Kirk H. Michaelian

Photoacoustic IR Spectroscopy

Related Titles

Everall, N., Griffiths, P. R., Chalmers, J. M. (eds.)

Vibrational Spectroscopy of Polymers
Principles and Practice

2007
ISBN: 978-0-470-01662-6

Telle, H. H., Urena, A. G., Donovan, R. J.

Laser Chemistry
Spectroscopy, Dynamics and Applications

2007
ISBN: 978-0-471-48571-1

Bordo, V. G., Rubahn, H.-G.

Optics and Spectroscopy at Surfaces and Interfaces

2005
ISBN: 978-3-527-40560-2

Kirk H. Michaelian

Photoacoustic IR Spectroscopy

Instrumentation, Applications and
Data Analysis

Second, Revised and Enlarged edition

WILEY-VCH Verlag GmbH & Co. KGaA

The Author

Kirk H. Michaelian
Natural Resources Canada
Devon, Alberta
Canada
Kirk.Michaelian@NRCan-RNCan.gc.ca

Cover Photo Spieszdesign GbR, Neu-Ulm

All books published by **Wiley-VCH** are carefully produced. Nevertheless, authors, editors, and publisher do not warrant the information contained in these books, including this book, to be free of errors. Readers are advised to keep in mind that statements, data, illustrations, procedural details or other items may inadvertently be inaccurate.

Library of Congress Card No.: applied for

British Library Cataloguing-in-Publication Data
A catalogue record for this book is available from the British Library.

Bibliographic information published by the Deutsche Nationalbibliothek
The Deutsche Nationalbibliothek lists this publication in the Deutsche Nationalbibliografie; detailed bibliographic data are available on the Internet at <http://dnb.d-nb.de>.

© 2010 WILEY-VCH Verlag GmbH & Co. KGaA, Boschstr. 12, 69469 Weinheim, Germany

All rights reserved (including those of translation into other languages). No part of this book may be reproduced in any form – by photoprinting, microfilm, or any other means – nor transmitted or translated into a machine language without written permission from the publishers. Registered names, trademarks, etc. used in this book, even when not specifically marked as such, are not to be considered unprotected by law.

Composition Toppan Best-set Premedia Limited, Hong Kong
Printing and Binding Fabulous Printers Pte. Ltd., Singapore
Cover Design Adam Design, Weinheim

Printed in Singapore
Printed on acid-free paper

ISBN: 978-3-527-40900-6

For Diane

Contents

Preface to the Second Edition *XI*
Preface to the First Edition *XIII*

1 Introduction *1*
1.1 Single- and Multiple-Wavelength PA Spectroscopies *2*
1.2 Scope *3*
1.3 Other Sources of Information *4*
     References *5*

2 History of PA Infrared Spectroscopy *7*
2.1 Early History *8*
2.2 Multiple-Wavelength PA Infrared Spectroscopy *11*
2.3 Arrival of PA FT-IR Spectroscopy *15*
     References *22*

3 Instrumental Methods *25*
3.1 Dispersive PA Infrared Spectroscopy *25*
3.1.1 Near-Infrared PA Spectroscopy *25*
3.1.2 Mid-Infrared PA Spectroscopy *26*
3.1.3 Instrumentation *28*
3.2 Rapid-Scan PA FT-IR Spectroscopy *29*
3.2.1 Normalization of Rapid-Scan Spectra *32*
3.3 Step-Scan PA FT-IR Spectroscopy *34*
3.4 Mid-Infrared Laser PA Spectroscopy *38*
3.4.1 CO_2 Laser PA Spectroscopy *38*
3.4.2 CO Laser PA Spectroscopy *40*
3.4.3 External-Cavity Quantum-Cascade Laser PA Spectroscopy *41*
3.5 Quartz-Enhanced PA Spectroscopy *42*
3.6 Cantilever Enhanced PA Spectroscopy *47*
3.7 Piezoelectric Detection *51*
3.8 Photothermal Beam Deflection Spectroscopy *56*
3.8.1 Reverse Mirage Detection *59*

Photoacoustic IR Spectroscopy. 2nd Ed., Kirk H. Michaelian
Copyright © 2010 WILEY-VCH Verlag GmbH & Co. KGaA, Weinheim
ISBN: 978-3-527-40900-6

3.9	Optothermal Window Spectroscopy	62
3.9.1	Research of Bićanić's Group	63
3.9.2	Research of McQueen's Group	64
	References	65

4 Signal Recovery 71
4.1 DSP Demodulation 71
4.2 Lock-in Demodulation 74
References 77

5 Experimental Techniques 79
5.1 Amplitude Modulation 79
5.2 Phase Modulation 82
5.3 Synchrotron Infrared PA Spectroscopy 87
5.4 PA Infrared Microspectroscopy 92
5.5 Quantitative Analysis 96
5.5.1 Quantification in PA Near-Infrared Spectroscopy 97
5.5.2 Quantification in PA Mid-Infrared Spectroscopy 98
5.5.3 Quantitative Analysis at Higher Concentrations 102
5.6 Depth Profiling 103
5.6.1 Amplitude Modulation 107
5.6.2 Phase Modulation 111
5.6.3 The Phase Spectrum 112
5.6.4 Phase Rotation 113
5.6.5 Generalized Two-Dimensional (G2D) Correlation 117
5.6.6 G2D Correlation in the Study of Layered Polymer Systems 118
5.6.7 G2D Correlation in the Study of Foods 118
References 120

6 Numerical Methods 127
6.1 Averaging of PA Data 127
6.1.1 Single-Sided Interferograms: Averaging of Interferograms and Spectra 128
6.1.2 Double-Sided Interferograms: The Average Modulus Spectrum 130
6.2 Spectrum Linearization 132
6.3 Phase Analysis 137
References 141

7 Applications 143
7.1 Carbons 143
7.1.1 Research of Low's Group 143
7.1.2 Other Groups 151
7.1.3 Hydrocarbon Cokes 154
7.2 Hydrocarbon Fuels 156
7.2.1 Coals 157
7.2.1.1 Early PA Infrared Spectra of Coals 157

7.2.1.2	Near-Infrared PA Spectra of Coals	*159*
7.2.1.3	Study of Coal Oxidation by PA Infrared Spectroscopy	*161*
7.2.1.4	Depth Profiling of Oxidized Coals	*162*
7.2.1.5	Numerical Analysis of PA Spectra of Coals	*163*
7.2.2	Liquid Fuels	*165*
7.2.3	Other Samples	*168*
7.3	Organic Chemistry	*171*
7.3.1	Mid-Infrared PA Spectra of Aromatic Hydrocarbons	*173*
7.3.2	Far-Infrared PA Spectra of Aromatic Hydrocarbons	*174*
7.4	Inorganic Chemistry	*176*
7.4.1	Carbonyl Compounds	*176*
7.4.2	Semiconductors	*179*
7.4.3	Superconductors	*182*
7.4.4	Other Materials	*182*
7.5	Biology and Biochemistry	*182*
7.6	Medical Applications	*187*
7.7	Polymers	*194*
7.7.1	Research of Urban's Group	*199*
7.7.2	Other Groups	*201*
7.8	Catalysts	*202*
7.8.1	Early Work	*202*
7.8.2	Surface Characterization and Reactions	*206*
7.8.3	Research of Ryczkowski's Group	*208*
7.9	Gases	*209*
7.9.1	CO Laser Radiation	*210*
7.9.2	CO_2 Laser Radiation	*213*
7.9.3	Other Radiation Sources	*218*
7.9.4	FT-IR PA Spectra of Gases	*225*
7.10	Wood and Paper	*227*
7.11	Food Products	*237*
7.11.1	Early PA Studies of Foods	*237*
7.11.2	A Typical Example: Milk Chocolate	*239*
7.11.3	Agricultural Grains	*239*
7.11.4	The PA Infrared Spectrum of Flour	*240*
7.11.5	Further PA Studies of Food Products	*241*
7.11.6	Research of Irudayaraj's Group	*243*
7.11.7	Food Production	*245*
7.12	Clays and Minerals	*247*
7.13	Textiles	*257*
	References	*263*

Appendix 1: Glossary *293*
Appendix 2: Literature Guide – Solids and Liquids *295*
Appendix 3: Literature Guide – Gases *355*
Index *383*

Preface to the Second Edition

The second edition of this text on photoacoustic (PA) infrared spectroscopy was written with several objectives in mind. Comparison of the titles of the two editions[1] reveals a clear shift in emphasis. The first edition attempts a general review and summary of the subject: individual chapters discuss its history and evolution, experimental methods, depth profiling, numerical methods, and quantitative chemical analysis. By contrast, the second edition focuses primarily on instrumental methods (some of which did not exist when the first edition was published), experimental techniques, signal recovery, and numerical methods of data analysis. Both works include reviews on applications of PA infrared spectroscopy and appendices that summarize much of the published scientific literature on this subject. Thus the reader should not anticipate a close correlation between the first and second editions, despite their fundamental relationship.

Another goal of the current work is to incorporate material published during the noticeable interval (more than seven years) that separates the first edition from the second. The rapid growth of the scientific literature in this field (which, of course, is typical of much modern research) can be roughly quantified by comparing the number of published references to PA infrared spectra of solids and liquids: Appendix 2 cites 543 publications in the first edition, and 883 in the second! It is hoped that these literature summaries facilitate access to the primary scientific literature. Consistent with this reasoning, Appendix 3 – a similar guide to publications on gas-phase PA infrared spectroscopy – was added to the second edition. This task was undertaken despite the fact that the PA research conducted by the author for more than two decades has been almost exclusively confined to the study of condensed-phase materials.

A third objective that was addressed in the second edition involves the correction of numerous minor errors. While it cannot be realistically claimed that the present edition is entirely free from these errors, it is anticipated that it contains fewer mistakes than the earlier work.

The acknowledgments to the first edition remain in effect despite a perceptible passage of time. To these appreciations, I add my thanks to colleagues from the

---

1) One could (humorously) denote these as PA-IR(n), $n=1,2$.

Photoacoustic IR Spectroscopy. 2nd Ed., Kirk H. Michaelian
Copyright © 2010 WILEY-VCH Verlag GmbH & Co. KGaA, Weinheim
ISBN: 978-3-527-40900-6

Canadian Light Source (*Centre canadien de rayonnement synchrotron*) at the University of Saskatchewan. Dr. Qing Wen read the entire manuscript and provided valuable suggestions for its improvement. This work was written for Diane and in her memory.

November, 2010

Kirk H. Michaelian
Edmonton, Alberta, Canada

Preface to the First Edition

'Science is concerned with the rational correlation of experience rather than the discovery of fragments of absolute truth.'
                Sir Arthur Eddington, *The Philosophy of Physical Science*

The continued growth of the primary scientific literature and the increased level of specialization that characterize modern physical science would both appear to require that forthcoming entries in the Chemical Analysis Series be constrained to topics somewhat narrower than those in previous volumes. The subject of photoacoustic (PA) spectroscopy is certainly not an exception to this trend. To be specific, A. Rosencwaig surveyed this then-emergent field in Volume 57, *Photoacoustics and Photoacoustic Spectroscopy*, a little more than two decades ago. The preparation of a similar treatise today would undoubtedly be a daunting task – one that might be expected to yield several texts. This assertion is corroborated by the 1994 publication of *Air Monitoring by Spectroscopic Techniques* (Volume 127, edited by M. W. Sigrist), followed in 1996 by *Photothermal Spectroscopy Methods for Chemical Analysis* (Volume 134, by S. E. Bialkowski). The present text, which is dedicated to the subject of photoacoustic infrared spectroscopy, is the newest volume in this continuing Chemical Analysis Series on photoacoustic and photothermal spectroscopy.

It is never inappropriate to question the need for a book on a particular analytical method. Two factors are especially pertinent with regard to the current submission. First, while several excellent review articles and book chapters on photoacoustic Fourier transform infrared (FTIR) spectroscopy already exist, much of this work is concerned with specific sample classes (e.g., polymers), numerical methods (mostly dealing with the phase of the PA signal), or instrumentation. The present volume adopts a perspective that is alternately historical and contemporary by reviewing both early and recent literature on an array of applications of PA infrared spectroscopy.

Moreover, a slightly different technological viewpoint is adopted here: This text discusses various implementations of PA spectroscopy at near-, mid- and far-infrared wavelengths, and is not restricted to the topic of 'PA FTIR' spectroscopy. For example, spectra of gases obtained with CO or CO_2 laser radiation at discrete wavelengths, without the use of an interferometer, are included. Similarly, dispersive near- and mid-infrared PA spectra of condensed-phase materials are discussed in several contexts. This approach to the subject was taken so as to minimize the effects of the seemingly arbitrary description of PA infrared spectra that is based on the specific apparatus used to acquire them.

The second justification for the present work arises from the very significant advances in, and expansion of, PA and photothermal science in recent years. Much of the early work in this discipline consisted of research in PA spectroscopy, albeit primarily at wavelengths shorter than those considered here. On the other hand, the entire field of PA and photothermal research has now grown and diversified to the point where it is justifiable – and perhaps necessary – that its major specializations be summarized more or less independently. The ever-increasing extent of the pertinent scientific literature implies that PA infrared spectroscopy certainly merits this treatment, as do several other related topics. It is hoped that this overview and synthesis of the PA infrared literature will prove useful to readers who may require detailed information on this subject.

This book has several specific objectives. Even though PA infrared spectroscopy is well established, many infrared spectroscopists tend to regard the method as unconventional or marginal and therefore ascribe second-class status to the technique. This situation is partly attributable to the low intensities associated with many early PA spectra, a factor that has been minimized in recent years by numerous advances in instrumentation and data analysis. It also arises from the usual lack of familiarity associated with any developing method. Hence the first objective of the present work is to clearly demonstrate the viability of PA infrared spectroscopy. This is accomplished by showing its widespread acceptance and reviewing its successful utilization in a variety of disciplines.

As the relevant literature was assembled, it became apparent that a significant number of proven experimental and numerical PA infrared techniques exist. It was noted above that some of these methods have already been discussed in one or more review articles or book chapters. However, there does not appear to be a single source that attempts a more comprehensive review of the relevant literature. This goal has now been addressed: The second objective of the current work was to assemble a reference text on PA infrared spectroscopy, which discusses most of its important applications and provides the interested reader with a point of entry into the original literature on the topic.

Because this is a volume in a series on Chemical Analysis, a nonmathematical approach has been adopted where possible. On the other hand, when equations are employed, the notation of the original authors has generally been followed; this ensures that the reader will not encounter unnecessary changes when reading both the present work and the primary literature. Because symbol conventions inevitably vary, the reader may notice a few minor examples of alternative usage within the present work. Insofar as mathematical symbols are defined upon their introduction in a particular context, this situation should not cause any significant confusion.

The organization of this text is based on the straightforward division of the published literature into several categories. To begin, Chapters 1 and 2 provide an introduction and historical review of PA infrared spectroscopy, with particular emphasis on early work. After this perspective is established, Chapter 3 discusses a series of seven different experimental PA methods that have been employed by researchers during the last two decades.

One of the most important capabilities of PA infrared spectroscopy–depth profiling–is described in Chapter 4. In this discussion the principles that underlie several relevant experimental techniques are explained and typical results are presented. Chapter 5 then outlines three important numerical methods that pertain to PA spectroscopy. The phase of the PA signal (which is distinct from the instrumental phase that affects all FTIR spectra) plays a key role in several of these manipulations.

A thorough reading of the appropriate scientific literature confirms that the disciplines in which PA infrared spectroscopy has been applied are particularly diverse. Many analysts may be familiar with one or more specific uses of the technique; in all likelihood, very few will be acquainted with the entire array of research topics in which it has been successfully utilized. Chapter 6, which describes a total of 15 different applications of PA infrared spectroscopy, was written to address this situation. Each section is meant to convey a sense of the research carried out in a specific area and to guide the reader who may wish to pursue the subject in more detail. Typical spectra (reproduced from the published literature or measured in the author's laboratory) are presented in each section of this chapter.

As a spectroscopic technique matures, its use for quantitation generally becomes more viable. The status of PA infrared spectroscopy with regard to quantitative analysis is reviewed in Chapter 7. Indeed, while the majority of the literature on PA spectra of solids and liquids describes qualitative or semi-quantitative studies, this discussion shows that many successful experiments in quantitative analysis have also been carried out.

Finally, Chapter 8 discusses two emerging techniques: PA infrared microspectroscopy and synchrotron infrared PA spectroscopy. Very recent experiments thus demonstrate that the field of PA infrared spectroscopy continues to advance after several decades of active research by a particularly wide community of spectroscopists.

The main part of the book is followed by two appendices. Appendix 1 provides definitions of a number of relevant terms used in the PA literature. Appendix 2 is an alphabetical listing of publications (including journal articles, book chapters, and conference proceedings) on PA infrared spectra of solids and liquids; the particular application(s) discussed in each article are indicated using a classification scheme similar to that in Chapter 6. This information has been collected as an aid to the reader who wishes to consult the original literature on PA infrared spectroscopy of condensed-phase samples.

It is usual to conclude preliminary remarks such as these with a series of acknowledgments. In this regard, the support of Natural Resources Canada during this research and the preparation of this manuscript is greatly appreciated. Similarly, I have benefited considerably from continued cooperation and collaboration with Bruker Optics in the implementation of PA infrared spectroscopy at CANMET during the last two decades. My entry into this field came about as the result of a prescient suggestion approximately 20 years ago. The use of rather tentatively worded acknowledgments is often effected stylistically or deferentially; in this case, however, it is literally necessary. Specifically, I would like to thank Dr. J. C. Donini for ensuring my involvement in PA spectroscopy, but cannot; his sudden passing five years ago has made this impossible.

November, 2002

Kirk H. Michaelian
Edmonton, Alberta, Canada

1
Introduction

This book discusses the experimental technique known as photoacoustic (PA) infrared spectroscopy. Research and applications in this field have enjoyed more or less continual development since its emergence over 30 years ago: a substantial body of literature – comprising more than one thousand publications – on PA infrared spectra exists today. The present work attempts to review and synthesize this literature, a principal objective being to summarize the current status of the technique. Recent advances in this field are also described. The assembled information will allow spectroscopists and researchers in specific disciplines to determine whether the method is appropriate for their needs.

PA infrared spectroscopy can be viewed from several perspectives. For example, it can be regarded as one of a large number of photoacoustic and photothermal (PT) methods used by physicists, chemists, and other researchers to characterize condensed matter and gases. It should be noted that both optical and thermophysical properties of materials can be investigated by these methods. From this viewpoint, PA infrared spectroscopy is a specialization within a much broader field, made possible by the development and application of transducers and optical instrumentation that operate in the infrared region of the electromagnetic spectrum. Of course some might suggest that this (obviously valid) interpretation tends to ascribe a secondary status to PA infrared spectroscopy.

Vibrational (infrared and Raman) spectroscopists[1] will almost certainly approach PA infrared spectroscopy in a different way. For these scientists, PA detection of infrared spectra can be described as an enabling technology, significantly increasing the number and type of samples for which viable data can be obtained. The reader will soon recognize that the viewpoint of the infrared spectroscopist is adopted in this book. In fact PA detection of infrared absorption spectra, using modern equipment and radiation sources, offers several well-known advantages. The most important of these are the following:

- Minimal sample preparation is required.
- The technique is suitable for opaque materials.
- Depth profiling can be effected for inhomogeneous or layered solids.
- PA spectroscopy is nondestructive (the sample is not consumed).

1) The author of this text is included in this group.

Photoacoustic IR Spectroscopy. 2nd Ed., Kirk H. Michaelian
Copyright © 2010 WILEY-VCH Verlag GmbH & Co. KGaA, Weinheim
ISBN: 978-3-527-40900-6

It is not an exaggeration to assert that these characteristics are critical in many circumstances: for example, samples exhibiting various problematic characteristics may be encountered in industrial laboratories on a daily basis. These include, but are not limited to, viscous liquids, semisolids, and dispersions; metal powders, carbonaceous solids, and granular materials; polymers and layered solids with physical structures or chemical compositions that may be altered by grinding. Traditional infrared sample preparation methods are often inappropriate or have deleterious effects on these substances. Hence the minimization of sample preparation in PA infrared spectroscopy can be considered its most important attribute. Examination of the scientific literature confirms that the majority of spectroscopists using this technique implicitly agree with this statement. The use of PA infrared spectroscopy for the characterization of problematic, even 'intractable', samples is discussed throughout this book.

Notwithstanding these statements, the capacity for PA depth profiling is an almost equally important feature: this experimental technique has been utilized extensively to study layered polymers, adsorption on substrates, and surface oxidation of hydrocarbon fuels or other species. Thus the PA spectroscopist also possesses the capability for analysis of surface and subsurface layers (in this context, 'surface' implies depths on the order of micrometers, while 'subsurface' regions extend tens of micrometers), a goal that surely can be said to be the dream of many chemists and physicists.

PA spectroscopy – frequently referred to by the acronym PAS – is sometimes described as an 'unconventional' infrared technique. The very significant number of publications in the primary scientific literature reporting research-quality PA spectra belie this somewhat pejorative description. It is hoped that the present account adequately demonstrates the wide-ranging applicability of the technique and makes a convincing argument for its increased future use.

1.1
Single- and Multiple-Wavelength PA Spectroscopies

PA spectroscopy can be divided into two broad categories. The first can be described as single-wavelength spectroscopy, since only one wavelength of light impinges on the sample of interest. Signal generation in a gas-microphone cell can be used to illustrate this technique. Three steps can be identified. First, modulated radiation from a laser or other suitable source impinges on the condensed-phase sample; second, the absorbed radiation is converted to heat by radiationless processes; and third, the heat generated within the sample is transferred to its cooler surroundings. Periodic heating of the boundary layer of carrier gas adjacent to the warm surface creates a pressure (acoustic) wave that is detected by the transducer (microphone). This experiment can be extended to include measurement of wavelength (wavenumber) dependence of optical absorption by systematically changing the wavelength of the incident radiation to build up a PA spectrum. Sequential observation of PA signals can be effected by selecting different lines from a multiple-wavelength laser or by use of an optical filter, such as a grating monochromator, in conjunction with broad-band radiation. These techniques were used to obtain PA infrared spectra by several research groups, particularly in the 1970s and early 1980s when PA spectroscopy enjoyed a rapid increase in popularity. Currently, multi-line gas (CO_2 and CO) and solid-state mid- and near-infrared lasers are used for specific PA applications, an important example being trace gas detection. This implementation of single-wavelength PA infrared spectroscopy is discussed in later chapters.

The second category is multiple-wavelength (multiplex) PA spectroscopy, as practiced with Fourier Transform infrared (FT-IR) spectrometers. Most readers already know, of course, that these spectrometers have attained very wide acceptance in analytical, research, and teaching laboratories during the last four decades. In the present context, the most important attribute of an FT-IR spectrometer is its capability for simultaneous measurements at a range of wavelengths; spectral coverage is determined mainly by the optical characteristics of the beamsplitter, the window material in the sample accessory, and the detector. An optical detector is not required in conventional PA FT-IR spectroscopy, and the accessible wavelength interval depends only on the beamsplitter and the window fitted on the gas-microphone cell. This technique has been used extensively for about three decades and is the source of the majority of the literature discussed in this book. Signal generation in the PA FT-IR experiment can be described in terms similar to those in the previous paragraph. Modulation is provided by the moving mirror in the interferometer or by use of an external device such as a chopper. This is discussed in more detail in Chapter 3.

1.2
Scope

As noted in the previous section, PA infrared spectroscopy has long been practiced with lasers, scanning monochromators, and FT-IR spectrometers. Indeed, the ongoing use of many types of instrumentation in PA spectroscopy demonstrates the breadth of the field. Although many workers today naturally associate infrared spectroscopy with FT-IR spectrometers, it should be emphasized at the outset that PA FT-IR spectroscopy is, in fact, a specialization within the broader discipline of PA infrared spectroscopy. This book adopts the wider definition of the field and examines a number of relevant PA infrared techniques from both historical and modern perspectives.

Spectroscopists are well aware that definitions of wavelength regions tend to differ for reasons that may be either historical, technological, or a combination of the two. Near-, mid-, and far-infrared PA spectroscopies are discussed in this book. Unless otherwise stated, these regions are demarcated as follows: near-infrared, 4000–12 500 cm^{-1} (wavelengths 2.5–0.8 µm); mid-infrared, 400–4000 cm^{-1} (25–2.5 µm); and far-infrared, 50–400 cm^{-1} (200–25 µm). It should be noted that this division of the electromagnetic spectrum, although not uncommon, is somewhat arbitrary; for example, the low-wavenumber limit given for the far-infrared region is based on the few published PA far-infrared spectra rather than more conventional far-infrared limits that encompass lower wavenumbers and longer wavelengths.

1.3
Other Sources of Information

Many review articles and conference proceedings dealing with PA spectroscopy have been published during the last three decades. Some literature reviews provide an overview of PA methods and emphasize work at shorter wavelengths (ultraviolet and visible) that is not discussed in this book. Mid-infrared PA spectroscopy receives very limited attention in most of these articles; some, however, discuss near-infrared spectroscopy and are therefore relevant to specific sections of this text. The early work of Adams (1982) is a typical example. Similarly, Vargas and Miranda (1988) published a detailed summary of PA and PT techniques that contains a short section on PA spectroscopy in the near- and mid-infrared regions. Further references discuss PA spectroscopy and its relationship to PA and PT methods (Pao, 1977; Rosencwaig, 1978, 1980; Tam, 1986; Almond and Patel, 1996).

Initial work in mid-infrared PA spectroscopy was summarized by Vidrine (1982) and Graham, Grim, and Fateley (1985). McClelland (1983) discussed several aspects of signal generation and instrumentation in an important survey of PA spectroscopy that emphasized the infrared region. The latter article is considered to be authoritative and continues to be cited by many practitioners who utilize PA infrared spectroscopy. Numerical methods, specifically phase correction and signal averaging, were discussed a few years later by the present author (Michaelian, 1990).

Two research groups made major contributions to the advancement of PA infrared spectroscopy and should be particularly noted with regard to review publications. R. A. Palmer of Duke University, together with many students and other collaborators, published extensively on research topics including PA infrared spectra of polymers, the role and significance of the PA phase, and the development of step-scan FT-IR PA spectroscopy. Consistent with this research effort, a review on PA spectroscopy of polymers by Dittmar, Palmer, and Carter (1994) contains a useful summary of the history and principles of PA infrared spectroscopy. Numerous other publications by this research group are referred to in later chapters.

Similarly, J. F. McClelland and co-workers at Iowa State University made a very considerable contribution to PA infrared spectroscopy during the last three decades. These investigators have a substantive history in instrumentation that culminated in the successful manufacture of commercial sample accessories for FT-IR spectrometers. McClelland and his colleagues published several detailed review articles, including a summary of the PA FT-IR technique that discusses signal generation and demonstrates a series of qualitative and quantitative applications (McClelland *et al.*, 1992). Other reviews discussed sample handling in PA FT-IR spectroscopy (McClelland *et al.*, 1993) and the implementation of PA spectroscopy with step-scan and rapid-scan spectrometers (McClelland *et al.*, 1998). PA FT-IR spectroscopy was reviewed in Volume 2 of *Handbook of Vibrational Spectroscopy* (McClelland, Jones, and Bajic, 2002) with particular reference to signal generation, instrumentation and sampling. These publications will be of considerable use to investigators who require an introduction to PA infrared spectroscopy.

References

Adams, M.J. (1982) Photoacoustic spectroscopy. *Prog. Analyt. Atom. Spectrosc.*, **5**, 153–204.

Almond, D. and Patel, P. (1996) *Photothermal Science and Techniques*, Chapman & Hall, London.

Dittmar, R.M., Palmer, R.A., and Carter, R.O. (1994) Fourier transform photoacoustic spectroscopy of polymers. *Appl. Spectrosc. Rev.*, **29** (2), 171–231.

Graham, J.A., Grim, W.M., and Fateley, W.G. (1985) Fourier transform infrared photoacoustic spectroscopy of condensed-phase samples, in *Fourier Transform Infrared Spectroscopy*, vol. **4** (eds J.R. Ferraro and L.J. Basile), Academic Press, New York, pp. 345–392.

McClelland, J.F. (1983) Photoacoustic spectroscopy. *Anal. Chem.*, **55** (1), 89A–105A.

McClelland, J.F., Luo, S., Jones, R.W., and Seaverson, L.M. (1992) A tutorial on the state-of-the-art of FTIR photoacoustic spectroscopy, in *Photoacoustic and Photothermal Phenomena III* (ed. D. Bićanić), Springer-Verlag, Berlin, pp. 113–124.

McClelland, J.F., Jones, R.W., Luo, S., and Seaverson, L.M. (1993) A practical guide to FTIR photoacoustic spectroscopy, in *Practical Sampling Techniques for Infrared Analysis* (ed. P.B. Coleman), CRC Press, Boca Raton, pp. 107–144.

McClelland, J.F., Bajic, S.J., Jones, R.W., and Seaverson, L.M. (1998) Photoacoustic spectroscopy, in *Modern Techniques in Applied Molecular Spectroscopy* (ed. F.M. Mirabella), John Wiley & Sons, Inc., New York, pp. 221–265.

McClelland, J.F., Jones, R.W., and Bajic, S.J. (2002) Photoacoustic spectroscopy, in *Handbook of Vibrational Spectroscopy*, vol. 2 (eds J.M. Chalmers and P.R. Griffiths), John Wiley & Sons, Ltd, Chichester, pp. 1231–1251.

Michaelian, K.H. (1990) Data treatment in photoacoustic FT-IR spectroscopy, in *Vibrational Spectra and Structure*, vol. 18 (ed. J.R. Durig), Elsevier Science Publishers B.V., Amsterdam, pp. 81–126.

Pao, Y.-H. (ed.) (1977) *Optoacoustic Spectroscopy and Detection*, Academic Press, New York.

Rosencwaig, A. (1978) Photoacoustic spectroscopy. *Adv. Electron. Electron Phys.*, **46**, 207–311.

Rosencwaig, A. (1980) *Photoacoustics and Photoacoustic Spectroscopy*, Chem. Anal., vol. **57**, John Wiley & Sons, Inc. (Interscience), New York.

Tam, A.C. (1986) Applications of photoacoustic sensing techniques. *Rev. Mod. Phys.*, **58** (2), 381–431.

Vargas, H. and Miranda, L.C.M. (1988) Photoacoustic and related photothermal techniques. *Phys. Rep.*, **161** (2), 43–101.

Vidrine, D.W. (1982) Photoacoustic Fourier transform infrared spectroscopy of solids and liquids, in *Fourier Transform Infrared Spectroscopy*, vol. 3 (eds J.R. Ferraro and L.J. Basile), Academic Press, New York, pp. 125–148.

2
History of PA Infrared Spectroscopy

This chapter is an account of the historical development of PA infrared spectroscopy. Relevant literature published in the period beginning in 1968 and ending in 1981 is reviewed in the following pages. The reader should appreciate that these dates are not entirely arbitrary: the former year marked the appearance of a groundbreaking article on PA detection of infrared absorption in gases, while the latter is the approximate point at which PA infrared spectroscopy first achieved a measure of maturity and acceptance by the community of vibrational spectroscopists (Figure 2.1). This acceptance is, of course, indicated by the publication of a significant number of articles on the subject; because of its size, this emergent body of literature is not very amenable to the chronological perspective adopted here to place the early work in context. Hence this second date is where the historical review comes to an end, and discussions based on experimental techniques and applications of PA infrared spectroscopy logically begin.

Researchers who utilized PA infrared spectroscopy in these formative years may well question the criteria used to determine which articles are included in this discussion. Inevitably, some publications have been excluded. There are three primary reasons why a few works are omitted. First, some publications may not be widely available, or have not come to the attention of the author; second, the current selection of articles was deliberately chosen so as to illustrate the evolution of the field; and finally, some publications were intentionally omitted from this chapter and deferred for a fuller discussion later in this book.

This chapter is divided into three main sections. The first deals with work published between 1968 and 1978; the latter date was selected because it corresponds to publication of the first article that reported PA FT-IR spectra. The second section describes early multiple-wavelength PA infrared spectroscopy, which can be thought of as a precursor to multiplex (FT-IR) PA spectroscopy. Finally, the third section discusses the emergence of PA FT-IR spectroscopy, a process that took place between 1978 and 1981. Further work published after 1981 is reviewed in subsequent chapters that are organized according to subject matter, rather than year of publication.

Photoacoustic IR Spectroscopy. 2nd Ed., Kirk H. Michaelian
Copyright © 2010 WILEY-VCH Verlag GmbH & Co. KGaA, Weinheim
ISBN: 978-3-527-40900-6

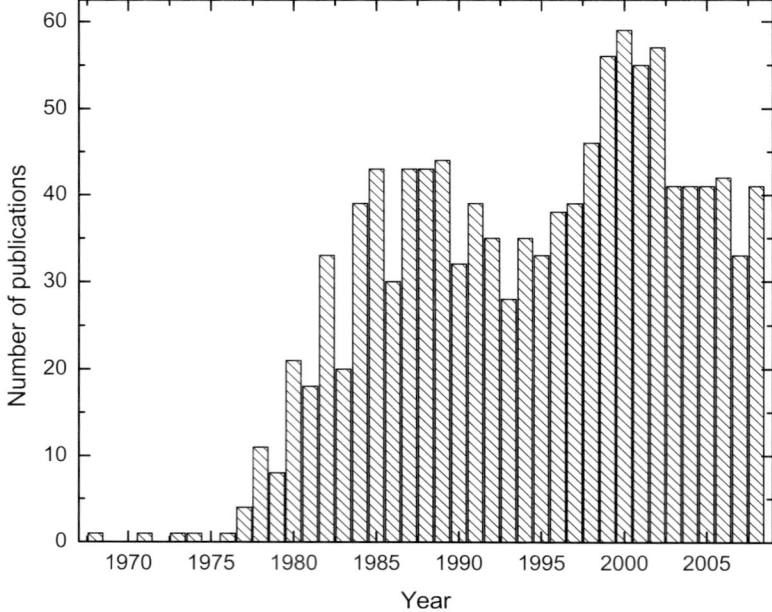

Figure 2.1 Plot of the number of publications on PA infrared spectroscopy as a function of the year of publication.

2.1
Early History

PA infrared spectroscopy is sometimes said to originate with the work of Busse and Bullemer (1978), who obtained a PA spectrum of methanol vapor using a commercial FT-IR spectrometer and an absorption cell fitted with a microphone. This equipment allowed comparison of an ordinary mid-infrared absorption spectrum with a PA spectrum obtained under the same conditions. To put this work in proper context and describe the development of the field more accurately, it should be emphasized that this was, in fact, the first successful PA FT-IR experiment described in the primary scientific literature; nevertheless, a number of articles on PA infrared spectroscopy using other instrumentation were actually published prior to this work. These articles are briefly summarized in this section.

Table 2.1 lists some of these early publications on PA infrared spectroscopy. All of these works appeared before that of Busse and Bullemer, that is, prior to 1978; it can be noted that the majority describe PA spectra of gases. Some of these references are mentioned again in Chapter 7, which deals with spectra of gases and a number of other specific applications.

In 1968, Kerr and Atwood described an 'absorptivity spectrophone' designed to measure weak absorptivities at visible and infrared wavelengths. In the latter case a CO_2 laser, operating at 9.6 μm (1042 cm^{-1}) rather than the more familiar 10.6-μm

Table 2.1 Photoacoustic infrared spectroscopy: early publications.

Author	Subject
Kerr and Atwood (1968)	CO_2 laser absorptivity spectrophone
Kreuzer (1971)	PA spectroscopy of CH_4, CH_3OH
Rosengren (1973)	Theory of optoacoustic gas detector
Max and Rosengren (1974)	Optoacoustic detection of NH_3 in air
Kanstad and Nordal (1977)	Grazing incidence PA spectroscopy of metal surfaces
Nordal and Kanstad (1977a)	PA spectroscopy of $(NH_4)_2SO_4$, glucose
Nordal and Kanstad (1977b)	PA spectroscopy of $(NH_4)_2SO_4$, hexachlorobenzene
Patel, Kerl, and Burkhardt (1977)	Optoacoustic spectroscopy of excited-state NO

wavelength, was used to detect CO_2 in CO_2/N_2 mixtures. The resulting signal from a differential pressure transducer was observed to vary linearly with CO_2 concentration. An absorptivity of 5.8×10^{-7} was measured at a concentration of 300 parts per million (ppm), the approximate value for CO_2 in air. Moreover, the authors predicted that an absorptivity as low as 1.2×10^{-7} could be measured with the use of a 10-W CO_2 laser. Thus the high sensitivity achievable in gas-phase PA infrared spectroscopy was demonstrated at a very early date.

Three years later, Kreuzer (1971) discussed the theory of the absorption of infrared radiation by a gas with particular reference to coherent radiation sources. PA spectra of CH_4 and CH_3OH vapor in a narrow section of the C–H stretching region were obtained by tuning an HeNe laser through a series of emission lines between 2947.87 and 2947.93 cm^{-1}. Despite the extremely limited wavenumber range covered in this work, several absorption bands were observed. The calculated sensitivity was about 10^{-2} ppm CH_4 in air, although Kreuzer estimated that this might eventually be improved by a factor as great as 10^5. A subsequent theoretical model for this experiment (Rosengren, 1973) rationalized and confirmed the minimum detectable molecular CH_4 density reported by Kreuzer. This theoretical analysis concluded that the highest sensitivity predicted in the experimental study, which is equivalent to 10^{-2} ppm, was indeed achievable.

Trace gas detection was studied further by Max and Rosengren (1974), who constructed a resonant PA gas cell. These authors showed that the minimum detectable gas concentration is determined by microphone and preamplifier noise; in particular, acoustic noise picked up by the microphone was found to be the most important factor. A CO_2 laser was used to detect trace amounts of NH_3 in air in this investigation. For a laser power of 280 mW, the lower concentration limit giving rise to a measurable signal was about 0.003 ppm. It was again suggested that a lower detection limit of about 10^{-4} ppm should be achievable with this PA technique.

Early PA infrared spectra were also obtained for condensed phases. Nordal and Kanstad (1977a, 1977b) noted recent successes in PA spectroscopy at ultraviolet, visible, and near-infrared wavelengths, and successfully extended the technique into the mid-infrared region. A tunable CO_2 laser with lines in three wavelength intervals (approximately 9.2–9.7, 10.1–10.3, and 10.5–10.7 μm) was used as the infrared radiation source, and the PA cell was constructed in the authors' laboratory. Ammonium sulfate (both solid and in aqueous solution), glucose, and hexachlorobenzene were successfully investigated in these experiments. Insensitivity to the morphology of solid samples was noted as a particular advantage; indeed, the appearance of the totally symmetric sulfate v_1 band, normally observed only in Raman spectra, was conjectured to arise from reduced symmetry in crystallite surface layers that would have been destroyed by grinding. Although this reasoning was subsequently modified by these workers, it nicely demonstrates that a very real advantage may be afforded by the minimal sample preparation required in PA spectroscopy. The capability for quantitative analysis was also shown in this work; PA intensities for $(NH_4)_2SO_4$ solutions varied linearly over a wide concentration range. However, the intensity tended to saturate at very high concentrations, a phenomenon that had already been observed in spectra obtained at shorter wavelengths.

Another article published at about the same time by these authors (Kanstad and Nordal, 1977) described grazing incidence PA spectroscopy on metal surfaces, an application of a technique termed PA reflection-absorption spectroscopy (PARAS). This experiment also utilized a tunable CO_2 laser and was designed to study optical absorption in thin surface layers. The layout of the PA apparatus is shown schematically in Figure 2.2. The straightforward measurement principle is that radiation polarized in the plane of incidence may be absorbed by the sample, whereas perpendicularly polarized radiation may not; the geometry of the experiment was carefully controlled to ensure that only light with parallel polarization impinged on the sample. Experiments were performed on aluminum oxide layers and iron surfaces exposed to SO_2. For the first group of samples, a strong absorption near 10.55 μm (950 cm^{-1}) was observed and attributed to a longitudinal Al–O vibration. Results for the iron surface, although less satisfactory, were tentatively ascribed to the presence of $FeSO_4 \cdot 7H_2O$. Importantly, Kanstad and Nordal (1977) concluded that aluminum oxide could be detected at surface coverages equivalent to a fraction of a monolayer; the putative sulfate species was thought to have a detection limit of several layers. Thus very high sensitivity was also demonstrated in PA infrared spectra of solids using radiation from a CO_2 laser.

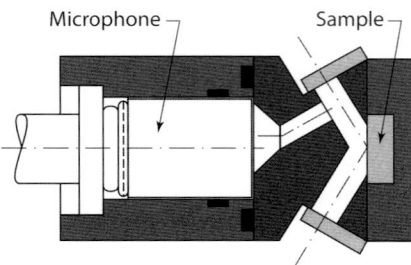

Figure 2.2 Experimental layout for PA reflection-absorption spectroscopy (PARAS). (Reproduced from Kanstad, S.O. and Nordal, P.-E., *Int. J. Quantum Chem.* **12** (Suppl. 2): 123–130, by permission of John Wiley & Sons, Inc.; copyright © 1977.)

The final work mentioned in this short review involves PA spectra of gases, but its perspective is significantly different from those in the investigations mentioned above. Patel, Kerl, and Burkhardt (1977) obtained high-resolution excited-state PA spectra of nitric oxide (NO), specifically utilizing PA detection because of its high sensitivity. In this experiment, NO was first excited using CO laser emission at 1917.8611 cm^{-1} and then probed with tunable radiation from a spin-flip Raman laser. The PA signal was measured as a function of the magnetic field of the Raman laser, which is equivalent to variation of infrared wavenumber. Zeeman spectra of excited-state NO were also obtained using PA detection. This investigation provides a different type of information than that in the previously mentioned studies, but is nevertheless relevant since it shows the versatility of PA spectroscopy in an important application at infrared wavelengths.

2.2
Multiple-Wavelength PA Infrared Spectroscopy

The second phase in the development of PA infrared spectroscopy can be described as the period in which the pioneering single-wavelength measurements described above gave way to multiple-wavelength experiments, performed with scanning spectrometers or by the use of multiple infrared laser lines. The broader spectral coverage in these later PA experiments greatly increased the quantity and diversity of information that could be obtained, making it possible to use PA spectroscopy as a tool for the analysis of samples of interest to chemists and biologists; thus it was no longer strictly necessary to restrict PA studies to systems that absorb radiation at particular wavelengths dictated by the available monochromatic sources. Both mid- and near-infrared dispersive PA spectra are included in the second group of experiments. Results obtained using CO_2 laser radiation by Nordal and Kanstad, as well as Perlmutter and coworkers, are also discussed in this section. All of this multiple-wavelength work was published between 1978 and 1980, a short interval during which several significant developments in PA FT-IR spectroscopy also occurred. The latter topic is discussed separately in the next section of this chapter.

Dispersive PA infrared spectra are discussed first. In the mid-infrared region, Low and Parodi (1978) successfully used PA spectroscopy as a means to characterize surfaces and adsorbates. The goal of their work was to study species adsorbed on catalysts or silica powder. The spectrometer available to these researchers incorporated a LiF prism, which made it possible to record infrared spectra from 2000 to 4000 cm^{-1} but precluded observation of low-wavenumber bands due to heavy atoms such as silicon and oxygen. In their first experiment, PA spectra of a chromia-alumina catalyst exhibited bands due to OCH_3 and OH groups, the former appearing after exposure of the catalyst to $(CH_3O)_3CH$ vapor and the latter arising from surface hydroxyls and adsorbed water. Similarly, the PA infrared spectrum of treated silica exhibited bands due to the CH_3 functionality in Si–OCH_3 groups; after exposure to $HSiCl_3$ vapor, a new band characteristic of the chemisorbed silane was observed. Low and Parodi noted that similar results could, in principle, have been observed in transmission measurements on thin pellets. By contrast, PA spectroscopy offers a decisive advantage in the study of whole catalyst pellets, for which transmittance is obviously very low.

This initial demonstration of dispersive mid-infrared PA spectroscopy was soon followed by an entire series of publications by the same authors (Low and Parodi, 1980a, 1980b, 1980c, 1980d, 1980e). This work is described in more detail in the first part of Chapter 3. In these investigations, the authors pursued their interest in the characterization of surfaces and opaque solids (Low and Parodi, 1980c, 1980d) and various carbonaceous samples (Low and Parodi, 1980a, 1980b). With regard to the latter subject, it is relevant to note that PA spectra of carbons are frequently utilized to normalize spectra of less strongly absorbing samples; this topic is also mentioned in the next chapter.

Several articles on dispersive near-infrared PA spectroscopy are also relevant to this discussion. Three of these publications describe the work of Kirkbright and his collaborators. In the first, Adams, Beadle, and Kirkbright (1978) described instrumentation and applications of PA spectroscopy at wavelengths between approximately 0.8 and 2.7 μm (12 500–3700 cm^{-1}). Experimental apparatus included a quartz-halogen lamp, chopper, monochromator, home-built PA cell, and lock-in amplifier. Spectra were reported for two rare-earth oxides and aliphatic and aromatic hydrocarbons; the hydrocarbon spectra are discussed in Chapter 7 of this book. The bands observed for the rare-earth oxides arise from low-energy electronic transitions and were sufficiently well defined for the authors to suggest their use as wavelength standards.

In a related investigation, Kirkbright (1978) presented a brief overview of PA spectroscopy, mentioning an entire series of advantages and applications of the technique: freedom from light scattering interference, the potential for depth profiling, complementary measurements of thermal conductivity and sample thickness, accessibility to wide spectral ranges, capability for the determination of luminescence quantum efficiencies, the capacity for study of very weakly absorbing materials, and, finally, the use of an inert carrier gas, which facilitates the study of air-sensitive materials. This discussion was followed by near-infrared PA results for a wide variety of samples, including blood, proteins, clays, and minerals, and the aforementioned hydrocarbons. This article illustrates the considerable diversity of PA near-infrared spectroscopy and the detailed information obtainable by this technique.

In subsequent research carried out by this group (Castleden, Kirkbright, and Menon, 1980) near-infrared PA spectroscopy was used to determine the moisture content in single-cell protein. The spectra reported in this work spanned the wavelength range from about 1.3 to 2.3 μm (roughly 7700–4350 cm^{-1}). A band at 1.9 μm (5265 cm^{-1}) was observed to vary with the amount of moisture present in the samples and used for quantification. Importantly, the intensity of this band was first corrected by ratioing it against an N–H band at 1.55 μm (6450 cm^{-1}); the latter feature, which was due to protein, was known to be very similar from one sample to the next. The authors found that quantitative analysis was not feasible until samples were separated into a series of size fractions, since PA intensities are greater for smaller particle sizes. When spectra for specific size ranges were compared, it was observed that corrected PA intensities at 1.9 μm increased linearly with moisture content. In fact, these results were superior to those obtained by low-resolution NMR spectroscopy: NMR was less selective than PA spectroscopy because it responded to all protons, rather than those in particular functional groups.

Other research teams were also investigating dispersive near-infrared PA spectroscopy during this period. For example Blank and Wakefield (1979), working at Gilford Instrument Laboratories, described a double-beam PA spectrometer that functioned in the ultraviolet, visible, and near-infrared spectral regions. This article contains near-infrared PA spectra of the rare-earth oxides Ho_2O_3, Nd_2O_3 and Pr_6O_{11}, which exhibit a number of well-defined bands. A wide variety of other samples was also mentioned by these authors, although corresponding spectra were not shown. Thus, as was the case in the work of Kirkbright's group, the studies of Blank and Wakefield demonstrated that PA near-infrared spectroscopy is a particularly versatile method for chemical analysis. It can also be noted that this article described a divided PA cell which was subsequently adapted for mid-infrared PA spectroscopy.

The final example of dispersive near-infrared PA spectroscopy to be mentioned in this section is the work of Lochmüller and Wilder (1980a, 1980b). These researchers used a Princeton Applied Research spectrometer equipped with an Xe lamp to obtain ultraviolet, visible, and near-infrared PA spectra. The first article by these authors (1980a) discussed the qualitative characterization of chemically modified silica surfaces, while the second (1980b) emphasized quantification of the degree of surface coverage. In the former study, samples were prepared by reacting microparticulate silica gel with various organomethyldichlorosilanes so as to introduce cyclic hydrocarbons onto the silica surface. Spectra of a series of silicas in which cyclohexyl, cyclohexenyl, or phenyl groups are bonded to the surface contained features due to the second overtone of the strongest aliphatic C–H stretching band at 1.186 μm (8430 cm^{-1}) and a shoulder at 1.173 μm (8525 cm^{-1}). These were attributed to cyclohexyl methylene groups and the adjacent methyl group on the organomethyldichlorosilane, respectively. Another band at 1.14 μm (8770 cm^{-1}) arises from *cis* hydrogens on sp^2 carbons and consequently was most intense for the phenyl system. These results demonstrate the capability of PA near-infrared spectroscopy for qualitative analysis. In the second investigation, it was observed that coverage with various alkyl, phenyl, and aminoalkyl compounds could be monitored using either the first overtone of the strongest aliphatic C–H stretching band, which occurs at 1.71 μm (5850 cm^{-1}), or the second overtone that was mentioned above. Because the first overtone – which is naturally the stronger of the two – is superimposed on a sloping baseline, the authors chose the other alternative for quantification. A linear relationship was established between PA signal amplitude and carbon content.

Three publications by Nordal and Kanstad were described in the first section of this chapter. These authors used several CO_2 laser lines in their PA studies and were able to identify a number of important infrared bands. Six additional articles by this group, published between 1978 and 1980, are pertinent to the current discussion of multiple-wavelength PA spectroscopy. This later research was based on the use of emission lines from both $^{12}CO_2$ and $^{13}CO_2$ lasers, which effectively increased the number of available wavelengths where PA intensities could be measured.

In the first of these publications, Kanstad and Nordal (1978a) described an 'open-membrane' PA cell in which the sample remained outside the acoustic chamber and was suspended on a membrane that was used in place of the more familiar infrared-transparent window. In this experiment, heat generated as a consequence of absorption of the incident radiation by the sample was transported through the membrane into the chamber. The resulting acoustic wave was detected by a microphone. This approach is a variation of the so-called 'open PA cell' and was eventually adapted by other workers. The technique is mentioned again in Chapter 3. Kanstad and Nordal (1978a, 1978b) used the method to obtain PA spectra of an Al_2O_3 film, a polychlorotrifluoroethylene oil film, and hexachlorobenzene crystallites. Measurements were made at more than 20 wavelengths in four regions between 9.2 and 10.7 µm. Although the CO_2 laser radiation utilized in this work was not continuously tunable, the number of lines available was sufficient to enable characterization of several absorption bands of particular interest.

PA infrared spectra of ammonium sulfate were obtained in the initial investigations by these authors that were described in Section 2.1. This work was extended to include the study of single crystals of $(NH_4)_2SO_4$ in two subsequent publications (Kanstad and Nordal, 1979, 1980a). These later results are noteworthy because they include one of the first examples of polarized PA infrared spectra in the literature. Indeed, the single-crystal study revealed that the v_1 sulfate band at 975 cm^{-1} is comparatively strong when the electric vector of the infrared radiation is parallel to the a_0 crystal axis, but practically absent when the vector is polarized along b_0. This result led to the conclusion that the appearance of the v_1 peak in the PA spectra of $(NH_4)_2SO_4$ powder obtained by these authors (Nordal and Kanstad, 1977a, 1977b) was not due to surface properties since the band was also observed for single crystals. The PA infrared spectrum of $(NH_4)_2SO_4$ is discussed again in the next section of this chapter.

PA reflection-absorption spectroscopy, briefly mentioned in the previous section, was considered in greater detail by these authors (Nordal and Kanstad, 1978; Kanstad and Nordal, 1979, 1980b). This technique was developed for studying thin films, particularly oxides, on metal surfaces. In addition to the open-membrane spectrophone discussed above, the authors described two other PARAS techniques, using either a modified closed-chamber spectrophone or piezoelectric detection. In the former experiment, the CO_2 laser radiation impinges on the sample at an angle of 60°, with the reflected beam being allowed to exit the spectrophone chamber (Figure 2.2). A microphone detects the acoustic wave that arises from absorption and radiationless deexcitation by the sample. Spectra are plotted as the quantity $\Delta = \delta R_\parallel - \delta R_\perp$, where $\delta R_\parallel$ and $\delta R_\perp$ are the fractional changes in reflectivity for light polarized parallel and perpendicular, respectively, to the plane of incidence. A 10.55-μm (948-cm^{-1}) band appeared only in the parallel-polarized spectrum for an aluminum oxide layer. Moreover, the authors noted that this band shifts by a few wavenumbers when the oxide layer is thicker, and that band positions observed using PARAS differ slightly from published values obtained by other techniques.

In the second experiment, the piezoelectric transducer (PZT) was affixed directly to the back surface of the sample, while the laser beam impinged on the front surface. This arrangement removed any requirement for sample modification or the use of a gas-microphone cell. This technique was also utilized by Kanstad and Nordal (1979, 1980b) to obtain spectra of oxide layers on aluminum. Piezoelectric detection is discussed in more detail in Chapter 3.

The next publication on multiple-wavelength PA infrared spectroscopy to be discussed also describes the use of CO_2 lasers. Perlmutter, Shtrikman, and Slatkine (1979) reported the PA detection of trace amounts of ethylene in the presence of interfering gases. The authors pointed out three limiting factors in these measurements: (i) additive (microphone) electrical noise; (ii) modulation noise, which arises primarily from variations in laser power; and (iii) the accuracies with which the absorption coefficients of the interfering gases are known. The third factor is by far the most important, typically amounting to levels of a few per cent. Perlmutter *et al.* used both $^{12}CO_2$ and $^{13}CO_2$ lasers in their work, detecting C_2H_4 concentrations of 130 ppb in an urban environment. They estimated a detection limit of about 5 ppb, but noted that the presence of 1% CO_2 would increase this limit by a factor of about 10. PA detection of ethylene is discussed further in Chapter 7.

The final example in this section utilized a different laser as a radiation source. Monchalin *et al.* (1979) obtained PA infrared spectra of chrysotile asbestos in the hydroxyl stretching region between about 3400 and 3800 cm^{-1}, using a series of emission lines near 2.7 μm from an HF laser. This form of asbestos, a hydrated magnesium silicate, was of interest to these authors as an environmental pollutant. About a dozen HF laser lines were available in this experiment; these were sufficient to allow observation of a doublet at 3645 and 3688 cm^{-1}, the second peak being the stronger of the two. These bands arose from hydroxyl groups in two different environments and displayed an intensity ratio consistent with their expected relative abundances. In the second part of this investigation, Monchalin *et al.* measured PA spectra for two orientations of asbestos fibers relative to the laser polarization; as expected, the PA intensity was much lower when the polarization was parallel to the fibers than when it was in the perpendicular orientation.

2.3
Arrival of PA FT-IR Spectroscopy

A third stage in the development of PA infrared spectroscopy occurred between 1978 and 1981. During this period, the enhanced capabilities and widespread availability of FT-IR spectrometers led an increasing number of vibrational spectroscopists to utilize these instruments, rather than traditional scanning infrared spectrometers, on a routine basis. At the same time, recent advances in dispersive near- and mid-infrared PA spectroscopy, in addition to multiple-wavelength laser-based methods, helped to create and define the need for multiplex mid-infrared PA spectroscopy. These parallel developments made the emergence of PA FT-IR spectroscopy in the late 1970s and early 1980s predictable and, perhaps, inevitable.

As mentioned earlier in this chapter, Busse and Bullemer (1978) are generally recognized as the first workers to obtain a viable PA infrared spectrum using an FT-IR spectrometer. In addition, several other groups developed productive research programs in PA FT-IR spectroscopy during the next two to three years. Interestingly, virtually all of the latter efforts pertained to condensed matter (primarily solids) rather than gases; in other words, these investigators generally dealt with PA intensities that were weaker than those in the first successful PA FT-IR experiment. The most important publications from several of these groups in the years up to and including 1981 are reviewed in this section. Work published after this date is discussed in later chapters, which are organized according to subject matter rather than date of publication.

Table 2.2 lists selected publications on PA FT-IR spectroscopy that appeared between 1979 and 1981. These articles presented historically important early work in the field. The research groups mentioned are those of Rockley and his collaborators at Oklahoma State University; Mead, Vidrine and their colleagues at Nicolet Instrument Corporation; and finally, Royce, Teng, and coworkers at Princeton University. The publication references from each of these three sources are grouped together in Table 2.2 and discussed in the same sequence in the following paragraphs.

This history of PA FT-IR spectroscopy begins with the work of M. G. Rockley and his group, who had already utilized PA spectroscopy at shorter wavelengths. The purported first publication of a PA FT-IR spectrum of a solid material presented a result extending from 4000 to $800\,\text{cm}^{-1}$ for polystyrene film (Rockley,

Table 2.2 Photoacoustic FT-IR spectroscopy: early publications.

Author	Subject
Rockley (1979)	Polystyrene
Rockley (1980a)	$(NH_4)_2SO_4$
Rockley (1980b)	Charcoal; polystyrene; aspirin
Rockley and Devlin (1980)	Coal
Rockley, Richardson, and Davis (1980)	Naphthalene; benzophenone; KNO_3; silica
Mead et al. (1979)	Coal
Mead, Lowry, and Anderson (1981)	LiF; Ge; Si; GaAs; near-, mid-, far-infrared
Vidrine (1980)	PVC; polyurethane; lecithin; plastic; coal
Vidrine (1981)	Catalysts; conducting polymers; depth profiling
Laufer et al. (1980)	AgCN
Royce, Teng, and Enns (1980)	Polystyrene; acrylates; coal; zeolites

1979). The PA cell used in this work, originally developed for use at ultraviolet and visible wavelengths, was fitted with an NaCl window for this exploratory infrared study. This strategy was not uncommon in the early days of PA infrared spectroscopy, and illustrates how PA spectroscopy was gradually adapted for work at longer wavelengths. Several C–H stretching bands are visible in the uncorrected polystyrene spectrum reported by Rockley, with absorption by ambient CO_2 and H_2O also being quite significant. This PA spectrum is noisy and also contains spikes due to ground loops, but is noteworthy simply because it is the first of its kind. The author correctly predicted that PA FT-IR spectroscopy would eventually be used to study samples with rough surfaces, catalysts, and gases. Many examples of these and other applications are discussed in later chapters of this book.

The use of PA FT-IR spectroscopy for characterization of ammonium sulfate, which is of interest as an atmospheric aerosol, was described one year after the initial work on polystyrene (Rockley, 1980a). As discussed above, Nordal and Kanstad (1977a, 1977b) had already identified the v_1 band of the sulfate ion in PA infrared spectra of solid $(NH_4)_2SO_4$ obtained with a CO_2 laser; Rockley (1980a) confirmed the existence of this band in the corresponding PA FT-IR spectrum. However, the more recent work showed that this band occurred as a shoulder rather than a well-defined peak in the FT-IR spectrum of this compound. Rockley attributed this observation to local field-induced distortions that broaden the stronger v_3 band near $1100 \, cm^{-1}$, causing the v_1 peak to appear as a shoulder. Figure 2.3 shows that the v_1 sulfate band is indeed barely visible in the PA infrared spectrum of polycrystalline ammonium sulfate obtained with modern instrumentation.

Rockley (1980b) next described PA FT-IR spectra of several opaque solids. Charcoal, polystyrene beads used as catalyst supports, and an aspirin tablet were studied in this work. Carbon black and finely divided lead were considered as reference materials. The author noted that absorption by atmospheric H_2O and CO_2 tends to complicate PA infrared spectra: when these gases are present in the optical path between the interferometer and the PA cell, they absorb part of the incident radiation and give rise to transmission-like features in the spectra. Further, Rockley found that these bands are only approximately canceled by ratioing the PA spectrum of a sample to that of a reference material, since the amounts of CO_2 and water vapor in the instrument frequently differ from one experiment to the next. Despite this complication, useful PA spectra of polystyrene and aspirin were obtained in this investigation.

PA infrared spectroscopy has been used extensively to characterize coals during the last three decades. Rockley and Devlin (1980) were among the first to report PA FT-IR spectra of coals, describing spectra of aged and freshly exposed coal surfaces. These authors obtained research-quality PA spectra of sub-bituminous, bituminous, and anthracitic coals at a time when the use of FT-IR spectrometers in PA spectroscopy was barely under way – a rather impressive accomplishment. As shown in Figure 2.4, their investigation revealed noteworthy differences among the spectra of the three coals, the most striking being the relative weakness of the bands due to CH_2 and CH_3 groups in the anthracite spectrum. Significant

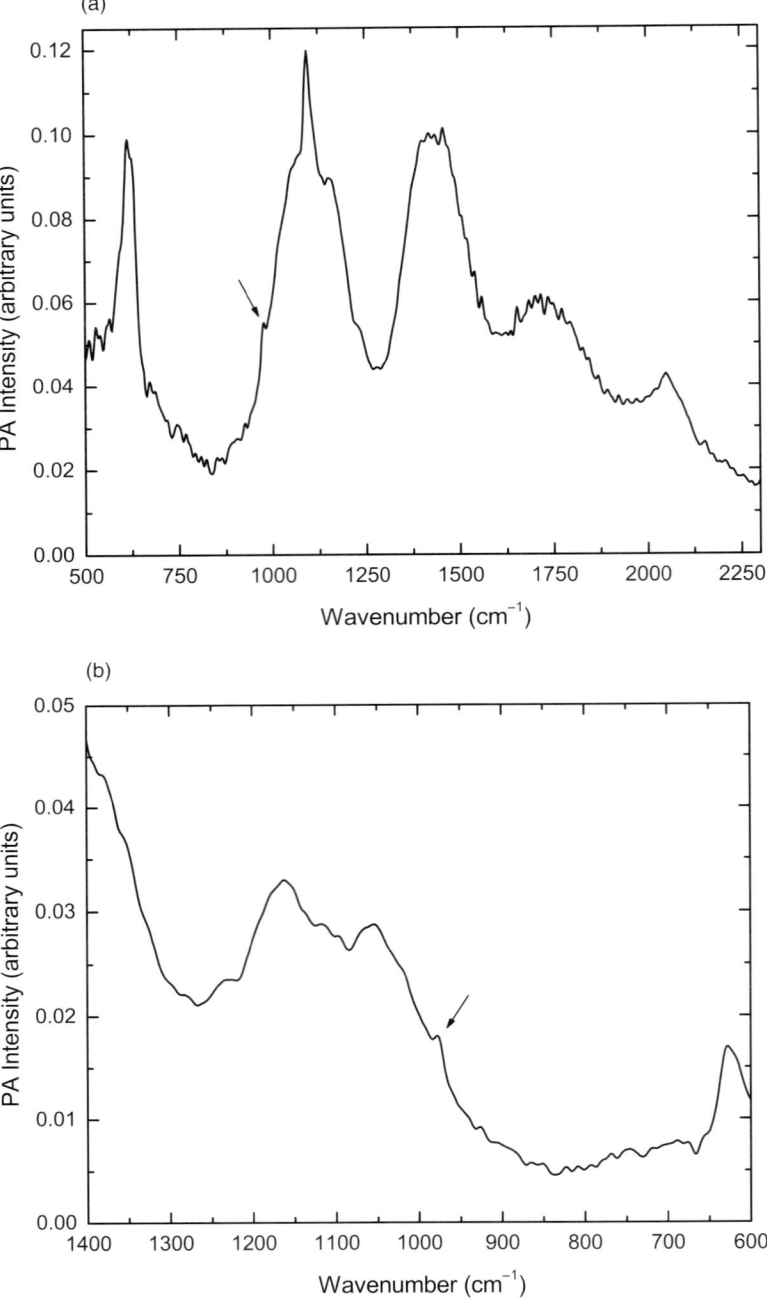

Figure 2.3 (a) PA FT-IR spectrum of $(NH_4)_2SO_4$, measured at a resolution of $6 cm^{-1}$ using a Bruker IFS 88 FT-IR spectrometer and MTEC 200 PA cell. Arrow indicates the 975-cm^{-1} band. (b) Expanded plot, with a wavenumber scale similar to that used by Rockley (1980a).

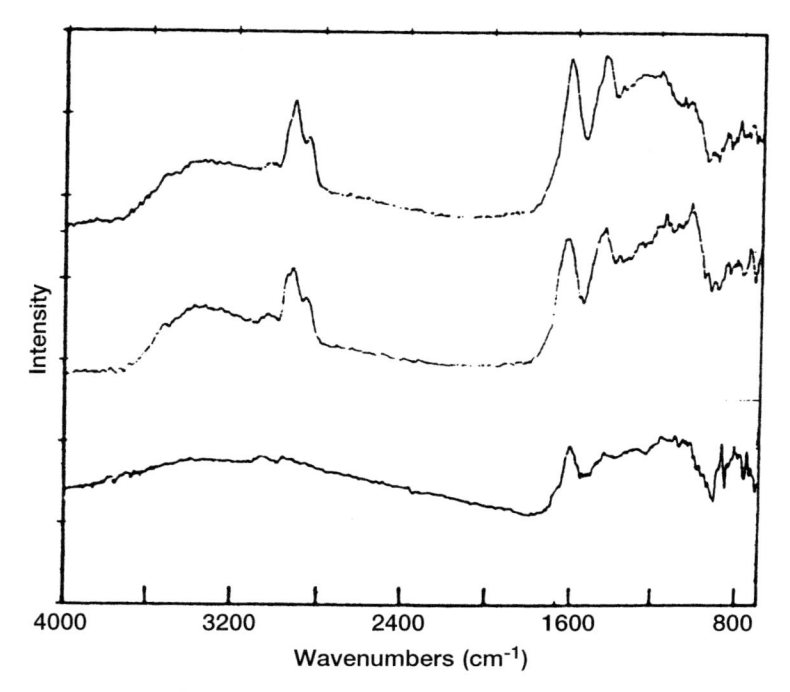

Figure 2.4 PA infrared spectra of three coals. Upper curve, Illinois No. 6; middle curve, Pittsburgh bituminous; bottom curve, Reading anthracite. (Reproduced from Rockley, M.G. and Devlin, J.P., *Appl. Spectrosc.* **34**: 407–408, by permission of the Society for Applied Spectroscopy; copyright © 1980.)

differences were also observed among the PA spectra of the aged coal surfaces. These results are discussed in more detail in Section 7.2.

The final publication by Rockley's group to be discussed in this historical review reports early work on PA FT-IR spectroscopy of solids, with particular reference to the capabilities of the technique for quantitative analysis (Rockley, Richardson, and Davis, 1980). The first attempt at quantification by these authors utilized previously melted binary mixtures of naphthalene and benzophenone; this system did not display band intensities that were proportional to the concentrations of the components, possibly because of the formation of eutectic mixtures. However, the second example considered in this work (mixtures of $K^{14}NO_3$ and $K^{15}NO_3$) did exhibit a linear relationship between intensity and concentration. This result led Rockley *et al.* to conclude that PA spectroscopy is, in fact, suitable for quantitative analysis. This topic is considered in more detail in Chapter 5.

In the same publication, Rockley, Richardson, and Davis (1980) carried out one of the first investigations of the effect of particle size on PA infrared spectra. Silica powder with particle sizes ranging up to 60 μm was separated into five different size fractions. For particles up to about 25 μm, the ratio of PA intensity at $1100\,cm^{-1}$ to that at $2000\,cm^{-1}$ decreased as particle size increased; on the other hand, PA intensity was approximately constant between 25 and 60 μm. The results were interpreted as an indication that analyte bands become visible in PA spectra only when particle sizes are similar to or less than the wavelength of the incident light—a model that does not seem to have been invoked in subsequent research.

Important exploratory work in PA infrared spectroscopy was carried out at Nicolet Instrument Corporation (now known as Thermo Nicolet) in the period between 1979 and 1981 by Mead, Vidrine, Lowry, and other researchers. Some of the earliest articles written by this group are listed in Table 2.2. A review by Vidrine that also dates from this period was mentioned in Chapter 1.

In a brief—but important—conference abstract Mead *et al.* (1979) compared PA and transmission (KBr pellet) FT-IR spectra of a Japanese coal. There are significant differences between the two spectra. For example, the authors noted that the PA spectrum clearly displayed bands from aliphatic and aromatic C–H groups, whereas the transmission spectrum was especially sensitive to hydroxyl oxygen content. In the latter case, the infrared spectrum is affected by the interaction between the KBr matrix and the coal; this well-known phenomenon arises from the tendency for small quantities of water retained by KBr to form hydrogen bonds with oxygen-containing functional groups in coals. This phenomenon is completely eliminated when PA infrared spectra are obtained for neat (undiluted) coal samples. Hence this preliminary result illustrates an important advantage of PA spectroscopy regarding coal characterization. Other problems associated with the use of KBr pellets are discussed elsewhere in this book.

Mead, Lowry, and Anderson (1981) next described much more extensive PA FT-IR studies of a variety of samples, including LiF, Ge, Si, and GaAs. Significantly, near-, mid- and far-infrared data were presented in this work; both powders and crystals were studied. The effects of dopants on near-infrared band edges were considered for the latter three samples, all of which are semiconductors. By contrast, mid- and far-infrared PA spectra were recorded for LiF. A broad peak near 600 cm^{-1} in the spectrum of LiF powder exhibited possible fine structure, which the authors attributed to surface vibrational modes and a change in the usual selection rules that govern the infrared spectrum of LiF. This, in turn, led to the suggestion that PA spectroscopy might be used to study surface vibrational modes, a point discussed above with regard to the work of Nordal and Kanstad (1977a, 1977b).

Two additional articles by D. W. Vidrine published during this period are relevant to this discussion. In the first, Vidrine (1980) discussed PA FT-IR spectroscopy of solids, particularly emphasizing the fact that traditional sample preparation is not required in this technique. To demonstrate the advantages arising from the use of PA spectroscopy during the analysis of difficult samples, the author acquired PA FT-IR spectra for a series of commercial materials, including plastic, polyurethane chips, phenoxy pellets, liquid lecithin, and a pharmaceutical tablet. The relative insensitivity of PA spectroscopy to sample morphology is demonstrated by the high quality of these spectra; a comparison of PA spectra of a nitrile plastic—examined as a powder, with sawn or smooth surfaces, and as pellets—affirmed this conclusion. PA spectra of coals were also investigated in this work. In fact, PA spectra of both low- and high-aromatic coals were shown to display better band definition and signal/noise ratios than those observed in the diffuse reflectance spectra of the same coals.

This article also discussed the use of a carbon black spectrum as a reference for normalizing the PA FT-IR spectrum of a sample measured under similar conditions, a procedure still widely used by many investigators. Noise levels in PA spectra were characterized; the root mean square (RMS) noise in a PA spectrum was observed to be independent of acoustic frequency, and to be negligible in most circumstances. Coherent noise was also found to be rather weak, except at very low wavenumbers where throughput is limited. The PA cell used in this work exhibited typical frequency response, the intensity diminishing as modulation frequency (f) increased. Although cell resonance was not discussed in detail, the results in this work indicate a resonance frequency of approximately 1.1 kHz. Similar resonance frequencies are observed in many other gas-microphone cells.

In a publication based on a lecture given at the 1981 International Conference on Fourier Transform Infrared Spectroscopy, Vidrine (1981) presented a general description of PA FT-IR spectroscopy of solids. The surface sensitivity of the technique was mentioned in this work. The phenomenology of the PA effect is explained in this paper, and the relative weakness of PA signals for solids – typically at least one order of magnitude smaller than those observed using ordinary infrared detectors – is stressed. The concept of depth of penetration (sampling depth) and its dependence on the thermal properties of the sample is discussed, although no estimates of this distance are given. The author stated that the PA and attenuated total reflectance (ATR) sampling methods are the only two surface-sensitive techniques for infrared spectroscopy, a generalization that could be disputed by workers who utilize other reflectance techniques.

The PA spectra illustrated by Vidrine (1981) in this article are noteworthy. For example, spectra of a catalyst were obtained at a series of different mirror velocities, providing one of the first examples of depth profiling in rapid-scan PA FT-IR spectroscopy. This experiment is mentioned again in Chapter 5. PA infrared spectra were also obtained for conducting polymers, which can be difficult to analyze because of high absorptivity at infrared wavelengths. The PA spectroscopy of gases is also mentioned in this work, as is the use of He as a carrier gas in the gas-microphone PA cell. Thus Vidrine addressed a number of significant issues in PA FT-IR spectroscopy in this early work, many of which retain their importance in current research.

The Princeton University group of Royce, Teng, and coworkers was also among the first to utilize PA FT-IR spectroscopy in the study of solid materials. In their first work, Laufer et al. (1980) described PA spectra of powdered AgCN. This material did not yield satisfactory transmission spectra when pellets or undiluted powder were examined: the Christiansen effect produced sloping baselines that depend on the refractive index of the matrix material. By contrast, PA spectra were obtained for a neat sample, and this undesirable effect was therefore absent. The authors observed that the C≡N stretching band in the PA spectrum of AgCN is much wider than the corresponding Raman peak; an increase in f did not yield a change in the width or shape of the PA band, leading to the conclusion that saturation was not the source of band broadening.

This work on PA infrared spectra of solids was expanded in another investigation by these researchers (Royce, Teng, and Enns, 1980). In addition to AgCN, polystyrene, acrylates, coal, and zeolites were included in this second work. Several features of FT-IR spectrometers particularly relevant to PA infrared spectroscopy were noted; most important, of course, are the well-known multiplex (Fellgett) and throughput advantages. It was also pointed out that a PA interferogram qualitatively resembles the interferogram in an emission experiment, rather than the more familiar transmission result where much information is confined to the narrow zero-path-difference region; the latter result is a consequence of the width of the blackbody spectrum. In most circumstances, a PA interferogram displays more intensity in the 'wings' farther from the centerburst: the dynamic range of a typical PA interferogram is much smaller than that when an optical detector is used. This implies that a single gain setting can be used in a PA FT-IR measurement, in contrast with many other applications where the gain is increased to improve accuracy away from the interferogram centerburst.

References

Adams, M.J., Beadle, B.C., and Kirkbright, G.F. (1978) Optoacoustic spectrometry in the near-infrared region. *Anal. Chem.*, **50** (9), 1371–1374.

Blank, R.E. and Wakefield, T. (1979) Double-beam photoacoustic spectrometer for use in the ultraviolet, visible, and near-infrared spectral regions. *Anal. Chem.*, **51** (1), 50–54.

Busse, G. and Bullemer, B. (1978) Use of the opto-acoustic effect for rapid scan Fourier spectroscopy. *Infrared Phys.*, **18** (5–6), 631–634.

Castleden, S.L., Kirkbright, G.F., and Menon, K.R. (1980) Determination of moisture in single-cell protein utilising photoacoustic spectroscopy in the near-infrared region. *Analyst*, **105** (1256), 1076–1081.

Kanstad, S.O. and Nordal, P.-E. (1977) Grazing incidence photoacoustic spectroscopy of species and complexes on metal surfaces. *Int. J. Quantum Chem.*, **12** (Suppl. 2), 123–130.

Kanstad, S.O. and Nordal, P.-E. (1978a) Open membrane spectrophone for photoacoustic spectroscopy. *Opt. Commun.*, **26** (3), 367–371.

Kanstad, S.O. and Nordal, P.-E. (1978b) Photoacoustic and photothermal spectroscopy – novel tools for the analysis of particulate matter. *Powder Technol.*, **22** (1), 133–137.

Kanstad, S.O. and Nordal, P.-E. (1979) Infrared photoacoustic spectroscopy of solids and liquids. *Infrared Phys.*, **19** (3–4), 413–422.

Kanstad, S.O. and Nordal, P.-E. (1980a) Photoacoustic and photothermal spectroscopy. *Phys. Technol.*, **11** (4), 142–147.

Kanstad, S.O. and Nordal, P.-E. (1980b) Photoacoustic reflection-absorption spectroscopy (PARAS) for infrared analysis of surface species. *Appl. Surf. Sci.*, **5** (3), 286–295.

Kerr, E.L. and Atwood, J.G. (1968) The laser illuminated absorptivity spectrophone: a method for measurement of weak absorptivity in gases at laser wavelengths. *Appl. Opt.*, **7** (5), 915–921.

Kirkbright, G.F. (1978) Analytical optoacoustic spectrometry. *Optica Pura y Aplicada*, **11**, 125–136.

Kreuzer, L.B. (1971) Ultralow gas concentration infrared absorption spectroscopy. *J. Appl. Phys.*, **42** (7), 2934–2943.

Laufer, G., Huneke, J.T., Royce, B.S.H., and Teng, Y.C. (1980) Elimination of dispersion-induced distortion in infrared absorption spectra by use of photoacoustic spectroscopy. *Appl. Phys. Lett.*, **37** (6), 517–519.

Lochmüller, C.H. and Wilder, D.R. (1980a) Qualitative examination of chemically-modified silica by near-infrared photoacoustic spectroscopy. *Anal. Chim. Acta*, **116** (1), 19–24.

Lochmüller, C.H. and Wilder, D.R. (1980b) Quantitative photoacoustic spectroscopy of chemically-modified silica surfaces. *Anal. Chim. Acta*, **118** (1), 101–108.

Low, M.J.D. and Parodi, G.A. (1978) Infrared photoacoustic spectra of surface species in the 4000–2000 cm^{-1} region using a broad band source. *Spectrosc. Lett.*, **11** (8), 581–588.

Low, M.J.D. and Parodi, G.A. (1980a) Infrared photoacoustic spectra of solids. *Spectrosc. Lett.*, **13** (2–3), 151–158.

Low, M.J.D. and Parodi, G.A. (1980b) Carbon as reference for normalizing infrared photoacoustic spectra. *Spectrosc. Lett.*, **13** (9), 663–669.

Low, M.J.D. and Parodi, G.A. (1980c) Infrared photoacoustic spectroscopy of surfaces. *J. Mol. Struct.*, **61**, 119–124.

Low, M.J.D. and Parodi, G.A. (1980d) Infrared photoacoustic spectroscopy of solids and surface species. *Appl. Spectrosc.*, **34** (1), 76–80.

Low, M.J.D. and Parodi, G.A. (1980e) An infrared photoacoustic spectrometer. *Infrared Phys.*, **20** (5), 333–340.

Max, E. and Rosengren, L.-G. (1974) Characteristics of a resonant opto-acoustic gas concentration detector. *Opt. Commun.*, **11** (4), 422–426.

Mead, D.G., Lowry, S.R., and Anderson, C.R. (1981) Photoacoustic infrared spectroscopy of some solids. *Int. J. Infrared Millimeterf Waves*, **2** (1), 23–34.

Mead, D.G., Lowry, S.R., Vidrine, D.W., and Mattson, D.R. (1979) Infrared spectroscopy using a photoacoustic cell, Fourth international conference on infrared and millimeter waves and their applications, p. 231.

Monchalin, J.-P., Gagné, J.-M., Parpal, J.-L., and Bertrand, L. (1979) Photoacoustic spectroscopy of chrysotile asbestos using a cw HF laser. *Appl. Phys. Lett.*, **35** (5), 360–363.

Nordal, P.-E. and Kanstad, S.O. (1977a) Photoacoustic spectroscopy on ammonium sulphate and glucose powders and their aqueous solutions using a CO_2 laser. *Opt. Commun.*, **22** (2), 185–189.

Nordal, P.-E. and Kanstad, S.O. (1977b) Infrared photoacoustic spectroscopy for studying surfaces and surface-related effects. *Int. J. Quantum Chem.*, **12** (Suppl. 2), 115–121.

Nordal, P.-E. and Kanstad, S.O. (1978) Photoacoustic reflection-absorption spectroscopy (PARAS) of thin oxide films on aluminium. *Opt. Commun.*, **24** (1), 95–99.

Patel, C.K.N., Kerl, R.J. and Burkhardt, E.G. (1977) Excited-state spectroscopy of molecules using opto-acoustic detection. *Phys. Rev. Lett.*, **38** (21), 1204–1207.

Perlmutter, P., Shtrikman, S., and Slatkine, M. (1979) Optoacoustic detection of ethylene in the presence of interfering gases. *Appl. Opt.*, **18** (13), 2267–2274.

Rockley, M.G. (1979) Fourier-transformed infrared photoacoustic spectroscopy of polystyrene film. *Chem. Phys. Lett.*, **68** (2–3), 455–456.

Rockley, M.G. (1980a) Reasons for the distortion of the Fourier-transformed infrared photoacoustic spectroscopy of ammonium sulfate powder. *Chem. Phys. Lett.*, **75** (2), 370–372.

Rockley, M.G. (1980b) Fourier-transformed infrared photoacoustic spectroscopy of solids. *Appl. Spectrosc.*, **34** (4), 405–406.

Rockley, M.G., and Devlin, J.P. (1980) Photoacoustic infrared spectra (IR-PAS) of aged and fresh-cleaved coal surfaces. *Appl. Spectrosc.*, **34** (4), 407–408.

Rockley, M.G., Richardson, H.H., and Davis, D.M. (1980c) Fourier-transformed infrared photoacoustic spectroscopy, the technique and its applications, in Ultrasonics Symposium Proceedings, vol. 2, pp. 649–651.

Rosengren, L.-G. (1973) A new theoretical model of the opto-acoustic gas concentration detector. *Infrared Phys.*, **13**, 109–121.

Royce, B.S.H., Teng, Y.C., and Enns, J. (1980) Fourier transform infrared photoacoustic spectroscopy of solids, in Ultrasonics Symposium Proceedings, vol. 2, pp. 652–657.

Vidrine, D.W. (1980) Photoacoustic Fourier transform infrared spectroscopy of solid samples. *Appl. Spectrosc.*, **34** (3), 314–319.

Vidrine, D.W. (1981) Photoacoustic Fourier transform infrared (FTIR) spectroscopy of solids. *Proc. SPIE Int. Soc. Opt. Eng.*, **289**, 355–360.

3
Instrumental Methods

3.1
Dispersive PA Infrared Spectroscopy

The discussion of multiple-wavelength PA infrared spectroscopy in Chapter 2 mentioned a number of early investigations that were carried out using dispersive (scanning) spectrometers. This work, which can be regarded as the logical extension of previous research in ultraviolet and visible PA spectroscopy, began in the near-infrared region and later expanded to include mid-infrared wavelengths. Dispersive PA mid-infrared spectroscopy, in turn, was eventually displaced by multiplex (FT-IR) PA spectroscopy. The use of scanning spectrometers for the acquisition of mid-infrared PA spectra is now therefore primarily of historical interest. By contrast, dispersive near-infrared PA spectroscopy continues to be utilized by a number of research groups, often in conjunction with work in the near-ultraviolet and visible regions. Dispersive PA infrared spectroscopy is briefly reviewed in this section, both for completeness and to place this technique in context with the discussion of other experimental PA methods in later sections of this chapter.

3.1.1
Near-Infrared PA Spectroscopy

The rapid growth of interest in PA and photothermal science in the 1970s and early 1980s prompted numerous workers to initiate research programs that involved dispersive near-infrared PA spectroscopy. Representative publications from four different groups are listed among the references at the end of this chapter. The reader should understand that this list is not meant to be comprehensive: the articles were chosen to illustrate the variety of different applications of the technique and to recognize some of the most well-known research groups that were active in this area at that time.

G. F. Kirkbright and his colleagues at Imperial College, London, published several articles describing their work in dispersive PA near-infrared spectroscopy before turning their efforts to multiplex spectroscopy and longer wavelengths. For example, Kirkbright (1978) presented a very useful general review of dispersive

Photoacoustic IR Spectroscopy. 2nd Ed., Kirk H. Michaelian
Copyright © 2010 WILEY-VCH Verlag GmbH & Co. KGaA, Weinheim
ISBN: 978-3-527-40900-6

PA spectroscopy, with analytical applications that were drawn from a variety of different fields. This reference is mentioned several times in this book. Some of the results that it describes were taken from other important articles by this group, which are discussed in more detail in Chapter 7. Related subsequent publications on dispersive PA near-infrared spectroscopy by these researchers discussed aromatic hydrocarbons (Adams, Beadle, and Kirkbright, 1978), single-cell protein (Castleden, Kirkbright, and Menon, 1980), and pharmaceutical drugs (Castleden, Kirkbright, and Long, 1982). These investigators made many significant early contributions to PA infrared spectroscopy, although they generally did not specialize in particular analytical applications.

Several PA near-infrared spectrometers were commercially available during this period. Blank and Wakefield (1979) described the design and construction of a dual-beam spectrometer at Gilford Instrument Laboratories. Both solid and liquid samples were successfully analyzed with this apparatus. Similarly, Lochmüller and Wilder (1980a, 1980b) utilized a Princeton Applied Research spectrometer to study the bonding of hydrocarbons to silica gel. The results of their work are described more fully in Chapters 2 and 7. (As an aside, it can be noted that the PA cells in these Princeton spectrometers were subsequently modified for use in FT-IR spectrometers. One such cell in the author's laboratory has remained operational for more than two decades.) In addition, a number of other research groups successfully utilized dispersive PA spectrometers manufactured by EDT Research.

Finally, the work of P. S. Belton and his collaborators at the AFRC Institute of Food Research, Norwich, on PA near- and mid- infrared spectra of food products should be specifically mentioned. This team utilized both dispersive and FT spectrometers in their research. Examples of their findings in dispersive near-infrared PA spectroscopy are given in works by Belton and Tanner (1983), and Belton, Wilson, and Saffa (1987). This group made numerous important contributions with regard to the quantitative analysis of solid mixtures and to the PA infrared spectroscopy of foods; their work in these areas is summarized in Chapters 5 and 7, respectively.

3.1.2
Mid-Infrared PA Spectroscopy

As mentioned above, dispersive mid-infrared PA instrumentation was employed for a limited period before the evolution and widespread adoption of PA FT-IR spectroscopy. The two well-known positive attributes of FT-IR spectrometers (the Fellgett and Jacquinot advantages) probably made this progression inevitable. Nevertheless, viable research was conducted using dispersive mid-infrared PA spectroscopy for a number of years. Some important aspects of this work are mentioned in the following paragraphs.

The work of M. J. D. Low and collaborators at New York University is discussed in several places in this book. These researchers published an extensive literature on the PA infrared spectra of solids and surface species, one of their major

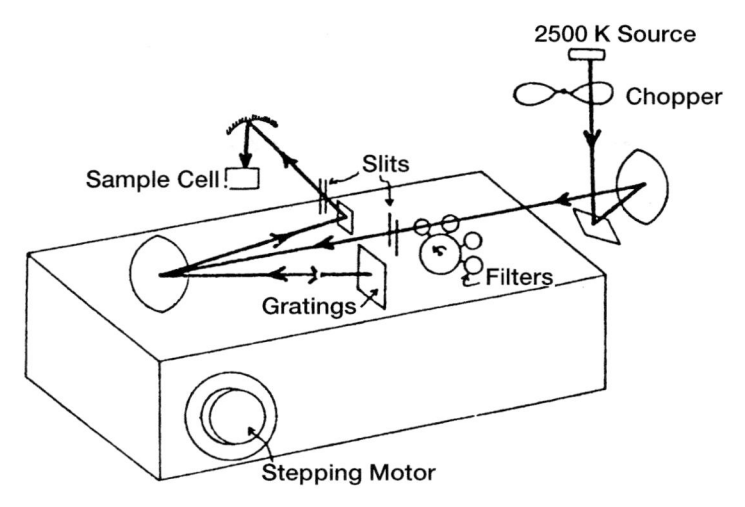

Figure 3.1 Optical system for dispersive mid-infrared PA spectroscopy. (Reprinted from Infrared Phys. **20**, Low, M.J.D. and Parodi, G.A., An infrared photoacoustic spectrometer, 333–340, copyright © 1980, with permission from Elsevier Science.)

contributions being a long series of studies of carbons and chars. In fact, much of the early work of this group was performed using dispersive spectrometers. Some representative publications are listed among the references at the end of this chapter. It should be noted that studies carried out using photothermal beam deflection (mirage) spectroscopy are discussed separately later in this chapter.

Several relevant publications by this group are already mentioned in the historical review in Chapter 2. Among these, an important series of studies by Low and Parodi (1978, 1980a, 1980b, 1980c, 1980d, 1980e) describe the dispersive PA mid-infrared spectra of various solids. A drawing of the apparatus used in some of these experiments is shown in Figure 3.1 (Low and Parodi, 1980c). In another configuration of this equipment, LiF and CaF_2 prisms were used instead of a grating to disperse the infrared radiation (Low and Parodi, 1978, 1980a, 1980d). Three different infrared spectrometers were successfully utilized in this project (Low and Parodi, 1982). A wide variety of surface species, including adsorbents, catalysts, coals, carbons, and corrosion products, were studied with this equipment. However, despite the success of these pioneering studies, this group increasingly relied on FT-IR spectrometers in their later PA work.

It appears that very little, if any, research in PA mid-infrared spectroscopy has been performed using dispersive spectrometers in recent years. Although low-cost scanning mid-infrared spectrometers are still manufactured, the many features of modern rapid- and step-scan FT-IR spectrometers make these instruments the logical choice for the present-day PA infrared spectroscopist. Hence, the vast majority of contemporary PA infrared spectra of condensed substances are obtained using FT-IR spectrometers. This situation is reflected in the published scientific literature that is discussed throughout the rest of this book.

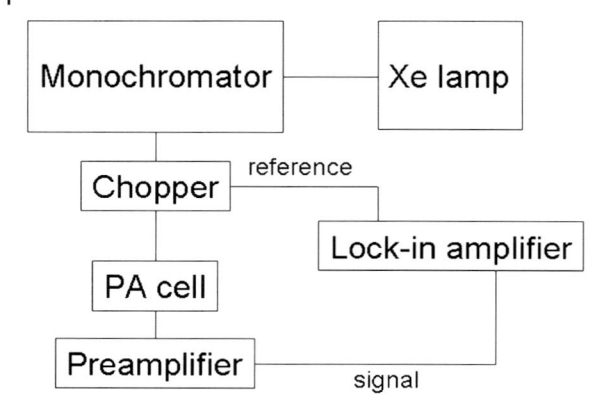

Figure 3.2 Schematic diagram of a dispersive near-infrared PA spectrometer.

3.1.3
Instrumentation

Many researchers have constructed dispersive PA spectrometers in the last three decades. As noted above, these systems are often designed for work in the ultraviolet and visible wavelength regions, although some also allow access to the near-infrared. A schematic diagram of a typical system is shown in Figure 3.2. Broadband radiation from an Xe lamp is filtered with a monochromator fitted with one or more gratings; near-infrared wavelengths are accessed using a grating blaze wavelength near 1500 nm (~6670 cm^{-1}). Wavelength-selected radiation emerging from the monochromator is chopped before it impinges on the PA cell. The preamplified signal is demodulated using a lock-in amplifier, which yields the real and imaginary (in-phase and quadrature) components. These two signals are used to calculate the PA amplitude and phase spectra. A typical near-infrared PA amplitude spectrum for carbon black, which contains a series of bands due to Xe emission lines, is displayed in Figure 3.3. Dias *et al.* (2002) utilized similar instrumentation to study cross-linking in polymers; other research groups have also designed and assembled spectrometers of this type.

Two interesting examples of dispersive PA systems can be mentioned here. First, Kapil, Joshi, and Rai (2003) developed a spectrometer for use with a resonant PA cell at temperatures down to −40 °C; spectra were acquired up to 1100 nm (~9000 cm^{-1}) using an Xe source. Lines from the lamp were observed in the PA spectrum of carbon black at wavelengths greater than 800 nm (12 500 cm^{-1}). Second, Guo *et al.* (2007) inserted a prism between the monochromator and chopper, dividing the radiation into two beams at wavelengths λ and $\lambda + \Delta\lambda$ before it reached the PA cell. This arrangement produced first-derivative spectra, which can be used for background suppression or improvement of the visibility of weak bands that occur as shoulders to stronger features. Direct observation of derivative spectra is expected to facilitate the analysis of strongly scattering or opaque substances.

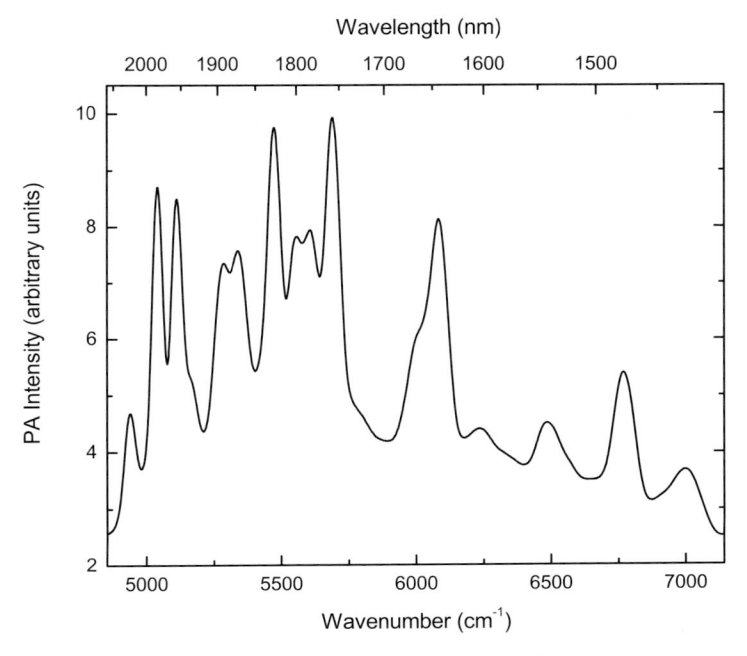

Figure 3.3 PA spectrum of carbon black obtained using the apparatus in Figure 3.2. The bands arise from Xe emission lines.

3.2
Rapid-Scan PA FT-IR Spectroscopy

The initial development of rapid-scan PA FT-IR spectroscopy was outlined in Chapter 2. At the time of writing, this technique is about 30 years old and has been the subject of more than 900 publications (see Appendices 2 and 3). Much of this literature is discussed in the remaining chapters of this book. The present section is restricted to a few general points regarding rapid-scan PA spectra of condensed-phase materials.

The major difference between rapid- or continuous-scan PA infrared spectroscopy and the dispersive method described in the preceding section exists in the fact that external modulation is not required in the rapid-scan experiment. In simple terms, the moving mirror in the interferometer provides the audio-frequency modulation necessary for the generation of a PA signal. Gas-microphone PA cells that are compatible with rapid-scan FT-IR spectrometers have been readily available for three decades; the acquisition of a PA spectrum with this equipment is a straightforward task in many circumstances. Therefore the non-specialist can approach rapid-scan PA FT-IR spectroscopy in a manner that might be appropriate for a variety of other infrared measurement techniques (e.g., diffuse reflectance, attenuated total reflectance (ATR), transmission). Qualitative – and sometimes

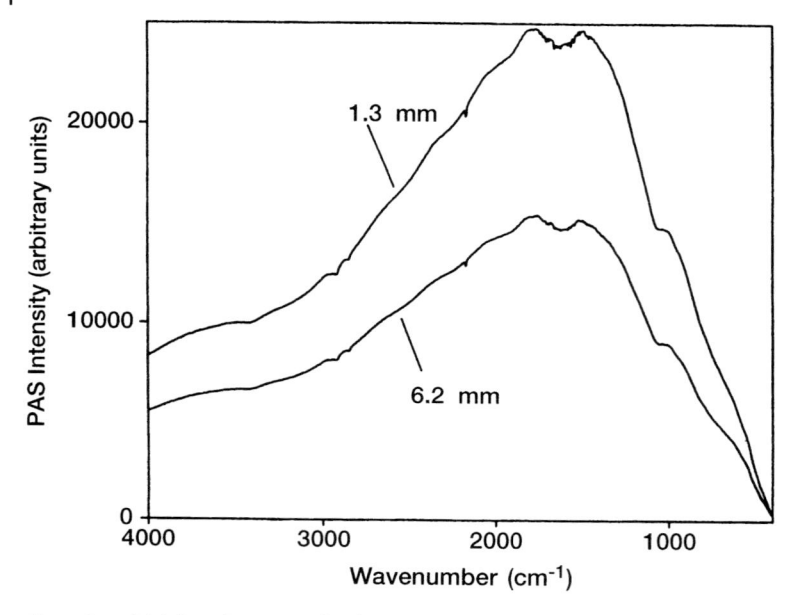

Figure 3.4 PA infrared spectra of carbon black-filled rubber at two sample-window distances. The mirror velocity was 0.08 cm s^{-1}, corresponding to f values ranging from 64 to 640 Hz in the mid-infrared. (Reproduced from Carter, R.O. and Wright, S.L., *Appl. Spectrosc.* **45**: 1101–1103, by permission of the Society for Applied Spectroscopy; copyright © 1991.)

quantitative – PA infrared spectra of a wide variety of solids and liquids are readily obtainable in this way.

The PA spectroscopist will eventually be faced with another set of less obvious issues. Some are specific to the use of the gas-microphone cell: these include effects due to limited sample quantities, the sizes of the sample cups, and perhaps the method by which they are filled. A related question discussed by Carter and Wright (1991) pertains to the position of the sample within the PA cell. As shown in Figure 3.4, these authors found that the PA signal was maximal when the distance between the sample surface and the cell window was smaller than the boundary layer (μ_g, the thermal diffusion length of the carrier gas). They also concluded that powdered carbon black was the most suitable reference material for normalizing PA infrared spectra of finely divided solids, whereas carbon-filled rubber was more appropriate for non-porous materials (Carter and Paputa Peck, 1989). The normalization of PA infrared spectra is discussed below and in Chapter 7.

The dependence of the PA signal on sample position was examined in more detail by Jones and McClelland (2001). The magnitude of the signal from the reference (glassy carbon), its dependence on modulation frequency, and its phase all vary with the separation between the reference material and the cell window. The values of the PA magnitude and phase at low f (low wavenumbers in rapid-scan experiments) differ significantly from those observed at high f. These observations were put down to the fact that the walls of the cell act as a heat sink in these experiments. The cell window and microphone also contribute to the observed non-ideal behavior. The sample position should be kept constant for accurate work, thereby reducing the number of variables that affect PA intensities.

When modern FT-IR spectrometers became widely available in the 1970s, many infrared spectroscopists were pleasantly surprised at the very high signal/noise ratios that these instruments afforded under many circumstances. The excellent performance of these spectrometers led to greatly reduced acquisition times and encouraged implementation of numerical techniques such as deconvolution, curve-fitting, and chemometric methods, all of which are adversely affected by noise in the spectra to be analyzed.

Despite the very good performance of these spectrometers, many early rapid-scan PA FT-IR spectra were badly corrupted by noise. In fact, typical signals from some PA cells were weaker than those obtained using optical detectors by one to two orders of magnitude; even weaker signals were observed for some samples. Random noise in these spectra was partially eliminated by signal averaging, although at the expense of greatly increased measurement times. Environmental noise was reduced by interrupting the flow of purge gas during data acquisition, thereby improving signal/noise ratios in rapid-scan PA spectra (Donini and Michaelian, 1988). Moreover, considerable improvement was achieved with acoustic isolation of the PA cell (Duerst and Mahmoodi, 1984; Mahmoodi, Duerst, and Meiklejohn, 1984; Duerst, Mahmoodi, and Duerst, 1987). However, it should be noted that the environmental isolation required for some PA cells of older design may not be necessary with modern accessories.

Generation of the PA signal in a rapid-scan FT-IR experiment was discussed in detail by McClelland *et al.* in several reviews cited earlier in this book and briefly mentioned in Chapter 1. The phenomenology of the experiment as outlined by McClelland *et al.* (1992) is summarized as follows. Modulated infrared radiation emerges from the interferometer and impinges on the PA gas-microphone cell containing a sample of interest. Part of this radiation is absorbed at wavelengths dictated by the optical properties of the sample. Absorption induces temperature oscillations in a layer at a depth x and thickness dx that are proportional to $P_o(1 - R)\beta e^{-\beta x}$, where P_o is the incident power, R is the fraction of incident light reflected by the surface of the sample, and β is the absorption coefficient. The temperature oscillations generate thermal waves that decay according to $\mu_s = (\alpha/\pi f)^{1/2}$, where μ_s is the thermal diffusion length and α is the thermal diffusivity of the sample (these and other relevant terms are defined in Appendix 1). A small fraction of each thermal wave is transferred to the layer of carrier gas immediately adjacent to the sample. The resulting pressure wave is detected by a sensitive microphone (transducer) in the PA cell. This microphone produces a voltage that is amplified and then input to the FT-IR electronics for further processing. The signal from the microphone thus replaces that from an optical detector which might be used in a transmission, diffuse reflectance, or ATR experiment.

3.2.1
Normalization of Rapid-Scan Spectra

Because PA FT-IR spectra are measured in single-beam experiments, it is generally desirable that they be corrected for wavenumber-dependent instrumental effects through division by a reference ('background') spectrum. The reference spectrum can be regarded as an energy curve or a representation of the optical throughput for an 'empty instrument'; in other words, its profile is dictated by the optical and electronic characteristics of the spectrometer and detector. In PA infrared spectroscopy, reference spectra are commonly obtained using powdered carbon black, glassy carbon, and carbon-filled polymers. These reference data are used to normalize PA spectra of ordinary samples.

The normalization of a typical PA infrared spectrum of a clay mineral is illustrated in the next four figures. Figure 3.5 depicts the data used in the calculations, which were obtained using a Bruker IFS 113v FT-IR and an MTEC 300 PA cell. The upper curve – the carbon black reference spectrum – is featureless except for broad beamsplitter absorption at about 1000–1300 cm^{-1} and minor bands arising from ambient water vapor and CO_2. The lower curve shows the PA spectrum measured under similar conditions for kaolin, a common layer silicate with empirical composition $Al_4Si_4O_{10}(OH)_8$. This is the spectrum to be corrected.

Division of the kaolin spectrum by the carbon black spectrum yields the result in Figure 3.6. It is evident that this calculation has its most noticeable effects near the limits of the mid-infrared region, where the intensity of the incident radiation is lower than that in the central part (roughly 700–2500 cm^{-1}). This ratioed spec-

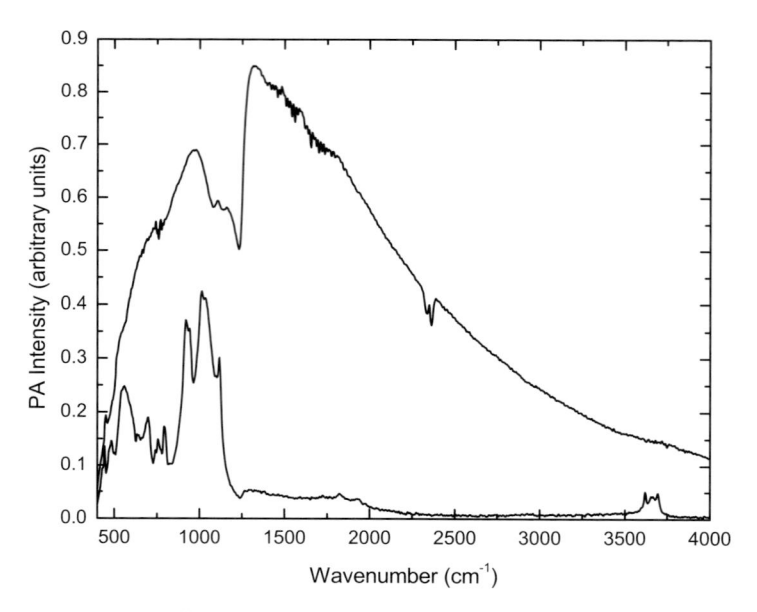

Figure 3.5 PA infrared spectra of carbon black (upper curve) and kaolin (lower curve).

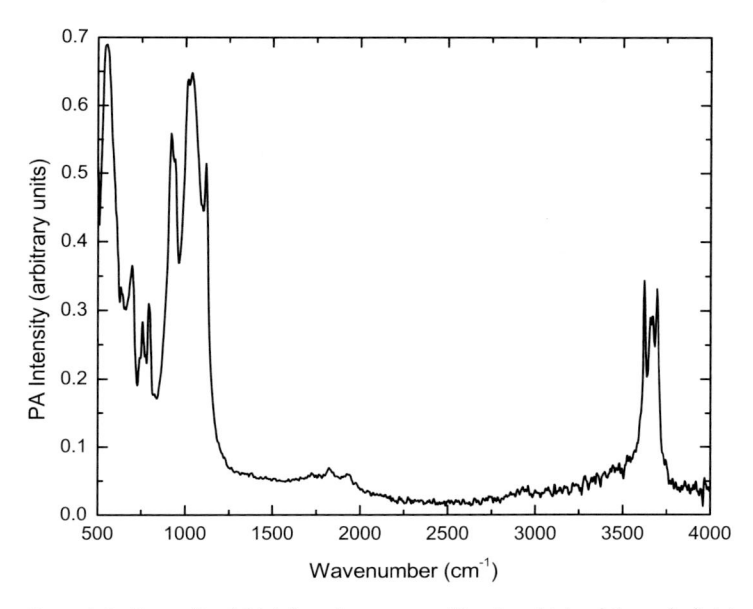

Figure 3.6 Normalized PA infrared spectrum of kaolin, obtained through division of the lower spectrum in Figure 3.2 by the upper spectrum.

Table 3.1 Rosencwaig-Gersho theory: Dependence of PA signal on $f^{a)}$.

		Frequency dependence
Thermally thin ($\mu_s > l$)	$\mu_s > \mu_\beta$ or $\mu_s < \mu_\beta$	f^{-1}
Thermally thick ($\mu_s < l$)	$\mu_s > \mu_\beta$	
	$\mu_s < \mu_\beta$	$f^{-3/2}$

a) $\mu_s = (\alpha/\pi f)^{1/2}$, thermal diffusion length; $\mu_\beta = 1/\beta$, optical absorption length; l = sample dimension.

trum would generally be considered adequate for most qualitative and semi-quantitative work.

The observant reader may recognize that the hydroxyl stretching bands in the 3600–3700 cm^{-1} region in Figure 3.6 are somewhat weaker relative to the low-wavenumber bands than those in published infrared spectra of this clay, most of which were obtained by transmission measurements on alkali halide pellets. There is one obvious reason for this discrepancy. According to Rosencwaig and Gersho (1976), PA intensity is proportional to $f^{-3/2}$ for an optically opaque, thermally thick solid. By contrast, PA intensity varies as f^{-1} for an optically opaque, thermally thin solid such as carbon black (Table 3.1). The second relationship is demonstrated in Figure 3.7, which shows the frequency dependence of the PA intensity for carbon black in a simple experiment performed with an HeNe laser, mechanical chopper, and lock-in amplifier.

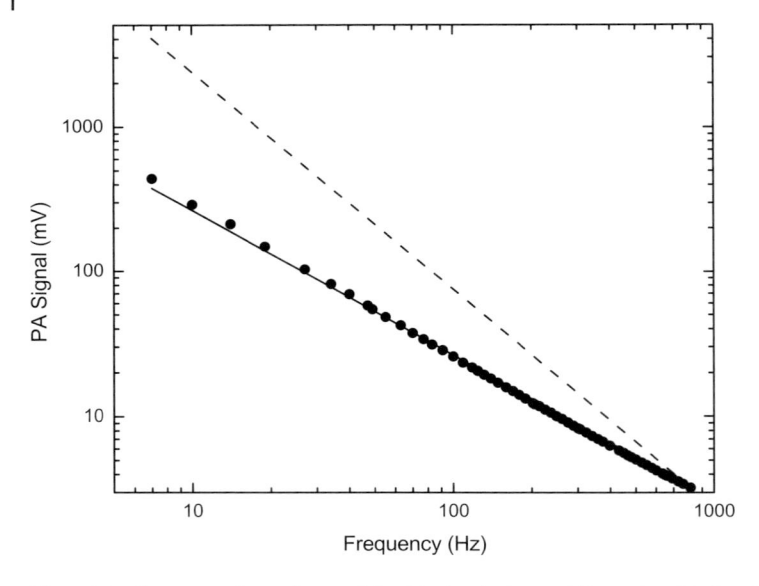

Figure 3.7 Frequency dependence of the PA signal observed for carbon black using an MTEC 300 cell, an HeNe laser, mechanical chopper, and lock-in amplifier. Solid and dashed lines have slopes of −1 and −1.5, respectively.

In rapid-scan PA FT-IR spectroscopy, $f \propto \nu$ (ν = wavenumber). Therefore the ratioed intensity should be multiplied by $\nu^{1/2}$ at each spectral ordinate to correct for the different frequency dependencies of the sample and reference intensities. This procedure yields the result plotted in Figure 3.8. As expected, the redistribution of intensities effected by this calculation produces a PA spectrum that agrees much more closely with published absorbance spectra of this clay. Although this correction strategy is not widely utilized, its simplicity and effectiveness suggest that it be considered when a qualitative comparison of PA and absorption spectra is required.

Glassy carbon is frequently used as an alternative to carbon black powder for measurement of a PA reference spectrum. The single-wavelength (632.8-nm) experiment mentioned above was repeated for the glassy carbon reference provided with the MTEC 300 PA cell, yielding a frequency dependence close to $f^{-3/2}$. This implies that correction of a ratioed PA spectrum through multiplication by $\nu^{1/2}$ is unnecessary when this reference material is used.

3.3
Step-Scan PA FT-IR Spectroscopy

The many applications of step-scan PA FT-IR spectroscopy described throughout this book attest to the considerable importance of this method. The reader will

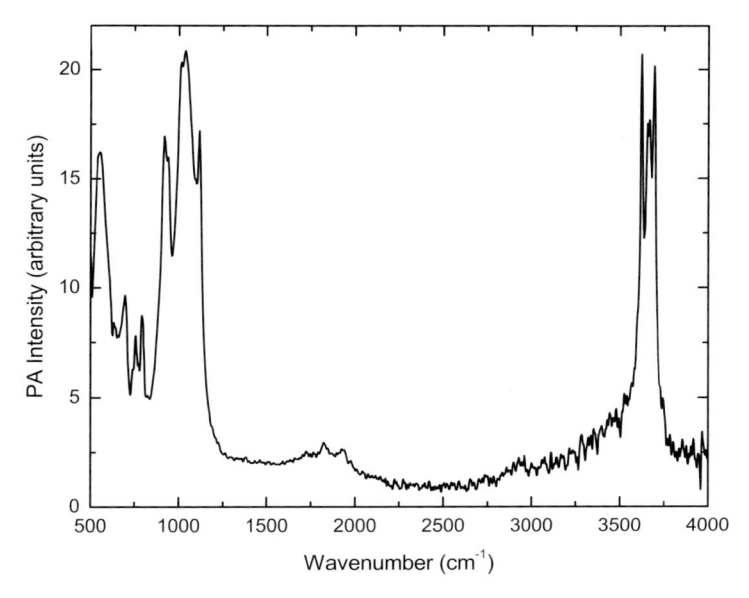

Figure 3.8 Normalized PA infrared spectrum of kaolin, calculated by multiplying the spectrum in Figure 3.3 by $v^{1/2}$.

recognize that the predominant use of step-scan spectra documented here has been in depth profiling experiments. In addition, because step-scan measurements involve the use of a constant modulation frequency, spectra obtained in this way do not exhibit the wavenumber-dependent frequency effects that affect rapid-scan PA infrared spectra. Consequently, step-scan spectra are a closer approximation to ordinary absorption spectra in many situations.

Initial step-scan PA depth profiling results, discussed in Chapter 5, were obtained using amplitude (chopper) modulation in the late 1980s. Phase modulation was successfully implemented in PA step-scan experiments at about the same time and has proven to be the more popular technique during the last two decades; this is mainly due to its ready availability and ease of use. Because of this circumstance, a considerable literature on applications of step-scan phase modulation PA spectroscopy has developed. This literature is reviewed in Chapter 5. Several references of a more general nature on step-scan spectroscopy are discussed in this section.

As a preliminary comment, it is appropriate to mention that the capability for rapid-scan movement of the mobile interferometer mirror was a key feature of the new generation of FT-IR spectrometers that became commercially available in the 1970s. In fact, many older interferometers utilized step-scan mirror movement: the detector signal was digitized and recorded at each resting position of the movable mirror for later processing. This data acquisition strategy was reliable but slow, and infrared spectroscopists generally welcomed the availability of continuous ('rapid') scanning. However, researchers interested in PA FT-IR spectroscopy soon recognized that the rapid-scan technique presented its own set of problems because each spectral ordinate is associated with a different f. It is well known that identification of saturation effects and depth profiling experiments are both complicated by the existence of a range of frequencies in a single spectrum. The reintroduction of step-scan spectrometers eliminated these problems and was therefore welcomed by PA spectroscopists. Of course, these step-scan instruments are also utilized by researchers who investigate time-dependent phenomena.

Before commercial step-scan FT-IR spectrometers became available, a few individual research laboratories developed purpose-built instruments that operated at shorter wavelengths. Lloyd et al. (1980) described a step-scan spectrometer for PA spectroscopy in the visible region. The interferometer incorporated an inchworm translator that controlled the moving mirror; the other mirror in the interferometer provided phase modulation. A lock-in amplifier was employed to demodulate the signal from a piezoelectric transducer. Spectra were obtained in the 400–700 nm region with this equipment.

Pioneering work in this field was also performed by Débarre, Boccara, and Fournier (1981), who constructed a step-scan, or step-and-integrate, interferometer for visible and near-infrared PA spectroscopy. These authors also utilized internal (phase) modulation, noting that a mechanical chopper can give rise to a coherent stray signal. Moreover the derivative of the direct current (DC) interferogram, which exhibits a mean value of zero, is measured with this technique. This allows more precise measurement of weak signals. Spectra were obtained for carbon black and rare-earth oxides with this apparatus. In-phase and quadrature spectra were compared, allowing identification of the surface and bulk contributions to the PA spectra.

The Bruker IFS 88 FT-IR was introduced as the first modern commercial spectrometer with both rapid- and step-scan capabilities in 1987. During the next few years, several other instrument manufacturers developed similar instrumentation, and today's PA spectroscopist can choose from a number of viable options. Signal demodulation has also undergone major advances. Whereas a lock-in amplifier was originally required for signal recovery in a step-scan PA experiment, DSP demodulation now makes it possible to perform signal recovery entirely in software. The simultaneous detection of PA signals at different frequencies or phase angles further facilitates depth profiling measurements.

The research group led by R. A. Palmer at Duke University modified several conventional rapid-scan FT-IR instruments to enable alternative step-scan operation. For example, Smith et al. (1988) described a number of changes made to an IBM Instruments IR/44 spectrometer. The position of the movable mirror was controlled by a feedback loop that monitored the interference pattern of the HeNe laser. The length of each step was 1/4 of the HeNe laser wavelength, that is, about 0.158 µm; this yielded a wide free spectral range that included the near-infrared and visible regions as well as the far- and mid-infrared. Both amplitude and phase modulation PA experiments were carried out with this instrument. The authors

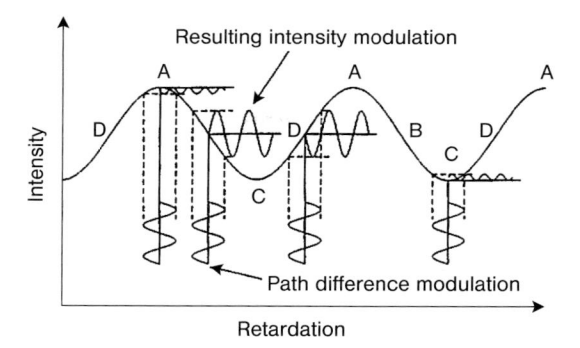

Figure 3.9 Simplified representation of modulation in a step-scan spectrometer. (Reproduced from Manning, C.J., Palmer, R.A. and Chao, J.L., *Rev. Sci. Instrum.* **62**: 1219–1229; used with permission. Copyright © American Institute of Physics, 1991.)

noted that phase modulation measures only the alternating current (AC) component of the interferogram; by contrast, amplitude modulation also produces a large DC background. This point was noted above in the work of Débarre, Boccara, and Fournier (1981) and is discussed further in Chapter 5.

This instrument was described in greater detail in a subsequent publication (Manning, Palmer, and Chao, 1991). The modulation scheme in this spectrometer is depicted schematically in Figure 3.9. Particular emphasis was placed on instrumentation and the mirror position control circuit in this second article. To demonstrate the use of this instrument, step-scan phase modulation PA spectra were obtained for carbon black-filled elastomers and adhesive tape. Step-scan phase spectra of the elastomers exhibited details at carbon loadings as high as 60%, whereas rapid-scan amplitude spectra were already saturated at lower carbon content. Moreover, spectra of the adhesive and backing of the tape were retrieved in a successful depth-profiling experiment. The potential use of step-scan spectrometers for the study of various time-dependent phenomena was also mentioned in this work.

This group next modified a Nicolet 800 FT-IR spectrometer to allow step-scan operation (Gregoriou *et al.*, 1993). The performance of this second instrument was generally superior to that of the first spectrometer described above; importantly, larger phase-modulation amplitude (up to 10 λ_{HeNe}, where $\lambda_{HeNe} = 0.6328\,\mu m$) was potentially available in the second instrument. The significance of this capability with regard to far-infrared PA spectroscopy is discussed later in this book. The authors went on to utilize this instrument for the measurement of PA spectra of polymer films, demonstrating use of the phase angle rotation method for depth profiling.

Two relevant review articles were published by this group at about the same time that the modification of the second instrument was completed. In the first, Palmer *et al.* (1993) discussed dynamic (time- and phase-resolved, i.e., time- and frequency-domain) vibrational spectroscopy and several applications of step-scan

FT-IR spectrometers. The following advantages were noted for step-scan PA spectroscopy: (i) results at near-infrared and visible wavelengths are improved because the high Fourier frequencies normally associated with these regions are replaced by much lower f values; (ii) the use of the phase spectrum extends the dynamic range, making it possible to detect weakly absorbing components within a strongly absorbing matrix; and (iii) the depth profiling capability of step-scan PA spectroscopy is superior to that in rapid-scan spectroscopy. In fact, a depth resolution of about 1 μm is attainable in step-scan spectroscopy under favorable circumstances. The second review (Palmer, 1994) surveyed time- and phase-resolved techniques in vibrational spectroscopy. PA spectroscopy is mentioned briefly, together with a number of other infrared techniques. No PA spectra are presented in this work.

At the time of writing, a number of commercial research-grade FT-IR spectrometers incorporating step-scan mirror movement are available. These spectrometers operate at far-, mid- and near-infrared wavelengths. Hence, researchers in PA infrared spectroscopy can select from instruments offered by many instrument manufacturers, with no requirement for optical or electronic equipment development. This situation encourages the continued and expanded implementation of step-scan PA infrared spectroscopy.

3.4
Mid-Infrared Laser PA Spectroscopy

The use of lasers as radiation sources in mid-infrared PA spectroscopy was reviewed in a historical context in Chapters 1 and 2. Although a large fraction of modern mid-infrared PA research is carried out using FT-IR spectrometers, laser-based PA spectroscopy continues to be the preferred alternative in some circumstances. PA investigations of solids and liquids using mid-infrared lasers are described in this section, while gas-phase PA spectroscopy is discussed in Chapter 7.

3.4.1
CO_2 Laser PA Spectroscopy

As might be expected, the CO_2 laser is the most common choice among PA spectroscopists who investigate either condensed matter or gases. Pioneering work by Kanstad and Nordal was mentioned in the first two chapters of this book. A number of additional research teams have performed similar investigations during the last three decades. Representative work is briefly outlined in this section.

Studies of adsorbed species on silver surfaces using CO_2 laser PA spectroscopy were described by two groups. First, Chuang, Coufal, and Träger (1983) examined SF_6 and NH_3 adsorption on silver films that were confined in an ultrahigh vacuum (UHV) chamber; a piezoceramic disc served as the transducer. Spectra of adsorbed

SF$_6$, acquired at 90 K, displayed broad maxima near 940 cm^{-1} that varied slightly with fractional surface coverage: this experiment demonstrated submonolayer sensitivity. Similar results were obtained for NH$_3$, where the intensity maximum occurred near 1080 cm^{-1}. Second, Wang et al. (1994) reported apparent torsional-rotational structure in spectra of ethanol adsorbed on a polycrystalline silver substrate; seven peaks were observed in the 1045–1054 cm^{-1} region and attributed to internal torsional rotation of the ethanol adsorbates. This work was also performed while maintaining samples at low temperature and under high vacuum. Sensitivity to 1 Langmuir (1 × 10^{-6} Torr s) exposure and a spectral resolution of 1 cm^{-1} were achieved in this work. Piezoelectric detection, discussed later in this chapter, was employed in this investigation.

CO$_2$ lasers have also been used to obtain PA spectra of liquids under ambient conditions. Bićanić et al. (1989) described a 'reverse mirage' experiment in which absorption coefficients (β) were determined for methanol at CO$_2$ laser wavelengths. This technique is discussed in some detail later in this chapter. Results for β at 10.26 and 10.59 µm agreed well with expected values at these wavelengths. Turning to a greatly different application, Harris et al. (2000) described remote detection of liquids by pulsed indirect PA spectroscopy. This technique involves detection of acoustic pulses at large distances (up to several centimeters) from a sample that is not enclosed in a gas-microphone cell. PA signal intensities were observed to vary linearly with laser pulse energies in this experiment. Although a small number of laser wavelengths were utilized, the available PA intensities for an oil sample were consistent with an ordinary absorption spectrum of the oil.

More traditional spectroscopic studies were carried out by R. L. Prasad and coworkers at Banaras Hindu University and other locations. These researchers used CO$_2$ laser PA spectroscopy to analyze three different narcotics (Prasad, Thakur, and Bhar, 2002a) and two explosives (Prasad et al., 2002b). Data were obtained in the ~920–990 and ~1020–1080 cm^{-1} regions in both studies. Calculations of molecular geometries and normal mode frequencies were also included in these studies. Despite the limited spectral coverage, approximately eight bands were identified for each material. These bands were assigned to particular vibrational modes, one of the primary objectives of infrared spectroscopists.

A previously unpublished demonstration of CO$_2$ laser PA spectroscopy for a sample of fine derivatized polystyrene beads is shown in Figure 3.10.[1] The CO$_2$ laser data in this figure (open circles connected by solid lines) were obtained using a chopper at $f = 10$ Hz and a lock-in amplifier. Power levels were about 10–20 mW for most laser lines (9P and 10R branches) in this experiment. For comparison, an FT-IR PA spectrum of the same sample is also shown in this figure (dashed line). The data from the two experiments are generally in good agreement; interestingly, a band near 1025 cm^{-1} is more well defined in the spectrum obtained using the laser as the infrared source. Results obtained for this sample using a different laser are discussed below.

1) The author is grateful to Dr. M. Petryk for his assistance with these measurements.

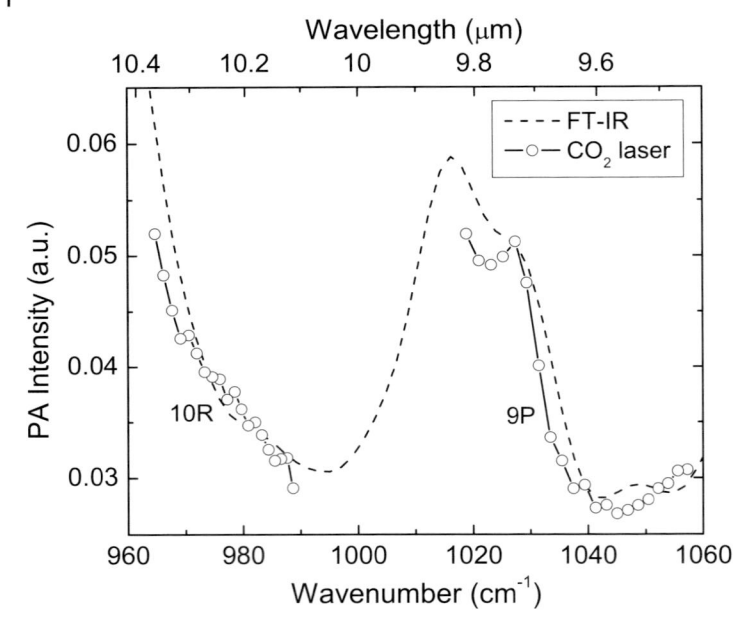

Figure 3.10 PA spectra of polystyrene beads obtained using a CO_2 laser, chopper and lock-in amplifier (solid line) and FT-IR spectrometer (dashed line). The chopper frequency was 10 Hz.

3.4.2
CO Laser PA Spectroscopy

CO lasers have been used much less extensively than the CO_2 lasers mentioned above for acquisition of mid-infrared PA spectra of solids and liquids. Two publications on CO laser PA spectroscopy by D. Bićanić at Wageningen Agricultural University and coworkers at several institutions can be mentioned in this context. In the first, Bićanić et al. (1995) used a CO laser for optothermal window (OW) experiments on oleic acid and water. The OW technique is described in detail later in this chapter. This PT method yields signals proportional to β in the weak-absorption case. Bićanić et al. determined β for oleic acid and H_2O in the 1700–1900 cm^{-1} region, where the CO laser emits more than 30 lines. The PT results agreed well with conventional absorption spectra.

A CO laser was also used for PA measurements of carbonate contents in different Hungarian soil samples (Dóka et al., 1998). A purpose-built gas-microphone PA cell was employed in this work. Measurements were performed only at 1801 cm^{-1}, a location where carbonates absorb weakly; in other words, PA spectra were not acquired in this study. The single-wavelength PA intensities varied linearly with carbonate content (determined by a chemical method) up to 30%, with a lower detection limit of 2%.

3.4.3
External-Cavity Quantum-Cascade Laser PA Spectroscopy

The external-cavity quantum-cascade laser (EC-QCL) has become an important radiation source in several types of mid-infrared spectroscopy (e.g., direct absorption, cavity ring-down, PA) during the last decade. Use of an external cavity dramatically extends QCL tuning ranges – from a few wavenumbers to intervals greater than $100\,cm^{-1}$ in some cases. EC-QCLs, which can be designed for cw or pulsed operation and are available commercially, have been successfully utilized in gas-phase PA infrared spectroscopy by a number of research groups. This work on gases is discussed in Chapter 7.

An EC-QCL has also been recently employed in PA experiments on solid samples (Wen and Michaelian, 2008). The pulsed laser used in this exploratory work had a nominal wavelength of $9.7\,\mu m$ and was tunable from about 9.30 to $10.10\,\mu m$ ($990–1075\,cm^{-1}$). Typical pulse widths ranged from 0.04 to $0.5\,\mu s$ and emission linewidths were less than $1\,cm^{-1}$. A gas-microphone PA cell and lock-in amplifier were included in this setup; no spectrometer was required for the measurements, which consisted of series of single-wavelength observations for each sample and the reference (carbon black). PA intensity was found to be proportional to the laser pulse width, that is, intensity varied linearly with the incident energy.

Figure 3.11 shows a typical EC-QCL PA spectrum (solid line) obtained with this equipment for the same acetyl polystyrene beads mentioned above. The laser pulse

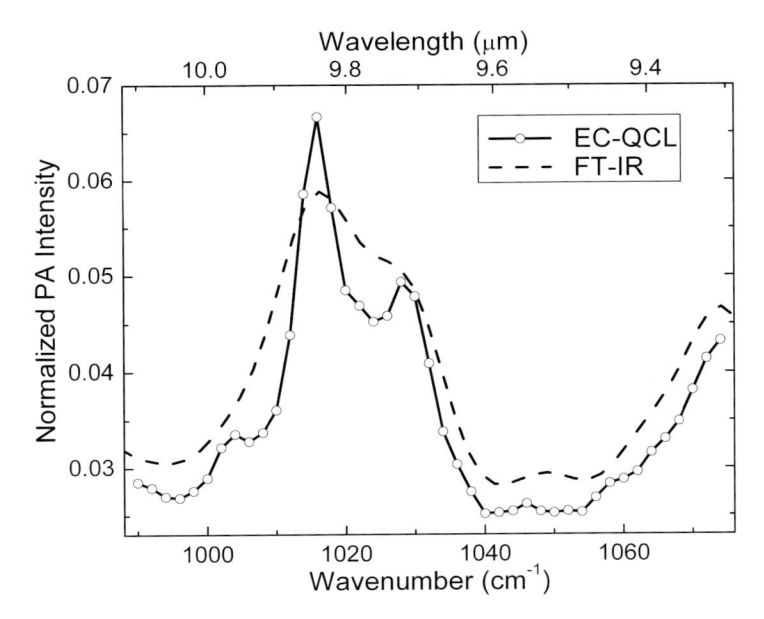

Figure 3.11 PA spectra of acetyl polystyrene beads (75–150 μm) obtained using an external cavity quantum cascade laser (solid line) and FT-IR spectrometer (dashed line). The pulse (modulation) frequency was 475 Hz in both experiments.

width and current were set at 0.5 µs and 1650 mA, respectively. Three maxima are visible between 1000 and 1040 cm^{-1} in this spectrum. The good quality of this result, which was normalized through division by a carbon black reference spectrum, demonstrates the viability of this novel method. It can be noted that the EC-QCL PA spectrum exhibits better band definition than that in the PA spectrum measured with an FT-IR spectrometer at a resolution of 6 cm^{-1} (dashed line): this result is reminiscent of the CO_2 laser PA spectrum in Figure 3.10. Improved band definition was also observed in EC-QCL PA spectra of other samples. This phenomenon may arise from reduced PA saturation due to the use of pulsed excitation or another, unknown, cause.

3.5
Quartz-Enhanced PA Spectroscopy

The research team of Kosterev, Tittel and coworkers at Rice University introduced the technique of quartz-enhanced photoacoustic spectroscopy (QEPAS) in a key publication several years ago (Kosterev *et al.*, 2002). This method is based on the use of a quartz tuning fork (QTF, the mass-produced timing device used in watches and clocks) as a transducer in PA experiments on gases. In less than a decade, QEPAS has been used to obtain spectra of approximately 15 different gases. Various mid- and near-infrared lasers, as well as an optical parametric oscillator, were employed as radiation sources in research described in a long series of publications. Quantification of trace amounts of gases such as CH_4, CO_2, and NH_3 has been a primary objective of the Tittel group, as have instrumentation and sensor development. Related work has also been performed at Pacific Northwest National Laboratory in the United States, the Life Science Trace Gas Facility in Nijmegen, University of Twente, and several other laboratories and manufacturing facilities.

The key aspect of QEPAS is the accumulation of absorbed energy in the detector, rather than a resonant acoustic mode in a conventional PA gas-microphone cell. In fact there is no need to confine the gas to be analyzed in QEPAS, unless it is necessary to isolate the sample from the environment. The most important advantages of QEPAS include the following:

- The high resonance frequency of the tuning fork (~32.8 kHz) and its narrow bandwidth (a few hertz) yield very high Q factors, typically on the order of ~10^4.
- The detector is not spectrally sensitive, and is practically unaffected by environmental noise.
- QEPAS can be implemented over a wide range of gas pressures.
- The volume of gas analyzed (specifically, the volume between the forks of the QTF) is less than 1 mm^3.

Insensitivity to environmental noise in QEPAS can be put down to two causes. First, because the wavelengths associated with this noise (typically, one centimeter or more) are large with respect to the dimensions of the QTF, these acoustic waves normally impinge on both forks in the same direction. The resulting deflection is not piezoelectrically active and, in contrast with the antisymmetric vibration, does not yield a detectable signal. Second, ambient noise is generally low above 10 kHz, and therefore has minimal effect on the QEPAS signal. In other words, little environmental noise falls within the analyzed bandwidth.

Table 3.2 summarizes the most important articles in the QEPAS literature for 2002–2009, listing the gases studied together with the corresponding radiation

Table 3.2 QEPAS literature, 2002–2009.

Author	Gas	Radiation source[a]	Wavenumber (cm^{-1})[b]	Wavelength (μm)[b]
Kosterev et al. (2002)	CH_4	DFB diode laser	5997–6001	1.66
Horstjann et al. (2004)	H_2CO	ICL	2832	3.53
Kosterev and Tittel (2004)	NH_3	DFB diode laser	6529	1.53
Kosterev et al. (2004)	CO	DFB QCL	2195–2198	4.55
Weidmann et al. (2004)	CO_2 H_2O C_2H_2 NH_3	SG-DBR diode laser	6514 6541 6529 6529	1.53
Kosterev, Bakhirkin, and Tittel (2005)	N_2O, CO	DFB QCL	2195–2198	4.55
Kosterev, Mosely, and Tittel (2006)	HCN	DFB diode laser	6539	1.53
Wojcik et al. (2006)	Freon 134a[c]	QCL	1189	8.41
Wysocki, Kosterev, and Tittel (2006)	CO_2	DFB diode laser	4991	2
McCurdy et al. (2007)	NH_3 acetone, Freon 125[d]	diode laser, EC-QCL	6529 1130–1250	1.53 8.4
Ngai et al. (2007)	C_2H_6	OPO	2983, 2990	3–4
Kosterev et al. (2008a)	Freon 125[d] CO_2	EC-QCL	1120–1250 4991	8.42
Kosterev et al. (2008b)	CH_4	DFB diode laser	6057	1.65
Wysocki et al. (2008)	NO C_2H_5OH N_2O	EC-QCL	1903 1830–1980 1175	5.3 5.3 8.4
Liu et al. (2009)	H_2O	DFB diode laser	7166	1.4
Petra et al. (2009)	C_2H_2 NH_3	Diode laser	6524–6529	1.53
Schilt, Kosterev, and Tittel (2009)	C_2H_4	DFB diode laser	6177–6190	1.62

a) DFB = distributed feedback; ICL = interband cascade laser; QCL = quantum cascade laser; EC = external cavity; SG-DBR = sample-grating distributed Bragg reflector; OPO = optical parametric oscillator.
b) Wavelength and wavenumber values are approximate.
c) 1,1,1,2-tetrafluoroethane (C2H2F4).
d) Pentafluoroethane (C2HF5).

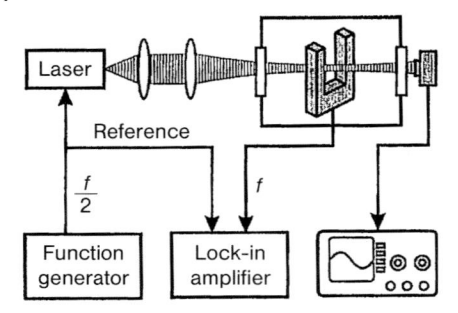

Figure 3.12 Experimental setup for QEPAS. Laser current is modulated at $f/2$, where f is the resonant frequency of the quartz tuning fork. (Reproduced from Kosterev, A.A., Bakhirkin, Yu.A., Curl, R.F. and Tittel, F.K., Quartz-enhanced photoacoustic spectroscopy, *Opt. Lett.* **27**: 1902–1904, Fig. 2, by permission of the Optical Society of America, copyright © 2002.)

sources and their characteristic wavenumbers and wavelengths. The publications that describe this work are organized chronologically.

Kosterev et al. (2002) first demonstrated the technique using a 1.66-μm (~6000-cm^{-1}) DFB diode laser to detect overtone carbon-hydrogen stretching bands in methane. Figure 3.12 shows a schematic drawing of the experimental layout. The laser current was sinusoidally modulated at half the QTF resonant frequency, yielding wavelength and amplitude modulation at the same frequency. This caused the laser emission to cross the absorption band twice in each period, producing a PA signal at the resonant frequency. Alternatively, the emission wavelength was tuned by changing the temperature of the laser. The lock-in amplifier signals resembled second-derivative curves. At a pressure of 375 torr and a laser power of about 2 mW, the detected QEPAS signal for ambient air containing 6.7% CH_4 ranged up to about 100 μV.

The stronger absorption bands observed in the mid-infrared region naturally led to the implementation of QEPAS at these longer wavelengths. Horstjann et al. (2004) used a 3.53-μm DFB interband cascade laser (ICL) to observe fundamental C–H bands in formaldehyde. This laser was operated at liquid nitrogen temperature and yielded a single-mode power of 12 mW. The wavenumber of the laser emission was adjusted between 2831.8 and 2833.7 cm^{-1} by tuning the current from 3 to 60 mA. The QTF signal varied linearly up to 20 μV as the H_2CO concentration was increased to about 20 ppm. A detection sensitivity of 0.6 ppm was estimated for a laser power of 3.4 mW and a time resolution of 10 s. The fundamental noise of the QEPAS spectrometer is determined by the QTF; since the noise is not related to laser power, the detection limit can – in principle – be improved by utilizing a more powerful laser.

Other mid-infrared experiments were performed using a 4.55-μm DFB quantum-cascade laser (QCL). This wavelength is well suited for the study of CO and N_2O. First, Kosterev et al. (2004) used this QCL to detect CO at ppm levels in propylene. Two glass tubes, 2.45 mm in length and 0.32 mm in diameter, were arranged perpendicular to the QTF plane, forming an acoustic microresonator. A similar setup was discussed by these authors in their initial QEPAS publication (Kosterev et al., 2002). The signal exhibited the familiar second-derivative shape when the lock-in phase was properly adjusted. By contrast, side lobes appeared and the central feature was almost masked when the phase setting was not optimized. Experimental phase lags arose from delays between the function generator and the laser current modulation, the current and wavelength modulation, and the phase shift between energy input and pressure. It was observed that vibration-translation (V–T) relaxation rates can be studied using this technique.

Both N_2O and CO were analyzed in a second set of experiments using a 4.55-μm QCL (Kosterev, Bakhirkin, and Tittel, 2005). The QEPAS signal of ambient N_2O (approximate concentration 315 ppb) increased as SF_6 was added to air at levels of a few per cent, promoting V–T relaxation in the target gas. Although V–T energy transfer in $CO:N_2$ was too slow to produce a QEPAS response, rotational relaxation produced a signal at a CO concentration of 280 ppb. This sensitivity is too low to monitor atmospheric amounts of this gas, but could be improved through the use of higher laser power. As an aside, the authors also noted that a shift of the QTF resonance frequency to 10 kHz – which could be achieved by adding weights to the ends of the tuning fork – would increase QEPAS sensitivity by a factor of 6. This advantage obtains because the signal/noise ratio varies as $f^{-1.5}$.

The research of this group gradually expanded to include a number of additional light sources and was directed toward the study of an increasing number of gases. Kosterev and Tittel (2004) used a 1.53-μm DFB telecommunication laser to detect NH_3, finding that the QEPAS signal (transimpedance amplifier output) varied linearly with concentration from 0 to 100 ppm. The sensitivity of the apparatus was judged lower than the best that was currently obtainable using wavelength-modulation PA spectroscopy, but better than that in amplitude-modulation experiments. Importantly, a faster operational amplifier in the transimpedance amplifier circuit improved sensitivity with regard to the first demonstration of QEPAS. As in other work, higher multiples of the laser modulation frequency f were employed: the lock-in amplifier detected the QEPAS signal at $2f$, while the zero-crossing of the $3f$ signal from a reference cell was used to lock the laser wavelength to the band of interest.

The 1.53-μm laser was later used to study the effect of humidity on the QEPAS signal amplitude for HCN (Kosterev, Mosely, and Tittel, 2006). Humidified samples were found to yield significantly higher PA signals, because of a reduction in the V–T relaxation time by three orders of magnitude. A detection limit of 155 ppb for HCN was determined in this work. More recently, Petra et al. (2009) presented a theoretical analysis of a QEPAS sensor and validated their model using a similar laser to detect NH_3 and C_2H_2. A QTF with a resonance frequency of 4.25 kHz was also evaluated in this study.

A different telecommunication diode laser was used for CO_2 trace gas detection by Wysocki, Kosterev, and Tittel (2006). This laser, with an emission wavelength near 2 μm, was tuned by varying both temperature and current. A CO_2 combination band at 4991 cm^{-1} was monitored in this investigation. The QEPAS signal for CO_2 increased concomitantly with water vapor concentration, prompting the suggestion that the instrumentation could be used for the simultaneous monitoring of CO_2 and H_2O. V–V and V–T processes in the CO_2–N_2–H_2O system were studied in this work.

The parallel use of QEPAS for the analysis of a number of gases obviously requires the radiation source to emit over a wide wavelength range. Thus Weidmann et al. (2004) employed a tunable near-infrared diode laser with sample-grating (SG) distributed Bragg reflectors for the detection of CO_2, H_2O, C_2H_2, NH_3, CO, and HCN. Absorption bands in the 6514–6541 cm^{-1} region were analyzed in this work. CO_2 and H_2O were measured in human breath samples. The noise-equivalent sensitivity of about $8 \times 10^{-9} cm^{-1} W^{-1} Hz^{-1/2}$ confirmed that QEPAS performed at a level similar to that in other types of PA spectroscopy.

As mentioned in the preceding section, major developments in the technology associated with quantum-cascade lasers (QCLs) have taken place in the last decade. In the present context, the commercial availability of EC QCLs is particularly relevant because of the wide tunability of these devices. Wysocki et al. (2008) discussed the use of QCLs in the context of high-resolution spectroscopy of gases. Tuning ranges of 155 and 182 cm^{-1} were attained for 5.3- and 8.4-μm lasers, respectively. The shorter wavelength was used to detect NO, while the 8.4-μm region is suitable for the analysis of N_2O, CH_4, SO_2, and NH_3. Kosterev et al. (2008a) reviewed the use of QCLs for trace gas sensing in detail, including QEPAS, cavity ring down spectroscopy, absorption, and integrated cavity output spectroscopy. A DFB QCL can be tuned by about 4 cm^{-1} by changing the injection current; variation of temperature increases this range to about 20 cm^{-1}. By contrast, a 5.3-μm EC QCL was tuned from about 5.05 to 5.5 μm (1820–1980 cm^{-1}). The performance of these lasers is continually being improved.

An acoustic microresonator consisting of glass tubes was described above. Kosterev et al. (2008b) evaluated an alternative device, formed from aluminum blocks with 0.35-mm holes. While the Al version also increased the PA signal, the Q factor was degraded as compared with the glass tubes. CH_4 was analyzed in humidifed and dry nitrogen using a 1.65-μm diode laser in this study.

The development of gas sensors based on QEPAS is a recent emphasis of the research team at Rice University. McCurdy et al. (2007) reviewed laser spectroscopy methods for the analysis of exhaled breath, including all of the techniques mentioned by Kosterev et al. (2008a). QEPAS results for acetone showed that a concentration of 70 ppb, appropriate for breath analysis, could be detected using a 1-min time constant. Schilt, Kosterev, and Tittel (2009) developed an ethylene sensor for monitoring fruit quality. A DFB diode laser (1.62 μm, 6177 cm^{-1}) was used to obtain QEPAS spectra in this work. Liu et al. (2009), of the Anhui Institute of Optics and Fine Mechanics, described a QEPAS-based sensor based on a 1.396-μm fiber-coupled DFB diode laser. H_2O in ambient air was detected at 0.64%, corresponding to a relative humidity of 26% at 21 °C.

The literature reviewed so far in this section is based on the use of lasers in QEPAS. The reader may well suggest that other suitable radiation sources could be employed in similar experiments. For example, Ngai et al. (2007) used the 3–4 μm idler output of an optical parametric oscillator (OPO) for trace gas detection by QEPAS. Spectral coverage was 16.5 cm^{-1}. C_2H_6 and H_2O were monitored at 2990 and 2994 cm^{-1}, respectively; the detection limit for C_2H_6 was 25 ppb. Mixtures of CH_4, C_2H_6, and H_2O were also analyzed.

Two review articles nicely summarize QEPAS research up to 2007. Tittel et al. (2007a) discussed detection of trace gases by QEPAS, reviewing the main attributes of the technique and several representative results. Other techniques, such as direct absorption and cavity-enhanced spectroscopy, were also described. Similarly, Tittel et al. (2007b) presented a detailed discussion of the use of mid- and near-infrared lasers in QEPAS. A useful table listing 11 trace gases, the locations of their absorption bands, and their normalized noise-equivalent absorption coefficients, is included in this work.

Finally, an article by Wojcik et al. (2006) at Pacific Northwest National Laboratory should be mentioned. These researchers utilized an 8.41-µm Fabry Pérot QCL for QEPAS detection of Freon 134a. An important aspect of this work was the use of amplitude modulation at the QTF resonance frequency, rather than the usual wavelength-modulation approach. The QTF signal varied linearly with absorption coefficient over three orders of magnitude with this instrumentation. The authors developed the theory of the QTF in detail, treating the near-field interaction between the acoustic gradient and the transducer as a driven damped harmonic oscillator.

3.6
Cantilever Enhanced PA Spectroscopy

An important alternative to the common gas-microphone cell has been developed in recent years by J. Kauppinen and co-workers at the University of Turku. In this technique a cantilever-type optical microphone, rather than a conventional capacitative (condenser or electret) device, is utilized to obtain PA spectra of gases. This approach is motivated by the fact that the cantilever (fabricated from silicon) is not subject to several limitations that restrict the performance of capacitative microphones: these include damping due to air flow between the flexible membrane and the metal electrode, relatively low sensitivity, and poor thermal stability (Koskinen et al., 2006). Moreover, because a cantilever bends but does not stretch, the movement at its free end is up to two orders of magnitude greater than that at the middle of the tightened membrane in a capacitative microphone. The larger movement of the cantilever yields enhanced sensitivity. This displacement can be measured interferometrically and is linear over a wide dynamic range (Kauppinen et al., 2004). The cantilever can therefore detect very low gas concentrations, rivaling those observed by PA methods that use long-path cells and powerful lasers.

Methane (CH_4) gas was detected and quantified using cantilever-enhanced PA spectroscopy in a number of experiments performed by this group. In their first publication on this technique, Kauppinen et al. (2004) described the use of an optical filter to select the 2600–3400 cm^{-1} region, which contains the prominent C–H stretching band. The chopping frequency (118 Hz) was significantly lower

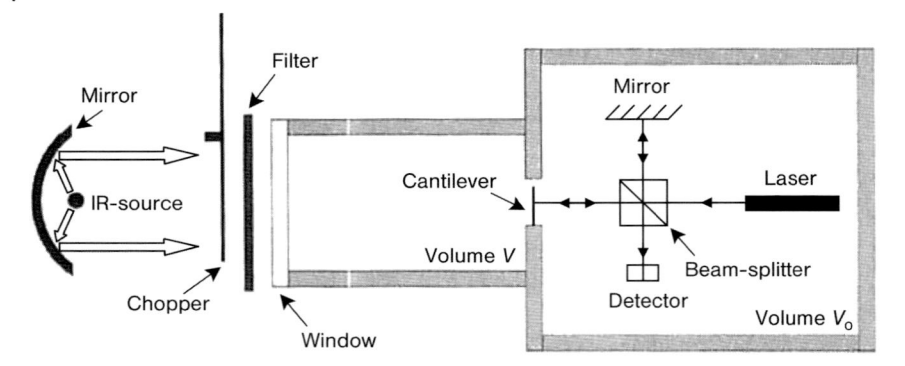

Figure 3.13 PA gas-detection system using an interferometric cantilever as an optical microphone.

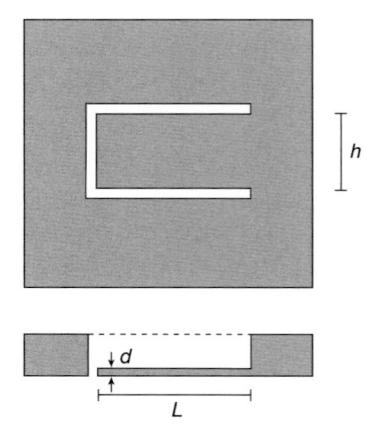

Figure 3.14 Structure of cantilever and frame. Typical dimensions are given in the text. (Reproduced from Kuusela, T. and Kauppinen, J., Photoacoustic gas analysis using interferometric cantilever microphone, *Appl. Spectrosc. Rev.* **42**: 443–474, by permission of Taylor & Francis, copyright © 2007.)

than the cell resonance, found to occur at 470 Hz. Although chopper irregularity produced side bands, a very low detection limit of 0.2 ppb was calculated for CH_4. The experimental layout is shown in Figure 3.13. Mathematical expressions for the amplitude of the PA signal and the resonance frequency were given in this article, and the cell resonance was modeled successfully. The cantilever dimensions were 4 mm (length) × 2 mm (width) × 5 µm (thickness). The structure of the cantilever appears in Figure 3.14.

Uotila, Koskinen, and Kauppinen (2005) next described a differential gas-detection system that employed broadband infrared radiation and a cantilever sensor. PA and long-path absorption spectroscopies were effectively combined in this approach; the infrared radiation was divided into two parallel beams that impinged on cells containing the sample and nonabsorbing reference (N_2) gases, respectively. It was found that the pressure of the analyte affected signal intensity; the normalized difference signal S_{B-A}/S_B increased when the pressure was raised from 20 to 400 mbar, but a further increase did not produce a larger signal. A detection limit of 13 ppb was calculated for a 0.37-s measurement time. Selectivity was studied using CH_4, CO_2, and butane. As might be expected, no cross-interference was observed for CH_4/CO_2, while the similar absorption bands of CH_4 and butane led to considerable interference. The use of long-path cells lowered the detection limit in this work. Ambient noise was not found to be problematic.

Selective differential detection of gases using the cantilever method was demonstrated by Koskinen et al. (2006). The first resonance frequency for a cantilever with a rectangular cross-section was given as

$$\omega_{0,cant} = \frac{(1.875)^2}{l^2} \sqrt{\frac{Et^2}{12\rho}} \quad (3.1)$$

where E is Young's modulus, ρ is the density, l is the length, and t is the thickness of the cantilever. The amplitude of the cantilever is

$$A(\omega) = \frac{F_0}{\sqrt{(\omega_0^2 - \omega^2)^2 + (\omega\beta/m)^2}} \quad (3.2)$$

where F_0 is the external force and β and m are the damping constant and mass, respectively.

The selective differential method relies on the difference in PA signals observed for the sample and nonabsorbing gases, this difference occurring when there is a cross-correlation between the gas in the sample cell and the reference gas in a (third) PA cell. Importantly, no optical filters are required in this method. The sensitivity and linearity of this detector were much better than those of a capacitative microphone; the detection limit for CH_4 with the cantilever apparatus was comparable to that obtained with resonant PA cells and lasers.

The cantilever sensor was also applied to tunable diode laser-based PA spectroscopy (TDLPAS). Laurila et al. (2005) employed a DFB diode laser and a cantilever to detect CO_2 at 1572 nm (6361 cm^{-1}). Wavelength modulation was effected at $f = 163$ Hz by adding a sinusoidal component to the DC driving current of the laser. The PA signal was detected at 2f. The Fourier transform of the signals yielded noisy curves containing narrow peaks at f and 2f, as well as a broader peak at 400 Hz due to cell resonance. The f peak, which was more prominent at higher concentration, contained contributions from the background as well as sample absorption. The 2f signal varied linearly with laser power and CO_2 concentration. Energy transfer from CO_2 to N_2 affected the PA signal, sometimes leading to nonlinear concentration dependence.

The work carried out by this group in the first three years of research on cantilever-enhanced PA spectroscopy was summarized in a detailed review by Kuusela and Kauppinen (2007). A number of topics were discussed with regard to the theory of cantilever operation. These included cantilever dynamics, cell volume effects, heating of the sample gas, effects of the cantilever-frame gap, gas flow due to pressure difference between cells, energy transfer caused by temperature differences, and conversion of the absorbed light to heat. Four types of noise were identified: acceleration noise (due to external low-frequency vibrations), acoustical noise (arising from leakage of acoustic waves into the cell), Brownian noise, and electrical noise. Experiments were performed with an infrared source, a tunable diode laser, and an FT-IR spectrometer. A system incorporating sample and (nonabsorbing) reference cells and a differential PA cell containing a high concentration of the analyzed gas was also shown. In addition to this comprehensive review of cantilever-enhanced PA spectroscopy, the authors also briefly discussed condenser and electret microphones and Helmholtz resonance. Most subjects were treated mathematically in this useful article.

A further review of cantilever-enhanced PA spectroscopy of gases was presented by Koskinen *et al.* (2008). Many of the topics mentioned in the previous paragraph were discussed again. Cell resonance varied from about 340 to 410 Hz for mixtures of CH_4 in N_2 or undiluted CO_2, a result that was modeled successfully. The resonance frequency increased with pressure when CH_4 was mixed with N_2 or Ar, but hardly varied when the pressure of undiluted CO_2 was changed. Cantilever-enhanced detection was used together with an FT-IR spectrometer to detect CH_4 in the presence of H_2O vapor. In general, cantilever-enhanced PA spectroscopy was concluded to be at least an order of magnitude more sensitive than other PA techniques.

This research group next constructed a PA cell capable of gas, solid, or liquid analysis in an FT-IR spectrometer (Uotila and Kauppinen, 2008). This desirable implementation of cantilever detection enabled the acquisition of research-quality spectra of sunflower oil, polymer chips, carbon black, and CH_4. The gas-phase result was comparable to that obtainable by conventional absorption spectroscopy with a 6-cm cell and a standard optical detector. The signal/noise ratio obtained using the cantilever cell was about five times that observed with the popular MTEC gas-microphone cell. The development of this new cell could be significant because of the widespread use of PA infrared spectroscopy for the characterization of solid and liquid samples.

Recently, Fonsen *et al.* (2009) utilized an electrically modulated broadband infrared source with a dual cantilever apparatus for CH_4 detection. An optical filter was used to select the wavenumber interval of interest, which was centered at 2950 cm^{-1}. Although the cantilever resonance occurred at 288 Hz, the source was completely modulated at 20 Hz. The dual cantilever arrangement enabled suppression of acceleration noise, which was 180° out of phase with respect to the PA signal. For 100 ppm CH_4 the relative noise was 0.30%, about the same as the noise due to source intensity fluctuations. This dual-cantilever device offers good sensitivity at a low equipment cost. This group has also utilized mid-infrared LEDs as sources in another cost-saving refinement (Kuusela *et al.*, 2009).

3.7
Piezoelectric Detection

Photoacoustic detection of the absorption of modulated infrared radiation can be effected in several ways. The most common, of course, involves the use of a microphone to monitor the sound waves in an enclosed cell. Alternatively, it is possible to detect the elastic wave induced in a piezoelectric substrate in intimate contact with the sample. The reader is probably aware of the fact that the majority of the applications discussed in this book are based on the former technique. Nevertheless, it is important to recognize the viability of piezoelectric measurement of PA infrared spectra – indeed, piezoelectric PA infrared spectroscopy has been known for approximately the same length of time as its gas-microphone counterpart. One could speculate that the rather limited use of piezoelectric detection is at least partly due to the fact that it was never developed and commercialized in the same manner as the gas-microphone cell, various models of which have been widely available for about three decades. Piezoelectric detection in PA infrared spectroscopy is briefly outlined in this section.

An important series of early articles by Kanstad and Nordal was discussed in Chapter 2. These authors used CO_2 laser radiation as the infrared source in their PA experiments, together with several means of detection. Some of their publications deal with piezoelectric PA infrared spectroscopy and are summarized in the following paragraphs.

A scheme for the measurement of both piezoelectric and gas-microphone PA spectra, referred to as PA reflection-absorption spectroscopy, was developed by Kanstad and Nordal (1980a, 1980b). The experimental layout is sketched in Figure 3.15. It can be noted that piezoelectric measurement is quite simple: the piezoelectric transducer (PZT,[2] also referred to as a piezotransducer) is affixed to the rear surface of the sample, with no need for a sample chamber or other similar enclosure. Absorption of the CO_2 laser radiation by the sample leads to the development of a voltage by the PZT, which can be detected using a lock-in amplifier; in the configuration shown in Figure 3.15, the PZT signal is ratioed against that from a reference spectrophone, or gas-microphone cell. Kanstad and Nordal (1980b) obtained a PA spectrum of an aluminum oxide layer on a 0.04-mm thick Al foil in this way, and observed the prominent longitudinal optic (LO) mode of the Al–O stretching vibration at 10.49 μm (953 cm^{-1}). An uncompensated single-beam experiment would have involved measurement of the PZT signal alone.

The piezoelectric PA technique described in the previous paragraph was subsequently used to obtain spectra of skin lipids by Kanstad et al. (1981). Source compensation was achieved with the use of a blackened reference transducer in this investigation. This work is discussed in more detail in the section on medical applications of PA infrared spectroscopy in Chapter 7.

An early demonstration of piezoelectric PA infrared spectroscopy involving the use of an FT-IR spectrometer was performed by Lloyd, Yeates, and Eyring (1982).

---

2) PZT is also used as an acronym for lead zirconate titanate, a ceramic perovskite that exhibits the piezoelectric effect.

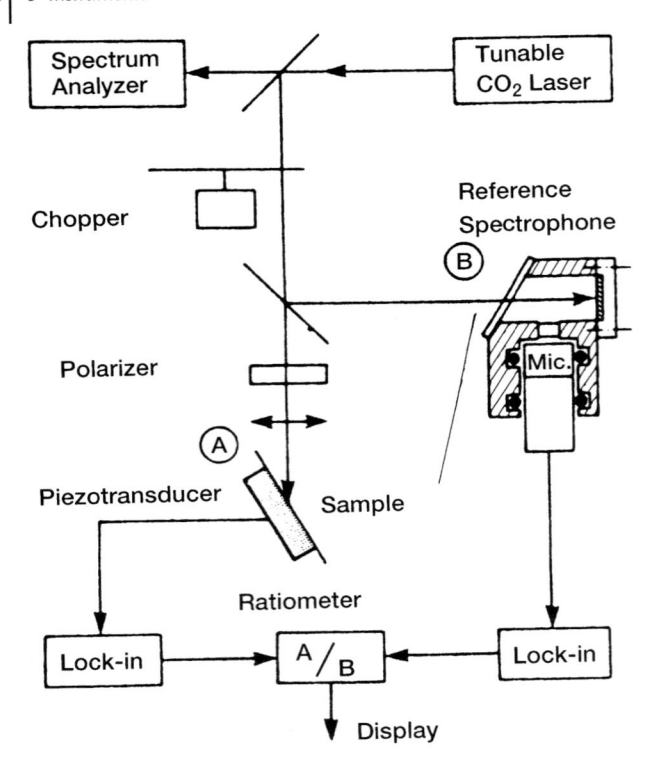

Figure 3.15 Apparatus for piezoelectric PA spectroscopy, using a spectrophone as a reference for source compensation. (Reprinted from *Appl. Surf. Sci.* **5**, Kanstad, S.O. and Nordal, P.-E., Photoacoustic reflection-absorption spectroscopy (PARAS) for infrared analysis of surface species, 286–295, copyright © 1980, with permission from Elsevier Science.)

These authors obtained both gas-microphone and piezoelectric PA spectra of samples such as tetraphenylcyclopentadienone on TLC plates. The object of this work was to develop methods for acquiring spectra of separated compounds without removing them from the plates. Neat sample pellets were also examined in this research. In the piezoelectric PA FT-IR experiments, the PZT signal was directed to a current-sensitive preamplifier and lock-in amplifier. The DC output of the lock-in was input to the FT-IR electronics.

Several noteworthy results were obtained in this investigation. For example, a pellet of the ketone that was attached to the PZT yielded a transmission-like spectrum rather than the expected absorptive spectrum. This result was put down to the fact that the PZT and sample were not rigidly coupled, so that absorption of the incident light did not deform the PZT; instead, radiation transmitted by the sample reached the detector, producing a piezoelectrically detected transmission spectrum. By contrast, coupling between an alumina TLC plate and the PZT was more effective, and an absorption-like spectrum was obtained. Another, more general, observation from this study was that the elimination of a carrier gas in piezoelectric PA spectroscopy offers a useful simplification with respect to the familiar gas-microphone technique.

Piezoelectric detection of PA spectra can, of course, be employed at wavelengths much shorter than the mid-infrared. Moreover, there may be no need to employ an FT spectrometer in these experiments. This was the case in the work of Manzanares, Blunt, and Peng (1993), who studied fundamental and overtone spectra for the C–H stretching vibrations of *cis-* and *trans-*3-hexene. For the fourth ($v = 5$) and fifth ($v = 6$) overtones, a dye laser, acousto-optic (AO) modulator, and piezoelectric detector were used. The AO modulator interrupted the Ar^+ pump laser at a frequency of about 80 kHz, two orders of magnitude greater than typical f values employed in PA FT-IR spectroscopy. The wavelength of the incident light was scanned by means of a birefringent filter in the dye laser, while the signal from the PZT was demodulated using a lock-in amplifier. This article emphasizes curve fitting, band assignments, and calculation of local-mode frequencies and anharmonicities, rather than the use of PA detection.

All of the publications described so far in this section refer to experiments in which a lock-in amplifier was used for signal recovery at specific wavelengths. In rapid-scan PA multiplex (FT-IR) spectroscopy, modulation exists, of course, because of the moving mirror in the interferometer. A spectroscopist might suggest that this modulation – albeit at lower frequencies than those normally associated with PZT detection – could be utilized directly in a piezoelectric PA FT-IR experiment. In fact PA infrared spectra have been successfully obtained in this way. Zhang, Michaelian, and Burt (1997) recorded piezoelectric PA infrared spectra of mica in an arrangement where the PZT detector was in direct contact with the rear surface of the mica sheet, while the infrared radiation impinged on the front surface. The adsorption of stearic acid on mica was also investigated in this experiment. The relatively small voltage from the PZT was input directly to the electronics of an FT-IR spectrometer, yielding weak but stable interferograms. However, at the low mirror velocities normally used with gas-microphone cells, these interferograms lacked centerbursts at or near the zero-path-difference (ZPD) point: conventional FT-IR software therefore misidentified this location and calculated unrealistic spectra. By contrast, viable spectra were calculated when ZPD was fixed at the expected location and the interferogram phase was constrained to the first and fourth quadrants (Figure 3.16). Interestingly, these piezoelectric PA spectra resemble transmission spectra, a result reminiscent of that observed by Lloyd, Yeates, and Eyring (1982) and discussed above. It should also be pointed out that the phase correction problem encountered with these PA data is similar to that in some types of differential spectroscopy, where the presence of both positive and negative bands leads to the absence of a central peak in the interferogram. Improper phase correction wrongly converts the negative-going features in these spectra into positive bands.

Thin films are particularly amenable to the measurement of piezoelectric PA spectra because it is possible to ensure good contact between sample and detector. Some of these films may be of greater interest to materials scientists and physicists than to chemists; this is the case in several examples from the literature that are reviewed below. The section on inorganic materials in Chapter 7 can be consulted for more information.

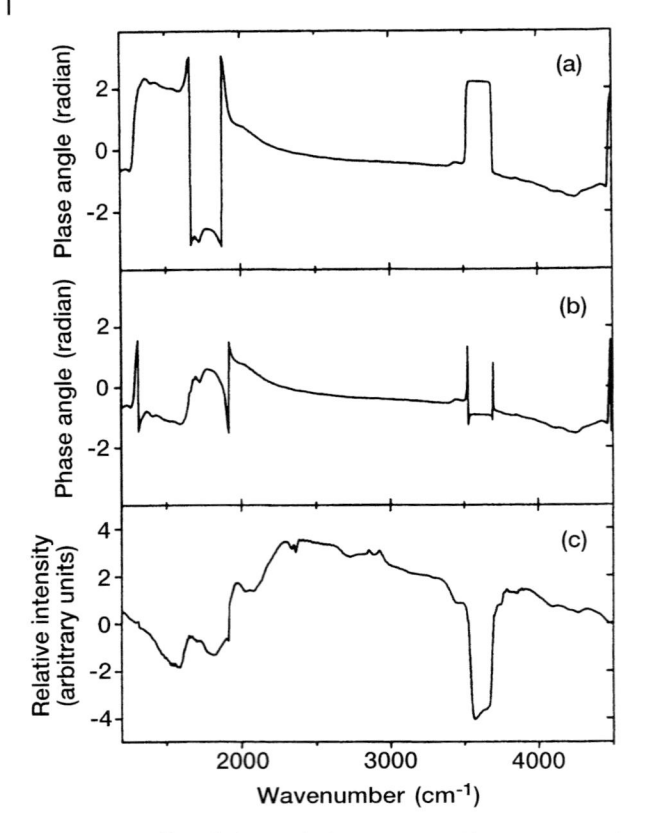

Figure 3.16 Effect of phase calculation on piezoelectric PA infrared spectrum of mica. (a) phase calculated in all four quadrants; (b) phase calculated in first and fourth quadrants; (c) PA spectrum corresponding to the phase in (b). (Reproduced from Zhang, S.L. et al., Phase correction in piezoelectric photoacoustic Fourier transform infrared spectroscopy of mica, *Opt. Eng.* **36**: 321–325, by permission of the International Society for Optical Engineering, copyright © 1997.)

Yoshino *et al.* (1999) reported piezoelectric PA spectra of the I–III–VI$_2$ chalcopyrite semiconductor CuInSe$_2$, grown on an oriented GaAs substrate by molecular beam epitaxy. PA spectra were obtained in the near-infrared region (0.6–1.8 eV, or 4840–14 520 cm^{-1}) at both liquid nitrogen and room temperatures using a disc-shaped PZT attached to the rear surface of the sample. The object of this work was to investigate defect levels and the bandgap energy of CuInSe$_2$, which is of interest for possible solar cell applications. Spectra were obtained before and after quenching by irradiation at a photon energy greater than 1.1 eV (8870 cm^{-1}). Figure 3.17 shows that the spectrum is dramatically affected by quenching. The positions of the bands give the activation energies of the lattice defects.

In related work, Fukuyama *et al.* (2000) investigated near-infrared piezoelectric PA spectra of GaAs. This detection method was chosen because it directly monitors nonradiative relaxation and is sensitive to very weak absorption in highly

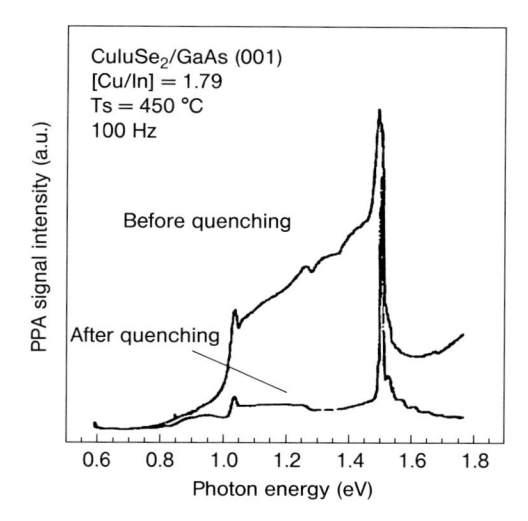

Figure 3.17 Piezoelectric PA spectra of CuInSe$_2$ film on GaAs substrate, before and after quenching at 77 K by irradiation at 1.1 eV. (Reprinted from *Thin Solid Films* **343–344**, Piezoelectric photoacoustic spectra of CuInSe$_2$ thin film grown by molecular beam epitaxy, 591–593, copyright © 1999 with permission from Elsevier Science.)

transparent samples. The PA spectra, which are rather featureless except for two peaks superimposed on a continuum that extends between 0.8 and 1.5 eV (6450–12 100 cm^{-1}), changed as the sample was quenched by 1.12-eV (9035-cm^{-1}) illumination at 80 K. This suggested the contribution of an acceptor-like deep level that changed the activation energy during the photoquenching process.

Piezoelectric PA spectra of n-type phosphorus-doped and p-type boron-doped silicon were measured near the energy gap by Kuwahata, Muto, and Uehara (2000). The effect of carrier concentration was investigated for these spectra, which extended from visible to near-infrared wavelengths and exhibited two very broad bands in the 500–1200 nm region. PA intensities at energies higher than the energy gap decreased for the n-type samples and increased for the p-type samples because of the increase in free electrons at the bottom of the conduction band and the increase in holes at the top of the valence band, respectively. Variation of the carrier concentration suggested that free electrons suppress thermoelastic transfer.

Sakai *et al.* (2002) reported piezoelectric PA spectra for Co-doped ZnO in the region from about 0.7 to 3.5 eV. The temperature dependence of the spectra and the effect of varying Co concentration were studied. A disk-shaped PZT was attached to the back of a pressed pellet using silver conducting paste, and samples were held in a cryostat. The signal was generated by nonradiative carrier recombination.

Nonresonant multiphoton PA spectroscopy (NMPPAS), a method developed for subsurface tissue diagnostics by Chandrasekharan, Gonzales, and Cullum (2004), utilizes a PZT to detect the PA signal arising from the absorption of radiation emitted by an optical parametric oscillator. Solutions under study were contained in quartz cuvettes, the wall of the cuvette being in contact with the detector. The signal was amplified and recorded using a digital sampling oscilloscope.

The technique was demonstrated with dye solutions at visible and near-infrared wavelengths.

3.8
Photothermal Beam Deflection Spectroscopy

An alternative to the popular gas-microphone PA technique is discussed in this section. Photothermal beam deflection spectroscopy (PBDS) – which is based on the so-called "mirage effect" – has been used by a number of research groups to acquire near- and mid-infrared spectra of solids, liquids, and gases during the last three decades. In fact this detection scheme actually predates the development of modern PA FT-IR spectroscopy, having originally been developed for studies in the visible and near-infrared wavelength regions. The adaptation of the method to the mid-infrared occurred in the early 1980s, at about the same time as many other experimental advances that are described throughout this book. Unlike the gas-microphone cell, which has been commercially available for many years, PBDS instrumentation has generally been purpose-built. As noted above for piezoelectric PA spectroscopy, this factor has limited the number of infrared spectroscopists who utilize the technique.

For condensed-phase materials, beam deflection arises from the thermally induced refractive index gradient that exists near a surface as a consequence of the absorption of modulated light. An extensive theoretical treatment was given by Aamodt and Murphy (1981). In contrast with the gas-microphone method, the sample need not be confined in a cell in PBDS. Instead, a beam of light from a probe laser is made to pass immediately above the sample, through the periodically heated gas. Because the refractive index of this gas is modulated, the probe beam is periodically deflected from its original path. This phenomenon is similar to the mirage effect, and was named accordingly by early workers in this area. Thermal diffusivity can also be measured for opaque materials using this effect (Gendre, 1987).

The deflection of the beam can be monitored in various ways, a simple example being the use of a knife-edge and position detector. Many workers in this field have utilized bicells or various types of linear position detectors. The resulting signal can be demodulated using a lock-in amplifier and then analyzed in a manner similar to that for other PA data. The major advantage of this technique is that the sample need not be enclosed: this makes it possible to obtain spectra of large or irregularly shaped objects. An illustration of this versatility occurs in the *in vivo* measurement of spectroscopic and photochemical properties of intact leaves (Havaux, Lorrain, and Leblanc, 1989).

The research of M. J. D. Low and his colleagues on the PA infrared spectra of carbons, catalysts, and other solids is discussed in several places in this book. This group used the PBD technique in many of their studies. A majority of their publications tended to emphasize the interpretation of the infrared spectra rather than the technique by which they were acquired; these articles are not discussed here.

3.8 Photothermal Beam Deflection Spectroscopy | 57

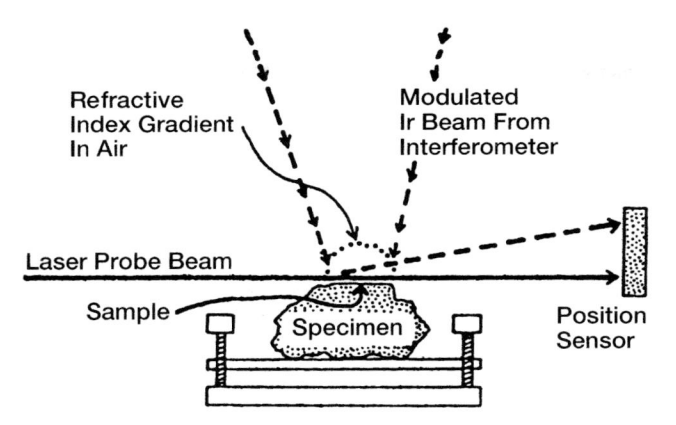

Figure 3.18 A schematic drawing of a simple experimental layout for PBDS. (Reproduced from Low, M.J.D. and Tascon, J.M.D., *Phys. Chem. Minerals* **12**: 19–22, Fig. 1, by permission of Springer-Verlag, copyright © 1985.)

On the other hand, several papers that describe the PBD method in some detail were also published by these researchers (Low and Lacroix, 1982; Low, Lacroix, and Morterra, 1982a; DeBellis and Low, 1987, 1988). Some additional pertinent references are mentioned in the next paragraph.

A simple drawing of an experimental setup for PBDS is reproduced in Figure 3.18 (Low and Tascon, 1985). As mentioned above, the position sensor can be used in conjunction with a knife-edge that is arranged to partially block the probe laser beam. A dual detector [measuring $(A + B)$ or $(A - B)/(A + B)$, where A and B are the intensities detected in the two channels] or a double-beam dual detector arrangement can also be employed (Low, Lacroix, and Morterra, 1982b). The effects of vibrational noise on this experiment can, of course, be quite deleterious. Low (1986) demonstrated that some sources of noise in the spectra were not random; extensive averaging therefore reduced, but did not eliminate, the noise in the spectra. Although irregularly shaped samples can be analyzed by this technique, powders were found to yield spectra superior to those for coarse solids (Low and Morterra, 1987). This result was put down to the increased surface area and scattering caused by the smaller particles. The geometry of the experiment is also better defined for a well-packed finely divided solid, or a film, than for a coarse solid.

Some of the most important early PBDS experiments were carried out by Fournier, Boccara, and their collaborators. Three publications by this group will be mentioned here. In the first, Boccara *et al.* (1980) described the use of PBD for measurement of absorption in optical coatings at visible wavelengths. Absorption by C_6H_6 near 610 nm was also studied. The deflection amplitude was shown to be proportional to the absorption coefficient in the weak-absorption case. The method was distinguished from the thermal lens effect, where the pump and probe beams are collinear. At approximately the same time, Fournier *et al.* (1980) described the measurement of gas-phase spectra by means of the photothermal deflection technique. A CO_2 laser, with emission lines in the 9.4- and 10.4-µm (~1080- and

970-cm^{-1}) regions served as the infrared source in this experiment. The optical layout was such that the probe beam from an HeNe laser was nearly collinear with the CO_2 laser beam. The photothermal signal detected for ethylene varied linearly with concentration over four orders of magnitude. An impressive detection limit of 5 ppb was estimated in this work.

Another elegant experiment by this group demonstrated the high sensitivity of PBDS in two additional applications (Fournier, Boccara, and Badoz, 1982). This technique was shown to be more sensitive than conventional PA spectroscopy by three orders of magnitude at near-infrared and visible wavelengths. Hence the authors were able to obtain absorption and circular dichroism data that were completely inaccessible by other methods.

Several authors utilized the PBD technique in subsequent years. Bain, Davies, and Ong (1992) studied the C–H stretching region in spectra of organic monolayers on both smooth and rough silicon surfaces. The authors described an enclosed cell that could be filled with perfluorodecalin or CCl_4 to increase the magnitude of the beam deflection. While this strategy increases the sensitivity of the method, it is suitable only for infrared-transparent liquids; this explains the choice of liquids for the deflecting medium. Similarly, Zhou et al. (1996) used silicone oil as a medium in near-infrared and visible measurements on solid C_{60}.

Near-infrared PBDS was used to characterize non-oxidized polythiophenes with different alkyl side chains by Einsiedel et al. (1998). These materials are of interest with regard to nonlinear optical waveguide applications. PBDS detected several weak near-infrared absorption bands, even though the optical density varied by as much as five units. Finally, Bouzerar et al. (2001) obtained visible and near-infrared PBD spectra of amorphous hydrogenated carbon thin films (a-C:H). No well-defined absorption bands are visible in the spectra of these samples and experimental details were not given.

Deng et al. (1996) designed an electrochemical cell that enabled the measurement of PBD spectra. Figure 3.19 depicts a test cell used for setup and optimiza-

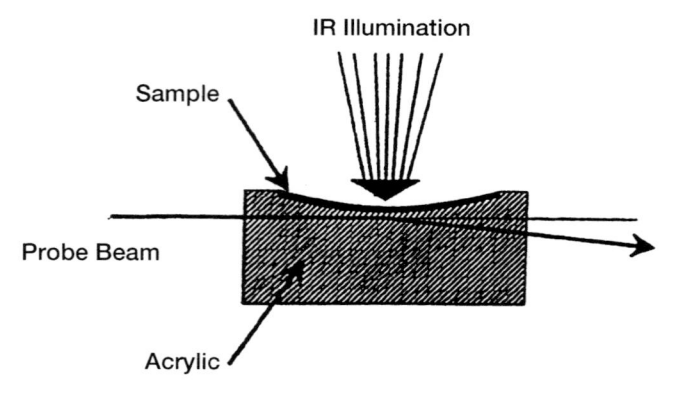

Figure 3.19 PBD test cell. An electrochemical cell was also constructed using this design. (Reproduced from Deng, Z. et al., *J. Electrochem. Soc.* **143**: 1514–1521, by permission of the Electrochemical Society, Inc.; copyright © 1996.)

tion in this experiment. Infrared illumination was provided by tunable diode lasers that emit in the 8–16 μm (1250–660 cm^{-1}) region. The probe beam passed through the acrylic block that supported the sample. Infrared absorption caused a deflection of this beam, which was monitored using a position detector. This system was used to obtain infrared spectra between 600 and 1000 cm^{-1} for lithium surface films after exposure to air and for an electrolyte after both discharging and recharging. Although the spectral coverage afforded by the lasers was relatively narrow, the results demonstrated the sensitivity and versatility of this novel technique.

3.8.1
Reverse Mirage Detection

An arrangement of the PBDS experiment that incorporates a liquid deflection medium was described above. This concept is also utilized in reverse mirage PA spectroscopy. For PBDS, submersion of the sample in a liquid increases the magnitude of beam deflection relative to that obtained in air, thereby increasing sensitivity; however, the accessible spectral region may be limited because of absorption of the infrared beam by the liquid. Because this beam does not traverse the liquid in the reverse mirage arrangement, this problem is obviated.

Figure 3.20 depicts the experimental setup. An infrared-transparent window, such as Si, Ge, or GaAs, is partially submerged in the liquid, which acts as both sample and deflection medium. Modulated infrared radiation passes through the window and into the liquid. The transverse probe beam passes through the liquid immediately adjacent to the window. Absorption of the incident radiation is followed by the production of heat. The resulting change in refractive index causes a deflection of the probe beam that can be monitored with a position-sensitive photodiode. The primary advantage of this reverse mirage arrangement is that the liquid need transmit only the probe beam (usually from an HeNe laser), not the

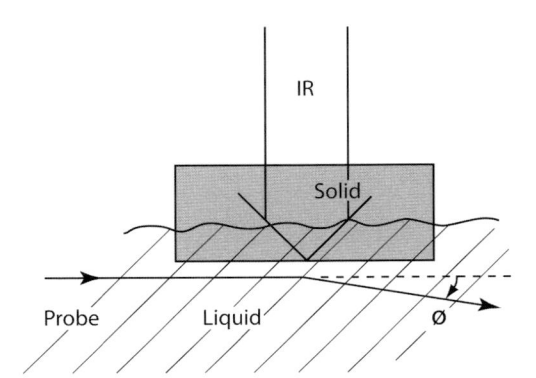

Figure 3.20 Sketch of the layout for reverse mirage spectroscopy. The probe beam is deflected by the angle φ due to the absorption of the infrared radiation by the liquid. (Reproduced from Palmer, R.A. and Smith, M.J., *Can. J. Phys.* **64**: 1086–1092, by permission of NRC Research Press; copyright © 1986.)

infrared radiation. The method is suitable for studying absorption spectra of liquids and species at the solid-liquid interface.

The reverse mirage technique can be implemented in an FT-IR spectrometer, or alternatively it can be used to detect the absorption of infrared laser radiation. Pioneering FT-IR experiments were described by Palmer and Smith (1986), who obtained a research-quality spectrum of acetonitrile using this method. These authors noted similarities between reverse mirage spectroscopy and the phenomenon of thermal lensing, the latter being the basis of the mirage effect in air. Moreover, reverse mirage spectroscopy is reminiscent of ATR infrared spectroscopy, where spectra of strongly absorbing liquids and solid-liquid interfaces are acquired. In fact, optical materials used as ATR crystals are also suitable as windows for reverse mirage spectroscopy.

Smith and Palmer (1987) continued their reverse mirage investigations of acetonitrile and observed a somewhat disconcerting result: the distance between the probe beam and the surface of the Si window had a dramatic effect on the spectra. Specifically, strong bands were often distorted while weaker features exhibited more conventional shapes. The distortion increased when the probe beam was farther from the window. These results were rationalized as follows. Unlike the situation in the ordinary mirage effect, the thermal gradient in the reverse mirage effect does not arise from a localized source; instead, it is distributed within the absorbing medium. When absorption is strong, the optical penetration depth is shallow and the heat source is close to the solid/liquid interface. By contrast, the heat source is deeper within the sample for weaker bands. Thus the optimal location of the probe laser beam is not the same in the two cases. Smith and Palmer (1987) showed that the thermal gradient is proportional to absorbance only within the first $15\,\mu m$ of the liquid; at greater distances from the window, the thermal gradient passes through a maximum and tends towards zero. Hence the probe beam must be situated so that it literally grazes the window if strong bands are to be observed reliably by reverse mirage spectroscopy. When this is done, the proportionality between thermal gradient and absorbance extends up to absorption coefficients of about $1000\,cm^{-1}$.

A few years after these initial investigations, Palmer's research group published several additional papers dealing with reverse mirage spectroscopy. In a study of propylene carbonate, Manning *et al.* (1992a) again noted distortions of stronger absorption bands, even though the magnitude of the deflection signal varied monotonically at lower absorptivities. The distortion of the stronger bands arises from attenuation of the thermal gradient in the liquid, leading to the appearance of notched bands.

This investigation was carried out with a step-scan instrument, making it possible to calculate both phase and amplitude spectra. Manning *et al.* (1992a) observed that the phase spectrum was actually a close approximation to the 'true' spectrum (obtained in an ATR experiment). The reverse mirage phase saturates at higher absorptivity than does the amplitude, effectively extending the dynamic range of the technique. The authors noted that the phase, rather than the in-phase and quadrature spectra, varies monotonically with absorptivity. However, caution must still be used when interpreting the phase data, because the calculation can be adversely affected when the phase angle changes from one quadrant to the next in a region of strong absorption. The relationship between the PA phase and absorptivity is also used in spectrum linearization, a numerical method discussed in Chapter 6.

Another, more theoretical, paper by Manning et al. (1992b) presented a detailed mathematical model of the reverse mirage effect. This treatment explicitly included the thermal influence of the window, and derived expressions for the thermal gradient at various distances within this influence (about three times the thermal diffusion length). The effect of the window can be ignored at greater distances. The Gaussian profile of the probe beam was incorporated through the use of numerical integration. An important result of this treatment is the prediction that the natural logarithm of the magnitude of the reverse mirage signal varies linearly with probe distance: the proportionality constant relating these quantities is the absorptivity. Thus absolute absorptivity can be determined in the infrared using reverse mirage spectroscopy. It was estimated that absorptivity values up to $3000\,\text{cm}^{-1}$ could be measured using the phase response obtained from step-scan reverse mirage data. Absorptivities calculated for 11 different bands of acetonitrile agreed fairly well with those obtained by ATR spectroscopy (Manning et al., 1992c).

As mentioned above, reverse mirage spectroscopy can also be used to study absorption at specific wavelengths. An example of this approach occurs in the work of Bićanić et al. (1989), who studied reverse mirage spectra of liquid methanol at CO_2 laser wavelengths. According to these authors, the amplitude of the deflection signal is given by

$$S = \eta(L/n)(dn/d\theta)(d\theta/d\chi) \tag{3.3}$$

where η is the transducer conversion factor (voltage per unit angle), L is the interaction length, n is the refractive index of the liquid, θ is the amplitude of the AC temperature rise, and x is the distance between the probe beam and the window. The absorption coefficient is obtained from

$$\beta = (\Delta)^{-1} \ln[S_n(\chi)/S_n(\chi + \Delta)] \tag{3.4}$$

where Δ is the difference between two x values and the subscript n indicates that the signals are normalized. This relationship is expected to hold at probing distances up to approximately $20\,\mu\text{m}$ for absorption coefficients as large as $10^5\,\text{m}^{-1}$. These limiting values are similar to those reported by Smith and Palmer (1987), which were mentioned earlier.

To demonstrate this method, Bićanić et al. (1989) carried out reverse mirage measurements on methanol using two strong CO_2 laser lines. For the 10P(20) and 10R(20) lines at 10.59 and $10.26\,\mu\text{m}$ (944 and $975\,\text{cm}^{-1}$) they obtained absorption coefficients of 0.122×10^5 and $0.162 \times 10^5\,\text{m}^{-1}$, respectively. These results are in good agreement with the expected values of 0.12×10^5 and $0.175 \times 10^5\,\text{m}^{-1}$.

3.9
Optothermal Window Spectroscopy

The final experimental technique to be described in this chapter is referred to as optothermal window (OW) spectroscopy. In contrast with many of the methods discussed so far, the OW technique has received somewhat limited attention in the literature. Nevertheless, it is important to include it in this survey of experimental methods: it bears a clear relationship to the other PA techniques mentioned and, importantly, has been utilized in both medical applications and the study of food products. These applications are discussed in other chapters in this book.

The use of the term 'optothermal window' is, in fact, quite instructive: this method is based on the detection of the temperature increase in an optical window that is in intimate contact with an absorbing sample. As is the case in other PA experiments, the absorption of light results in the production of a thermal wave; in OW spectroscopy the heat is transferred to the window, where it is detected piezoelectrically or by means of a thermistor. Hence the OW scheme is essentially a photothermal experiment, although an important aspect of the technique is also reminiscent of PA spectroscopy with PZT detection (see above). A related technique, based on inverse photopyroelectric detection with a PVDF film, has also been described briefly (Bićanić et al., 1989).

The basis of OW spectroscopy was discussed by Helander (1993) and Bićanić et al. (1995). For a thermally thick, optically opaque sample and a thermally thick window, the amplitude ψ of the optothermal signal is given by

$$\psi = \beta_s \mu_s \left[(1+\beta_s \mu_s)^2 + 1\right]^{-1/2} (1 + e_s/e_w)^{-1} \tag{3.5}$$

where the subscripts s and w denote sample and window, respectively, and e is thermal effusivity (see Appendix 1). The phase is given by

$$\varphi = -\tan^{-1}(1+\beta_s \mu_s)^{-1} \tag{3.6}$$

Equation 3.5 can be rearranged to yield an expression for the dimensionless optothermal signal amplitude S:

$$S = \beta_s \mu_s \left[(1+\beta_s \mu_s)^2 + 1\right]^{-1/2} \tag{3.7}$$

$$S = \psi(1 + e_s/e_w) \tag{3.8}$$

The first expression for S was also given by Lima et al. (2003) in consideration of the situation where the sample is thermally thick and the window is thermally thin. For the important case where $\beta_s \mu_s \ll 1$, the simple expressions $S = \beta_s \mu_s/2^{1/2}$ and $\varphi = -45°$ obtain. Thus the optothermal signal is proportional to the sample absorbance; in principle an absorption spectrum can be obtained using this technique when the sample thickness is equal to μ_s. By contrast, the phase is saturated and does not provide any useful information under most conditions. For strongly absorbing samples μ_s is constrained to low values, which implies that a high modulation frequency must be used. This naturally yields reduced signal intensity, for a reason analogous to that which affects both rapid- and step-scan PA FT-IR spectra.

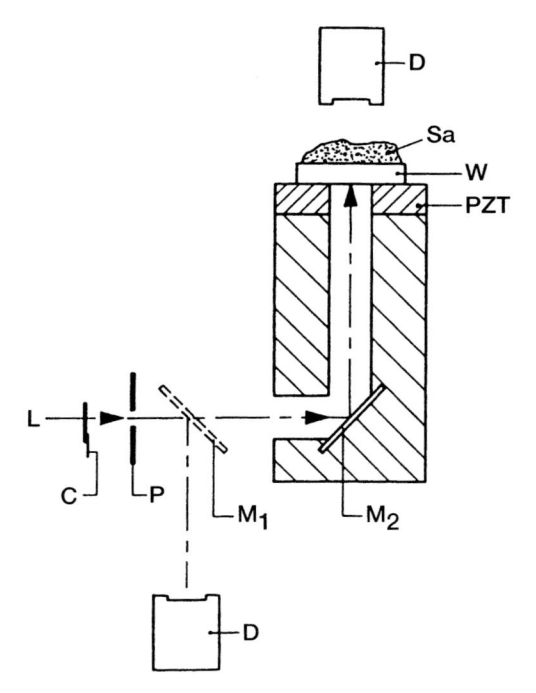

Figure 3.21 Experimental arrangement for OW spectroscopy. L, laser beam; C, chopper; P, diaphragm; M_1, M_2, mirrors; D, pyroelectric detector; PZT, piezoelectric detector; W, sapphire window; Sa, sample. A CO laser was used in this experiment. (Reproduced from Bićanić, D. et al., *Appl. Spectrosc.* **49**: 1485–1489, by permission of the Society for Applied Spectroscopy; copyright © 1995).

A simplified view of the experimental apparatus used by Bićanić et al. for the observation of OW spectra of liquids is shown in Figure 3.21. In this arrangement, a CO laser was used as the source of infrared radiation; obviously, a broadband near- or mid-infrared thermal source and appropriate optical filters could instead have been utilized to select radiation in the wavelength region of interest. The CO_2 and quantum-cascade lasers, as well as an OPO, are other infrared sources that are suitable for this technique. A creative experimentalist might arrange to perform OW measurements with the use of an FT-IR spectrometer, so that an entire mid-infrared spectrum could be acquired. The output voltage from the PZT could be preamplified and directed to the ADC of the spectrometer or detected with a lock-in amplifier in such an experiment.

3.9.1
Research of Bićanić's Group

Bićanić et al. (1995) used the OW apparatus shown in Figure 3.21 to obtain discrete-wavelength (sequentially measured) PA infrared spectra of oleic acid

($C_{18}H_{34}O_2$) and water. A CO laser, with about 30 emission lines between 1710 and 1890 cm^{-1}, was used as the infrared source in these experiments. This interval was sufficiently wide to allow the detection of a band near 1720 cm^{-1} due to C=O groups in the spectrum of oleic acid; however, only the high-frequency wing of the water v_2 band was accessible in this experiment. The intensity data were calibrated using the previously known absorption coefficient of water at the 1781-cm^{-1} CO laser frequency. This permitted calculation of β at each line position and the construction of plots of β vs wavenumber for both liquids.

This initial investigation was followed by several studies on food products. These publications are discussed in greater detail in Chapter 7. It should be noted that a CO_2 laser, which emits radiation between about 9 and 11 μm (1100–900 cm^{-1}), was used as the infrared source in these investigations.

The OW technique was compared with several other analytical methods with regard to the detection of trans fatty acids (TFAs) in margarines (Favier et al., 1996) and unsaturated vegetable oils (Bićanić et al., 1999). These acids exhibit a characteristic absorption band at 966 cm^{-1}, a wavenumber that fortuitously coincides with the CO_2 laser 10R6 line. TFA contents that ranged from about 4 to 60 per cent were successfully measured in these studies. Similarly, Favier et al. (1998) used OW spectroscopy to detect the adulteration of extra-virgin olive oil by safflower and sunflower oils. These contaminants were analyzed using the 10P10 and 9P26 lines (953 and 1041 cm^{-1}, respectively).

This group also used OW spectroscopy in a medical application. Annyas, Bićanić, and Schouten (1999) determined total body water in human blood serum samples by measuring absorption intensities at a wavelength of 4 μm. Heavy water was used as a tracer in this investigation. An impressive detection limit of 30 ppm was demonstrated; this could be further improved by replacing the thermal source (lamp) with a suitable laser.

3.9.2
Research of McQueen's Group

Optothermal spectroscopy has also been used to analyze foods by D. H. McQueen at the Chalmers University of Technology and his colleagues at several other institutions. The principal results of this work are summarized in Chapter 7, and the present discussion is restricted to a few salient points.

McQueen et al. (1995a) compared optothermal near-infrared spectroscopy and ATR mid-infrared spectroscopy with regard to their capabilities for the analysis of cheese samples. No spectra are illustrated in this study. Measurements were carried out in three relatively broad wavelength regions that were selected with the use of bandpass filters. The central wavelengths (1740, 1935 and 2180 nm) were chosen so as to yield information on the contents of protein, fat, and moisture in the cheeses. The simplicity of this experiment and the quality of the results it provided led the authors to conclude that the optothermal technique was superior to ATR spectroscopy for this particular application.

The authors also noted several limitations of the optothermal method as implemented in their laboratory. These include the following: (i) instrument reproducibility was about 1.5% of signal strength; (ii) good thermal contact between the sample and the window was essential; and (iii) temperature variations tended to affect spectral sensitivities and the transport of heat from the cheese samples to the window. These points were reiterated in a review article on the analysis of foods by near- and mid-infrared PA spectroscopy (McQueen et al., 1995b).

References

Aamodt, L.C. and Murphy, J.C. (1981) Photothermal measurements using a localized excitation source. *J. Appl. Phys.*, **52** (8), 4903–4914.

Adams, M.J., Beadle, B.C., and Kirkbright, G.F. (1978) Optoacoustic spectrometry in the near-infrared region. *Anal. Chem.*, **50** (9), 1371–1374.

Annyas, J., Bićanić, D., and Schouten, F. (1999) Novel instrumental approach to the measurement of total body water: optothermal detection of heavy water in the blood serum. *Appl. Spectrosc.*, **53** (3), 339–343.

Bain, C.D., Davies, P.B., and Ong, T.H. (1992) Vibrational spectroscopy of monolayers by pulsed photothermal beam deflection (PBD), in *Photoacoustic and Photothermal Phenomena III* (ed. D. Bićanić), Springer-Verlag, Berlin, pp. 158–160.

Belton, P.S. and Tanner, S.F. (1983) Determination of the moisture content of starch using near infrared photoacoustic spectroscopy. *Analyst*, **108** (1286), 591–596.

Belton, P.S., Wilson, R.H., and Saffa, A.M. (1987) Effects of particle size on quantitative photoacoustic spectroscopy using a gas-microphone cell. *Anal. Chem.*, **59** (19), 2378–2382.

Bićanić, D., Krüger, S., Torfs, P., Bein, B., and Harren, F. (1989) The use of reverse mirage spectroscopy to determine the absorption coefficient of liquid methanol at CO_2 laser wavelengths. *Appl. Spectrosc.*, **43** (1), 148–153.

Bićanić, D., Jalink, H., Chirtoc, M., Sauren, H., Lubbers, M., Quist, J., Gerkema, E., van Asselt, K., Miklós, A., Sólyom, A., Angeli, Gy.Z., Helander, P., and Vargas, H. (1992) Interfacing photoacoustic and photothermal techniques for new hyphenated methodologies and instrumentation suitable for agricultural, environmental and medical applications, in *Photoacoustic and Photothermal Phenomena III* (ed. D. Bićanić), Springer-Verlag, Berlin, pp. 20–27.

Bićanić, D., Chirtoc, M., Chirtoc, I., Favier, J.P., and Helander, P. (1995) Photothermal determination of absorption coefficients in optically dense fluids: application to oleic acid and water at CO laser wavelengths. *Appl. Spectrosc.*, **49** (10), 1485–1489.

Bićanić, D., Fink, T., Franko, M., Močnik, G., van de Bovenkamp, P., van Veldhuizen, B., and Gerkema, E. (1999) Infrared photothermal spectroscopy in the science of human nutrition. *AIP Conf. Proc.*, **463**, 637–639.

Blank, R.E. and Wakefield, T. (1979) Double-beam photoacoustic spectrometer for use in the ultraviolet, visible, and near-infrared spectral regions. *Anal. Chem.*, **51** (1), 50–54.

Boccara, A.C., Fournier, D., Jackson, W., and Amer, N.M. (1980) Sensitive photothermal deflection technique for measuring absorption in optically thin media. *Opt. Lett.*, **5** (9), 377–379.

Bouzerar, R., Amory, C., Zeinert, A., Benlahsen, M., Racine, B., Durand-Drouhin, O., and Clin, M. (2001) Optical properties of amorphous hydrogenated carbon thin films. *J. Non-Cryst. Solids*, **281** (1–3), 171–180.

Carter, R.O. and Paputa Peck, M.C. (1989) Photoacoustic detection of rapid-scan Fourier transform infrared spectra from low surface-area solid samples. *Appl. Spectrosc.*, **43** (3), 468–473.

Carter, R.O. and Wright, S.L. (1991) Evaluation of the appropriate sample position in a PAS/FT-IR experiment. *Appl. Spectrosc.*, **45** (7), 1101–1103.

Castleden, S.L., Kirkbright, G.F., and Menon, K.R. (1980) Determination of moisture in single-cell protein utilising photoacoustic spectroscopy in the near-infrared region. *Analyst*, **105** (1256), 1076–1081.

Castleden, S.L., Kirkbright, G.F., and Long, S.E. (1982) Quantitative assay of propranolol by photoacoustic spectroscopy. *Can. J. Spectrosc.*, **27** (1), 245–248.

Chandrasekharan, N., Gonzales, B., and Cullum, B.M. (2004) Non-resonant multiphoton photoacoustic spectroscopy for noninvasive subsurface chemical diagnostics. *Appl. Spectrosc.*, **58** (11), 1325–1333.

Chuang, T.J., Coufal, H., and Träger, F. (1983) Infrared laser photoacoustic spectroscopy of adsorbed species. *J. Vac. Sci. Technol. A*, **1** (2), 1236–1239.

Débarre, D., Boccara, A.C., and Fournier, D. (1981) High-luminosity visible and near-IR Fourier- transform photoacoustic spectrometer. *Appl. Opt.*, **20** (24), 4281–4286.

DeBellis, A.D. and Low, M.J.D. (1987) Dispersive infrared photothermal beam deflection spectroscopy. *Infrared Phys.*, **27** (3), 181–191.

DeBellis, A.D. and Low, M.J.D. (1988) Dispersive infrared phase angle photothermal beam deflection spectroscopy. *Infrared Phys.*, **28** (4), 225–237.

Deng, Z., Spear, J.D., Rudnicki, J.D., McLarnon, F.R. and Cairns, E.J. (1996) Infrared photothermal deflection spectroscopy: a new probe for the investigation of electrochemical interfaces. *J. Electrochem. Soc.*, **143** (5), 1514–1521.

Dias, D.T., Medina, A.N., Baesso, M.L., Bento, A.C., Porto, M.F., and Rubira, A.F. (2002) The photoacoustic spectroscopy applied in the characterization of the cross-linking process in polymeric materials. *Braz. J. Phys.*, **32** (2B), 523–530.

Dóka, O., Bićanić, D., Szücs, M., and Lubbers, M. (1998) Direct measurement of carbonate content in soil samples by means of CO laser infrared photoacoustic spectroscopy. *Appl. Spectrosc.*, **52** (12), 1526–1529.

Donini, J.C. and Michaelian, K.H. (1988) Low-frequency photoacoustic spectroscopy of solids. *Appl. Spectrosc.*, **42** (2), 289–292.

Duerst, R.W. and Mahmoodi, P. (1984) IR-PAS chamber for signal-to-noise enhancement. *Prepr. Am. Chem. Soc., Div. Polym. Chem.*, **25** (2), 194–195.

Duerst, R.W., Mahmoodi, P., and Duerst, M.D. (1987) IR-PAS studies: signal-to-noise enhancement and depth profile analysis, in *Fourier Transform Infrared Characterization of Polymers* (ed. H. Ishida), Plenum Press, New York, pp. 113–122.

Einsiedel, H., Kreiter, M., Leclerc, M., and Mittler-Neher, S. (1998) Photothermal beam deflection spectroscopy in the near IR on poly[3-alkylthiophene]s. *Opt. Mater.*, **10** (1), 61–68.

Favier, J.P., Bićanić, D., van de Bovenkamp, P., Chirtoc, M., and Helander, P. (1996) Detection of total trans fatty acids content in margarine: an intercomparison study of GLC, GLC + TLC, FT-IR, and optothermal window (open photoacoustic cell). *Anal. Chem.*, **68** (5), 729–733.

Favier, J.P., Bićanić, D., Cozijnsen, J., van Veldhuizen, B., and Helander, P. (1998) CO_2 laser infrared optothermal spectroscopy for quantitative adulteration studies in binary mixtures of extra-virgin olive oil. *J. Am. Oil Chem. Soc.*, **75** (3), 359–362.

Fonsen, J., Koskinen, V., Roth, K., and Kauppinen, J. (2009) Dual cantilever enhanced photoacoustic detector with pulsed broadband IR-source. *Vib. Spectrosc.*, **50** (2), 214–217.

Fournier, D., Boccara, A.C., Amer, N.M., and Gerlach, R. (1980) Sensitive *in situ* trace-gas detection by photothermal deflection spectroscopy. *Appl. Phys. Lett.*, **37** (6), 519–521.

Fournier, D., Boccara, A.C., and Badoz, J. (1982) Photothermal deflection Fourier transform spectroscopy: a tool for high-sensitivity absorption and dichroism measurements. *Appl. Opt.*, **21** (1), 74–76.

Fukuyama, A., Akashi, Y., Suemitsu, M., and Ikari, T. (2000) Detailed observation of the photoquenching effect of EL2 in semi-insulating GaAs by the piezoelectric photoacoustic measurements. *J. Cryst. Growth*, **210** (1), 255–259.

Gendre, D. (1987) Mesure par effet mirage de la diffusivité thermique jusqu'à 500 K de matériaux opaques. *Rev. Gen. Therm.*, **26** (301), 54–58.

Gregoriou, V.G., Daun, M., Schauer, M.W., Chao, J.L., and Palmer, R.A. (1993) Modification of a research-grade FT-IR spectrometer for optional step-scan operation. *Appl. Spectrosc.*, **47** (9), 1311–1316.

Guo, L., Tang, Z., He, Y., and Zhang, H. (2007) Characterization of a derivative photoacoustic spectrometer. *Rev. Sci. Instrum.*, **78** (2), 023104 (4 pages).

Harris, M., Pearson, G.N., Willetts, D.V., Ridley, K., Tapster, P.R., and Perrett, B. (2000) Pulsed indirect photoacoustic spectroscopy: application to remote detection of condensed phases. *Appl. Opt.*, **39** (6), 1032–1041.

Havaux, M., Lorrain, L., and Leblanc, R.M. (1989) In vivo measurement of spectroscopic and photochemical properties of intact leaves using the 'mirage effect'. *FEBS Lett.*, **250** (2), 395–399.

Helander, P. (1993) A method for the analysis of the optothermal and photoacoustic signals. *Meas. Sci. Technol.*, **4** (2), 178–185.

Horstjann, M., Bakhirkin, Y.A., Kosterev, A.A., Curl, R.F., Tittel, F.K., Wong, C.M., Hill, C.J., and Yang, R.Q. (2004) Formaldehyde sensor using interband cascade laser based quartz-enhanced photoacoustic spectroscopy. *Appl. Phys. B*, **79** (7), 799–803.

Jones, R.W. and McClelland, J.F. (2001) Phase references and cell effects in photoacoustic spectroscopy. *Appl. Spectrosc.*, **55** (10), 1360–1367.

Kanstad, S.O. and Nordal, P.-E. (1980a) Photoacoustic reflection-absorption spectroscopy (PARAS) for infrared analysis of surface species. *Appl. Surf. Sci.*, **5** (3), 286–295.

Kanstad, S.O. and Nordal, P.-E. (1980b) Photoacoustic and photothermal spectroscopy. *Phys. Technol.*, **11** (4), 142–147.

Kanstad, S.O., Nordal, P.-E., Hellgren, L., and Vincent, J. (1981) Infrared photoacoustic spectroscopy of skin lipids. *Naturwissenschaften*, **68** (1), 47–48.

Kapil, J.C., Joshi, S.K., and Rai, A.K. (2003) *In situ* photoacoustic investigations of some optically transparent samples lilke ice and snow. *Rev. Sci. Instrum.*, **74** (7), 3536–3543.

Kauppinen, J., Wilcken, K., Kauppinen, I., and Koskinen, V. (2004) High sensitivity in gas analysis with photoacoustic detection. *Microchem. J.*, **76** (1–2), 151–159.

Kirkbright, G.F. (1978) Analytical optoacoustic spectrometry. *Opt. Pura Apl.*, **11**, 125–136.

Koskinen, V., Fonsen, J., Kauppinen, J., and Kauppinen, I. (2006) Extremely sensitive trace gas analysis with modern photoacoustic spectroscopy. *Vib. Spectrosc.*, **42** (2), 239–242.

Koskinen, V., Fonsen, J., Roth, K., and Kauppinen, J. (2008) Progress in cantilever enhanced photoacoustic spectroscopy. *Vib. Spectrosc.*, **48** (1), 16–21.

Kosterev, A.A. and Tittel, F.K. (2004) Ammonia detection by use of quartz-enhanced photoacoustic spectroscopy with a near-IR telecommunication diode laser. *Appl. Opt.*, **43** (33), 6213–6217.

Kosterev, A.A., Bakhirkin, Yu.A., Curl, R.F., and Tittel, F.K. (2002) Quartz-enhanced photoacoustic spectroscopy. *Opt. Lett.*, **27** (21), 1902–1904.

Kosterev, A.A., Bakhirkin, Y.A., Tittel, F.K., Blaser, S., Bonetti, Y., and Hvozdara, L. (2004) Photoacoustic phase shift as a chemically selective spectroscopic parameter. *Appl. Phys. B*, **78** (6), 673–676.

Kosterev, A.A., Bakhirkin, Y.A., and Tittel, F.K. (2005) Ultrasensitive gas detection by quartz-enhanced photoacoustic spectroscopy in the fundamental molecular absorption bands region. *Appl. Phys. B*, **80** (1), 133–138.

Kosterev, A.A., Mosely, T.S., and Tittel, F.K. (2006) Impact of humidity on quartz-enhanced photoacoustic spectroscopy based detection of HCN. *Appl. Phys. B*, **85** (2–3), 295–300.

Kosterev, A., Wysocki, G., Bakhirkin, Y., So, S., Lewicki, R., Fraser, M., Tittel, F., and Curl, R.F. (2008a) Application of quantum cascade lasers to trace gas analysis. *Appl. Phys. B*, **90** (2), 165–176.

Kosterev, A.A., Bakhirkin, Y.A., Tittel, F.K., McWhorter, S., and Ashcraft, B. (2008b) QEPAS methane sensor performance for humidifed gases. *Appl. Phys. B*, **92** (1), 103–109.

Kuusela, T. and Kauppinen, J. (2007) Photoacoustic gas analysis using interferometric cantilever microphone. *Appl. Spectrosc. Rev.*, **42** (5), 443–474.

Kuwahata, H., Muto, N., and Uehara, F. (2000) Carrier concentration dependence of photoacoustic spectra of silicon by a piezoelectric transducer method. *Jpn. J. Appl. Phys., Part 1*, **39** (5B), 3169–3171.

Kuusela, T., Peura, J., Matveev, B.A., Remennyy, M.A., and Stus', N.M. (2009) Photoacoustic gas detection using a cantilever microphone and III–V mid-IR LEDs. *Vib. Spectrosc.*, **51** (2), 289–293.

Laurila, T., Cattaneo, H., Koskinen, V., Kauppinen, J., and Hernberg, R. (2005) Diode laser-based photoacoustic spectroscopy with interferometrically enhanced cantilever detection. *Opt. Express*, **13** (7), 2453–2458.

Lima, J.A.P., Massunaga, M.S.O., Cardoso, S.L., Vargas, H., de Melo Monte, M.B., Duarte, A.C.P., do Amaral, M.R., Jr., and de Souza-Barros, F. (2003) Contributions for soil treatment: the controlled release of biological phosphate monitored by the optothermal window technique. *Rev. Sci. Instrum.*, **74** (1), 773–775.

Liu, K., Li, J., Wang, L., Tan, T., Zhang, W., Gao, X., Chen, W., and Tittel, F.K. (2009) Trace gas sensor based on quartz tuning fork enhanced laser photoacoustic spectroscopy. *Appl. Phys. B*, **94** (3), 527–533.

Lloyd, L.B., Riseman, S.M., Burnham, R.K., Eyring, E.M., and Farrow, M.M. (1980) Fourier transform photoacoustic spectrometer. *Rev. Sci. Instrum.*, **51** (11), 1488–1492.

Lloyd, L.B., Yeates, R.C., and Eyring, E.M. (1982) Fourier transform infrared photoacoustic spectroscopy in thin-layer chromatography. *Anal. Chem.*, **54** (3), 549–552.

Lochmüller, C.H. and Wilder, D.R. (1980a) Qualitative examination of chemically modified silica surfaces by near-infrared photoacoustic spectroscopy. *Anal. Chim. Acta*, **116** (1), 19–24.

Lochmüller, C.H. and Wilder, D.R. (1980b) Quantitative photoacoustic spectroscopy of chemically modified silica surfaces. *Anal. Chim. Acta*, **118** (1), 101–108.

Low, M.J.D. (1986) Some practical aspects of FT-IR/PBDS. Part I: vibrational noise. *Appl. Spectrosc.*, **40** (7), 1011–1019.

Low, M.J.D. and Lacroix, M. (1982) An infrared photothermal beam deflection Fourier transform spectrometer. *Infrared Phys.*, **22** (3), 139–147.

Low, M.J.D. and Morterra, C. (1987) Some practical aspects of FT-IR/PBDS. Part II: sample handling procedures. *Appl. Spectrosc.*, **41** (2), 280–287.

Low, M.J.D. and Parodi, G.A. (1978) Infrared photoacoustic spectra of surface species in the 4000–2000 cm^{-1} region using a broad band source. *Spectrosc. Lett.*, **11** (8), 581–588.

Low, M.J.D. and Parodi, G.A. (1980a) Infrared photoacoustic spectroscopy of surfaces. *J. Mol. Struct.*, **61** (C), 119–124.

Low, M.J.D. and Parodi, G.A. (1980b) Infrared photoacoustic spectra of solids. *Spectrosc. Lett.*, **13** (2–3), 151–158.

Low, M.J.D. and Parodi, G.A. (1980c) An infrared photoacoustic spectrometer. *Infrared Phys.*, **20** (5), 333–340.

Low, M.J.D. and Parodi, G.A. (1980d) Infrared photoacoustic spectroscopy of solids and surface species. *Appl. Spectrosc.*, **34** (1), 76–80.

Low, M.J.D. and Parodi, G.A. (1980e) Carbon as reference for normalizing infrared photoacoustic spectra. *Spectrosc. Lett.*, **13** (9), 663–669.

Low, M.J.D. and Parodi, G.A. (1982) Dispersive photoacoustic spectroscopy of solids in the infrared range. *J. Photoacoustics*, **1** (1), 131–144.

Low, M.J.D. and Tascon, J.M.D. (1985) An approach to the study of minerals using infrared photothermal beam deflection spectroscopy. *Phys. Chem. Miner.*, **12** (1), 19–22.

Low, M.J.D., Lacroix, M., and Morterra, C. (1982a) Infrared photothermal beam deflection Fourier transform spectroscopy of solids. *Appl. Spectrosc.*, **36** (5), 582–584.

Low, M.J.D., Lacroix, M., and Morterra, C. (1982b) Infrared spectra of massive solids by photoacoustic beam deflection Fourier transform spectroscopy. *Spectrosc. Lett.*, **15** (1), 57–64.

McClelland, J.F., Luo, S., Jones, R.W., and Seaverson, L.M. (1992) A tutorial on the state-of-the-art of FTIR photoacoustic spectroscopy, in *Photoacoustic and Photothermal Phenomena III* (ed. D. Bićanić), Springer-Verlag, Berlin, pp. 113–124.

McCurdy, M.R., Bakhirkin, Y., Wysocki, G., Lewicki, R., and Tittel, F.K. (2007) *J. Breath Res.*, **1**, 014001 (12 pages).

McQueen, D.H., Wilson, R., Kinnunen, A., and Jensen, E.P. (1995a) Comparison of two infrared spectroscopic methods for cheese analysis. *Talanta*, **42** (12), 2007–2015.

McQueen, D.H., Wilson, R., and Kinnunen, A. (1995b) Near and mid-infrared photoacoustic analysis of principal components of foodstuffs. *TrAC, Trends Anal. Chem.*, **14** (10), 482–492.

Mahmoodi, P., Duerst, R.W., and Meiklejohn, R.A. (1984) Effect of acoustic isolation chamber on the signal-to-noise ratio in infrared photoacoustic spectroscopy. *Appl. Spectrosc.*, **38** (3), 437–438.

Manning, C.J., Palmer, R.A., and Chao, J.L. (1991) Step-scan Fourier-transform infrared spectrometer. *Rev. Sci. Instrum.*, **62** (5), 1219–1229.

Manning, C.J., Dittmar, R.M., Palmer, R.A., and Chao, J.L. (1992a) Use of step-scan FT-IR to obtain the photoacoustic/photothermal response phase. *Infrared Phys.*, **33** (1), 53–62.

Manning, C.J., Palmer, R.A., Chao, J.L., and Charbonnier, F. (1992b) Photothermal beam deflection using the reverse mirage geometry: theory and experiment. *J. Appl. Phys.*, **71** (5), 2433–2440.

Manning, C.J., Charbonnier, F., Chao, J.L., and Palmer, R.A. (1992c) Reverse mirage photothermal beam deflection: theory and experiment, in *Photoacoustic and Photothermal Phenomena III* (ed. D. Bićanić), Springer-Verlag, Berlin, pp. 161–164.

Manzanares, C., Blunt, V.M., and Peng, J. (1993) Vibrational spectroscopy of nonequivalent C–H bonds in liquid cis- and trans-3-hexene. *Spectrochim. Acta, Part A*, **49** (8), 1139–1152.

Ngai, A.K.Y., Persijn, S.T., Lindsay, I.D., Kosterev, A.A., Groß, P., Lee, C.J., Cristescu, S.M., Tittel, F.K., Boller, K.-J., and Harren, F.J.M. (2007) Continuous wave optical parametric oscillator for quartz-enhanced photoacoustic trace gas sensing. *Appl. Phys. B*, **89** (1), 123–128.

Palmer, R.A. (1994) Time-resolved and phase-resolved vibrational spectroscopy by use of step-scan FT-IR. *Proc. SPIE Int. Soc. Opt. Eng.*, **2089**, 53–61.

Palmer, R.A. and Smith, M.J. (1986) Rapid-scanning Fourier-transform infrared spectroscopy with photothermal beam-deflection (mirage effect) detection at the solid-liquid interface. *Can. J. Phys.*, **64** (9), 1086–1092.

Palmer, R.A., Chao, J.L., Dittmar, R.M., Gregoriou, V.G., and Plunkett, S.E. (1993) Investigation of time-dependent phenomena by use of step-scan FT-IR. *Appl. Spectrosc.*, **47** (9), 1297–1310.

Petra, N., Zweck, J., Kosterev, A.A., Minkof, S.E., and Thomazy, D. (2009) Theoretical analysis of a quartz-enhanced photoacoustic spectroscopy sensor. *Appl. Phys. B*, **94** (4), 673–680.

Prasad, R.L., Thakur, S.N., and Bhar, G.C. (2002a) CO_2 laser photoacoustic spectra and vibrational modes of heroin, morphine and narcotine. *Pramana–J. Phys.*, **59** (3), 487–496.

Prasad, R.L., Prasad, R., Bhar, G.C., and Thakur, S.N. (2002b) Photoacoustic spectra and modes of vibration of TNT and RDX at CO_2 laser wavelengths. *Spectrochim. Acta, Part A*, **58** (14), 3093–3102.

Rosencwaig, A. and Gersho, A. (1976) Theory of the photoacoustic effect with solids. *J. Appl. Phys.*, **47** (1), 64–69.

Sakai, K., Fukuyama, A., Toyoda, T., and Ikari, T. (2002) Piezoelectric photothermal spectra of Co doped ZnO semiconductor. *Jpn. J. Appl. Phys., Part 1*, **41** (5B), 3371–3373.

Schilt, S., Kosterev, A.A., and Tittel, F.K. (2009) Performance evaluation of a near infrared QEPAS based ethylene sensor. *Appl. Phys. B*, **95** (4), 813–824.

Smith, M.J. and Palmer, R.A. (1987) The reverse mirage effect: catching the thermal wave at the solid/liquid interface. *Appl. Spectrosc.*, **41** (7), 1106–1113.

Smith, M.J., Manning, C.J., Palmer, R.A., and Chao, J.L. (1988) Step scan interferometry in the mid-infrared with photothermal detection. *Appl. Spectrosc.*, **42** (4), 546–555.

Tittel, F.K., Bakhirkin, Y.A., Curl, R.F., Kosterev, A.A., McCurdy, M.R., So, S.G., and Wysocki, G. (2007a) Laser based chemical technology: recent advances and applications, in *Advanced Environmental Monitoring* (eds Y.J. Kim and U. Platt), Springer, pp. 50–63.

Tittel, F.K., Wysocki, G., Kosterev, A., and Bakhirkin, Y. (2007b) Semiconductor laser based trace gas sensor technology: recent advances and applications, in *Mid-Infrared Coherent Sources and Applications* (eds M. Ebrahim-Zadeh and I.T. Sorokina), Springer, pp. 467–493.

Uotila, J. and Kauppinen, J. (2008) Fourier transform infrared measurement of solid-, liquid-, and gas-phase samples with a single photoacoustic cell. *Appl. Spectrosc.*, **62** (6), 655–660.

Uotila, J., Koskinen, V., and Kauppinen, J. (2005) Selective differential photoacoustic method for trace gas analysis. *Vib Spectrosc.*, **38** (1), 3–9.

Wang, Z.Y., Ng, C.F., Lim, P.K., Leung, H.W., Wu, J.G., and Chen, H.F. (1994) The torsional-rotational spectral structure of ethanol molecules adsorbed on polycrystalline silver substrate. *J. Phys. Condens. Matter*, **6** (36), 7207–7215.

Weidmann, D.A., Kosterev, A.A., Tittel, F.K., Ryan, N., and McDonald, D. (2004) Application of a widely electrically tunable diode laser to chemical gas sensing with quartz-enhanced photoacoustic spectroscopy. *Opt. Lett.*, **29** (16), 1837–1839.

Wen, Q. and Michaelian, K.H. (2008) Mid-infrared photoacoustic spectroscopy of solids using an external-cavity quantum-cascade laser. *Opt. Lett.*, **33** (16), 1875–1877.

Wojcik, M.D., Phillips, M.C., Cannon, B.D., and Taubman, M.S. (2006) Gas-phase photoacoustic sensor at 8.41 µm using quartz tuning forks and amplitude-modulated quantum cascade lasers. *Appl. Phys. B*, **85** (2–3), 307–313.

Wysocki, G., Kosterev, A.A., and Tittel, F.K. (2006) Influence of molecular relaxation dynamics on quartz-enhanced photoacoustic detection of CO_2 at $\lambda = 2$ µm. *Appl. Phys. B*, **85** (2–3), 301–306.

Wysocki, G., Lewicki, R., Curl, R.F., Tittel, F.K., Diehl, L., Capasso, F., Troccoli, M., Hofler, G., Bour, D., Corzine, S., Maulini, R., Giovannini, M., and Faist, J. (2008) Widely tunable mode-hop free external cavity quantum cascade lasers for high resolution spectroscopy and chemical sensing. *Appl. Phys. B*, **92** (3), 305–311.

Yoshino, K., Fukuyama, A., Yokoyama, H., Meada, K., Fons, P.J., Yamada, A., Niki, S., and Ikari, T. (1999) Piezoelectric photoacoustic spectra of $CuInSe_2$ thin film grown by molecular beam epitaxy. *Thin Solid Films*, **343–344** (1–2), 591–593.

Zhang, S.L., Michaelian, K.H., and Burt, J.A. (1997) Phase correction in piezoelectric photoacoustic Fourier transform infrared spectroscopy of mica. *Opt. Eng.*, **36** (2), 321–325.

Zhou, W.-Y., Xie, S.-S., Qian, S.-F., Wang, G., and Qian, L.-X. (1996) Photothermal deflection spectra of solid C_{60}. *J. Phys. Condens. Matter*, **8** (31), 5793–5800.

4
Signal Recovery

Step-scan PA infrared spectroscopy was briefly introduced in the previous chapter and is discussed more fully later in this book. The use of amplitude or phase modulation in step-scan experiments creates an obvious requirement for PA signal recovery. Both digital and analog signal recovery (demodulation) methods have been successfully utilized since step-scan PA infrared experiments became feasible more than two decades ago. Signal recovery through digital signal processing (DSP), or alternatively by use of lock-in amplifiers, is described in this chapter. Spectrum computation, which requires particular strategies for some of the resulting interferograms, is also mentioned.

4.1
DSP Demodulation

Detailed accounts of DSP demodulation were given previously with particular reference to PA infrared spectroscopy (Drapcho et al., 1997) and in a more general context (Curbelo, 1998). In this procedure, two interferograms (real and imaginary) representing the sample response to a periodic modulation are constructed. This is possible because the stepping of the movable mirror, amplitude or phase modulation, and sampling rate of the analog-to-digital converter (ADC) are all synchronous. The real and imaginary interferograms correspond to in-phase and quadrature interferograms, respectively, that might be obtained using a lock-in amplifier. Curbelo (1998) showed that ADC resolution is improved by adding noise, oversampling, digital filtering, and decimation. DSP demodulation can be used in step-scan PA spectroscopy, circular dichroism (CD) experiments that employ a photo-elastic modulator to switch the polarization of the incident radiation before it impinges on a sample, and a number of other applications.

An important aspect of DSP demodulation is the capability for simultaneous recovery of signals at several harmonic multiples of f. For example, square-wave modulation yields measurable PA intensities at the lower odd harmonics of f; these intensities are recovered by Fourier transformation of the time-domain response to the modulation. Parallel observation of PA intensities at these harmonic frequencies precludes the need for repeated measurements and facilitates

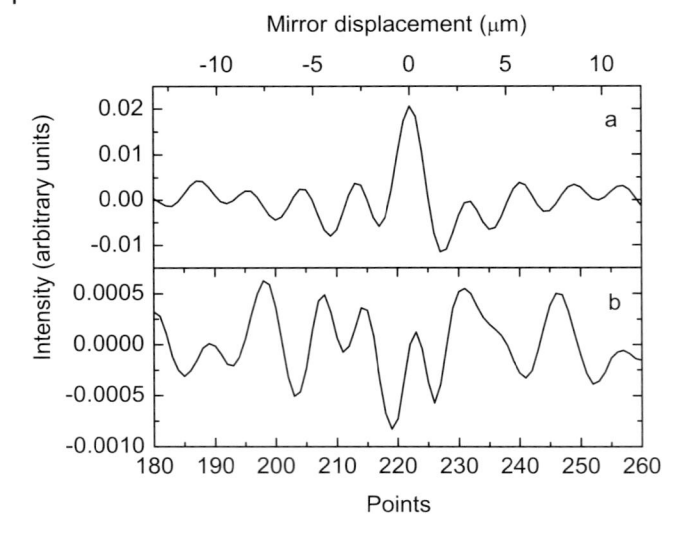

Figure 4.1 Central regions of (a) real and (b) imaginary interferograms obtained for kaolin in a step-scan phase modulation experiment at $f = 133$ Hz. Signals were recovered using DSP demodulation.

experiments such as depth profiling (Wang, Phifer, and Palmer, 1998). PA signals can be recovered at frequencies up to approximately $9f$ in this way.

A typical application of DSP demodulation in mid-infrared PA spectroscopy serves to illustrate the method.[1] Figure 4.1 shows the central regions of real (box a) and imaginary (box b) interferograms acquired for kaolin using a Bruker IFS 66v/S FT-IR spectrometer and an MTEC 300 PA cell; the phase modulation amplitude was 2λ at $f = 133$ Hz. These interferograms show that the real part of the signal is about 25 times as great as the imaginary component. Nevertheless, the signal/noise ratios in these two curves are comparable.

Calculation of spectra from these interferograms requires special consideration. The real data exhibit an obvious centerburst corresponding to the zero-path-difference (ZPD) location where the two beamsplitter–mirror distances in the interferometer are about equal. By contrast, an analogous central feature cannot be unambiguously identified in the imaginary interferogram. This means that Fourier transformation and spectrum computation based on the automatic recognition of ZPD, as implemented in standard FT-IR software, do not necessarily yield the correct imaginary spectrum. This situation is illustrated in Figure 4.2. The imaginary spectrum (lower curve, box a) does not show the expected bands in the 900–1200 and 3600–3700 cm^{-1} regions; moreover, the phase spectrum (box b) does not exhibit structure corresponding to the known bands in these regions.

Recalculation of the imaginary spectrum using the centerburst peak location and instrumental (spectrometer + PA cell) phase derived from the real interfero-

---

1) The author is grateful to Mr. C. Hyett for his assistance in the experiments described in this chapter. This work was performed at the Canadian Light Source.

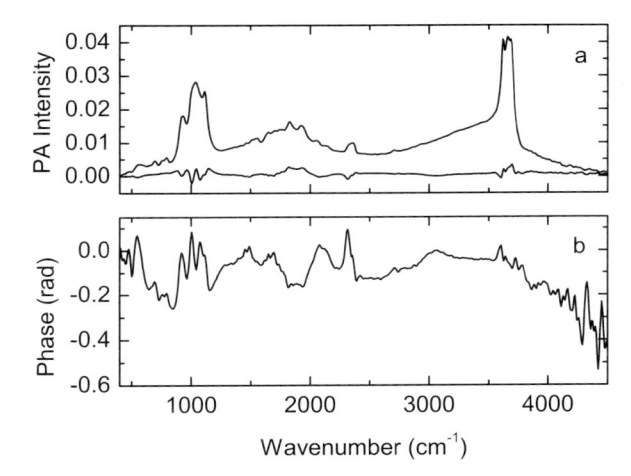

Figure 4.2 Spectra and phase calculated from the interferograms in Figure 4.1 using the standard procedure. The top and bottom curves in box a are the real and imaginary spectra, respectively; the PA phase calculated from these spectra is shown in box b.

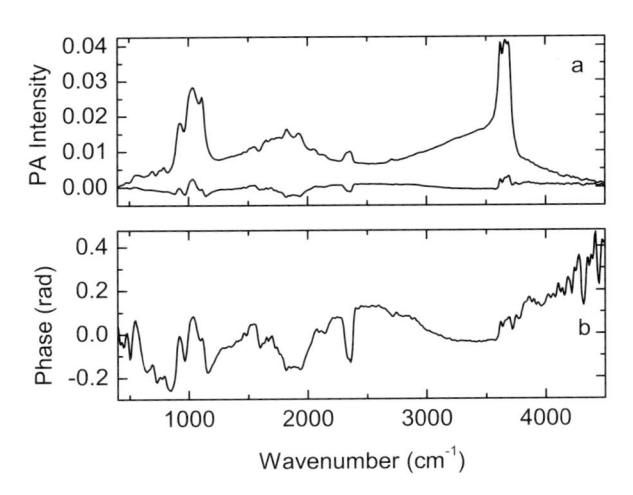

Figure 4.3 Spectra and phase calculated from the interferograms in Figure 4.1 using the stored-phase method. Upper and lower boxes display results corresponding to the curves in Figure 4.2.

gram yielded more satisfactory results. Figure 4.3 displays the real spectrum, the new imaginary spectrum (box b) obtained in this 'stored-phase' calculation, and the phase spectrum corresponding to this new imaginary spectrum (upper box). The phase spectrum is considerably improved with respect to that in Figure 4.2, with most of the kaolin bands being immediately recognizable. To summarize, the lack of a prominent centerburst in the imaginary interferogram is a consequence of the existence of regions of positive and negative intensity in the corresponding spectrum. Calculation of this spectrum requires a specialized strategy with regard to Fourier transformation and correction for the instrumental phase. Similar situations occur in several other types of FT spectroscopy.

DSP demodulation is commonly utilized in research-grade FT-IR spectrometers that are capable of step-scan operation. Many PA researchers use this means of signal recovery because of its convenience; as mentioned earlier, spectra at f and its harmonics are retrieved entirely in software. This procedure is satisfactory as long as digital signal recovery is accurate at both fundamental and harmonic frequencies (i.e., intensities are recovered at the correct frequencies). The analog alternative to DSP demodulation is described in the next section.

4.2
Lock-in Demodulation

Lock-in amplifiers have been used in a variety of signal recovery applications for many years. These instruments enable the measurement of in-phase and quadrature signals with high accuracies and sensitivities (down to nanoVolts) at arbitrary phase angles. Because these signals are recovered in narrow frequency intervals, considerable noise rejection is effected. Modern lock-in amplifiers – some using DSP in place of selected electronic components – allow signal detection at various harmonics of the reference (modulation) frequency, a capability relevant to PA infrared spectroscopy that was mentioned in the previous section. However, separate measurements must be performed at each harmonic frequency with a lock-in amplifier.

Signal recovery in step-scan PA infrared spectroscopy using a lock-in amplifier is demonstrated in this section. Experiments were performed in conjunction with the digital demodulation tests described above, utilizing the same spectrometer, PA cell, and sample. The step-scan operation of the spectrometer was also similar, except that the phase modulation was at $f = 77$ Hz.

A schematic layout of the lock-in demodulation experiment is shown in Figure 4.4 (Michaelian, 2007). The PA signal was demodulated by the lock-in amplifier and the DC output of this instrument was directed to the ADC in the FT-IR spectrometer; in other words, the lock-in amplifier was placed in series with the PA cell and FT-IR electronics. The phase angle of the lock-in was adjusted to maximize the in-phase PA signal, typically while the movable interferometer mirror was stopped near the ZPD position. This strategy is similar to that used a number of years ago in exploratory step-scan amplitude-modulation work (Michaelian, 1989). The experiment could be simplified by directly recording the digital data registered by the lock-in amplifier rather than converting these results to a DC signal; the present arrangement, which was chosen for convenience, allows data analysis using standard FT-IR software.

Data were recorded with the lock-in phase angle set to maximize the in-phase response at several different locations near ZPD. Figure 4.5 shows the central

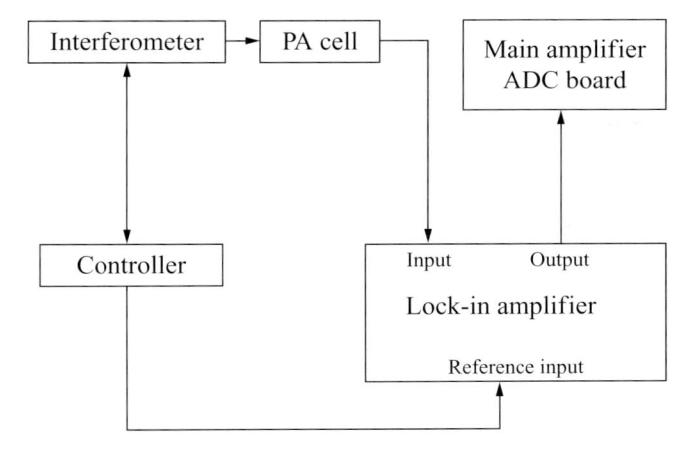

Figure 4.4 Experimental arrangement for lock-in demodulation in step-scan phase modulation PA FT-IR experiments.

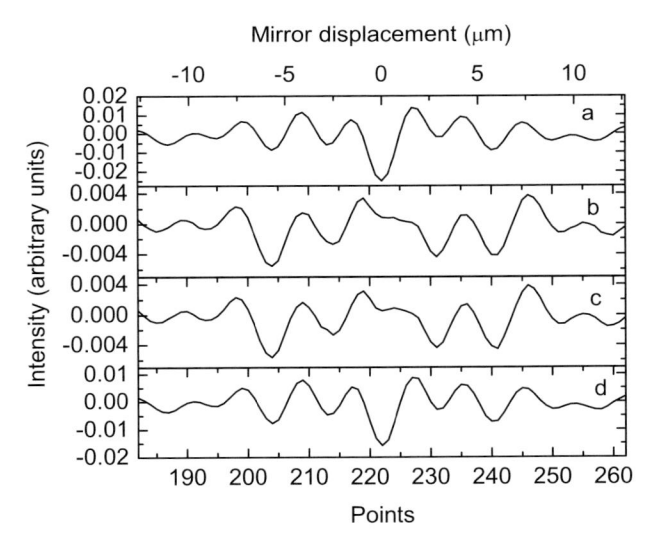

Figure 4.5 Central region of imaginary interferograms observed for kaolin using $f = 77\,\text{Hz}$ and lock-in demodulation. The lock-in phase angle was adjusted to maximize the intensity of the real interferogram at points (a) 226, (b) 224, (c) 222, and (d) 220.

regions of four quadrature (imaginary) interferograms obtained for kaolin when these phase angle adjustments were made at interferogram points 226, 224, 222 and 220 (top to bottom curves respectively). The spacing between consecutive points in these interferograms was $0.3164\,\mu\text{m}$, half the wavelength of the reference HeNe laser; this value is dictated by the bandwidth (free spectral range) in this experiment. It can be observed that imaginary interferograms exhibiting

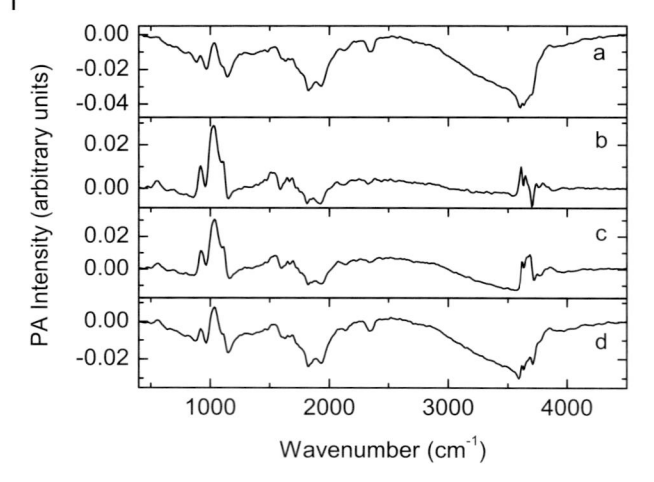

Figure 4.6 Spectra calculated from the interferograms in Figure 4.5. Curves (a–d) correspond to the interferograms with the same labels.

identifiable centerbursts were measured when the lock-in phase was set at positions a few points away from ZPD, which was situated near point 222; in these cases (curves a and d), a significant fraction of the PA intensity occurred in the quadrature channel. By contrast, less intensity was observed in the imaginary interferograms when the phase angle was adjusted closer to ZPD. The absence of clear centerbursts in curves b and c is reminiscent of the situation described earlier for the imaginary interferogram obtained with DSP demodulation; accordingly, stored-phase spectrum computation was also employed for the imaginary lock-in data.

Spectra calculated from the interferograms in Figure 4.5 are shown in Figure 4.6. The kaolin bands near $1000\,cm^{-1}$ are positive-going in all four spectra; the baseline in this region is negative in curves (a) and (d) but near zero in spectra (b) and (c). The hydroxyl bands above $3600\,cm^{-1}$ also evolve from one spectrum to the next, and reflect the grouping of the interferograms mentioned above. In other words, the imaginary spectra are grouped in the same way as the interferograms, as is expected.

Adjustment of the lock-in phase angle at different mirror positions causes a redistribution of the PA intensity between real and imaginary spectra, as mentioned above. Although the imaginary spectra differ with regard to intensity distribution among the individual bands when the lock-in phase is changed, the PA phase spectra calculated from each imaginary/real pair are essentially the same. This agreement implies that the lock-in signal recovery at each phase angle setting is accurate, and that the PA phase is correctly characterized (see below). Because the phase spectra can be utilized for band identification, the lock-in phase need not be set exactly at ZPD in all cases.

Comparison of DSP and lock-in demodulation is appropriate at this point in the discussion. An additional DSP demodulation experiment at 77 Hz provided results

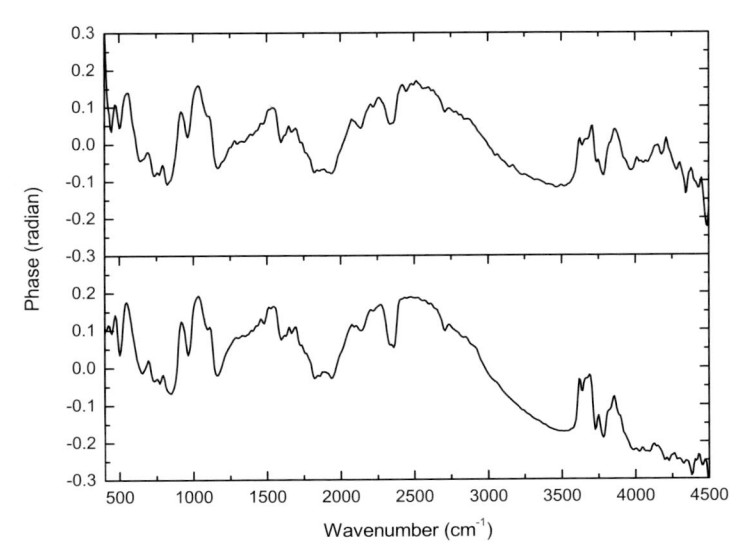

Figure 4.7 PA phase spectra of kaolin obtained at 77 Hz using DSP demodulation (lower box) and lock-in demodulation (upper box).

similar to the lock-in data discussed in the previous paragraphs. PA phase spectra, which have been shown to be independent of phase angle setting in either method, are suitable for such a comparison. Figure 4.7 illustrates the kaolin phase spectra obtained at 77 Hz using DSP (lower curve) and lock-in (upper curve) demodulation. These spectra are obviously quite similar up to 3500 cm^{-1}, exhibiting a series of maxima and minima corresponding to known absorption bands of this clay. A slight discrepancy between the two curves is visible above 4000 cm^{-1}, although no significant bands are expected in this region.

As a final note, it can be mentioned that when lock-in and DSP spectra acquired at higher f are compared, the lock-in spectra are often observed to exhibit better signal/noise ratios. In summary, lock-in demodulation was observed to yield equivalent or superior results as compared with DSP demodulation in step-scan PA infrared spectroscopy. Signal recovery in amplitude- and phase-modulation experiments at harmonic multiples of f using a lock-in amplifier is discussed in Chapter 5.

References

Curbelo, R. (1998) Digital signal processing (DSP) applications in FT-IR. Implementation examples for rapid and step scan systems. *AIP Conf. Proc.*, **430**, 74–83.

Drapcho, D.L., Curbelo, R., Jiang, E.Y., Crocombe, R.A., and McCarthy, W.J. (1997) Digital signal processing for step-scan Fourier transform infrared photoacoustic spectroscopy. *Appl. Spectrosc.*, **51** (4), 453–460.

Michaelian, K.H. (1989) Depth profiling and signal saturation in photoacoustic FT-IR spectra measured with a step-scan interferometer. *Appl. Spectrosc.*, **43** (2), 185–190.

Michaelian, K.H. (2007) Invited article: Linearization and signal recovery in photoacoustic infrared spectroscopy. *Rev. Sci. Instrum.*, **78** (1), 051301 (12 pages).

Wang, H., Phifer, E.B., and Palmer, R.A. (1998) Multi-frequency S2FTIR PA spectral depth profiling by use of sinusoidal phase modulation and harmonic demodulation up to 2250 Hz (10th harmonic). *Fresenius J. Anal. Chem.*, **362** (1), 34–40.

5
Experimental Techniques

This chapter describes six experimental techniques in PA infrared spectroscopy. Step-scan PA FT-IR spectroscopy, introduced in Chapter 3, is based on the use of amplitude and phase modulation; these techniques are discussed first, with particular reference to PA signal generation at various harmonics of the modulation frequency. Synchrotron infrared PA spectroscopy – an emerging topic – is the third method. PA infrared microspectroscopy, which has existed for more than a decade but not attained widespread popularity, is summarized next. By contrast the final two experimental techniques, quantitative analysis and depth profiling, have been widely used and discussed in the scientific literature. Numerical and experimental methods incorporated in these two techniques are introduced, and the literature is reviewed in this chapter.

5.1
Amplitude Modulation

As noted earlier, the modulation provided by the moving mirror in rapid-scan PA FT-IR spectroscopy ceases to exist when step-scan mirror motion is employed; this necessitates the use of external modulation.[1] This requirement slightly increases experimental complexity, but offers a significant advantage: all wavelengths of the available radiation are modulated at a single frequency. The thermal diffusion length is thus constant across the spectrum, an obvious advantage in depth-profiling experiments and other applications where the sampling depth is of particular interest.

Amplitude modulation will be discussed first. A chopper or shutter can be employed to provide this modulation, the chopper being a popular low-cost choice. To a good approximation, an ordinary chopper provides square-wave (Sq) amplitude modulation at a 50% duty cycle. Following Drapcho *et al.* (1997), Jiang *et al.* (1998), and Pichler and Sowa (2005a), modulation at angular frequency ω ($2\pi f$) is approximately equivalent to the sum of a series of odd harmonic terms:

---

1) This statement does not apply to the situation where the stepping rate approaches rapid- (continuous-) scan motion, providing a modulation similar to that in conventional rapid-scan PA spectroscopy.

$$Sq(\omega t) \approx \sum_{n=1}^{\infty} \frac{\sin(2n-1)\omega t}{2n-1} = \sin(\omega t) + \frac{1}{3}\sin(\omega t) + \frac{1}{5}\sin(5\omega t) + \ldots \quad (5.1)$$

Thus demodulation at each odd harmonic of the amplitude-modulation frequency is expected to yield a PA signal, and no observable intensity is predicted at the even harmonics. Demodulation can be effected at the harmonic frequencies using a lock-in amplifier. When spectra at several harmonics (f, $3f$, etc.) are required and a single lock-in is utilized, successive measurements must be performed. Alternatively, DSP (multifrequency) demodulation can be employed. In practice a chopper may not produce perfect square-wave modulation, causing a minor fraction of the PA intensity to occur at even harmonic frequencies.

Results obtained in mid-infrared amplitude-modulation experiments for carbon black and a common clay are illustrated in the next three figures. Spectra were acquired using a Bruker IFS 88 FT-IR spectrometer and an MTEC 200 PA cell. Dry nitrogen was used as the carrier gas in the cell, and ambient air remained in the instrument. A chopper provided 100-Hz modulation and a Signal Recovery lock-in amplifier was used to demodulate the PA signal at the first to fifth harmonic frequencies.

Figure 5.1 displays the spectra of carbon black calculated at the first, third and fifth harmonic frequencies. Virtually all of the PA intensity for this standard reference material occurs in the real spectra, that is, the signal appears only in the in-phase channel of the lock-in amplifier. Moreover, most of the PA intensity is

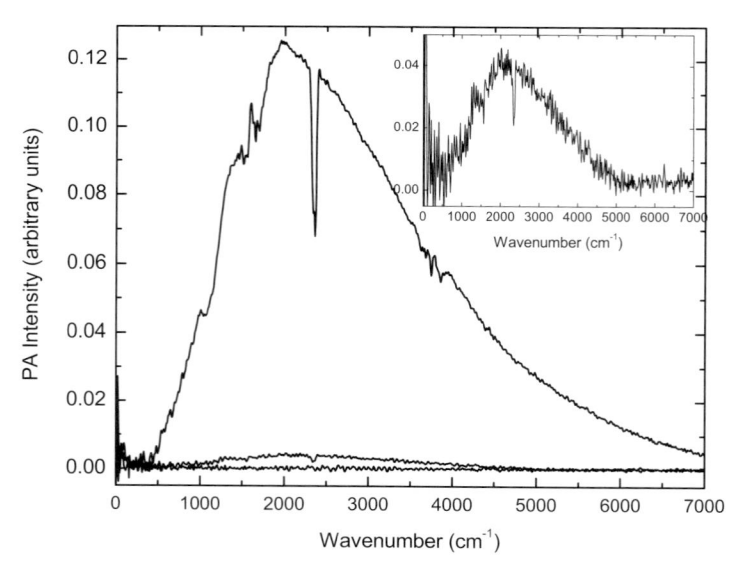

Figure 5.1 Amplitude-modulation PA spectra of carbon black (100 Hz) demodulated at the first, third, and fifth harmonic frequencies (top, middle, and bottom curves, respectively). Inset: third-harmonic spectrum multiplied by a factor of 10. The transmission-like features arise from water vapor and CO_2 in the spectrometer.

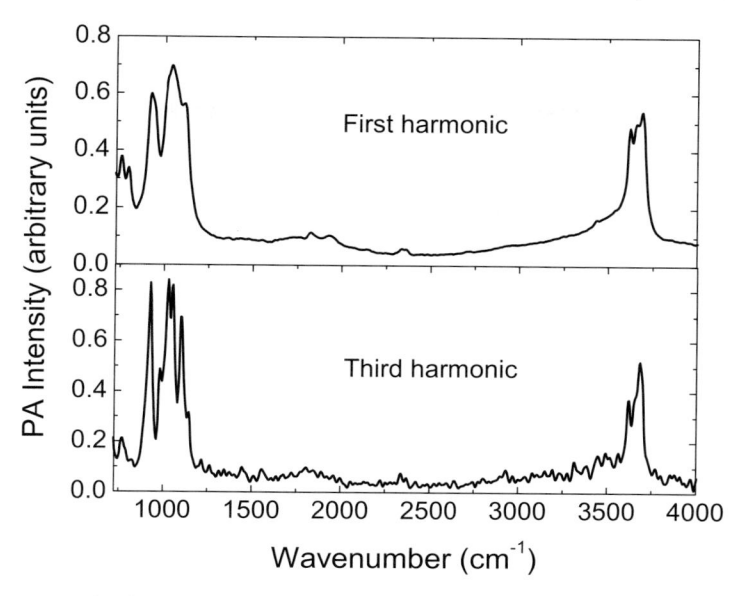

Figure 5.2 Amplitude-modulation PA spectra of kaolin (100 Hz), normalized through division by corresponding carbon black spectra.

detected at the fundamental frequency, with the $3f$ spectrum accounting for the remaining few per cent. As expected, second- and fourth-harmonic demodulation did not produce measurable signals.

Results obtained for kaolin, a layer silicate with empirical composition $Al_2Si_2O_5(OH)_4$, are somewhat more interesting. The first- and third-harmonic spectra, ratioed against the corresponding carbon black data, are plotted in Figure 5.2. The $1f$ spectrum (upper box) resembles published absorption spectra, except that the relatively low resolution used in this experiment (15 cm^{-1}) causes some loss of detail. Importantly, this spectrum exhibits a realistic relationship between the intensities of the band group below 1200 cm^{-1} and the hydroxyl stretching bands above 3600 cm^{-1}; this result is an improvement with regard to unratioed rapid-scan PA spectra of this material, where the intensities of the high-wavenumber bands are diminished because of the proportionality between modulation frequency and wavenumber. The $3f$ spectrum (lower box) shows similar relative intensities, but is adversely affected by noise in the 1200–4000 cm^{-1} region. It can also be noted that the reduced bandwidths in this spectrum actually allow identification of some of the unresolved features in the $1f$ spectrum.

PA intensity for kaolin occurs in both real and imaginary spectra, leading to a PA phase that varies with wavenumber. The phase spectrum calculated from the first-harmonic data is illustrated in Figure 5.3. Asterisks denote phase maxima and minima that correspond to known absorption bands. Phase spectra sometimes facilitate band identification, one example being the prominence of the combination bands at about 1825 and 1935 cm^{-1} in Figure 5.3.

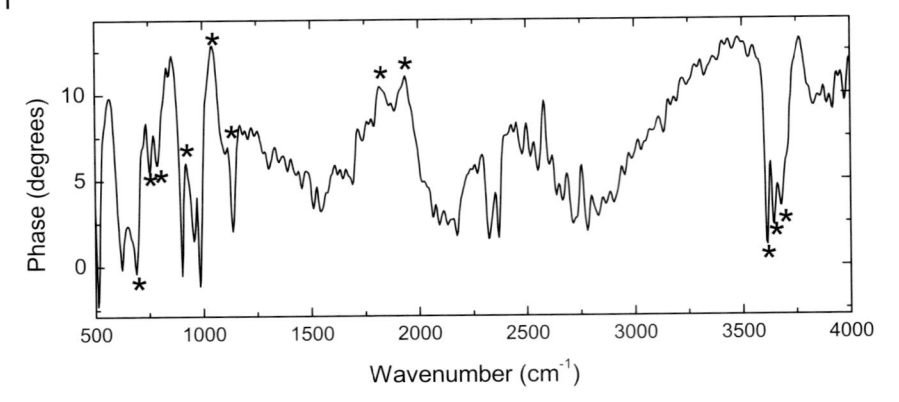

Figure 5.3 Phase spectrum of kaolin, calculated from the real and imaginary parts of the amplitude-modulation PA spectrum obtained at the fundamental frequency (100 Hz).

5.2
Phase Modulation

While amplitude modulation requires the use of a chopper or shutter, no additional equipment is needed for phase modulation in PA FT-IR spectroscopy. Instead, the movable mirror is 'dithered' at each stopping position during the scan, effectively yielding a derivative interferogram. Numerous workers have employed this technique in recent years, primarily because manufacturers of step-scan spectrometers generally incorporate the capability for phase modulation in their instruments. Moreover, demodulation is usually effected in software, so that the resulting PA amplitude spectra are calculated as readily as in conventional rapid-scan experiments. The real and imaginary spectra are also used to obtain the corresponding PA phase spectra. Measurement of both amplitude and phase spectra is illustrated in this section. Lock-in demodulation, an important alternative to the more common DSP demodulation, is emphasized.

As discussed by Débarre, Boccara, and Fournier (1981) and Wang, Phifer, and Palmer (1998), the intensity $I(\nu)$ in a sinusoidal phase-modulation experiment can be expanded in a harmonic series:

$$dI(\delta) = B(\nu)d\nu[1/2 + 1/2\, J_0(2\pi\nu\varepsilon)\cos(2\pi\nu\delta_0) - J_1(2\pi\nu\varepsilon)\sin(2\pi\nu\delta_0)\sin(\omega t)$$
$$+ J_2(2\pi\nu\varepsilon)\cos(2\pi\nu\delta_0)\cos(2\omega t) - J_3(2\pi\nu\varepsilon)\sin(2\pi\nu\delta_0)\sin(3\omega t) + \ldots]$$

(5.2)

where $\delta = \delta_0 + \varepsilon\sin(\omega t)$ is the optical path difference, ε and ν are the modulation amplitude and frequency, respectively, J_n are low-order Bessel functions, and $B(\nu)$ denotes the energy density at wavenumber ν. The terms in brackets correspond to the DC signal (not detected with AC coupling) and the PA signals at harmonic multiples of f. Demodulation at the fundamental frequency produces a spectrum approximately equal to $B(\nu)|J_1(2\pi\nu\varepsilon)|$. The equation shows that higher-frequency data can also be derived in the same experiment, simply by demodulating at the harmonics of the fundamental modulation frequency. Wang, Phifer, and Palmer (1998) employed this technique in a depth-profiling experiment, using DSP demodulation to obtain spectra up to the 10th harmonic of a 225-Hz modulation frequency.

This technique is illustrated in the following examples, which emphasize the far-infrared region. The occurrence of ε in the argument of J_1 causes the maximum of this function to shift to lower v as ε increases (Michaelian et al., 2009). Accordingly, large amplitudes are required if research-quality PA phase-modulation spectra are to be measured in the far infrared. This capability exists in a Bruker IFS 66v/S FT-IR spectrometer, where the movable mirror can be dithered at amplitudes up to 20 λ ($\lambda = 0.6328\,\mu m$, the HeNe laser wavelength). In these tests, this spectrometer was used in conjunction with an MTEC 300 PA cell. Ambient air remained in the instrument and also served as the carrier gas in the cell. A multi-layer mylar beamsplitter was employed and a polyethylene window was used to seal the sample chamber. These optical components restricted the accessible spectral region to about 100–700 cm^{-1}.

Phase-modulation PA spectra of carbon black obtained at $f = 5$ Hz and $\varepsilon = 15\,\lambda$ are considered first. Figure 5.4 shows spectra recovered by demodulating at the first five harmonics of the fundamental frequency using a lock-in amplifier. The transmission-like minima in these spectra are mainly due to beamsplitter absorption. The integrated intensities of these spectra, which can be thought of as energy

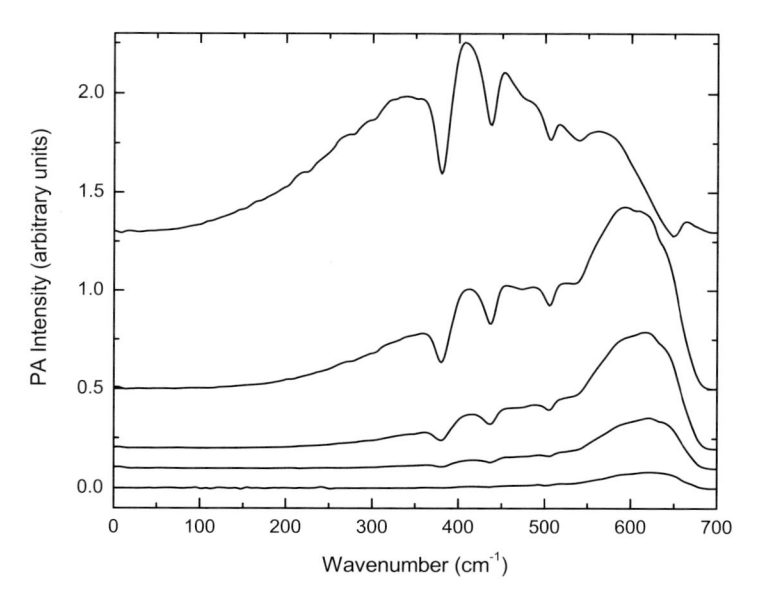

Figure 5.4 Phase-modulation PA spectra of carbon black (5 Hz, $\varepsilon = 15\,\lambda$) demodulated at the first five harmonic frequencies (top to bottom curves, respectively). Spectra are offset vertically for clarity.

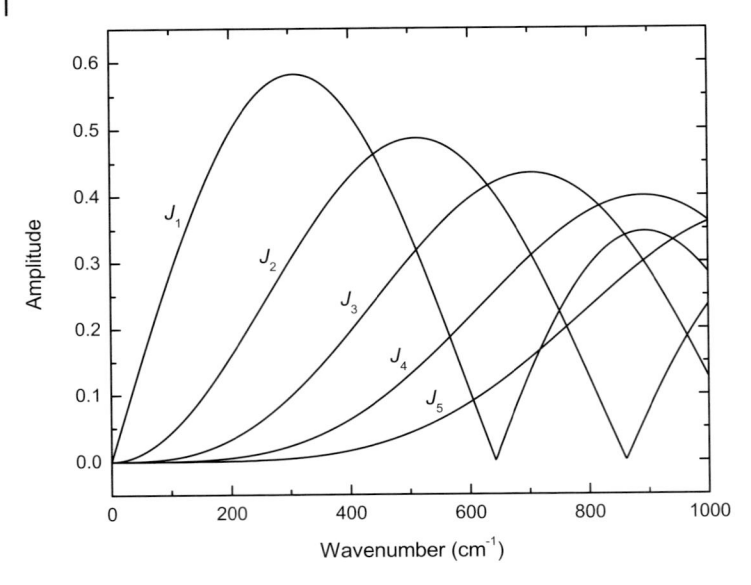

Figure 5.5 Bessel functions $|J_1|$ to $|J_5|$ for $\varepsilon = 15\ \lambda$.

curves, gradually decrease as the harmonic order increases; the first three harmonic spectra contain most of the total PA intensity. It can also be observed that the energy distribution is shifted toward higher wavenumbers in the four lower curves.

These trends are partially explained by differences among the Bessel functions $|J_n|$, which are plotted over a slightly wider wavenumber range in Figure 5.5. As the order n increases, the location of highest intensity shifts to higher ν. Because the widths of these functions increase with n, maximal amplitude decreases with increasing order. These curves help to explain the fact that a large fraction of the total PA intensity in the 0–700 cm^{-1} region occurs in the low-order (low-harmonic) spectra, and show that demodulation at the fundamental frequency affords the best opportunity for the observation of bands at very low wavenumbers.

The role of the Bessel functions is illustrated further in Figure 5.6, which presents results of modeling the energy curve at the fundamental frequency. An amplitude-modulation (constant f) spectrum of carbon black was multiplied by $|J_1|$ to calculate a model spectrum (solid line). The dashed line shows an experimental phase-modulation spectrum obtained at the same frequency. As might be expected, the simulated and measured spectra are in good agreement. To summarize, the first-harmonic spectrum is practically equal to $B(\nu)|J_1(2\pi\nu\varepsilon)|$, and the interval from about 100 to 650 cm^{-1} is readily accessible under the conditions employed in this experiment.

The variation in the locations of the maxima in Figures 5.4 and 5.5 can be put to good use. The first zero in $|J_1|$ occurs at about 640 cm^{-1}; indeed, $|J_2|$ displays greater intensity than $|J_1|$ in a significant region starting at approximately 440 cm^{-1}. Simi-

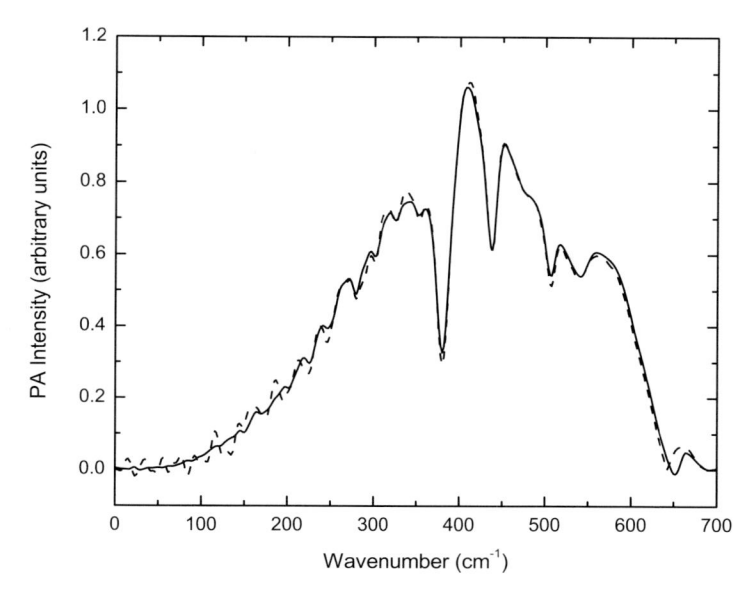

Figure 5.6 Phase-modulation PA spectra of carbon black (5 Hz, $\varepsilon = 15\ \lambda$) demodulated at the first harmonic frequency. Solid line, simulation; dashed line, experiment.

larly, $|J_3| > |J_2|$ above 630 cm^{-1}. These relationships suggest that particular higher-wavenumber regions may be studied advantageously by demodulating the PA signal at the second and third harmonic frequencies. In confirmation of this statement, Figure 5.7 shows unratioed PA amplitude spectra recovered at the first three harmonic frequencies for pyrene ($C_{16}H_{10}$), a four-ring aromatic hydrocarbon mentioned elsewhere in this book. The best results for the region extending from about 200 to 500 cm^{-1} are obviously obtained at the fundamental frequency. From 500 to 600 cm^{-1}, the second-harmonic spectrum provides equivalent or better data, particularly with regard to the neighboring bands near 580 and 590 cm^{-1}. The features in the 500–600 cm^{-1} region are also well defined in the third-harmonic curve, which additionally suggests the existence of several broad bands near the high-wavenumber limit of the plotted spectra. The bands above 475 cm^{-1} in all three spectra can also be observed in rapid-scan mid-infrared PA spectra of this compound.

The PA phase spectra, calculated from the real and imaginary amplitude spectra at the three respective harmonic frequencies, are shown in Figure 5.8. Asterisks are used to identify features, which can be either positive- or negative-going, that correspond to known absorption bands of pyrene. The phase is naturally subject to large uncertainty in regions of low intensity: this leads to the fluctuations above 600 cm^{-1} in the first-harmonic spectrum, and to similar behavior below 350 cm^{-1} in the $3f$ spectrum. It should also be noted that observation of bands in phase spectra sometimes facilitates peak identification, because the PA phase varies linearly with absorption over a wide range and phase spectra are less affected by saturation than amplitude spectra in regions of strong absorption.

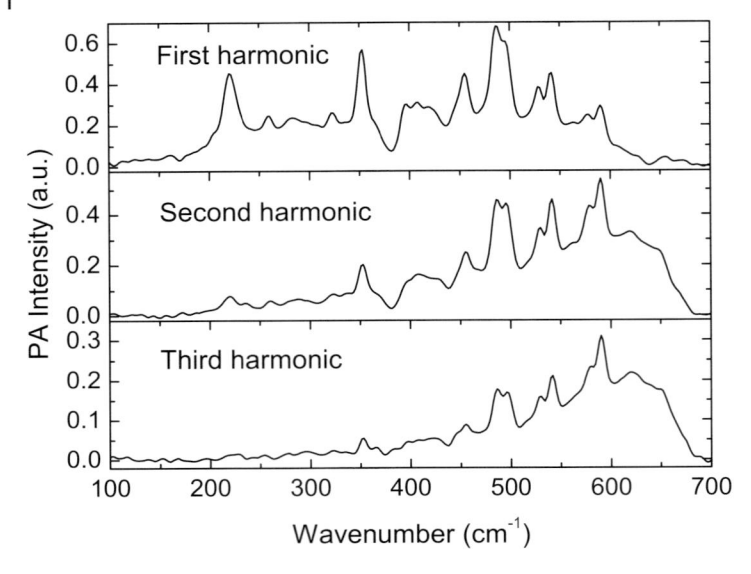

Figure 5.7 Single-beam phase-modulation PA spectra of pyrene (5 Hz, $\varepsilon = 15\ \lambda$) demodulated at the first three harmonic frequencies.

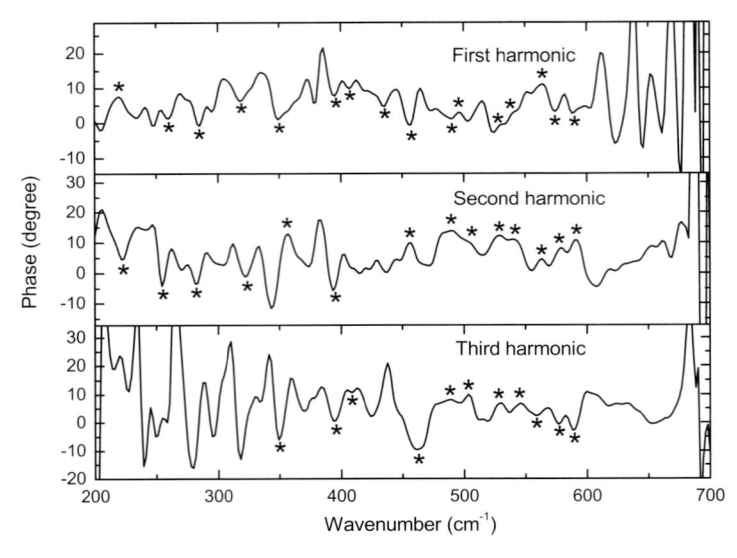

Figure 5.8 Phase spectra of pyrene, calculated from real and imaginary parts of the phase-modulation spectra (5 Hz, $\varepsilon = 15\ \lambda$) demodulated at the first three harmonic frequencies. The asterisks denote known bands in PA amplitude or absorption spectra.

5.3
Synchrotron Infrared PA Spectroscopy

Synchrotron radiation (SR), briefly defined as electromagnetic radiation emitted from an electron storage ring, is an attractive source for many optical experiments. The high brilliance (brightness) of SR and its continuous tunability across the entire electromagnetic spectrum, from X-ray wavelengths to the far-infrared, are currently leading to its increased utilization by physicists, materials scientists, biologists and other researchers. About 70 synchrotron facilities are presently in operation worldwide; many of these institutions operate mid- and far-infrared beamlines. Mid-infrared SR is most commonly used in spectromicroscopy, where high radiance (radiant power per unit area per unit solid angle) makes SR superior to an ordinary thermal source. This advantage enables near-diffraction-limited imaging, which is extensively employed in biomedical research and many other areas. By contrast, the primary application of far-infrared SR is to very high resolution gas-phase spectroscopy. In addition to these popular uses of infrared SR, mid- and far-infrared PA spectroscopy based on synchrotron sources has also been successfully demonstrated in the last decade. This topic is discussed in the following paragraphs.

The rationale for the utilization of SR in PA infrared spectroscopy requires elaboration. Indeed, the high brilliance of SR need not imply that these PA infrared spectra are superior to those acquired with a conventional thermal source. This qualification exists because PA spectra of condensed-phase materials are most commonly obtained for macroscopic (milligram-quantity) samples. Systematic comparisons of SR and thermal-source PA spectra were therefore undertaken, so that possible advantages of SR PA infrared spectroscopy could be identified.

Mid- and far-infrared SR PA spectroscopy was first demonstrated at the National Synchrotron Light Source, Brookhaven National Laboratory (Jackson, Michaelian, and Homes, 2001; Michaelian, Jackson, and Homes, 2001). An MTEC 100 PA cell, fitted with an appropriate window, was interfaced with a Bruker IFS 66v/S FT-IR spectrometer installed on a dedicated beamline. Spectra were obtained for carbon-filled rubber, glassy carbon, hydrocarbon coke, and clay samples. SR and thermal (globar) sources were employed in this study. It was readily observed that SR PA spectra were more intense than the corresponding thermal-source spectra in the far-infrared ($\sim$30–200 cm^{-1}) region; however, the relationship was reversed in the mid-infrared. The crossing wavenumber (v_c), where the SR and thermal-source spectra intersect, was near 200 cm^{-1} for unapertured ($\sim$5-mm) incident radiation. v_c gradually shifted to higher wavenumbers as the beam diameter was reduced to 1 mm. Thus comparison of the two radiation sources for PA spectroscopy depends on their relative intensities at specific wavelengths and on the beam size of the impinging radiation, as might be expected in SR work.

SR PA infrared spectroscopy has been investigated more fully at the Canadian Light Source, a third-generation synchrotron located at the University of Saskatchewan (Michaelian, May, and Hyett, 2008). Experiments were performed at the mid-infrared beamline using a Bruker IFS 66v/S spectrometer and an MTEC 300

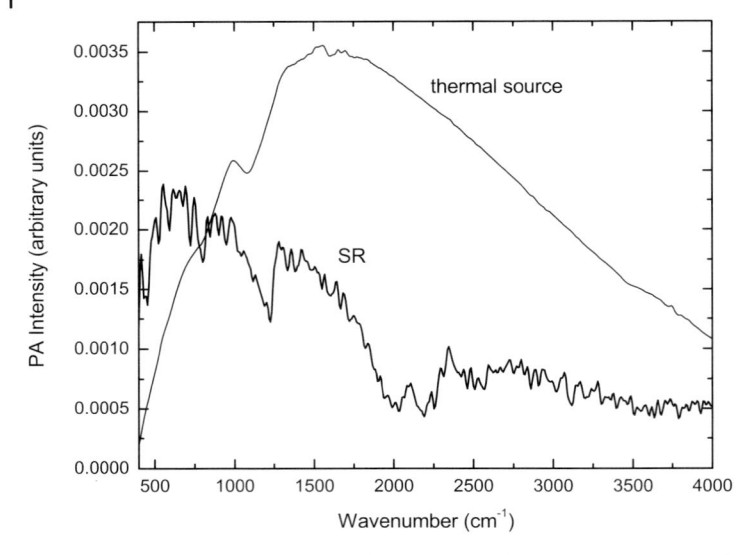

Figure 5.9 Mid-infrared PA spectra of carbon black obtained using a 1.5-mm beam. The SR spectrum is scaled to a beam current of 200 mA.

PA cell. Beam diameter tests were carried out by placing aluminum apertures above the sample cup inside the PA accessory. These apertures had outer diameters of 10 mm and inner diameters ranging from 0.7 to 5 mm. Figure 5.9 shows the mid-infrared region of SR and thermal-source PA spectra for carbon black obtained under like conditions using the 1.5-mm aperture. These spectra, which can be thought of as energy curves, intersect at $v_c \approx 800\,\text{cm}^{-1}$, somewhat higher than the corresponding position mentioned above.

The location of v_c shifts to higher wavenumbers as the aperture size is reduced because thermal-source PA intensity is attenuated more rapidly than SR PA intensity (Figure 5.10). Variation of v_c with aperture size is illustrated in Figure 5.11. Data acquired using two different spectrometer mirror velocities (HeNe laser modulation frequencies 1.6 and 2.2 kHz) are included in this figure. The experimental results were fitted using the relation

$$v_c = 3137 \exp(-x/0.810) + 343 \tag{5.3}$$

where x is aperture size in millimeters. Equation 5.3 suggests limiting values of 3480 and 343 cm^{-1} for very small and very large apertures, respectively. In other words, SR PA intensity is greater throughout most of the mid-infrared region for small aperture sizes; this is a manifestation of SR brilliance. By contrast, the thermal source should be employed for large apertures or sample sizes: SR offers no advantage in the acquisition of mid-infrared PA spectra for large samples.

The results in Figures 5.10 and 5.11 also provide means for estimation of the SR beam size in the PA cell. Equation 5.3 is of the general form

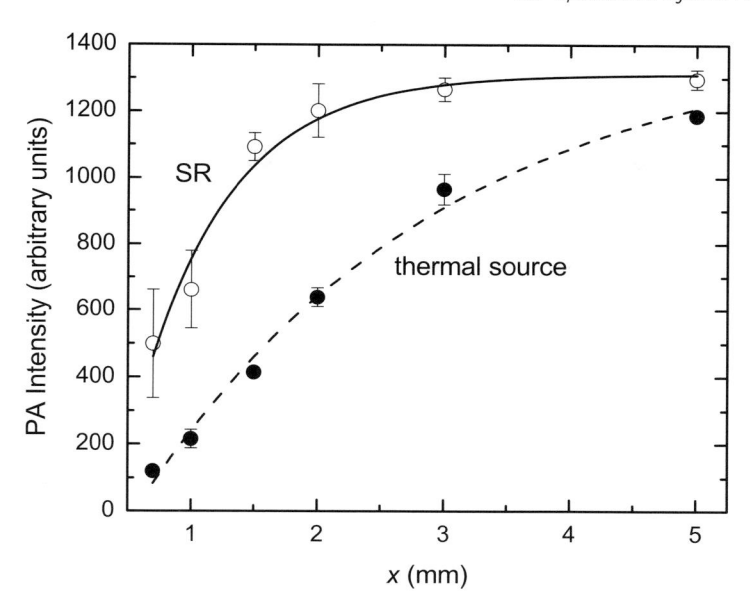

Figure 5.10 Variation of integrated PA intensity (interferogram centerburst amplitude) with aperture diameter. Open circles, SR; filled circles, thermal source. Error bars represent one standard deviation, and were calculated using spectra obtained at both 1.6 and 2.2 kHz. The lines are exponential fits to the two sets of experimental points.

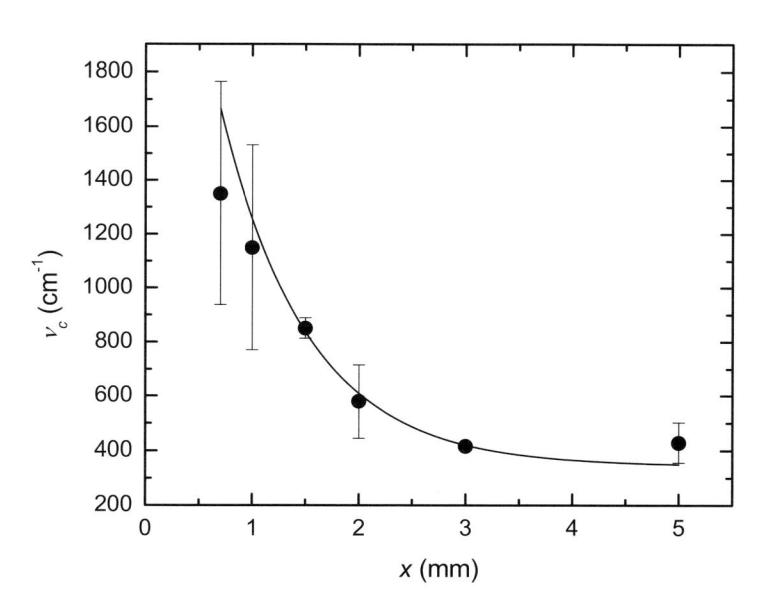

Figure 5.11 Dependence of v_c on aperture size. Error bars have a meaning similar to that in Figure 5.10. The line is an exponential fit to the data.

$$v_c = v_1 \exp(-x/x_0) + v_2 \qquad (5.4)$$

where v_1 and v_2 denote crossing wavenumbers at small and large apertures, respectively. x_0 can be taken as an estimate of the SR beam size in the FT-IR sample compartment. As shown in Equation 5.3, the data are fitted using $x_0 = 0.81$ mm. A second estimate is provided by fitting the SR data in Figure 5.10 to an exponential curve. This fit yields

$$I = -2290 \exp(-x/0.708) + 1310 \qquad (5.5)$$

where I is intensity. This expression can be rewritten more generally as

$$I = a_1 + a_2 \exp(-x/x_0) \qquad (5.6)$$

where a_1 and a_2 are constants. Hence the SR data in Figure 5.10 suggest $x_0 \approx 0.71$ mm, in reasonable agreement with the previous estimate. Both results are compatible with the ≤ 1-mm spot size observed visually when a Si/CaF_2 beamsplitter was installed in the FT-IR spectrometer. Moreover, the diameter of the SR beam was predicted to be 0.5–1.0 mm with the optical configuration used at this beamline. Of course, these SR beam-diameter estimates are all much lower than the ~4-mm size of the thermal-source beam in the PA cell.

The results in Figures 5.9 and 5.10 show that SR PA spectroscopy offers specific advantages at low wavenumbers (long wavelengths) and small beam (or sample) sizes. Far-infrared SR PA spectroscopy may prove to be particularly important, since traditional thermal radiation sources are weak in this region. Another, quite different, implementation of far-infrared SR is discussed in the following paragraphs.

In some applications, it may be desirable to extend SR PA infrared spectroscopy to very low wavenumbers. An alternative to the conventional operation of the electron storage ring, which has been implemented at several synchrotron facilities, leads to the production of far-infrared coherent synchrotron radiation (CSR). In this mode of operation the beam energy and current are reduced, allowing production of short electron bunches that emit coherently in the terahertz (THz) region. The power emitted by these bunches is typically several orders of magnitude greater than that in conventional SR. When the electron bunch length is less than 1 mm, emission occurs at wavelengths similar to this dimension. This emission can be analyzed using a far-infrared spectrometer.

CSR experiments were recently performed at the far-infrared beamline at the Canadian Light Source. Production of CSR was first verified using a Bruker IFS 125 FT-IR spectrometer equipped with a mylar beamsplitter and helium-cooled bolometer. Next, the coherent radiation was directed to the MTEC 300 PA cell mentioned above; the cell was fitted with a polyethylene window and installed in the spectrometer in the usual way. Figure 5.12 shows an interferogram (the two centerbursts are due to the 'forward-backward' scan) obtained for carbon black in this trial. The beam current was about 6 mA and the synchrotron frequency was 4.2 kHz during this measurement. The CSR PA spectrum derived from this

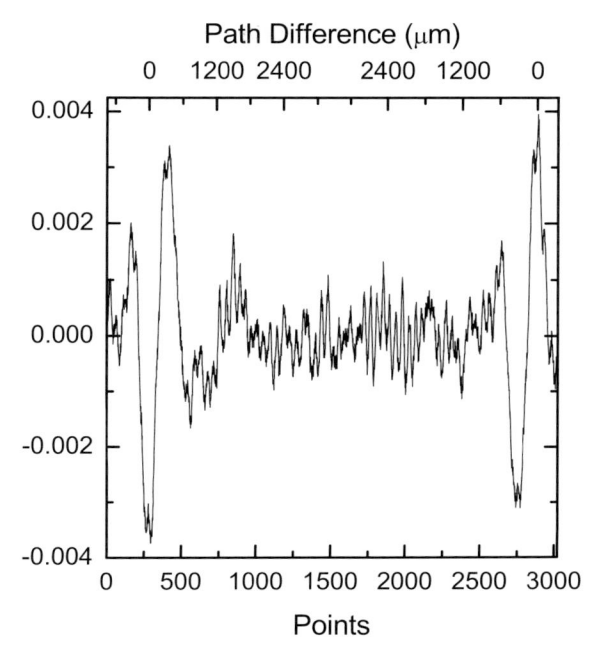

Figure 5.12 CSR PA 100-scan interferogram (resolution 6 cm^{-1}) for carbon black obtained at the Canadian Light Source using a Bruker IFS 125 spectrometer and an MTEC 300 PA cell. The upper abscissa shows the path difference for the forward (left-hand side) and backward (right-hand side) scans.

interferogram appears in Figure 5.13. Maximal intensity occurs near 14 cm^{-1} (700 μm), a much lower wavenumber than those in far-infrared energy curves obtained using conventional SR or thermal sources. Similar CSR energy curves have been reported for experiments performed at BESSY (Berliner Elektronenspeicherring-Gesellschaft für Synchrotronstrahlung m.b.H.) in Berlin (Abo-bakr et al., 2003). A bolometer was employed as the detector in that work. It is important to note that PA detection of CSR can be employed over a range of wavelengths, since no optical element is involved; this versatility could prove useful for synchrotron machine-development work. In summary, the results in Figures 5.12 and 5.13 clearly show the feasibility of CSR PA spectroscopy and demonstrate an important new capability for the acquisition of very-low wavenumber PA spectra of solids.

The work in this section should be placed in historical context. Specifically, it can be mentioned that Masujima et al. (1989) designed a PA detector for X-ray absorption spectroscopy using SR as the source. This detector was also modified for far-infrared investigations, and SR PA signals were detected for silver black and gold black. Importantly, PA signal intensity was shown to scale linearly with

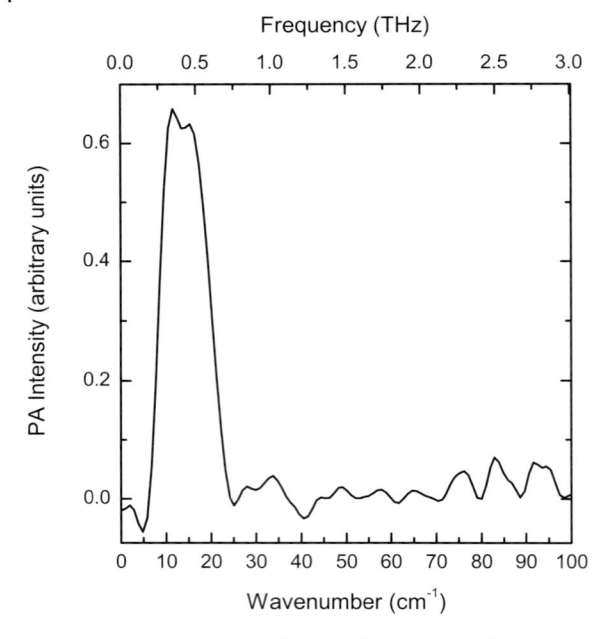

Figure 5.13 PA spectrum calculated from the interferogram in Figure 5.12.

beam current in the storage ring. No SR PA interferograms or spectra were included in the publication based on this research.

5.4
PA Infrared Microspectroscopy[2]

PA infrared spectroscopy is routinely implemented as a 'bulk' sampling technique. In typical gas-microphone cells, solid or liquid samples with cross-sectional areas on the order of 10–20 mm^2 are examined. A question sometimes asked of PA researchers pertains to the suitability of the method for much smaller samples; this raises the more fundamental issue of improved spatial resolution. Several articles relevant to this topic are briefly summarized in this section.

Jiang (1999) presented results obtained with a commercially available accessory in one of the first published reports on PA infrared microspectroscopy. In this pioneering work, single-particle and single-fiber microsamplers (Figure 5.14) were inserted in an MTEC 300 PA cell. PA FT-IR spectra were obtained for samples such as a coated polymer bead, grease-coated nylon, and human hair. The aims of the investigation were to demonstrate the microsampling accessory and to

---

2) 'Microspectroscopy' is used to distinguish the work from 'spectromicroscopy', where the emphasis is on imaging.

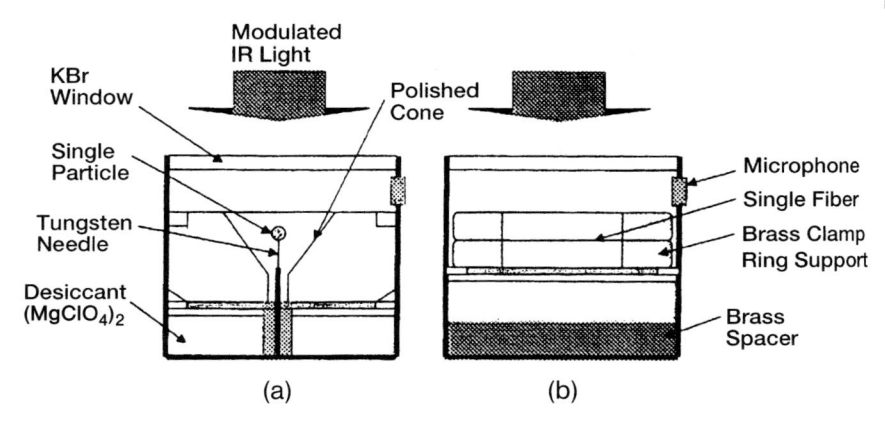

Figure 5.14 (a) Single-particle and (b) single-fiber microsamplers available from MTEC Photoacoustics. (Reproduced from Jiang, E.Y., *Appl. Spectrosc.* **53**: 583–587, by permission of the Society for Applied Spectroscopy; copyright © 1999.)

Table 5.1 Signal-to-noise ratios in PA infrared spectra of a 150-μm bead observed with various sampling methods (Jiang, 1999).

Sample accessory	Signal-to-noise ratio
Single-particle	6528
Small sample cup	2980
Large sample cup	2584

perform rapid- and step-scan depth profiling experiments. Sample areas as small as $25\,\mu m \times 25\,\mu m$ can be analyzed with this instrumentation.

Because both microsamplers reduce sample cup volume, concentrate the incident light on the sample, and provide better thermal isolation than ordinary sample cups, the PA spectra obtained with this accessory exhibit signal/noise ratios much higher than those in spectra of macroscopic samples. To illustrate this relationship, results reported by Jiang (1999) for a 150-μm polymer bead are summarized in Table 5.1. The single-particle accessory yielded a signal/noise ratio more than twice that obtained with the smaller (5-mm diameter) of two conventional sample cups, which contains approximately 70 mg when filled with an ordinary solid. A further reduction in signal was observed with a large (10-mm) cup. These results show that single-particle PA infrared spectroscopy may actually be preferable to the analysis of bulk samples in some circumstances.

Several depth profiling experiments were carried out in this investigation. For example, in rapid-scan PA tests on the coated bead, the use of lower mirror velocities (greater penetration depths) caused intensification of bands due to the bead, as expected. Step-scan phase-modulation experiments allowed variation of the sampling depth between about 9.5 and 25 μm for a silicon-greased wire-cored nylon fiber; these depths were calculated using $\alpha = 0.002\,\text{cm}^2\,\text{s}^{-1}$ for grease. Similarly, frequency- and phase-resolved step-scan experiments were performed for human hair coated with a commercial gel. DSP demodulation separated the various harmonics, and the higher-frequency spectra displayed enhanced contributions from surface species. Surface and bulk absorption were also separated using the phase spectrum and phase-rotation methods described earlier. The excellent data presented in this article show that the microsample accessory can play an important role in the acquisition of PA infrared spectra of solid materials. A review of the literature reveals that very few researchers have taken advantage of this fact.

Ordinarily, the best spatial resolution achievable in infrared spectromicroscopy corresponds to the diffraction limit, that is, the smallest detectable sample dimension is approximately equal to the infrared wavelength. The reader may be aware that much higher (up to several orders of magnitude) spatial definition is routinely achieved in scanning probe microscopy: if similar resolution could be obtained in the infrared, the diffraction limit would effectively be overcome. This was the motivation in an innovative investigation by Hammiche *et al.* (1999), who used a miniature Wollaston wire thermometer to record infrared spectra of polymers by detecting photothermally induced temperature fluctuations at the sample surface. In this work infrared spectra were obtained at a spatial resolution determined by the size of the contact between probe and sample; at a few hundred nanometers, this is well below the diffraction limit in the infrared region. The experiment performed by Hammiche *et al.* (1999) can be described as photothermal (PT), rather than PA, spectroscopy. While it did not involve the use of a gas-microphone cell or piezoelectric detection, it is based on PT detection of infrared spectra and is therefore related to several topics discussed elsewhere in this book.

The experimental arrangement is depicted in Figures 5.15 and 5.16. As the sample absorbed infrared radiation, the resulting temperature rise was detected by the probe. This induced changes in electrical resistance that were amplified and used to produce an interferogram. Fourier transformation produced an absorptive, or positive-going, spectrum. Spatial resolution was affected by a number of factors related to temperature distribution and contact area. The interferogram was quite weak and contained contributions from harmonics of the line frequency. PT spectra of polymers were compared with attenuated total reflectance (ATR) spectra: in some cases, major differences were observed. The exact reasons why the two types of spectra differed are unknown; however, it can be noted that the bands in the PT spectra are rather broad, suggesting partial saturation. PT spectra naturally reflect both optical and thermal characteristics of the sample, and hence this method could, in principle, be utilized to study both properties.

This experiment was refined and improved in two subsequent works. Hammiche *et al.* (2000) described the use of cantilever-type probes for the characteriza-

5.4 PA Infrared Microspectroscopy

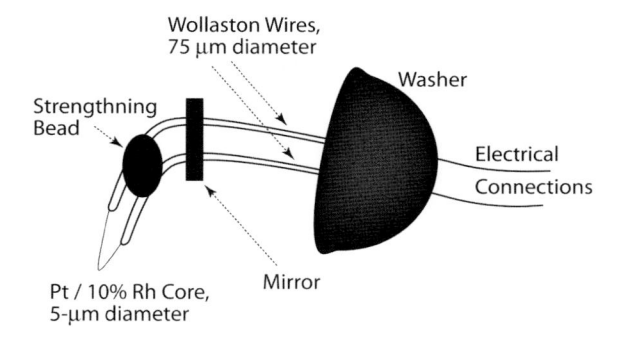

Figure 5.15 Schematic diagram of thermal probe used for photothermal infrared spectroscopy.

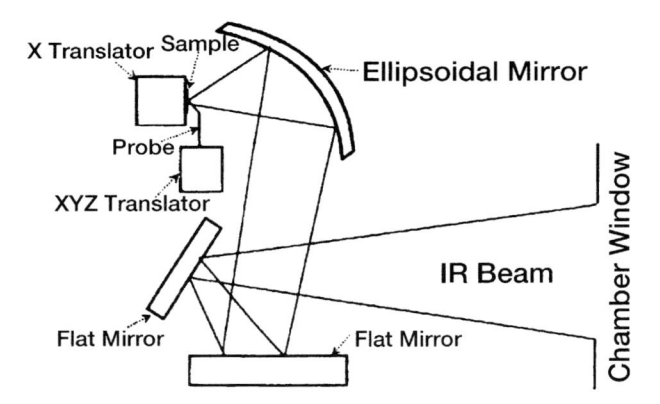

Figure 5.16 Probe and sample arrangement for photothermal infrared spectroscopy. (Reproduced from Hammiche, A., et al., *Appl. Spectrosc.* **53**: 810–815, by permission of the Society for Applied Spectroscopy; copyright © 1999.)

tion of polymers, and presented PT spectra of single and bilayer polymer samples. These spectra are of good quality, although the bands were again broadened with respect to those in analogous ATR spectra. In this and subsequent work (Bozec et al., 2001) a bilayer system (polyisobutylene on polystyrene) was studied. The case in which the top layer is optically transparent and thermally thin was examined in detail, and an equation relating the amplitude of the signal from the bottom layer to the thickness of the top layer was derived. A fit of the experimental data to this expression yielded $\mu_s = 8.9\,\mu m$ for the top layer, consistent with typical values for polymers at the f used in this experiment.

The intensities of the bands in the polystyrene spectrum obtained by this technique are undoubtedly affected by saturation. As is often the case in PA spectra obtained with a gas-microphone cell, relative band intensities are significantly different from those in other types of infrared spectra (transmission, diffuse reflectance, etc.). When the polyisobutylene overlayer thickness was increased to the point

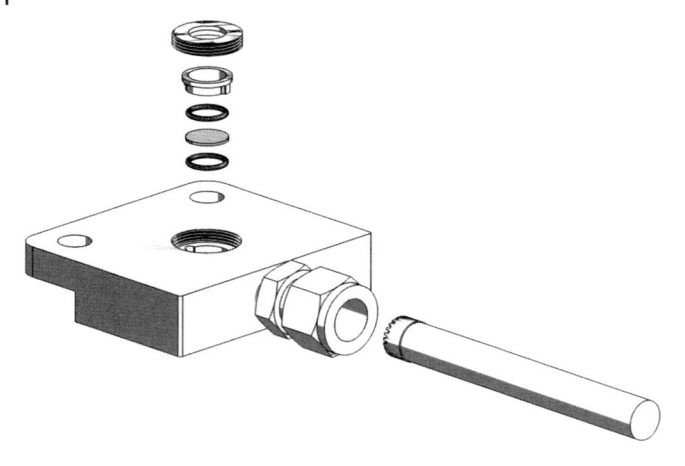

Figure 5.17 Schematic drawing of prototype micro-PA cell.

that it equaled the thermal diffusion length, the perturbation in the intensities of the polystyrene bands was reduced but not entirely eliminated. This improvement can be put down to the fact that only the surface of the polystyrene is probed in this experiment. A summary of this work is provided in a detailed review of microthermal analysis and microspectroscopy by Pollock and Hammiche (2001).

To conclude this discussion of PA infrared microspectroscopy, a schematic drawing of a prototype PA gas-microphone cell designed for use with an infrared microscope is shown in Figure 5.17. At the time of writing, this cell is under development at the Canadian Light Source. Single-particle PA spectra will be acquired with this equipment. Moreover, since it is compatible with a standard microscope stage, mapping of samples (spectromicroscopy) will also be possible.

5.5
Quantitative Analysis

The two quantities of greatest interest in virtually any type of spectroscopy are, of course, band positions and intensities – the former generally conveying qualitative information, the latter quantitative. As regards PA infrared spectroscopy, the reader may have already concluded that a substantial portion of the published literature describes qualitative (or, alternatively, semi-quantitative) applications of the technique. However, a different perspective is assumed in this section: this narrative is given over to a description of the use of PA infrared spectroscopy for the quantitative analysis of condensed-phase materials. It should be noted that PA spectra of gases, which are reviewed in Chapter 7, are mentioned only briefly in this discussion.

5.5.1
Quantification in PA Near-Infrared Spectroscopy

The potentiality for the utilization of near-infrared PA intensities for quantitative analysis was examined in the pioneering investigations by G. F. Kirkbright and his collaborators referred to earlier. These authors apparently considered this topic to be as important as the reliable assignment of the bands in the PA spectra – a perspective illustrated by the discussion in the following paragraphs.

Adams, Beadle, and Kirkbright (1978) showed that a characteristic near-infrared band at 2.2 μm (4545 cm^{-1}) could be used to identify aromatic hydrocarbons in the presence of aliphatics. PA spectra of benzene/n-hexane mixtures were analyzed, and a linear variation in the intensity of this band with benzene concentration was established. Similar results were obtained when benzene was added to an Iranian crude oil at concentrations of 10 and 30%.

These findings are significant because they demonstrate that it is feasible to use near-infrared PA spectroscopy for the quantification of total aromatics in hydrocarbon fuels. It can also be noted that this particular analysis will become increasingly important as sources of conventional light crude oils are depleted and the use of heavier (more aromatic) crudes becomes more prevalent. FT-Raman spectroscopy can also be used to determine aromatics contents in fuels (Michaelian, Hall, and Bulmer, 2003a, 2003b); however, Raman spectra of hydrocarbon samples with high boiling points are often plagued by fluorescence. Fortunately, a similar limitation does not exist in PA infrared spectroscopy.

The familiar strong absorption by water in the mid- and near-infrared regions provides another opportunity for quantitative PA analysis. This is demonstrated by work on the determination of moisture content in single-cell protein samples and milk substitutes (Castleden, Kirkbright, and Menon, 1980; Jin, Kirkbright, and Spillane, 1982). As discussed in Chapter 7, quantitative determination of moisture in these samples was feasible at concentrations up to about 10%, provided that the samples were first sorted according to particle size. When this essential preliminary step was carried out, linear correlations were obtained between the intensities of the 1.9-μm (5265-cm^{-1}) band and the amounts of moisture in the samples. It should also be noted that this technique could be applied – at both near- and mid-infrared wavelengths – to other foods, grains, and various classes of industrial materials.

The final demonstration of quantitative analysis by near-infrared PA spectroscopy to be mentioned here involved the characterization of drug tablets (Castleden, Kirkbright, and Long, 1982). This study also gave very encouraging results: drug concentrations as high as 36% yielded linear correlations between the intensities of specific bands and the amount of drug present in the samples. Similar results were obtained using ultraviolet and visible PA spectra. To summarize these three articles, the research of Kirkbright's group demonstrated that near-infrared PA spectroscopy is well suited for quantitative studies when analyte concentrations are in the low-to-moderate regime.

5.5.2
Quantification in PA Mid-Infrared Spectroscopy

The evolution of a particular spectroscopic technique normally passes through several stages, including its initial demonstration, acceptance by the scientific community, and refinement; this is often followed by its subsequent use for qualitative and quantitative analysis. Thus, after the early successful demonstrations of the feasibility of PA infrared spectroscopy, many investigators became interested in the question as to whether the technique could realistically be used for quantitative measurements. It was already known that ultraviolet and visible PA spectra are sometimes subject to saturation: as this effect begins to influence the spectra, the likelihood of reliable quantitative analysis is diminished. On the other hand, saturation effects would generally be expected to be less predominant in the infrared, where absorption coefficients can be orders of magnitude smaller than those at shorter wavelengths. In light of this situation, numerous research groups examined the possibility of using PA infrared spectroscopy for quantitative analysis in a variety of applications. Some relevant publications on this subject are mentioned below, beginning with the earlier investigations.

A series of articles on PA FT-IR spectroscopy in the 1980s took up the issue of quantitative analysis. Attempts to create suitable samples were naturally an integral part of these studies. Rockley, Richardson, and Davis (1980) first proposed the use of homogeneous mixtures prepared from molten naphthalene and benzophenone. Perhaps surprisingly, the authors found that the relative contributions of the two components to the PA spectra of the mixtures were not proportional to known concentrations. This led to the suggestion that eutectic mixtures of the hydrocarbons had been inadvertently prepared.

A more ideal system, consisting of intimate mixtures of $K^{14}NO_3$ and $K^{15}NO_3$, was investigated next. The spectra of these two salts can be distinguished by a characteristic band shift from 825 to 800 cm^{-1} caused by the increase in nitrogen mass; this makes the contributions of the two compounds to the spectra easily identifiable. Mixtures containing from 1 to 99% $K^{15}NO_3$ were analyzed, and a linear relationship was observed between the relative contribution of this isotopomer to the spectrum and its abundance (Rockley, Davis, and Richardson, 1981). This encouraging result suggested the feasibility of the use of mid-infrared PA intensities for quantitative analysis. However, these findings were accompanied by a caveat: the plot of $K^{15}NO_3$ intensity against the percentage of this species in the samples did not pass through zero. This result may have arisen from inaccurate resolution of the two overlapping peaks or, alternatively, contamination of this salt with the more common $K^{14}NO_3$.

Shortly after publication of the work by Rockley, Davis, and Richardson (1981), E. M. Eyring and his coworkers at the University of Utah employed PA infrared spectroscopy for the quantitative analysis of catalytic surface adsorption sites. An object of this work was to determine the concentrations of Brønsted and Lewis acid sites on silica-alumina through analysis of the PA spectrum of adsorbed pyridine. The results were compared with those for adsorption on γ-alumina, which has only Lewis acid sites. Riseman et al. (1982) showed that strong low-wavenumber substrate bands could be used to normalize the spectra: this procedure corrected for particle size effects and other experimental factors. The scaled PA spectra showed that 20% of the surface hydroxyls in silica-alumina interacted with pyridine, a result corroborated by other data. In subsequent work (Gardella, Jiang, and Eyring, 1983) the amount of CO adsorbed onto an Ni/SiO_2 catalyst surface was calculated. A linear calibration up to a CO volume of 300 µL was obtained by plotting PA intensity as a function of the amount of CO gas introduced. However, PA saturation occurred when larger quantities of CO were used. PA gas intensities were again calculated by normalizing the spectra to constant silica band intensity in this work.

The occurrence of saturation in PA infrared spectra of adsorbed species was also investigated by Highfield and Moffat (1985). Sorption of NH_3 or pyridine on 12-tungstophosphoric acid, $H_3PW_{12}O_{40}$, was studied in this work. In one series of experiments, the 1420-cm^{-1} NH_3 band intensified linearly with the amount of added gas. By contrast, the characteristic 1537-cm^{-1} pyridine band exhibited a linear variation with the amount of analyte only at lower concentrations. The intensity of this band grew more gradually as increasing amounts of pyridine were adsorbed, suggesting that saturation was limiting the PA intensities. The authors used known thermal properties for compounds resembling $H_3PW_{12}O_{40}$ to calculate a value of 55 cm^{-1} for $(\alpha/\pi f)^{-1/2}$, the reciprocal of μ_s, for the pyridine band. The absorptivity of the 1537-cm^{-1} band is known to be approximately 60 $L\,mol^{-1}\,cm^{-1}$. The concentration of pyridine at the suspected onset of saturation was estimated as 1.9 $mol\,L^{-1}$, corresponding to an absorption coefficient of 114 cm^{-1}. Because this value is significantly greater than the μ_s^{-1} calculated in the PA infrared experiment, it was concluded that signal saturation was the likely cause of the nonlinear relationship between PA intensity and concentration.

The dependence of PA intensity on analyte concentration was also studied in an early single-wavelength PA experiment. Yokoyama et al. (1984) used the CO_2 laser 1081-cm^{-1} line to detect the sulfate v_3 band in $KAl(SO_4)_2 \cdot 12H_2O$, $CaSO_4 \cdot 2H_2O$ and K_2SO_4. These compounds were divided into several size fractions and studied as binary mixtures with NaCl at concentrations ranging from 0 to 100%. At very low concentrations (below one per cent) PA intensity increased linearly with the amount of sulfate present in each mixture. However, the intensification of the PA signal was more gradual when the amount of analyte was greater. Yokoyama et al. (1984) noted that light scattering by the diluent can also contribute to PA intensity, and attributed the nonlinear increase in PA intensity to a diminution of this scattering at higher analyte concentrations. Thermal interactions were also thought to play a role in the reduction of the PA intensity. Signal saturation was not mentioned in this article.

In subsequent work, several research groups compared PA FT-IR spectroscopy with other infrared techniques as quantitative analysis methods. For example, Rosenthal et al. (1988) used PA and transmission infrared spectroscopies to analyze mixtures of acetylsalicylic acid, salicylic acid, and filler or binder used in the preparation of commercial analgesic tablets. Partial least squares (PLS)

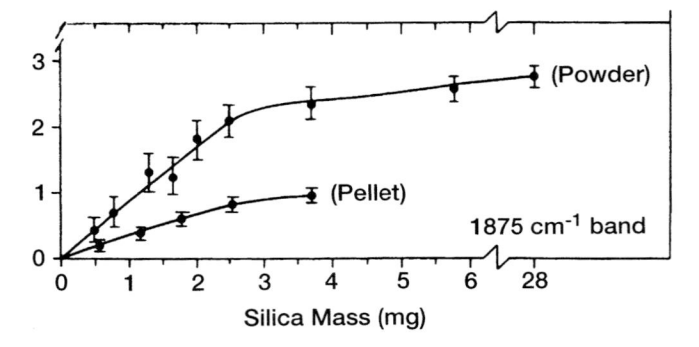

Figure 5.18 Variation of the intensity of the 1875-cm^{-1} silica band in the PA infrared spectra of powder and pellet forms of silica/kaolin mixtures. The nonlinear increase at higher concentrations was attributed to saturation. (Reproduced from Pandurangi, R.S. and Seehra, M.S., *Appl. Spectrosc.* **46**: 1719–1723, by permission of the Society for Applied Spectroscopy; copyright © 1992.)

analysis of the PA data gave correlation coefficients of ≥0.97 for calibration spectra, and at least 0.93 for validation samples. The latter results were obtained for concentrations between 1 and 5%. This article incorrectly stated that only one previous study on quantitative PA infrared spectroscopy existed at the time of publication.

Pandurangi and Seehra (1992) compared PA spectroscopy with diffuse reflectance and transmission for the determination of silica in silica-kaolin mixtures. Both powders and pellets were examined in the PA and diffuse reflectance experiments. As shown in Figure 5.18, the PA intensities of several different silica bands increased approximately linearly at low concentrations; however, this intensification was more gradual at higher concentrations. The latter results were attributed to saturation. Diffuse reflectance yielded a wider linear range than that in PA spectroscopy. It is relevant to mention that the onset of saturation for the weaker bands in the PA spectra occurred at higher concentrations.

Two examples of quantification in PA infrared spectroscopy are given in Chapter 7. The work of Gordon *et al.* (1990) on the growth of microorganisms on cellulose is discussed in the section on biology and biochemistry. Their results showed that the protein amide I band in spectra of *Saccharomyces cerevisiae* is less affected by saturation if the spectra are normalized using a polyacrylonitrile band as an internal standard. The saturation referred to in this case arises from layering of the yeast on the cellulose substrate, a phenomenon somewhat different than the effect described above for high concentrations in binary mixtures.

Saturation was much less troublesome in the PA infrared spectra of individual pea seeds (Letzelter *et al.*, 1995). Mixtures of starch in KBr first showed evidence of PA saturation at a concentration of about 50%, whereas protein-starch mixtures exhibited a linear variation in the intensity of the amide I band over an even wider concentration range. The authors concluded that the use of PA spectra for quantitative determination of the starch, lipid, and protein contents of pea seeds was quite feasible.

It is important to include the work on quantitative analysis by J. F. McClelland and his group at Iowa State University in this discussion. In a study of both qualitative and quantitative applications of PA infrared spectroscopy, McClelland *et al.* (1991) obtained spectra of vinyl acetate-polyethylene copolymers over a wide concentration range. After normalizing the PA spectra to constant integrated intensity for the C–H stretching region, the authors found that the vinyl acetate concentrations predicted by factor analysis of three different spectral regions showed excellent agreement with the concentrations measured by titration, an accepted method of analysis.

These experiments were described again in a wide-ranging review article by the same authors (McClelland *et al.*, 1992). Importantly, the latter work reported results of an even more impressive experiment in which a multicomponent system comprising coal and three of its commonly associated clays and minerals (kaolinite, pyrite, quartz) was examined using PA spectroscopy. The total inorganic concentration in each of the 15 samples studied was 30% by weight, with the amount of each component ranging from 0 to 30%. Excellent results were obtained in these experiments: factor analysis gave standard errors of prediction between 1 and 2% for all three components when single samples were successively treated as unknowns and the others were used as a learning set.

Norton and McClelland (1997) investigated the feasibility of using PA infrared spectroscopy for process control in an industrial setting where lime (CaO) is obtained by heating limestone ($CaCO_3$). The residual limestone from this commercial process is traditionally determined by an ignition test, which is relatively slow and susceptible to errors. PA spectroscopy was shown to be a viable alternative analysis method in this work: $CaCO_3$ can be identified by its characteristic 2513-cm^{-1} band in spectra of binary $CaCO_3$/CaO mixtures. Norton and McClelland obtained a linear calibration curve for limestone concentrations up to 5% in this way. An excellent detection limit of about 0.1% was thought to be achievable with this method.

Yang and Irudayaraj (2002) compared near-infrared, ATR, FT-Raman, diffuse reflectance, and mid-infrared PA spectroscopies for the determination of vitamin C in food and pharmaceutical products. All of the spectroscopic methods yielded satisfactory results. The 400–1728 and 2800–3718 cm^{-1} regions of the PA spectra – in other words, the fingerprint and C–H and O–H stretching regions – were observed to be correlated with vitamin C concentration. Both PA and diffuse reflectance spectroscopies are well suited to this analytical problem.

The above discussion can be briefly summarized through the statement that PA infrared spectroscopy has been successfully utilized for quantitative analysis by many research groups over the last three decades, and that a wide variety of samples are amenable to this approach. The greatest success has generally been achieved for analyte concentrations up to a few per cent; in favorable cases, much higher concentrations have been examined successfully. On the other hand, saturation limits the capability of PA infrared spectroscopy for quantitative analysis in some circumstances.

5.5.3
Quantitative Analysis at Higher Concentrations

Obviously, it is desirable to broaden the concentration range in which PA infrared spectroscopy can be used for quantitative analysis. A numerical method of achieving this aim was developed by Belton and coworkers in the 1980s. This procedure was subsequently implemented in the current author's laboratory and is summarized in this section.

As mentioned in the discussion on PA spectra of food products (Chapter 7), Belton and Tanner (1983) used near-infrared PA spectroscopy to determine the water content in starch. Their results showed that the intensity of the characteristic 1.9-μm (5265-cm^{-1}) water band initially increased linearly with concentration, but intensification proceeded more gradually as the amount of water in the samples increased further. This behavior–which resembles that in other work discussed in the present chapter–limited the suitability of PA spectroscopy for quantitative analysis, and led the authors to develop the theory described in the following paragraphs.

The general expression for PA intensity H resulting from the absorption of light by a thermally thick sample (Poulet, Chambron, and Unterreiner, 1980) may be written

$$H = AI_0 2^{1/2} \mu\beta / \left[(\mu\beta)^2 + (\mu\beta+2)^2\right]^{1/2} \tag{5.7}$$

where A is a composite term relating the thermal wave and PA signal intensities; I_o is the intensity of the incident light; and μ is thermal diffusion length. The absorption coefficient for a chromophore mixed with a diluent is given by

$$\beta = 2.303\varepsilon W_A \rho / MW_T \tag{5.8}$$

where M denotes the molecular weight of the absorbing species; W_A and W_T are the masses of the absorber and the total mixture, respectively; ε is molar absorptivity; and ρ is the density of the mixture.

It can be observed that Equation 5.7 simplifies to

$$H = AI_0 \mu\beta / 2^{1/2} \tag{5.9}$$

for $\mu\beta \ll 1$, and to

$$H = AI_0 2^{1/2} \mu\beta / (\mu\beta + 2) \tag{5.10}$$

for $(\mu\beta)^2 \ll (\mu\beta + 2)^2$. The first approximation corresponds to the weak-absorption case: the PA spectrum is proportional to a conventional absorption spectrum. PA intensity is predicted to vary linearly with concentration when Equation 5.9 obtains. Belton and Tanner (1983) plotted Equations 5.7, 5.9 and 5.10 against $\mu\beta$ and found

Table 5.2 Quantification of kaolin (per cent) in binary mixtures (Saffa and Michaelian, 1994a, 1994b). The infrared wavenumbers refer to the kaolin bands used in the calculations.

Kaolin/silica		Kaolin/KBr		
Actual value	PA result	Actual value	PA result	
	$550\,cm^{-1}$		$550\,cm^{-1}$	$924\,cm^{-1}$
32.7	37.3	14.6	12.4	12
64.1	71.3	80	75.4	71.5
76	78.6			

the latter equivalency to hold over a wide range of conditions. Moreover, Equation 5.10 can be rewritten to show the proportionality between H^{-1} and β^{-1}. This relationship is also valid at higher concentrations, where PA intensity no longer increases in direct proportion to analyte concentration.

The applicability of Equation 5.10 was investigated by Belton, Saffa, and Wilson (1987, 1988). These authors compared PA, ATR and diffuse reflectance infrared spectroscopies with regard to quantitative analysis of protein-starch mixtures and other food samples. They observed that the reciprocal of the sum of the amide I and II band intensities varied linearly with the reciprocal of the quantity of protein in the samples. Similar relations held for the other mixtures. Thus the range of concentrations where PA spectroscopy could be reliably used for quantitative analysis was considerably extended by use of this technique.

This method was subsequently used to analyze kaolin-silica and kaolin-KBr mixtures by Saffa and Michaelian (1994a, 1994b). Plots of $1/H$ vs W_T / W_A (the latter quantity being proportional to β^{-1}) were found to be linear over wide concentration ranges for both mixtures. Kaolin bands at 550 and $924\,cm^{-1}$ were used in this analysis; typical results are summarized in Table 5.2. Although kaolin concentrations as high as 80% were included in this work, the average error in the results was only 12%. It can therefore be suggested that this method may be generally applicable to quantitative analysis at high concentrations.

5.6
Depth Profiling

The ability to analyze solid samples at selected depths is generally recognized to be one of the most important features of PA spectroscopy. This section discusses depth profiling in PA infrared spectroscopy, a topic that has formed an active area of research for three decades. An examination of the relevant literature published in this period shows that some of the early attempts at characterization of layered or heterogeneous materials were adversely affected by low signal-to-noise ratios and other experimental limitations. Fortunately, the quality of the data acquired in these studies improved quickly, and excellent depth profiling results have been obtained in many laboratories for a number of years.

For typical samples, the PA signal arises from a superficial layer with a thickness approximately equal to the thermal diffusion length μ_s.[3] This dimension is given by the well-known expression $\mu_s = (\alpha/\pi f)^{1/2}$, where α denotes thermal diffusivity and f is modulation frequency. The mathematical definition of μ_s provides a basis for depth profiling: as f is changed, μ_s varies and different depths are analyzed. In general, this statement applies to both single-wavelength and multiplex PA experiments.

In a rapid-scan PA FT-IR experiment performed with a Michelson interferometer, f is determined by the mirror velocity V (cm s^{-1}) according to $f = 2V\nu$, where ν is infrared wavenumber. (In a Genzel interferometer, the different optical layout yields $f = 4V\nu$.) Consequently, the acquisition of PA spectra at different mirror velocities produces a change in both f and μ_s. It should be noted that some FT-IR spectrometers offer a limited choice of mirror velocities; depth profiling may not be feasible with such instruments.

Vidrine (1981), as well as Vidrine and Lowry (1983), were among the first to demonstrate PA depth profiling by changing interferometer mirror velocity. Numerous authors successfully applied this approach to a wide variety of samples in succeeding years (Donini and Michaelian, 1984; Muraishi, 1984; Ochiai, 1985; Yang et al., 1985; Urban and Koenig, 1986; Zerlia, 1986; Yang and Fateley, 1987; Yang, Bresee, and Fateley, 1987; Urban, 1987). These and other representative publications describing the results of PA infrared depth profiling experiments are listed in Table 5.3.

While the dependence of μ_s on mirror velocity enables depth profiling in rapid-scan FT-IR spectroscopy, the use of this principle is subject to an important qualification. Specifically, the dependence of μ_s on f implies a variation of sampling depth across the spectrum in a single PA spectrum. Taking the mid-infrared region as an example, the value of μ_s at 400 cm^{-1} is greater than that at 4000 cm^{-1} by a factor of $\sqrt{10} = 3.16$. This variation of depth with wavenumber need not influence the interpretation of spectra of homogeneous samples, but can be important when the composition is depth-dependent. For layered or inhomogeneous samples, a series of spectra obtained at different mirror velocities can be compared in a narrow spectral region, where the slight wavenumber dependence of μ_s can be neglected.

---

3) Many authors use μ_s and sampling depth interchangeably. Although related, they are not equal. As noted in Appendix 1, μ_s is defined as the dimension over which the amplitude of the thermal wave decays to $1/e$ (~0.37) of its original magnitude: the extent of μ_s does not define a boundary between 'sampled' and 'unsampled' regions. The *approximate* equivalence of μ_s and sampling depth can be used for convenience and simplicity.

Table 5.3 Examples of depth profiling in PA infrared spectroscopy.

Reference	Application/sample[a]
Dittmar, Chao, and Palmer (1991)	EVAc/PP, polyimide
Donini and Michaelian (1984)	PE/ink
Drapcho et al. (1997)	Polymer films
Gonon, Vasseur, and Gardette (1999)	Styrene/isoprene
Gonon et al. (2001)	Styrene/isoprene
Gregoriou and Hapanowicz (1996)	Polymer films
Gregoriou and Hapanowicz (1997)	Polymer laminate
Irudayaraj and Yang (2000)	Cheese wrapper
Irudayaraj and Yang (2002)	Starch/protein/polyethylene
Jiang, Palmer, and Chao (1995)	Polymer films
Jiang and Palmer (1997)	EVAc/PP
Jiang et al. (1997b)	Polymer films
Jiang et al. (1997a)	Polymer films
Jiang et al. (1998)	Polymer films
Jiang (1999)	Coated bead, grease/fiber, hair gel
Jones and McClelland (1996)	PET/polycarbonate
Jones and McClelland (2002)	PET/polycarbonate
Lerner et al. (1989)	Polymer laminate
McClelland et al. (1992)	Polymer films
McClelland et al. (1994)	PE, PS
McClelland et al. (1997)	PET/polycarbonate
McClelland et al. (2003)	Polymer films
Michaelian (1989)	Coal
Michaelian (1991)	Coal
Michaelian, Hall, and Kenny (2006)	Post-extraction oil sand
Muraishi (1984)	PVC/PVF
Noda et al. (1997)	Polymer films
Notingher et al. (2003)	Polymer films
Ochiai (1985)	PVC/PVF
Oh and Nair (2005)	Polycrystalline membranes
Palmer and Dittmar (1993)	Polymer films
Palmer et al. (1993)	EVAc/PP, Teflon/polyimide
Palmer, Jiang, and Chao (1994)	Polymer films
Palmer et al. (1997)	Polymer films
Pichler and Sowa (2005a)	Polymer films
Pichler and Sowa (2005b)	Polymer films
Pichler and Sowa (2005c)	Polymer films
Sawatari et al. (2005)	Polymer composites
Sowa and Mantsch (1994a)	Human tooth
Sowa and Mantsch (1994b)	Human tooth
Sowa et al. (1995)	Fingernails
Sowa et al. (1996)	PE sulfonation
Story, Marcott, and Noda (1994)	PDMS/PE/PS
Story and Marcott (1998)	Adhesive label
Stout and Crocombe (1994)	Polymer films
Szurkowski and Wartewig (1999a)	Olive oil/water
Szurkowski and Wartewig (1999b)	Olive oil/water

Table 5.3 Continued

Reference	Application/sample[a]
Szurkowski *et al.* (2000)	Olive oil/water
Teramae and Tanaka (1985)	Polymer films
Urban and Koenig (1986)	PVF$_2$ / PET
Urban (1987)	PVF$_2$ / PET
Urban *et al.* (1999)	Butyl acrylate/polyurethane latexes
Vidrine (1981)	Catalyst
Vidrine and Lowry (1983)	Catalyst
Wahls and Leyte (1998a)	PP/PET
Wahls and Leyte (1998b)	PP/PBA
Yamada *et al.* (1996)	Coals
Yamauchi *et al.* (2004)	Wood
Yang *et al.* (1985)	PET
Yang and Fateley (1987)	PVC
Yang, Bresee, and Fateley (1987)	PET
Yang *et al.* (1990)	PET
Yang and Irudayaraj (2000)	Protein and fat in cheese packaging
Yang, Irudayaraj, and Sakhamuri (2001)	Coatings, microorganisms on foods
Yang and Irudayaraj (2001)	Beef, pork
Zerlia (1986)	Coals
Zhang, Lowe, and Smith (2009)	Polymer coatings

a) EVAc, ethylene-vinylacetate; PDMS, polydimethylsiloxane; PE, polyethylene; PET, poly(ethylene terephthalate); PVC, polyvinyl chloride; PVF, polyvinyl fluoride; PVF$_2$, poly(vinylidene fluoride); PP, polypropylene; PS, polystyrene.

The development of modern step-scan FT-IR instruments in the 1980s and 1990s dramatically changed the situation with regard to depth profiling. In a step-scan spectrometer, the interferometer mirror movement is discrete rather than continuous; in effect, the scan velocity is reduced to zero, and some form of modulation must be supplied to give rise to the PA effect. For example, the infrared beam is periodically interrupted by a chopper or shutter in amplitude modulation. Obviously, all incident wavelengths are completely modulated (i.e., the beam is either 'on' or 'off') in this experiment. Alternatively, an interferometer mirror (usually the movable one) can be oscillated ('dithered') at each stopping position, producing phase modulation. Spectroscopists may recognize this as a derivative technique; for small modulation amplitudes, the phase modulation interferogram is equivalent to the first derivative of the amplitude-modulation or rapid-scan interferograms. Phase modulation naturally varies with wavelength. This technique is favored by many researchers because all of the available radiation from the interferometer is allowed to impinge on the PA cell, whereas half of the radiation is lost when an ordinary chopper is used for amplitude modulation. Moreover, phase-modulation amplitude and frequency are usually controlled by FT-IR software, reducing the complexity of these measurements.

5.6 Depth Profiling

Table 5.4 Principal depth profiling techniques in PA infrared spectroscopy.

Method	Salient features	Comments
Rapid-scan	Sampling depth varies inversely with mirror velocity	μ_s depends on wavenumber
Amplitude modulation	f dependence of PA intensity allows depth profiling	Incident intensity sacrificed
Phase modulation	Energy curve affected by Bessel functions	μ_s independent of wavenumber
Phase spectrum	Wavenumber dependence of PA phase indicates spatial origin	High depth resolution. Data acquired in a single scan
Phase rotation	Plot of PA intensity vs phase angle exhibits characteristic minimum	Complete extinction of bands possible. Used to determine layer thickness
G2D correlation	Cross peaks in synchronous plot have similar spatial origins; peaks in asynchronous plot indicate phase differences, enabling depth profiling	Magnitude and phase combined in a single qualitative representation. Aids identification of overlapping bands

The major depth profiling techniques used in PA infrared spectroscopy, which include amplitude and phase modulation, are summarized in Table 5.4. The principles involved in each method, as well as several advantages and limitations, are also listed in this table. Depth profiling in rapid-scan PA FT-IR spectroscopy, the first method, was briefly described above. The other five techniques in Table 5.3 are normally implemented with step-scan operation. These methods are discussed in the following paragraphs, with amplitude modulation being mentioned first. After this, phase modulation techniques are considered, and the uses of phase spectra and phase rotation plots are reviewed. Generalized two-dimensional correlation (which also has many applications outside of PA spectroscopy) is described in the last section.

5.6.1 Amplitude Modulation

The methodology utilized in depth profiling by amplitude modulation is relatively straightforward. As mentioned previously, the thermal diffusion length μ_s in a PA experiment is given by $\mu_s = (\alpha/\pi f)^{1/2}$. μ_s, the distance over which the thermal wave decays to a value equal to $1/e$ of its original amplitude, can be taken as an estimate of the sampling depth in PA spectroscopy. In an amplitude-modulation depth profiling experiment, the infrared radiation falling on the sample is periodically interrupted using a mechanical chopper or shutter. Variation of f results in the detection of PA signals from surface layers with different thicknesses. Comparison of these data makes it possible to characterize layered or inhomogeneous samples.

The reader is obviously aware that this book emphasizes PA FT-IR spectroscopy. On the other hand, it should be recognized that single-wavelength PA infrared depth profiling experiments may be adequate for some applications. For example, these measurements can be performed with pulsed or chopped laser radiation at a wavelength corresponding to an infrared absorption band of specific interest. Alternatively, single or multiple-wavelength PA depth profiling experiments can be carried out using a broadband source such as a lamp, the wavelength(s) being selected by means of suitable optical filters.

For most infrared spectroscopists, the depth profiling experiments of greatest interest will certainly be those carried out with FT-IR spectrometers. In a typical amplitude modulation experiment of this sort, the beam emerging from the interferometer is chopped before it impinges on the sample in the PA cell. The signal from the PA detector is demodulated using a lock-in amplifier, and the DC signal is communicated to the FT-IR electronics. All wavelengths are modulated at a single f, and μ_s is controlled by changing this parameter. Depth profiling is thus effected for a sample with a composition that varies with distance from the surface.

Amplitude-modulation depth profiling experiments were carried out on both fresh and oxidized coal samples shortly after the reintroduction of commercial step-scan FT-IR spectrometers about 20 years ago (Michaelian, 1989, 1991). Surface oxidation of coal, which has several negative effects, produces samples that are amenable to depth profiling by PA infrared spectroscopy. This oxidation creates carbonyl groups that absorb near 1750 cm^{-1}. One can use a simple model in which the infrared spectrum of oxidized coal contains contributions from the corresponding fresh coal and a generalized carbonyl band:

$$S_{ox}(v) = S_{fr}(v) + S_{C=O}(v) \tag{5.11}$$

where the subscripts ox, fr and C=O denote oxidized, fresh and carbonyl, respectively. The carbonyl band can be recovered through division of $S_{ox}(v)$ by $S_{fr}(v)$: this ratio is approximately equal to one, except in the region near 1750 cm^{-1} where $S_{C=O}(v)$ also contributes to the observed intensity (Figure 5.19).

In this experiment the intensity of the retrieved carbonyl band was measured as a function of f. At higher frequencies the signal arises mainly from the surface layer of the oxidized coal and the band is more predominant. As this frequency was reduced, μ_s increased and the interior of the coal particles contributed more to the PA signal. The C=O band was weaker in these spectra. These results are consistent with an intuitive model in which the surface of the coal particles is more extensively oxidized than the interior and demonstrate successful depth profiling using amplitude modulation.

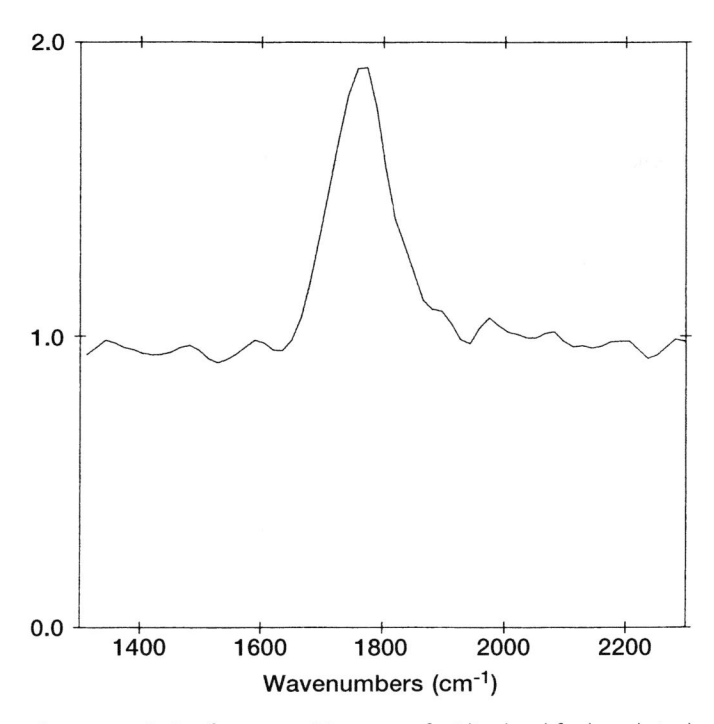

Figure 5.19 Ratio of step-scan PA spectra of oxidized and fresh coals in the carbonyl region. (Reproduced from Michaelian, K.H., *Appl. Spectrosc.* **43**: 185–190, by permission of the Society for Applied Spectroscopy; copyright © 1989.)

$S_{C=O}(\nu)$ can also be recovered by subtracting $S_{fr}(\nu)$ from $S_{ox}(\nu)$. When this approach is taken, the observed intensities must be corrected for the $\sim f^{-1}$ dependence of the PA signal (Michaelian, 1991). Depth profiling is then accomplished by measuring $S_{C=O}(\nu)$ at different frequencies (Figure 5.20). The results show that these intensities increase concomitantly with f up to about 180 Hz, and then remain approximately constant. If one assumes $\alpha = 0.1 \times 10^{-6}\,m^2\,s^{-1}$ for coal, the data suggest that an exterior layer about 12 μm thick is oxidized to approximately the same extent. Depth discrimination of this outer layer would be possible only for less extensively oxidized coals.

Because μ_s varies as $f^{-1/2}$, Figure 5.20 is not easily interpreted with regard to sampling depth. An alternative presentation of the data is given in Figure 5.21: μ_s was calculated at each f using the value of α given in the previous paragraph. Figure 5.21 illustrates an approximately linear relationship between carbonyl intensity and sampling depth; in other words, the extent of oxidation varies linearly with the distance from the surface of the coal particles in the ~13–26 μm region.

After these early amplitude-modulation depth profiling experiments, phase modulation gained popularity with many investigators. Several aspects of amplitude modulation contributed to this trend. These include the following: (i) the method requires the use of an external chopper; (ii) half of the available infrared

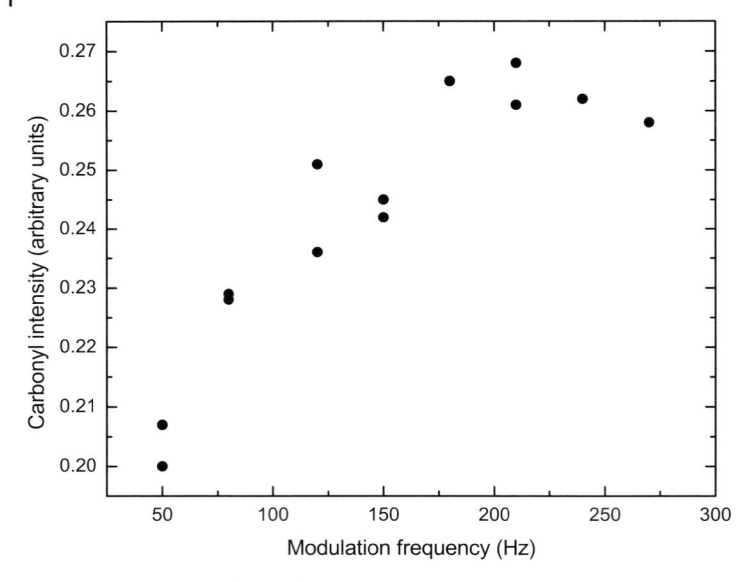

Figure 5.20 Depth profiling of oxidized coal: variation of carbonyl intensity with modulation frequency.

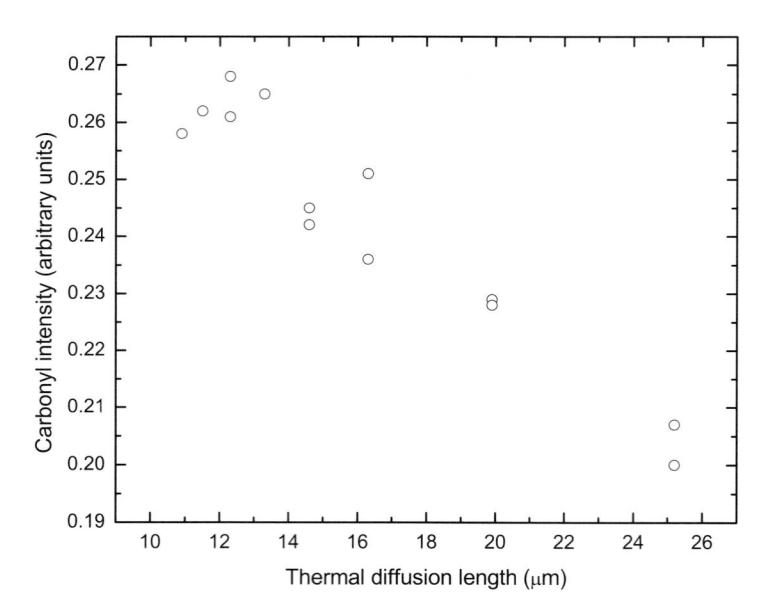

Figure 5.21 Depth profiling of oxidized coal: variation of carbonyl intensity with thermal diffusion length, μ_s.

radiation never reaches the sample and is sacrificed on the chopper blade; and (iii) a large DC background (which is independent of path difference, and can be observed in both rapid- and step-scan modes) reduces the dynamic range of the experiment. Moreover, it is essential that the infrared source be very stable since a change in source intensity can produce results similar to those caused by optical absorption. The last point may become important when depth profiling studies involve spectra obtained at significantly different times.

5.6.2
Phase Modulation

Depth profiling by phase modulation was demonstrated shortly after the amplitude modulation experiments described above. Phase modulation soon became the technique of choice for most workers in this field for several substantive reasons: (i) no external equipment is required; (ii) the experiment is conveniently controlled by FT-IR software; and (iii) in-phase and quadrature spectra are easily obtained in a single scan. (The last point also applies to amplitude modulation when a dual-phase lock-in amplifier is used.) Moreover, phase modulation is not subject to the limitations affecting amplitude modulation that were mentioned above.

The magnitude (M) and phase (θ) of the PA signal in phase modulation experiments can be utilized in depth profiling studies. These quantities are related by

$$M = [Q^2 + I^2]^{1/2} \tag{5.12}$$

and

$$\theta = \tan^{-1}[Q/I] \tag{5.13}$$

where Q and I are the quadrature and in-phase signals, respectively. The wavenumber dependence of these quantities is not shown in these equations.

Because all infrared wavelengths are modulated at a particular f in each phase modulation experiment, μ_s is constant. Therefore a change in f has a predictable effect on μ_s, and depth profiling is straightforward. Accordingly phase modulation magnitude spectra have been used in depth profiling studies by many research groups (Dittmar, Chao, and Palmer, 1991; Palmer and Dittmar, 1993; Szurkowski and Wartewig, 1999a, 1999b; Urban et al., 1999; Irudayaraj and Yang, 2000, 2002; Yang and Irudayaraj, 2000, 2001; Szurkowski et al., 2000; Gonon et al., 2001; Yang, Irudayaraj, and Sakhamuri, 2001; Jones and McClelland, 2002). The last reference in this list describes a numerical method for reducing saturation in spectra recorded at low f to a level similar to that at higher frequencies; this method can also be applied in rapid-scan experiments.

Another level of sophistication was introduced into this experiment through the use of digital signal processing (DSP). Digital demodulation allows simultaneous recovery of the PA signal at several harmonics of the phase modulation frequency, potentially providing all of the data necessary for depth profiling in one measurement. Obviously, considerable experiment time is saved in this way. Moreover, DSP detection was initially reported to yield signal/noise ratios superior to those obtained under similar conditions using a lock-in amplifier. While early experiments required a special 'DSP demodulator', the technique was soon implemented entirely in software (Drapcho et al., 1997). DSP detection was discussed further with regard to depth profiling by Story and Marcott (1998) and Jiang (1999). As many as nine harmonic frequencies can be examined at the same time in this way.

5.6.3
The Phase Spectrum

The depth profiling techniques discussed so far are based on the f dependence of the PA signal. On the other hand, the time required for the thermal wave to propagate to the surface produces a phase lag that also conveys depth-related information. A plot of θ vs wavenumber (a phase spectrum) may not be intuitive, since it usually does not resemble a conventional PA spectrum; however, the high phase resolution in such a plot is quite useful in applications such as depth profiling of very thin films.

The use of phase spectra for depth profiling layered polymer samples was pioneered in the early 1990s by R. A. Palmer and his collaborators. Since that time, the method has been utilized by other research groups in a variety of applications, some of which are described below. It is also relevant to point out that the phase spectrum yields results mathematically equivalent to those from phase rotation, discussed in the next section. These two approaches have been shown experimentally to yield equivalent results (Palmer, Jiang, and Chao, 1994; Jiang and Palmer, 1997).

The phase spectrum obtained by Jiang and Palmer (1997) for a sample consisting of an upper 12-μm ethylene-vinyl acetate copolymer layer and a 60-μm polypropylene substrate is shown in Figure 5.22. The letters in this figure denote magnitude-spectrum bands of particular interest. The larger phase angles of bands B, C_2, and C_3 imply deeper spatial origins, while the smaller angles for A_1, A_2, and C_1 indicate that the bands arise from a layer nearer to the surface. This demonstrates the use of phase angle spectra for depth profiling. In a series of related investigations, the method was applied to two-layer (Palmer, Jiang, and Chao, 1994, 1997), as well as three- and four-layer polymer samples (Jiang et al., 1997b).

Phase angle spectra have also been used in depth profiling studies on food products and food packaging materials. The phase angles for a model three-layer (polyethylene/protein/starch) sample confirmed the ordering of the layers in this system: the angles were smallest for the polyethylene, intermediate for protein, and greatest for starch (Irudayaraj and Yang, 2002). Similar results were obtained for two-layer (protein/apple, starch/apple) and three-layer (protein/starch/apple) samples (Yang, Irudayaraj, and Sakhamuri, 2001). These experiments are discussed below with respect to generalized two-dimensional correlation. Phase angle

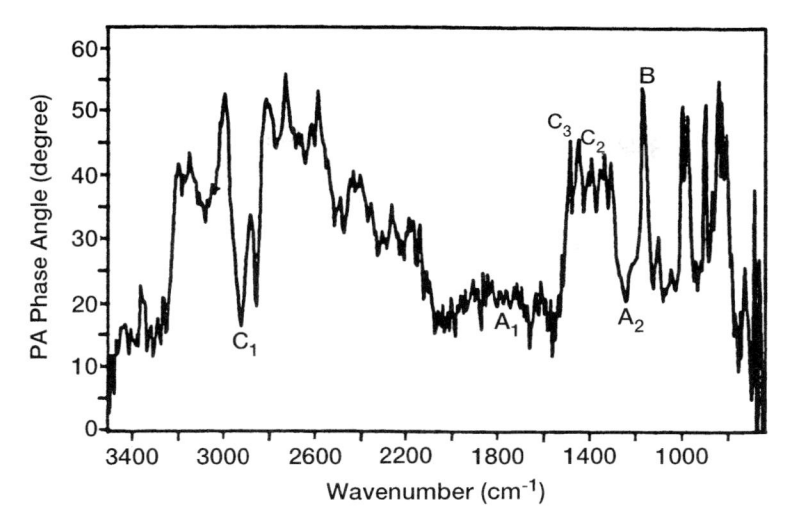

Figure 5.22 Phase angle spectrum of a two-layer polymer sample. The top layer was a 12-μm thick ethylene-vinyl acetate copolymer, while the substrate was a 60-μm thick layer of polypropylene. (Reprinted with permission from Jiang, E.Y. and Palmer, R.A., *Anal. Chem.* **69**: 1931–1935. Copyright © 1997 American Chemical Society.)

spectra were also used to confirm the migration of cheese components into polymeric packaging material (Irudayaraj and Yang, 2000).

5.6.4
Phase Rotation

Phase rotation was described in a series of articles by McClelland, Jones, and collaborators at Iowa State University, and was also implemented in other laboratories. It is mentioned here because the data are normally collected using step-scan phase modulation. The information utilized in phase rotation is essentially equivalent to that employed in the calculation of phase spectra, as discussed in the preceding section.

A phase rotation plot displays the variation of PA intensity with rotation angle. Data can be acquired by varying the detector phase (e.g., using a lock-in amplifier) to locate band extinction; alternatively, band intensities can be monitored as the phase angle is incremented in separate measurements. A third technique involves acquisition of PA spectra at two phase settings separated by 90°. The intensities at other angles are calculated by adding these two spectra vectorially.

It can be seen that phase angle plots are rather intuitive in appearance, an aspect that facilitates their interpretation. Phase rotation can be used in depth profiling studies to confirm the ordering of the layers in a sample. It also provides information regarding sampling depth in a PA experiment. Similarly, the technique can be used to determine layer thickness in appropriate circumstances. These applications are demonstrated in the following paragraphs.

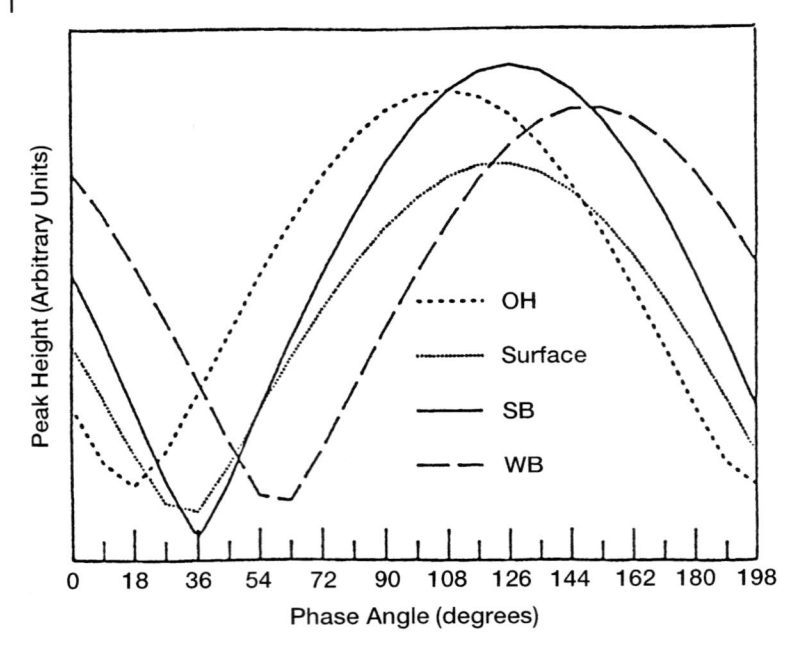

Figure 5.23 Phase rotation plot for chemically surface-treated polystyrene spheres. OH, hydroxyl groups; SB, strong band; WB, weak band (see text). (Reproduced from McClelland, J.F. et al., Depth profiling with step-scan FTIR photoacoustic spectroscopy, *SPIE* **2089**: 302–303, by permission of the International Society for Optical Engineering © 1994.)

Some capabilities of phase rotation are shown in a brief report on phase modulation PA infrared spectra of surface-treated polystyrene spheres (McClelland, Jones, and Ochiai, 1994). The magnitude spectrum for this sample contains a number of bands, among which four are of particular interest. These include a hydroxyl band, a feature due to the surface species, and two other bands simply described as weak and strong, respectively. Obviously, no information regarding the relative depths of the absorbing species can be derived from this single spectrum. By contrast, the variation of the peak heights with phase angle (Figure 5.23) is much more informative. According to the order of the phase angle minima in the 18–63° region, the curves arise from (i) the hydroxyl band, (ii) the peak from the surface species, (iii) the strong band, and (iv) the weak band. This sequence is consistent with the known ordering of the sample layers and confirms that depth profiling by phase rotation is indeed possible. The weak band exhibits a larger phase lag than the strong feature because the signal arises from the near-surface layer in the latter case (the incident radiation is attenuated sooner for the stronger band).

A second application of phase rotation was described by Bajic *et al.* (1995), who studied underground storage tank waste simulants. Aqueous mixtures of sodium nitrate and disodium nickel ferrocyanide, $Na_2NiFe(CN)_6$, were used to model the stored waste from nuclear fuel processes in this work. Solid samples for PA infrared analysis were prepared in three different ways (drying in air, oven drying, and freeze-drying), leading to the question as to whether the dried salts might form layers with varying thermal properties. Indeed, phase angle plots showed that the nitrate bands in the spectra of the air- and oven-dried samples exhibited earlier minima than those in the freeze-dried sample. This showed that the first two drying techniques caused the migration of the soluble sample to the surface, whereas freeze-drying yielded a more homogenous sample. $Na_2NiFe(CN)_6$ did not exhibit similar behavior: this material is insoluble and does not migrate during any of the drying processes. The phase rotation plots of the characteristic bands for the two salts also reflected the fact that the ferrocyanide absorption is the stronger of the two (its absorption coefficient maximizes at a smaller angle).

Jiang and Palmer (1997) used phase rotation to analyze PA data obtained for the two-layer sample mentioned above. Three bands due to the top layer exhibited minima in the phase rotation plot at angles of about 80°; two peaks with contributions from both layers showed a distinctly different minimum near 100°; and finally, the characteristic angle for a weak substrate band was about 118°. These results nicely demonstrate that phase rotation can be used for depth profiling in two-layer polymer samples. Jiang and Palmer (1997) also reported that the three respective angles determined by phase rotation differ by practically the same amounts as the angles deduced from the corresponding phase spectrum.

Phase rotation can also be used to calculate layer thickness. For the particular case of a transparent layer and an absorbing substrate, Adams and Kirkbright (1977a, 1977b) showed that the thermal wave at the surface lags behind that at the interface by

$$\Delta\theta = d/\mu_s \tag{5.14}$$

where d is the thickness of the upper layer and μ_s is its thermal diffusion length. In simple terms, the phase lag $\Delta\theta$ exists because of the time required for the thermal wave to propagate through the transparent layer and reach the upper surface of the two-layer sample.

Obviously, the upper layer is not transparent at all wavenumbers; in this example, it need only be transparent at the wavenumber where the substrate absorbs. The reference phase angle is established at a location where the upper layer absorbs strongly and should be independent of film thickness. The phase lag $\Delta\theta$ equals the difference between the reference and phase angles for the substrate band in a spectrum corresponding to a specific overlayer thickness.

This method was demonstrated by Jones and McClelland (1996), McClelland et al. (1997), and Wahls and Leyte (1998a). In the first two studies, thin films of polyethylene terephthalate (PET) on a much thicker polycarbonate substrate were examined. The reference phase angle was determined using the strong PET band at about $1725\,cm^{-1}$; as expected, this angle did not vary with film thickness. A polycarbonate band at $1771\,cm^{-1}$ was then studied in a series of spectra for samples with different PET film thicknesses. The differences between the phase

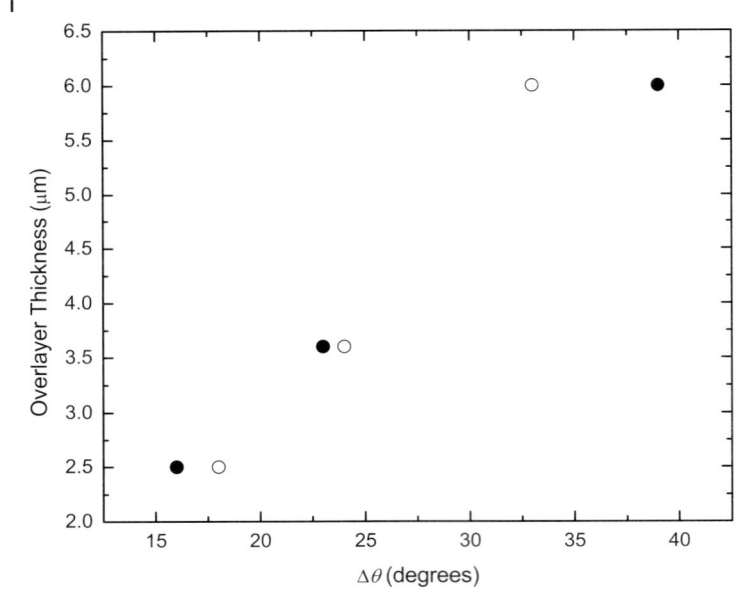

Figure 5.24 Determination of the thickness of PET films on a polycarbonate substrate by the phase rotation method (McClelland et al., 1997). Open circles, experimental results; filled circles, predicted values.

angles where the 1771-cm^{-1} band exhibited maximal intensity and the reference angle (McClelland et al., 1977) are plotted as open circles in Figure 5.24. The filled circles show values predicted by assuming $\alpha = 0.001\,\mathrm{cm^2\,s^{-1}}$ for PET. The good agreement in these results confirms that this method can be used to determine layer thickness. The data have been further analyzed using a numerical method known as expectation minimum analysis (Power and Prystay, 1995). Similar results were obtained in the other study of this two-layer system (Jones and McClelland, 1996).

It was mentioned above that PA spectra can be measured at a range of phase angles, with larger phase delays corresponding to greater sample depths. Alternatively, spectra can be calculated at various angles using the in-phase and quadrature components of the PA signal at a given f. This approach was taken by Sowa, Mantsch and their coworkers at National Research Council Canada in several biomedical investigations. Examples included an extracted human tooth, where an increasing protein contribution relative to the mineral (hydroxyapatite and phosphate) component was observed at increasing depths. Despite the complexity of tooth enamel, the distribution of these two components resembles a simple two-layer system (Sowa and Mantsch, 1994a, 1994b). Human finger nails were also analyzed successfully at selected depths using this principle (Sowa et al., 1995). This research is discussed in the section on medical applications in Chapter 7.

5.6.5
Generalized Two-Dimensional (G2D) Correlation

The final depth profiling method to be mentioned here is generalized two-dimensional (G2D) correlation. This procedure, originally developed by I. Noda and colleagues at the Procter and Gamble Company, has been used successfully by several groups in PA infrared experiments on layered samples, the most popular example being polymer laminates. Some applications of this technique are described below.

As discussed by Noda *et al.* (1997), G2D is an extension of the original method of two-dimensional infrared spectroscopy, where fluctuations in infrared intensities induced by small-amplitude sample deformations are studied by cross-correlation analysis (Noda, 1989, 1990). In G2D, the external perturbation applied to the sample does not have to be mechanical. Moreover, there is no requirement that the perturbation correspond to a sinusoidal variation; in fact, it need not be time dependent. The essential requirement is that it gives rise to selective changes in the band intensities in a spectrum.

The basis of G2D is that synchronous (asynchronous) correlations arise from similarities (dissimilarities) between variations of spectral intensities at wavenumbers v_1 and v_2. Peaks that occur along the diagonal of a synchronous correlation plot ('autopeaks') represent the magnitude of the changes in intensities during the perturbation; those in off-diagonal positions ('cross peaks') indicate the simultaneous intensification of two bands (positive) or, alternatively, opposite variations in band intensities (negative). A synchronous correlation map is symmetric with respect to its diagonal. In asynchronous 2D correlation, cross peaks appear when the changes in band intensities differ: a positive feature occurs when v_1 is affected before v_2, and a negative feature occurs in the opposite case. An asynchronous correlation map is antisymmetric with regard to the diagonal.

Because the information contained in a conventional (one-dimensional) spectrum is reorganized in a two-dimensional display, G2D can facilitate identification of weak features that are obscured by stronger neighboring bands in ordinary circumstances. This is particularly true with asynchronous correlation. However, Jiang and Palmer (1997) noted that artifactual peaks are occasionally created by this method. Hence the correlation results must be interpreted with care, as is the case with any numerical analysis of spectra.

In a PA step-scan depth profiling experiment, the external perturbation is the amplitude or phase modulation. For both types of modulation, bands arising from surface species exhibit greater relative intensities at higher f, while those due to the bulk are proportionately stronger at lower f. Synchronous cross peaks indicate similarity in the time signatures of the PA signals at the two wavenumbers in question. Because a time lag is associated with the propagation of the thermal wave through the sample, synchronicity implies that the signals arise from the same depth. The contours are positive because all band intensities decrease as f increases. On the other hand, in the asynchronous spectrum a positive feature at a given spectral ordinate implies that the first band arises from a shallower layer than that for the second peak. The depth relationship is reversed for a negative asynchronous cross peak. This principle enables depth profiling of heterogeneous or layered samples by G2D correlation analysis.

5.6.6
G2D Correlation in the Study of Layered Polymer Systems

Several articles have presented straightforward illustrations of G2D correlation for simple layered polymer samples. For example, Gregoriou and Hapanowicz (1997) examined a two-layer system consisting of a micrometer-thick acrylic polymer structure coated on a PET substrate several tens of micrometers thick. The synchronous correlation map for this sample confirmed that the PET carbonyl overtone was correlated with the 3055-cm^{-1} band that was also known to originate from PET. Similarly, Jiang et al. (1998) used G2D asynchronous correlation to analyze data for a sample consisting of a 1-µm polystyrene layer on a 100-µm mylar substrate. The signs of the contours in the correlation plot were consistent with the known ordering of the layers. High phase modulation frequencies were used in these experiments because the top layers were quite thin.

Jiang and Palmer (1997) used G2D correlation to depth profile the two-layer polymer sample mentioned above. The synchronous map (Figure 5.25) shows that the bands labeled A_1, A_2, C_2, and C_3 have similar phases, that is, their spatial origins (depths) are about the same. Bands A_1 and A_2 are due to the ethylene-vinyl acetate layer, while the other two (overlapping) bands arise mainly from polypropylene. By contrast, the asynchronous map (Figure 5.26) reveals that the cross peaks involving the much weaker band B (1169 cm^{-1}) and the other four bands are quite prominent: this confirms that band B arises from a greater sample depth (the polypropylene layer). Interpretation of these data takes both intensity and phase information into consideration.

The G2D technique was also applied to more complex layered systems. Jiang et al. (1997a) reported results for a four-layer sample consisting of polyethylene (10 µm), polypropylene (10 µm), PET (6 µm), and polycarbonate (6 mm). The ordering of the layers was confirmed from the signs of a series of asynchronous correlation contours involving various band-pair comparisons. Improved definition of partly overlapping bands was also observed in the 2D plots.

5.6.7
G2D Correlation in the Study of Foods

The literature on PA infrared spectra of food products is reviewed in Chapter 7. Several publications by the group of J. Irudayaraj at Pennsylvania State University can be mentioned in the context of the present discussion on depth profiling by G2D correlation analysis. The results of the G2D work are summarized in the next two paragraphs.

Yang and Irudayaraj (2000) used G2D in a study of the migration of cheese components into polymeric packaging material. PA spectra of cheese package wrappers were acquired after different storage periods and analyzed for evidence

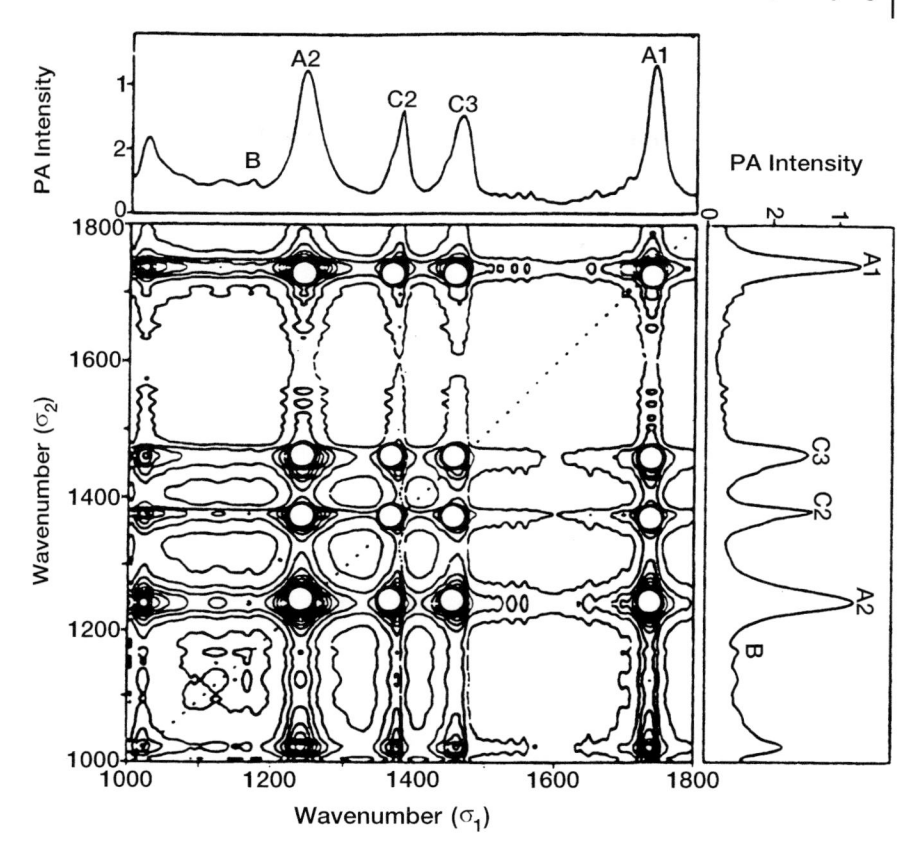

Figure 5.25 Synchronous 2D plot for two-layer polymer sample. The top layer was a 12-μm thick ethylene-vinyl acetate copolymer, and the substrate was 60-μm thick polypropylene. The modulation frequency was 200 Hz. (Reprinted with permission from Jiang, E.Y. and Palmer, R.A., *Anal. Chem.* **69**: 1931–1935. Copyright © 1997 American Chemical Society.)

of penetration of the cheese into the films. The synchronous G2D spectrum did not yield any information on the adherence of the cheese components to the package film, although it did facilitate identification of a C=C band at 1651 cm^{-1}. The asynchronous correlation spectrum was more informative, indicating that the amide I and II (protein) bands of the cheese originated from a shallower layer than several bands arising from CH_2 and CH_3 groups in the polymeric packaging material. This result confirmed the presence of the cheese near the surface of the package film. Moreover a 1461-cm^{-1} C–H band, not identifiable in the one-dimensional PA spectrum, was clearly revealed in the G2D correlation spectrum.

This group also studied edible coatings on fruit surfaces (Yang, Irudayaraj, and Sakhamuri, 2001). For a model two-layer sample consisting of a protein film on an apple skin, G2D was used to verify the known fact that the apple skin was beneath the protein film. Similarly, two three-layer samples were examined: G2D confirmed the starch/protein/apple and protein/starch/apple ordering of these

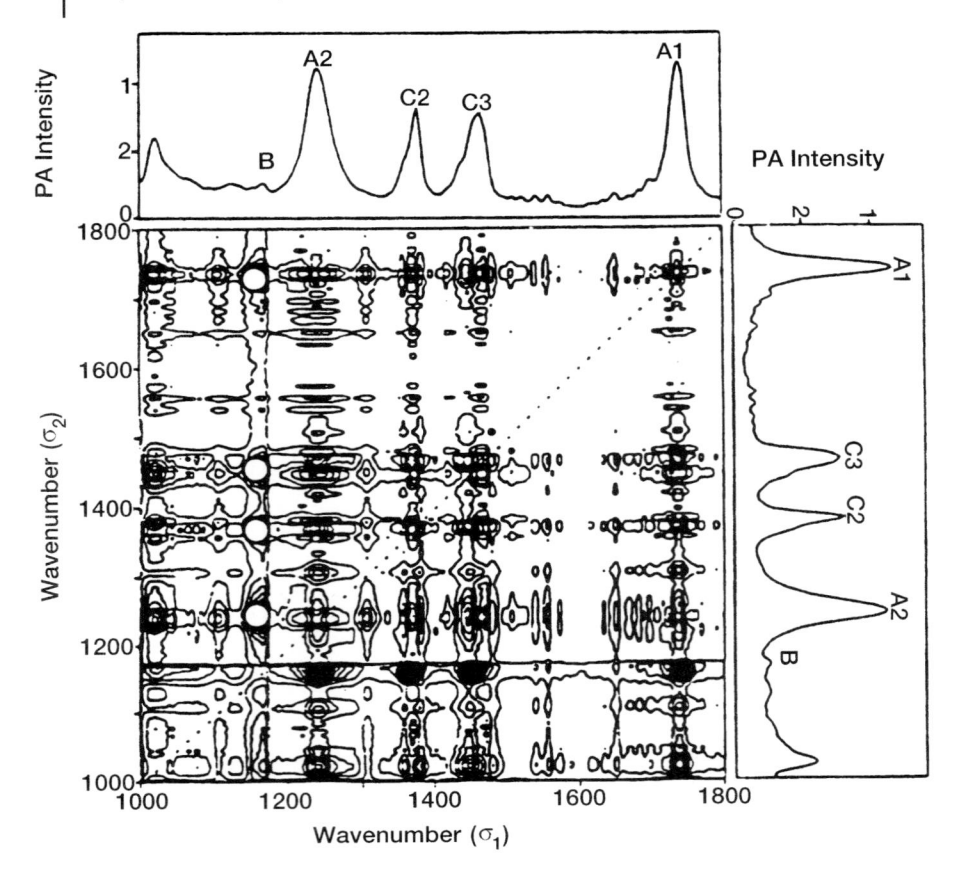

Figure 5.26 Asynchronous 2D plot for the two-layer polymer sample in Figure 5.25. Open circles represent positive peaks, whereas filled circles denote negative peaks. (Reprinted with permission from Jiang, E.Y. and Palmer, R.A., *Anal. Chem.* **69**: 1931–1935. Copyright © 1997 American Chemical Society.)

more complicated systems. Phase angle spectra yielded analogous results. A starch/protein/polyethylene three-layer system was also successfully analyzed by this technique (Irudayaraj and Yang, 2002). The ability of G2D to resolve overlapping peaks was noted as a particular advantage in this work. In summary, G2D can be utilized to detect the penetration of foods into package materials and to study the interactions between foods and coatings.

References

Abo-Bakr, M., Feikes, J., Holldack, K., Kuske, P., Peatman, W.B., Schade, U., Wüstefeld, G., and Hübers, H.-W. (2003) Brilliant, coherent far-infrared (THz) synchrotron radiation. *Phys. Rev. Lett.*, **90** (9), 094801 (4 pages).

Adams, M.J. and Kirkbright, G.F. (1977a) Analytical optoacoustic spectrometry. Part III. The optoacoustic effect and thermal diffusivity. *Analyst*, **102** (1217), 281–292.

Adams, M.J. and Kirkbright, G.F. (1977b) Thermal diffusivity and thickness measurements for solid samples utilising the optoacoustic effect. *Analyst*, **102** (1218), 678–682.

Adams, M.J., Beadle, B.C., and Kirkbright, G.F. (1978) Optoacoustic spectrometry in the near-infrared region. *Anal. Chem.*, **50** (9), 1371–1374.

Bajic, S.J., Luo, S., Jones, R.W., and McClelland, J.F. (1995) Analysis of underground storage tank waste by Fourier transform infrared photoacoustic spectroscopy. *Appl. Spectrosc.*, **49** (7), 1000–1005.

Belton, P.S. and Tanner, S.F. (1983) Determination of the moisture content of starch using near infrared photoacoustic spectroscopy. *Analyst*, **108** (1286), 591–596.

Belton, P.S., Saffa, A.M., and Wilson, R.H. (1987) Use of Fourier transform infrared spectroscopy for quantitative analysis: a comparative study of different detection methods. *Analyst*, **112** (8), 1117–1120.

Belton, P.S., Saffa, A.M., and Wilson, R.H. (1988) Quantitative analysis by Fourier transform infrared photoacoustic spectroscopy. *Proc. SPIE Int. Soc. Opt. Eng.*, **917**, 72–77.

Bozec, L., Hammiche, A., Pollock, H.M., Conroy, M., Chalmers, J.M., Everall, N.J., and Turin, L. (2001) Localized photothermal infrared spectroscopy using a proximal probe. *J. Appl. Phys.*, **90** (10), 5159–5165.

Castleden, S.L., Kirkbright, G.F., and Menon, K.R. (1980) Determination of moisture in single-cell protein utilising photoacoustic spectroscopy in the near-infrared region. *Analyst*, **105** (1256), 1076–1081.

Castleden, S.L., Kirkbright, G.F., and Long, S.E. (1982) Quantitative assay of propanolol by photoacoustic spectroscopy. *Can. J. Spectrosc.*, **27** (1), 245–248.

Débarre, D., Boccara, A.C., and Fournier, D. (1981) High luminosity visible and near-IR Fourier-transform photoacoustic spectrometer. *Appl. Opt.*, **20** (24), 4281–4286.

Dittmar, R.M., Chao, J.L., and Palmer, R.A. (1991) Photoacoustic depth profiling of polymer laminates by step-scan Fourier transform infrared spectroscopy. *Appl. Spectrosc.*, **45** (7), 1104–1110.

Donini, J.C. and Michaelian, K.H. (1984) Effect of cell resonance on depth profiling in photoacoustic FTIR spectra. *Infrared Phys.*, **24** (2–3), 157–163.

Drapcho, D.L., Curbelo, R., Jiang, E.Y., Crocombe, R.A., and McCarthy, W.J. (1997) Digital signal processing for step-scan Fourier transform infrared photoacoustic spectroscopy. *Appl. Spectrosc.*, **51** (4), 453–460.

Gardella, J.A., Jiang, D.-Z., and Eyring, E.M. (1983) Quantitative determination of catalytic surface adsorption sites by Fourier transform infrared photoacoustic spectroscopy. *Appl. Spectrosc.*, **37** (2), 131–133.

Gonon, L., Vasseur, O.J., and Gardette, J.-L. (1999) Depth profiling of photooxidized styrene-isoprene copolymers by photoacoustic and micro-Fourier transform infrared spectroscopy. *Appl. Spectrosc.*, **53** (2), 157–163.

Gonon, L., Mallegol, J., Commereuc, S., and Verney, V. (2001) Step-scan FTIR and photoacoustic detection to assess depth profile of photooxidized polymer. *Vib. Spectrosc.*, **26** (1), 43–49.

Gordon, S.H., Greene, R.V., Freer, S.N., and James, C. (1990) Measurement of protein biomass by Fourier transform infrared-photoacoustic spectroscopy. *Biotech. Appl. Biochem.*, **12** (1), 1–10.

Gregoriou, V.G. and Hapanowicz, R. (1996) Sub-micron resolution depth profiling of thin coatings using step-scan photoacoustic FT-IR spectroscopy. *Prog. Nat. Sci.*, **6** (Suppl.), S-10–S-13.

Gregoriou, V.G. and Hapanowicz, R. (1997) Applications of photoacoustic step-scan FT-IR spectroscopy to polymeric materials. *Macromol. Symp.*, **119**, 101–111.

Hammiche, A., Pollock, H.M., Reading, M., Claybourn, M., Turner, P.H., and Jewkes, K. (1999) Photothermal FT-IR spectroscopy: a step towards FT-IR microscopy at a resolution better than the diffraction limit. *Appl. Spectrosc.*, **53** (7), 810–815.

Hammiche, A., Bozec, L., Conroy, M., Pollock, H.M., Mills, G., Weaver, J.M.R.,

Price, D.M., Reading, M., Hourston, D.J., and Song, M. (2000) Highly localized thermal, mechanical, and spectroscopic characterization of polymers using miniaturized thermal probes. *J. Vac. Sci. Technol., B*, **18** (3), 1322–1332.

Highfield, J.G. and Moffat, J.B. (1985) The influence of experimental conditions in quantitative analysis of powdered samples by Fourier transform infrared photoacoustic spectroscopy. *Appl. Spectrosc.*, **39** (5), 550–552.

Irudayaraj, J. and Yang, H. (2000) Analysis of cheese using step-scan Fourier transform infrared photoacoustic spectroscopy. *Appl. Spectrosc.*, **54** (4), 595–600.

Irudayaraj, J. and Yang, H. (2002) Depth profiling of a heterogeneous food-packaging model using step-scan Fourier transform infrared photoacoustic spectroscopy. *J. Food Eng.*, **55** (1), 25–33.

Jackson, R.S., Michaelian, K.H., and Homes, C.C. (2001) Photoacoustic spectroscopy using a synchrotron source, in *Fourier Transform Spectroscopy, OSA Technical Digest*, Optical Society of America, Washington, DC, pp. 161–163.

Jiang, E.Y. (1999) Heterogeneity studies of a single particle/fiber by using Fourier transform infrared micro-sampling photoacoustic spectroscopy. *Appl. Spectrosc.*, **53** (5), 583–587.

Jiang, E.Y. and Palmer, R.A. (1997) Comparison of phase rotation, phase spectrum, and two-dimensional correlation methods in step-scan Fourier transform infrared photoacoustic spectral depth profiling. *Anal. Chem.*, **69** (10), 1931–1935.

Jiang, E.Y., Palmer, R.A., and Chao, J.L. (1995) Development and applications of a photoacoustic phase theory for multilayer materials: the phase difference approach. *J. Appl. Phys.*, **78** (1), 460–469.

Jiang, E.Y., McCarthy, W.J., Drapcho, D.L., and Crocombe, R.A. (1997a) Generalized two-dimensional Fourier transform infrared photoacoustic spectral depth-profiling analysis. *Appl. Spectrosc.*, **51** (11), 1736–1740.

Jiang, E.Y., Palmer, R.A., Barr, N.E., and Morosoff, N. (1997b) Phase-resolved depth profiling of thin-layered plasma polymer films by step-scan Fourier transform infrared photoacoustic spectroscopy. *Appl. Spectrosc.*, **51** (8), 1238–1244.

Jiang, E.Y., Drapcho, D.L., McCarthy, W.J., and Crocombe, R.A. (1998) Frequency-resolved, phase-resolved and time-resolved step-scan Fourier transform infrared photoacoustic spectroscopy. *AIP Conf. Proc.*, **430**, 381–384.

Jin, Q., Kirkbright, G.F., and Spillane, D.E.M. (1982) The determination of moisture in some solid materials by near infrared photoacoustic spectroscopy. *Appl. Spectrosc.*, **36** (2), 120–124.

Jones, R.W. and McClelland, J.F. (1996) Quantitative depth profiling of layered samples using phase-modulation FT-IR photoacoustic spectroscopy. *Appl. Spectrosc.*, **50** (10), 1258–1263.

Jones, R.W. and McClelland, J.F. (2002) Quantitative depth profiling using saturation-equalized photoacoustic spectra. *Appl. Spectrosc.*, **56** (4), 409–418.

Lerner, B., Perkins, J.H., Pariente, G.L., and Griffiths, P.R. (1989) Sample depth profiling by PAS/step-scanning interferometry: consideration of mirror positioning errors. *Proc. SPIE Int. Soc. Opt. Eng.*, **1145**, 476–477.

Letzelter, N.S., Wilson, R.H., Jones, A.D., and Sinnaeve, G. (1995) Quantitative determination of the composition of individual pea seeds by Fourier transform infrared photoacoustic spectroscopy. *J. Sci. Food Agric.*, **67** (2), 239–245.

Masujima, T., Yoshida, H., Kawata, H., Amemiya, Y., Katsura, T., Ando, M., Nanba, T., Fukui, K., and Watanabe, M. (1989) Photoacoustic detector for synchrotron-radiation research. *Rev. Sci. Instrum.*, **60** (7), 2318–2320.

McClelland, J.F., Luo, S., Jones, R.W., and Seaverson, L.M. (1991) FTIR photoacoustic spectroscopy applications in qualitative and quantitative analyses of solid samples. *Proc. SPIE Int. Soc. Opt. Eng.*, **1575**, 226–227.

McClelland, J.F., Luo, S., Jones, R.W., and Seaverson, L.M. (1992) A tutorial on the state-of-the-art of FTIR photoacoustic spectroscopy, in *Photoacoustic and Photothermal Phenomena III* (ed. D. Bićanić), Springer-Verlag, Berlin, pp. 113–124.

McClelland, J.F., Jones, R.W., and Ochiai, S. (1994) Depth profiling with step-scan FT-IR photoacoustic spectroscopy. *Proc. SPIE Int. Soc. Opt. Eng.*, **2089**, 302–303.

McClelland, J.F., Jones, R.W., Bajic, S.J., and Power, J.F. (1997) Depth profiling by FT-IR photoacoustic spectroscopy. *Mikrochim. Acta [Suppl.]*, **14**, 613–614.

McClelland, J.F., Jones, R.W., and Luo, S. (2003) Practical analysis of polymers with depth varying compositions using Fourier transform infrared photoacoustic spectroscopy (plenary). *Rev. Sci. Instrum.*, **74** (1), 285–290.

Michaelian, K.H. (1989) Depth profiling and signal saturation in photoacoustic FT-IR spectra measured with a step-scan interferometer. *Appl. Spectrosc.*, **43** (2), 185–190.

Michaelian, K.H. (1991) Depth profiling of oxidized coal by step-scan photoacoustic FT-IR spectroscopy. *Appl. Spectrosc.*, **45** (2), 302–304.

Michaelian, K.H., Jackson, R.S., and Homes, C.C. (2001) Synchrotron infrared photoacoustic spectroscopy. *Rev. Sci. Instrum.*, **72** (12), 4331–4336.

Michaelian, K.H., Hall, R.H., and Bulmer, J.T. (2003a) FT-Raman and photoacoustic infrared spectroscopy of Syncrude heavy gas oil distillation fractions. *Spectrochim. Acta, Part A*, **59** (4), 811–824.

Michaelian, K.H., Hall, R.H., and Bulmer, J.T. (2003b) FT-Raman and photoacoustic infrared spectroscopy of Syncrude light gas oil distillation fractions. *Spectrochim. Acta, Part A*, **59** (13), 2971–2984.

Michaelian, K.H., Hall, R.H., and Kenny, K.I. (2006) Photoacoustic infrared spectroscopy of Syncrude post-extraction oil sand. *Spectrochim. Acta, Part A*, **64** (3), 703–710.

Michaelian, K.H., May, T.E., and Hyett, C. (2008) Photoacoustic infrared spectroscopy at the Canadian Light Source: commissioning experiments. *Rev. Sci. Instrum.*, **79** (1), 014903 (5 pages).

Michaelian, K.H., Billinghurst, B.E., Shaw, J.M., and Lastovka, V. (2009) Far-infrared photoacoustic spectra of tetracene, pentacene, perylene and pyrene. *Vib. Spectrosc.*, **49** (1), 28–31.

Muraishi, S. (1984) Depth profile analysed by FT-IR/PAS spectrophotometry. *Bunko Kenkyu*, **33**, 269–270.

Noda, I. (1989) Two-dimensional infrared spectroscopy. *J. Am. Chem. Soc.*, **111** (21), 8116–8118.

Noda, I. (1990) Two-dimensional infrared (2D IR) spectroscopy: theory and applications. *Appl. Spectrosc.*, **44** (4), 550–561.

Noda, I., Story, G.M., Dowrey, A.E., Reeder, R.C., and Marcott, C. (1997) Applications of two-dimensional correlation spectroscopy in depth-profiling photoacoustic spectroscopy, near-infrared dynamic rheo-optics, and spectroscopic imaging microscopy. *Macromol. Symp.*, **119**, 1–13.

Norton, G.A. and McClelland, J.F. (1997) Rapid determination of limestone using photoacoustic spectroscopy. *Miner. Eng.*, **10** (2), 237–240.

Notingher, I., Imhof, R.E., Xiao, P., and Pascut, F.C. (2003) Spectral depth profiling of arbitrary surfaces by thermal emission decay–Fourier transform infrared spectroscopy. *Appl. Spectrosc.*, **57** (12), 1494–1501.

Ochiai, S. (1985) FT-IR spectroscopy analysis of depth profile of coating films. *Toso Kogaku*, **20**, 192–195.

Oh, W. and Nair, S. (2005) Spatially resolved *in situ* measurements of the transport of organic molecules in a polycrystalline nanoporous membrane. *Appl. Phys. Lett.*, **87** (15), 151912 (3 pages).

Palmer, R.A. and Dittmar, R.M. (1993) Step-scan FT-IR photothermal spectral depth profiling of polymer films. *Thin Solid Films*, **223** (1), 31–38.

Palmer, R.A., Chao, J.L., Dittmar, R.M., Gregoriou, V.G., and Plunkett, S.E. (1993) Investigation of time-dependent phenomena by use of step-scan FT-IR. *Appl. Spectrosc.*, **47** (9), 1297–1310.

Palmer, R.A., Jiang, E.Y., and Chao, J.L. (1994) Phase analysis and its application in step-scan FT-IR photoacoustic depth profiling. *Proc. SPIE. Int. Soc. Opt. Eng.*, **2089**, 250–251.

Palmer, R.A., Jiang, E.Y., and Chao, J.L. (1997) Step-scan FT-IR photoacoustic (S^2 FT-IR PA) spectral depth profiling of layered materials. *Mikrochim. Acta [Suppl.]*, **14**, 591–594.

Pandurangi, R.S. and Seehra, M.S. (1992) Quantitative analysis of silica in silica-kaolin mixtures by photoacoustic and diffuse reflectance spectroscopies. *Appl. Spectrosc.*, **46** (11), 1719–1723.

Pichler, A. and Sowa, M.G. (2005a) Independent component analysis of photoacoustic depth profiles. *J. Mol. Spectrosc.*, **229** (2), 231–237.

Pichler, A. and Sowa, M.G. (2005b) Blind phase projection as an effective means of recovering pure component spectra from phase modulated photoacoustic spectra. *Vib. Spectrosc.*, **39** (2), 163–168.

Pichler, A. and Sowa, M.G. (2005c) Blind source separation of photoacoustic depth profiles into independent components. *Appl. Spectrosc.*, **59** (2), 164–172.

Pollock, H.M. and Hammiche, A. (2001) Micro-thermal analysis: techniques and applications. *J. Phys. D: Appl. Phys.*, **34** (9), R23–R53.

Poulet, P., Chambron, J., and Unterreiner, R. (1980) Quantitative photoacoustic spectroscopy applied to thermally thick samples. *J. Appl. Phys.*, **51** (3), 1738–1742.

Power, J.F. and Prystay, M.C. (1995) Expectation minimum (EM): a new principle for the solution of ill- posed problems in photothermal science. *Appl. Spectrosc.*, **49** (6), 709–724.

Riseman, S.M., Massoth, F.E., Dhar, G.M., and Eyring, E.M. (1982) Fourier transform infrared photoacoustic spectroscopy of pyridine adsorbed on silica-alumina and γ-alumina. *J. Phys. Chem.*, **86** (10), 1760–1763.

Rockley, M.G., Richardson, H.H., and Davis, D.M. (1980) Fourier-transformed infrared photoacoustic spectroscopy, the technique and its applications, in 1980 Ultrasonics Symposium Proceedings, pp. 649–651.

Rockley, M.G., Davis, D.M., and Richardson, H.H. (1981) Quantitative analysis of a binary mixture by Fourier transform infrared photoacoustic spectroscopy. *Appl. Spectrosc.*, **35** (2), 185–186.

Rosenthal, R.J., Carl, R.T., Beauchaine, J.P., and Fuller, M.P. (1988) Quantitative applications of photoacoustic spectroscopy in the infrared. *Mikrochim. Acta*, **95** (1–6), 149–153.

Saffa, A.M. and Michaelian, K.H. (1994a) Quantitative analysis of clay mixtures by photoacoustic FT-IR spectroscopy. *Proc. SPIE Int. Soc. Opt. Eng.*, **2089**, 566–567.

Saffa, A.M. and Michaelian, K.H. (1994b) Quantitative analysis of kaolinite/silica and kaolinite/KBr mixtures by photoacoustic FT-IR spectroscopy. *Appl. Spectrosc.*, **48** (7), 871–874.

Sawatari, N., Fukuda, M., Taguchi, Y., and Tanaka, M. (2005) Composite polymer particles with a gradated resin composition by suspension polymerization. *J. Appl. Polym. Sci.*, **97** (2), 682–690.

Sowa, M.G. and Mantsch, H.H. (1994a) FT-IR step-scan photoacoustic phase analysis and depth profiling of calcified tissue. *Appl. Spectrosc.*, **48** (3), 316–319.

Sowa, M.G. and Mantsch, H.H. (1994b) Phase modulated–phase resolved photoacoustic FT-IR study of calcified tissues. *Proc. SPIE Int. Soc. Opt. Eng.*, **2089**, 128–129.

Sowa, M.G., Wang, J., Schultz, C.P., Ahmed, M.K., and Mantsch, H.H. (1995) Infrared spectroscopic investigation of in vivo and ex vivo human nails. *Vib. Spectrosc.*, **10** (1), 49–56.

Sowa, M.G., Fischer, D., Eysel, H.H., and Mantsch, H.H. (1996) FT-IR PAS depth profiling investigation of polyethylene surface sulfonation. *J. Mol. Struct.*, **379** (1–3), 77–85.

Story, G.M. and Marcott, C. (1998) Uniform depth profiling of multiple sample depth ranges in a single step-scanning FT-IR photoacoustic experiment. *AIP Conf. Proc.*, **430**, 513–515.

Story, G.M., Marcott, C., and Noda, I. (1994) Phase correction and two-dimensional correlation analysis for depth-profiling photoacoustic step-scan FT-IR spectroscopy. *Proc. SPIE Int. Soc. Opt. Eng.*, **2089**, 242–243.

Stout, P.J. and Crocombe, R.A. (1994) PAS/FT-IR spectroscopy with a step-scan spectrometer. *Proc. SPIE Int. Soc. Opt. Eng.*, **2089**, 300–301.

Szurkowski, J. and Wartewig, S. (1999a) Application of photoacoustic spectroscopy in visible and infrared regions to studies of thin olive oil layers on water. *AIP Conf. Proc.*, **463**, 618–620.

Szurkowski, J. and Wartewig, S. (1999b) Application of photoacoustic spectroscopy to studies of thin olive oil layers on water. *Instrument. Sci. Technol.*, **27** (4), 311–317.

Szurkowski, J., Pawelska, I., Wartewig, S., and Pogorzelski, S. (2000) Photoacoustic study of the interaction between thin oil layers with water. *Acta Phys. Pol., A*, **97** (6), 1073–1082.

Teramae, N. and Tanaka, S. (1985) Subsurface layer detection by Fourier transform infrared photoacoustic spectroscopy. *Appl. Spectrosc.*, **39** (5), 797–799.

Urban, M.W. (1987) Photoacoustic Fourier transform infrared spectroscopy: a new method for characterization of coatings. *J. Coatings Technol.*, **59** (745), 29–34.

Urban, M.W. and Koenig, J.L. (1986) Depth-profiling studies of double-layer PVF_2-on-PET films by Fourier transform infrared photoacoustic spectroscopy. *Appl. Spectrosc.*, **40** (7), 994–998.

Urban, M.W., Allison, C.L., Johnson, G.L., and Di Stefano, F. (1999) Stratification of butyl acrylate/polyurethane (BA/PUR) latexes: ATR and step-scan photoacoustic studies. *Appl. Spectrosc.*, **53** (12), 1520–1527.

Vidrine, D.W. (1981) Photoacoustic Fourier transform infrared (FTIR) spectroscopy of solids. *Proc. SPIE Int. Soc. Opt. Eng.*, **289**, 355–360.

Vidrine, D.W. and Lowry, S.R. (1983) Photoacoustic Fourier transform IR spectroscopy and its application to polymer analysis. *Adv. Chem. Ser. (Polym. Charact.)*, **203**, 595–613.

Wahls, M.W.C. and Leyte, J.C. (1998a) Step-scan FT-IR photoacoustic studies of a double-layered polymer film on metal substrates. *Appl. Spectrosc.*, **52** (1), 123–127.

Wahls, M.W.C. and Leyte, J.C. (1998b) Fourier transform infrared photoacoustic spectroscopy of polymeric laminates. *J. Appl. Phys.*, **83** (1), 504–509.

Wang, H., Phifer, E.B., and Palmer, R.A. (1998) Multi-frequency S^2FTIR PA spectral depth profiling by use of sinusoidal phase modulation and harmonic demodulation up to 2250 Hz (10th harmonic). *Fresenius J. Anal. Chem.*, **362** (1), 34–40.

Yamada, O., Yasuda, H., Soneda, Y., Kobayashi, M., Makino, M., and Kaiho, M. (1996) The use of step-scan FT-IR/PAS for the study of structural changes in coal and char particles during gasification. *Prepr. Pap.–Am. Chem. Soc. Div. Fuel Chem.*, **41** (1), 93–97.

Yamauchi, S., Sudiyani, Y., Imamura, Y., and Doi, S. (2004) Depth profiling of weathered tropical wood using Fourier transform infrared photoacoustic spectroscopy. *J. Wood Sci.*, **50** (5), 433–438.

Yang, C.Q. and Fateley, W.G. (1987) Fourier-transform infrared photoacoustic spectroscopy evaluated for near-surface characterization of polymeric materials. *Anal. Chim Acta*, **194** (C), 303–309.

Yang, C.Q., Ellis, T.J., Bresee, R.R., and Fateley, W.G. (1985) Depth profiling of FT-IR photoacoustic spectroscopy and its applications for polymeric material studies. *Polym. Mater. Sci. Eng.*, **53**, 169–175.

Yang, C.Q., Bresee, R.R., and Fateley, W.G. (1987) Near-surface analysis and depth profiling by FT-IR photoacoustic spectroscopy. *Appl. Spectrosc.*, **41** (5), 889–896.

Yang, C.Q., Bresee, R.R., and Fateley, W.G. (1990) Studies of chemically modified poly(ethylene terephthalate) fibers by FT-IR photoacoustic spectroscopy and X-ray photoelectron spectroscopy. *Appl. Spectrosc.*, **44** (6), 1035–1039.

Yang, H. and Irudayaraj, J. (2000) Depth profiling Fourier transform analysis of cheese package using generalized two-dimensional photoacoustic correlation spectroscopy. *Trans. Am. Soc. Agric. Eng.*, **43** (4), 953–961.

Yang, H. and Irudayaraj, J. (2001) Characterization of beef and pork using Fourier transform infrared photoacoustic spectroscopy. *Lebensm.-Wiss. u.-Technol.*, **34** (6), 402–409.

Yang, H. and Irudayaraj, J. (2002) Rapid determination of vitamin C by NIR, MIR and FT-Raman techniques. *J. Pharm. Pharmacol.*, **54** (9), 1247–1255.

Yang, H., Irudayaraj, J., and Sakhamuri, S. (2001) Characterization of edible coatings and microorganisms on food surfaces using Fourier transform infrared photoacoustic spectroscopy. *Appl. Spectrosc.*, **55** (5), 571–583.

Yokoyama, Y., Kosugi, M., Kanda, H., Ozasa, M., and Hodouchi, K. (1984) Infrared photoacoustic spectrometry of powdered samples. Calibration curves of sulfate compounds. *Bunseki Kagaku*, **33**, E1–E7.

Zerlia, T. (1986) Depth profile study of large-sized coal samples by Fourier transform infrared photoacoustic spectroscopy. *Appl. Spectrosc.*, **40** (2), 214–217.

Zhang, W.R., Lowe, C., and Smith, R. (2009) Depth profiling of coil coating using step-scan photoacoustic FTIR. *Prog. Org. Coat.*, **65** (4), 469–476.

6
Numerical Methods

6.1
Averaging of PA Data

Because PA spectra are often significantly weaker than spectra obtained using optical detectors, data averaging takes on considerable importance in PA FT-IR spectroscopy. In general, the acquisition of a large number of 'scans' (interferograms) is expected to yield satisfactory signal/noise ratios. This step can be repeated; for example, ten 50-scan PA interferograms might be recorded for a particular sample under a specific set of conditions (mirror velocity, resolution, etc.). 'Single-sided' interferograms are most commonly acquired. Two strategies can immediately be suggested for the treatment of these data:

a) The interferograms are averaged, with the result being used to obtain a single spectrum.
b) Spectra are calculated from these 10 interferograms and then averaged.

When double-sided interferograms are acquired, several approaches can be used for data averaging and spectrum calculation (Michaelian, 1987):

a) averaging of the interferograms and calculation of the modulus spectrum;[1]
b) averaging of the interferograms, followed by Fourier transformation and multiplicative phase correction (Mertz, 1965);[2]
c) calculation of individual modulus spectra, which are subsequently averaged;
d) calculation of individual spectra using multiplicative phase correction, followed by averaging;
e) calculation of the 'average modulus' spectrum (Birch, 1980).

It may be noted that strategies (b) and (d) in the second list are analogous to (a) and (b) in the first list, respectively. Averaging of PA FT-IR data is discussed in this section, beginning with experiments involving single-sided interferograms.

---

1) The modulus spectrum is commonly referred to as the power spectrum in FT-IR spectroscopy.
2) This is the default phase correction scheme in most FT-IR software.

Photoacoustic IR Spectroscopy. 2nd Ed., Kirk H. Michaelian
Copyright © 2010 WILEY-VCH Verlag GmbH & Co. KGaA, Weinheim
ISBN: 978-3-527-40900-6

6.1.1
Single-Sided Interferograms: Averaging of Interferograms and Spectra

As discussed above, it is common practice in PA FT-IR spectroscopy to acquire a series of n-scan interferograms under like conditions. The data must be averaged in some way to obtain a final PA spectrum. Two straightforward approaches are (i) the interferograms are averaged and a spectrum, designated $S_i(\nu)$, is calculated from the result, and (b) spectra are calculated from the individual interferograms and averaged, yielding $S_s(\nu)$. One might expect the spectra to agree, but the presence of noise in the data generally means that the two strategies do not produce equivalent spectra.

To illustrate these averaging strategies, we consider PA data obtained for carbon black several years ago at the Canadian Light Source. When these experiments were performed, low-frequency noise associated with the synchrotron radiation (SR) sometimes affected the PA spectra. Figure 6.1 shows the 0–100 Hz (~0–990 cm^{-1}) region of carbon black spectra calculated from 500 single-scan interferograms recorded at 1.6, 2.2 and 3.0 kHz (top to bottom boxes, respectively).[3] The abscissa in these spectra employs a frequency scale, rather than more familiar wavenumber units, to facilitate comparison of the results obtained at the three

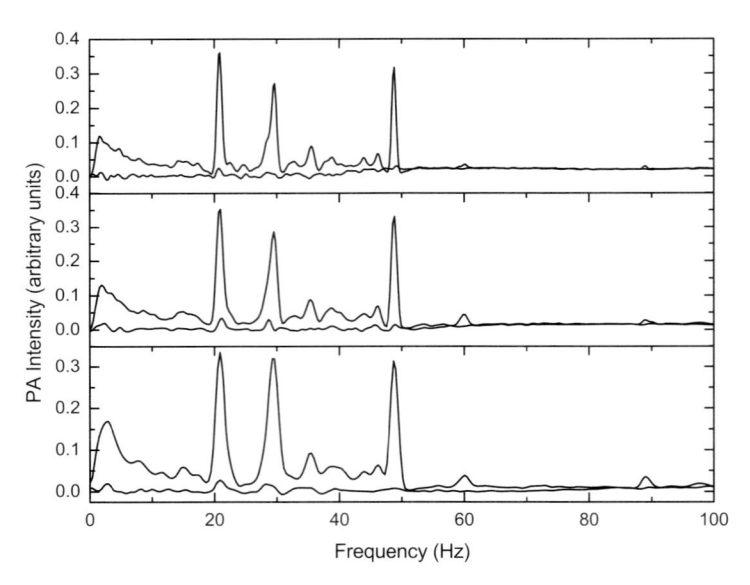

Figure 6.1 Low-frequency spectra of carbon black (1-scan data). Laser modulation frequencies as follows: upper box, 1.6 kHz; middle, 2.2 kHz; lower, 3.0 kHz. The upper curve in each box is S_s, while the lower is S_i.

3) It is standard practice to specify f at the He-Ne laser wavenumber (15 800 cm^{-1}) in rapid-scan FT-IR spectroscopy.

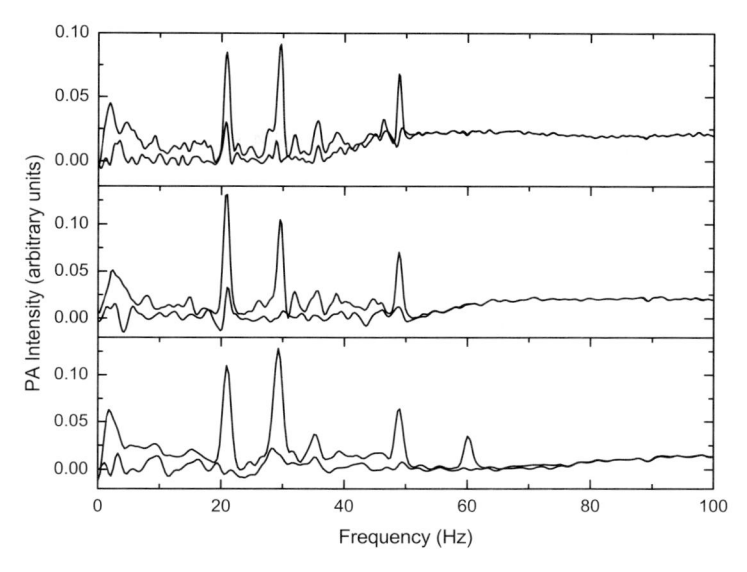

Figure 6.2 Low-frequency spectra of carbon black (10-scan data). Details are the same as in Figure 6.1.

mirror velocities. The prominent SR noise peaks in the upper curves [$S_s(v)$] occur at about 21, 29 and 49 Hz. These features are weak or absent in the lower spectra [$S_i(v)$], an important finding which favors use of the second averaging sequence.

Figure 6.2 reports similar spectra obtained when n was increased from 1 to 10. In this second trial 50 10-scan interferograms were acquired at each mirror velocity, that is, the total number of scans was maintained at 500. The spectra again clearly demonstrate that $S_s(v) \geq S_i(v)$: the noise features are significantly weaker when the interferograms are averaged prior to Fourier transformation. However, the $S_i(v)$ noise peaks are somewhat more prominent in Figure 6.2 than in Figure 6.1.

A simple model was put forward to account for these results (Michaelian, May, and Hyett, 2008). An SR noise peak corresponds to a cosine wave which adds to the PA interferogram due to the sample. Fourier transformation of a single-scan interferogram therefore yields a spectrum containing SR noise [$S_s(v)$, Figure 6.1]. If there is no fixed phase relationship between the cosine wave and the PA interferogram, the contribution from the SR noise will tend to cancel when a sufficient number of interferograms are averaged. The average interferogram therefore contains a reduced noise component, and the calculated spectrum [$S_i(v)$] is primarily that of the sample. The existence of minor noise peaks in $S_i(v)$ shows that the phase of the noise is not completely random; indeed, the enhanced $S_i(v)$ noise peaks in Figure 6.2 suggest that the SR noise was partially coherent. Similar differences between $S_s(v)$ and $S_i(v)$ were noted in an earlier (non-PA) SR infrared experiment by Bosch and Julian (2002).

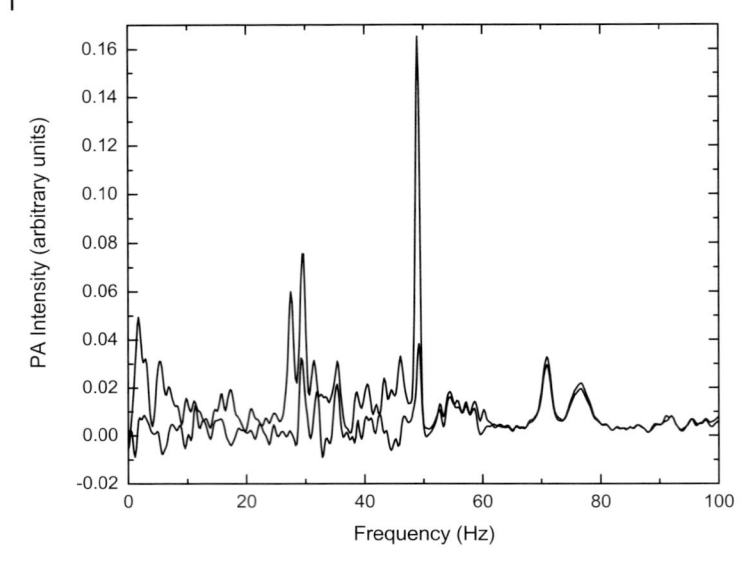

Figure 6.3 Low-frequency SR PA spectra of polystyrene particle obtained using microsample accessory. Upper curve, S_s; lower curve, S_i. Bands at 71 and 76 Hz are due to polystyrene; SR noise occurs in the 0–50 Hz region.

A practical application of the two averaging strategies is illustrated in Figure 6.3. Thirty-nine 50-scan SR PA interferograms were obtained for a 200-µm polystyrene particle at 1.6 kHz using a microsample accessory. SR noise appears below about 50 Hz (about 500 cm^{-1}), a region where minimal PA intensity is expected; suppression of this noise in $S_i(v)$ is obvious. Importantly, the broader polystyrene bands at about 71 and 76 Hz (705 and 755 cm^{-1}, respectively) exhibit comparable intensities in $S_s(v)$ and $S_i(v)$: this is an indication of their authenticities. Indeed, agreement between $S_s(v)$ and $S_i(v)$ has proven to be a reliable indicator of genuine bands in several other situations involving noisy PA spectra.

6.1.2
Double-Sided Interferograms: The Average Modulus Spectrum

The acquisition of double-sided interferograms creates additional alternatives for data averaging. It should be noted that interferogram peak locations must be consistent to ensure calculation of accurate spectra (Michaelian, 1987). The average modulus spectrum is described and compared with the average of individual modulus spectra in this section.

Birch (1980) expressed the complex spectrum $S(v)$ as

$$S(v) = c(v) + \varepsilon_c(v) + i\{s(v) + \varepsilon_s(v)\} \tag{6.1}$$

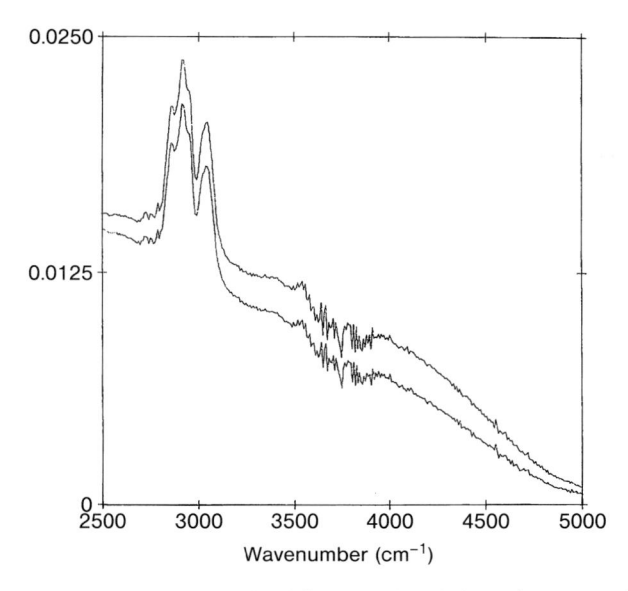

Figure 6.4 Spectra calculated from double-sided interferograms obtained for a coal sample. Upper curve, average of individual modulus spectra; lower curve, average modulus spectrum.

where $c(v)$ and $s(v)$ are the cosine and sine components of the spectrum, and $\varepsilon_c(v)$ and $\varepsilon_s(v)$ are the cosine and sine components of the noise spectrum, respectively. The average of the individually calculated modulus spectra [method (c)] is

$$\frac{1}{N}\sum\left[\{c(v)+\varepsilon_c(v)\}^2+\{s(v)\}^2\right]^{1/2} \qquad (6.2)$$

while the average modulus spectrum is

$$\frac{1}{N}\left\{\left[\sum\{c(v)+\varepsilon_c(v)\}\right]^2+\left[\sum\{s(v)+\varepsilon_c(v)\}\right]^2\right\}^{1/2} \qquad (6.3)$$

This alternative average spectrum was expected to lead to a reduction of noise. It can be shown that Equation 6.3 is equivalent to averaging interferograms prior to Fourier transformation.

Figure 6.4 compares spectra calculated according to Equations 6.2 and 6.3 for a typical western Canadian coal sample (Michaelian and Birch, 1991). The upper curve corresponds to the average of 10 modulus spectra, while the lower curve is the average modulus spectrum. It is obvious that the latter quantity is smaller at all wavenumbers, a result similar to those in Figures 6.1–6.3. However, it should be noted that the spectra in Figure 6.4 are also influenced by the fact that the interferogram peak locations varied within a range of two data points (about 1.26 µm) in this experiment.

The results presented here can be viewed in a somewhat different context. Gajić and Merkle (1988) assumed Gaussian interferogram noise and calculated expectation values of the intensities in $S_s(v)$ and $S_i(v)$ derived from double-sided interferograms. Their analysis showed that the signal level of the averaged modulus spectrum [or $S_s(v)$] is always higher than that of the spectrum obtained from the average interferogram. This relationship is similar to those depicted in Figures 6.1–6.4.

6.2
Spectrum Linearization

The object of linearization is to calculate a PA spectrum that resembles the result of a transmission experiment for a condensed-phase sample as closely as possible. In principle, either the amplitude or the phase of the PA signal can be used to quantitatively characterize sample absorption. However, the accuracies with which both quantities are known suffer from limitations: the amplitude is linear only at low absorptivities (McClelland, 1983), whereas the phase is subject to large uncertainties when the absorption is weak. Nevertheless, the amplitude and phase can be combined to calculate a linearized PA spectrum that is a closer approximation to the absorption spectrum than either individual quantity; this is possible because the upper limit of absorptivity in the PA phase spectrum is much higher than that in the amplitude spectrum. In fact, linearized PA spectra are proportional to absorption nearly three orders of magnitude above the onset of saturation observed in amplitude spectra (McClelland, Jones, and Bajic, 2002). This method was first developed and applied to the study of light-scattering samples in the ultraviolet and visible regions by Burggraf and Leyden (1981).

In typical experimental conditions the sample is thermally thick ($\mu_s < l$) and larger than the optical decay length ($\mu_\beta < l$). Combination of the expressions for the PA signal amplitude and phase yields the connection between the linearized amplitude spectrum q_l and the normalized amplitude spectrum q_n (Burggraf and Leyden, 1981; Carter, 1997):

$$q_l = 2^{1/2} \frac{q_n}{\sin(\psi - \pi/4)} \tag{6.4}$$

where ψ is the PA phase lag. Utilizing the trigonometric identity $\csc x = (\cot^2 x + 1)^{1/2}$, this equation can be rewritten

$$q_l = 2q_n \{[\cot^2(\psi - \pi/4) + 1]/2\}^{1/2} \tag{6.5}$$

The PA phase lag incorporates the related sample (s) and reference (r) quantities:

$$\psi = \psi_s - \psi_r + \pi/4 \tag{6.6}$$

ψ_s and ψ_r are obtained from

$$\sin\psi_j = I_j/(R_j^2 + I_j^2)^{1/2}$$
$$\cos\psi_j = R_j/(R_j^2 + I_j^2)^{1/2} \tag{6.7}$$

where I_j and R_j are the imaginary and real intensities, respectively, calculated at each wavenumber j. Carter (1992) rewrote Equation 6.5 as

$$q_l = q_n[\cot(\psi - \pi/4) + 1/2]^{1/2} = q_n[\cot(\psi_s - \psi_r) + 1/2]^{1/2} \tag{6.8}$$

The linearized spectrum can then be calculated from

$$q_l = q_s^2/(R_s I_r - I_s R_r) \tag{6.9}$$

where R and I denote real and imaginary spectra, $q_s = (R_s^2 + I_s^2)^{1/2}$ is the sample modulus (power) spectrum, and $q_n = q_s/q_r$. Thus one sample interferogram and one reference interferogram are measured, and the phase and intensity reference are obtained from the same data. It should also be noted that q_s is generally very similar to the spectrum calculated using conventional Mertz (multiplicative) phase correction, commonly implemented in FT-IR software. This analysis is based on the assumptions that the reference phase is equivalent to that measured for carbon black (Dittmar, Palmer, and Carter, 1994) and that the sample is homogeneous. Equation 6.9 is algebraically and numerically equivalent to Equation 6.5. The use of the reference interferogram in this calculation ensures that the linearized spectrum is normalized with respect to intensity. This approach is valid as long as sample and reference data are acquired under the same conditions, such that the instrumental phase is constant. Hence the sample interferogram should be measured immediately before or after the reference interferogram when spectrum linearization is to be performed.

Carter (1992) compared amplitude and linearized PA spectra of polymer and rubber samples, and found that linearization greatly reduced saturation effects, yielding a result that is similar to a conventional absorption spectrum. It was observed that the modulation frequency dependence of the sampling depth in rapid-scan PA spectroscopy persists in linearized PA spectra, although its effect is less dramatic than in amplitude spectra. It should also be noted that bands arising from surface species, such as thin films or oxide layers, are not affected by linearization in the same way as those arising from components that lie deeper within a sample. This is a consequence of the fact that the phase for surface species is close to $\pi/4$ (45°). The enhanced sensitivity to surface layers in linearized spectra was discussed further by Jiang, McCarthy, and Drapcho (1998) and McClelland, Jones, and Bajic (2002).

The linearization method has been used at National Research Council Canada in Winnipeg to facilitate analysis of PA infrared spectra of biomedical samples. For example, Sowa and Mantsch (1994) compared the linearized spectrum of tooth enamel with corresponding Mertz and modulus spectra. The linearized spectrum displayed somewhat better contrast between band maxima and background intensity than that observed in the other two spectra, which were actually superimposable. More recently, Pichler and Sowa (2004) compared linearized and amplitude spectra for layered polymer samples, and confirmed that linearization allows the detection of surface layers and correction of peak intensities after the onset of PA saturation. These authors explored the effects of sample interferogram shifts as large as ±9 points (this step is explained below). Optimal linearization results were obtained with the smallest ($\psi_r - \psi_s$) differences. As might be expected, only one or two particular shifts fulfilled this requirement.

Application of the linearization method is illustrated in the following figures. Double-sided PA interferograms were acquired at a resolution of 6 cm^{-1} for thimble solids (post-extraction oil sand, a mixture of clays and hydrocarbons obtained by distillation extraction) and carbon black using a Bruker IFS 66v/S spectrometer and an MTEC 300 PA cell. Each interferogram contains 9480 ordinates, with the centerburst occurring near (not necessarily at) point 4690. Calculations were performed using a spreadsheet program (Michaelian, 2007).

Figure 6.5 shows the central regions of the two interferograms. When these data are used in a linearization calculation based on Equation 6.9, unrealistic intense positive and negative features appear at several locations where bands are expected. This behavior occurs when the quantity $(R_s I_r - I_s R_r)$ is small or negative. Equation 6.9 can be rewritten

$$q_1 = q_s^2 / R_s R_r (\tan \psi_r - \tan \psi_s) \tag{6.10}$$

which shows that q_1 is badly behaved when $\psi_s \geq \psi_r$. Figure 6.6 shows that this condition obtains throughout much of the mid-infrared region.

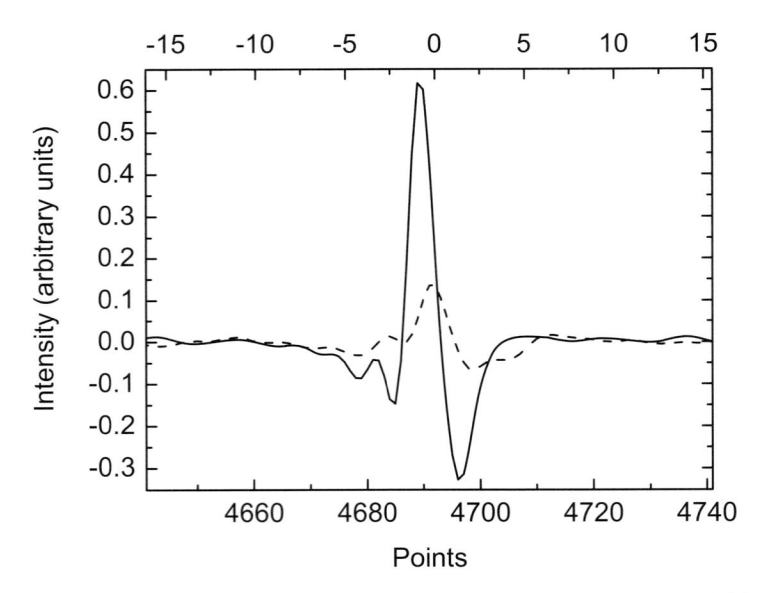

Figure 6.5 Central regions of measured interferograms for carbon black (solid line) and thimble solids (dashed line). The scale for the upper x axis refers to the weaker interferogram.

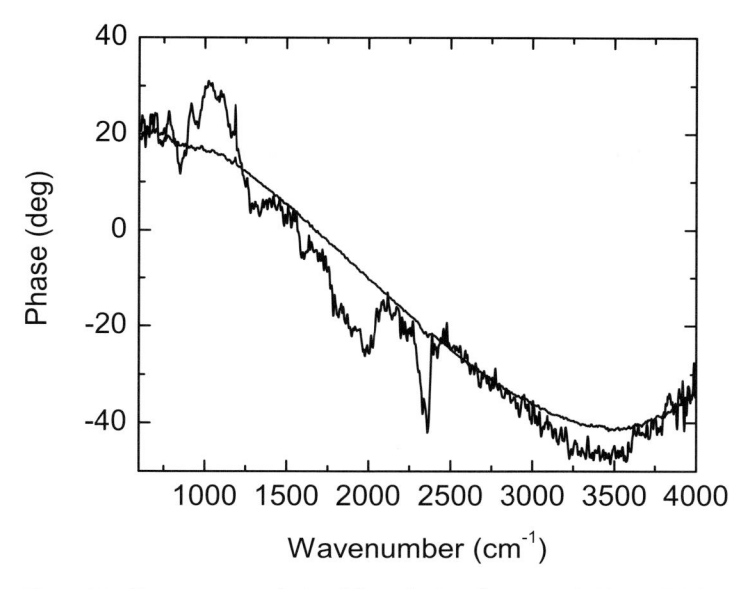

Figure 6.6 Phase spectra calculated from the interferograms in Figure 6.5. Noisy curve, thimble solids; smooth curve, carbon black.

This difficulty can be removed by slightly shifting one of the interferograms along the abscissa to bring about the relationship $\psi_s > \psi_r$ (McClelland, Jones, and Bajic, 2002; Pichler and Sowa, 2004).[4] It turns out that a shift of one or two data points – corresponding to a distance of less than one micrometer – is adequate in most cases. The carbon black phase curve is gradually displaced to greater values (larger phase angles) as the interferogram is shifted in the positive-path-difference direction (Figure 6.7). However, it should be noted that this modification of the phase is wavenumber dependent: the shift is relatively small at low wavenumbers, but increases significantly as the near-infrared region is approached. This trend affects the calculated band intensities in the linearized spectrum.

The result of the linearization calculation based on the two-point shift of the carbon black interferogram is shown in the upper box in Figure 6.8. It is evident that q_l exhibits band definition which is improved with respect to that in q_n (middle box), particularly in the important 'fingerprint' region below $2000\,cm^{-1}$. Moreover, the relative intensities of bands due to residual water vapor and CO_2 in the PA cell are diminished through the linearization calculation. Finally multiplication of q_l by wavenumber, an approximate scattering correction based on the original work of Burggraf and Leyden (1981), yields the spectrum displayed in the lowest box in Figure 6.8. This satisfactory result confirms the efficacy of spectrum linearization, showing that the effects of saturation can indeed be reduced numerically.

4) This strategy has also been pointed out by R. W. Jones (private communication).

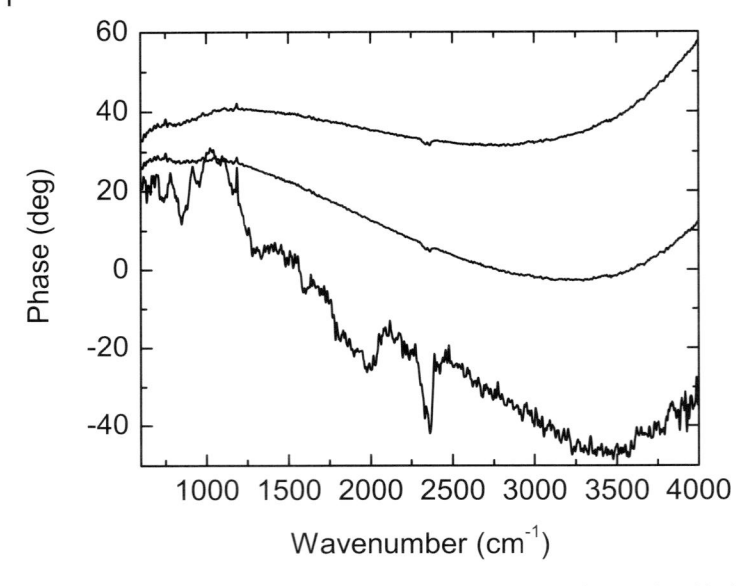

Figure 6.7 Phase spectra calculated after shifting the origin of the carbon black interferogram by one or two points in the positive-path-difference direction (middle and top curves, respectively). Identification of the curves is analogous to that in Figure 6.6.

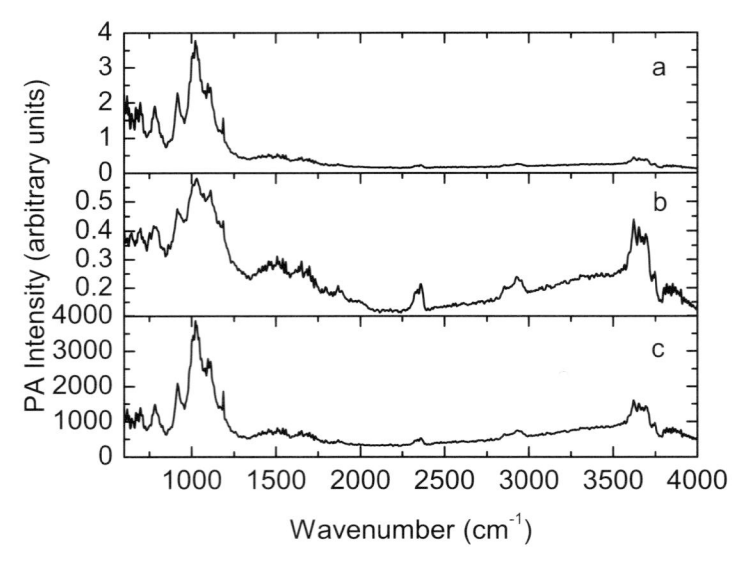

Figure 6.8 Comparison of normalized PA infrared spectra of thimble solids. (a) Linearized; (b) amplitude; (c) linearized and corrected for scattering through multiplication by wavenumber. The carbon black interferogram was shifted two points in the positive-path-difference direction, corresponding to the upper phase curve in Figure 6.7.

6.3
Phase Analysis

Another important numerical method relevant to PA infrared spectroscopy was developed by L. Bertrand and his collaborators at l'École Polytechnique de Montréal and several other institutions. The object of this method is to determine the absorption coefficient (β) of a sample as a function of wavenumber – in other words, acquisition of quantitative PA spectra rather than the qualitative data commonly reported by many authors. Elimination of saturation effects is another goal of these calculations. Moreover, because phase analysis is capable of distinguishing between surface and bulk absorption in some circumstances, it provides a depth profiling capability that augments those discussed in Chapter 5.

The PA phase can be obtained by two different approaches in FT-IR experiments. First, in what might be considered a conceptually simple method, the in-phase (0°) and quadrature (90°) components of the PA signal are directly observed using a lock-in amplifier. These measurements are ideally performed using a step-scan spectrometer, but could also be carried out with a very slow continuous scan and a high external modulation frequency. In the second approach, real and imaginary spectra calculated from a double-sided interferogram recorded in rapid-scan mode are utilized to obtain the PA phase. These spectra are available to the spectroscopist prior to the implementation of the conventional (Mertz) phase correction procedure in some commercial FT-IR data acquisition programs. However, the resolution of the phase spectra is restricted by the limited number of negative-path-difference points preceding the interferogram centerburst in many instruments. This limitation exists because many FT-IR spectrometers have been designed and constructed under the realistic assumption that low-resolution phase information is adequate in most (non-PA) applications. Fortunately, some research-level instruments allow the acquisition of full double-sided interferograms, thereby obviating this restriction.

Availability of the in-phase and in-quadrature, or real and imaginary, PA spectra is a prerequisite for the implementation of phase analysis. The basic principle of this technique was outlined by Bordeleau *et al.* (1986), Choquet, Rousset, and Bertrand (1985, 1986), and Bertrand (1988). These authors were particularly interested in the case of a thermally thin film adsorbed onto a thermally thick substrate, and showed that these two components of the PA signal are given by

$$\text{INPHAS} = C(1-\beta^s)\beta\mu/\left[(\beta\mu+1)^2+1\right] \tag{6.11}$$

and

$$\text{INQUAD} = C\left\{\beta^s + (1-\beta^s)(\beta\mu+1)\beta\mu/\left[(\beta\mu+1)^2+1\right]+D\right\} \tag{6.12}$$

where β^s is the (dimensionless) fraction of light absorbed by the surface layer, C is a constant that depends on the thermal properties of the sample, and D is a

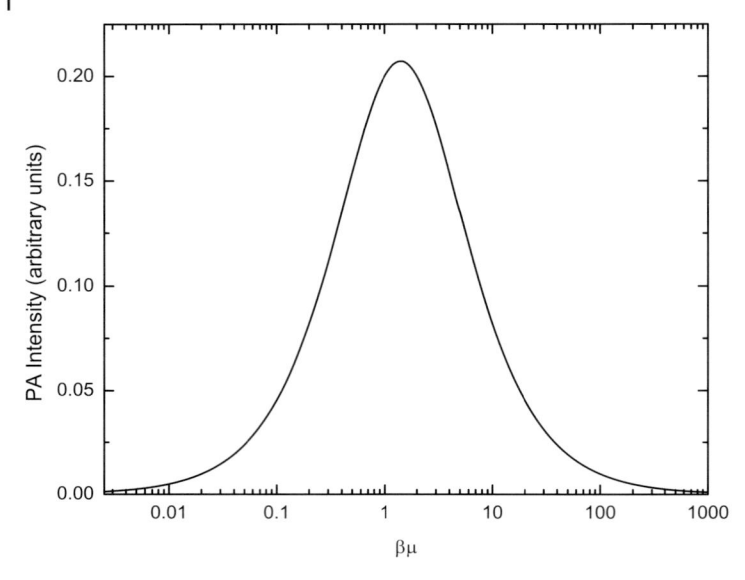

Figure 6.9 Variation of the in-phase (real) component of the PA intensity with the product $\beta\mu$.

thermal expansion term that is often negligible for solid samples. β and μ have their usual meanings.

In the limit where surface absorption is negligible and bulk absorption is large (i.e., $\beta^s = 0$ and $\beta\mu > 1$) these expressions can be rearranged to yield

$$\beta\mu = [\text{INQUAD} - \text{INPHAS}]/\text{INPHAS} \tag{6.13}$$

The variations of the in-phase and in-quadrature components of the PA intensity with $\beta\mu$ are readily calculated for the case where surface absorption can be neglected. These quantities are plotted in Figures 6.9 and 6.10, respectively.

On the other hand, when β^s is large and $\beta\mu \ll 1$, these equations reduce to

$$\text{INQUAD} - \text{INPHAS} = C\beta^s/2 + D \tag{6.14}$$

from which the surface absorption coefficient β^s can be easily determined. The wavenumber dependence of this quantity was shown to be small in PA data obtained for two polymers (Bordeleau et al., 1986).

The practicality of phase analysis has been illustrated in several publications. Choquet, Rousset, and Bertrand (1986) used the technique to correct spectra of asbestos (chrysotile), and demonstrated a significant reduction in saturation effects for this important mineral. The redistribution of intensity in the corrected PA spectrum with respect to an ordinary amplitude spectrum described by these authors qualitatively resembles that obtained when a rapid-scan PA spectrum is ratioed against a carbon black reference and corrected for modulation-frequency dependence through multiplication by $f^{1/2}$. Bertrand (1988) next applied the method to data for both asbestos and ethylene-propylene rubber (EPR); the results for the

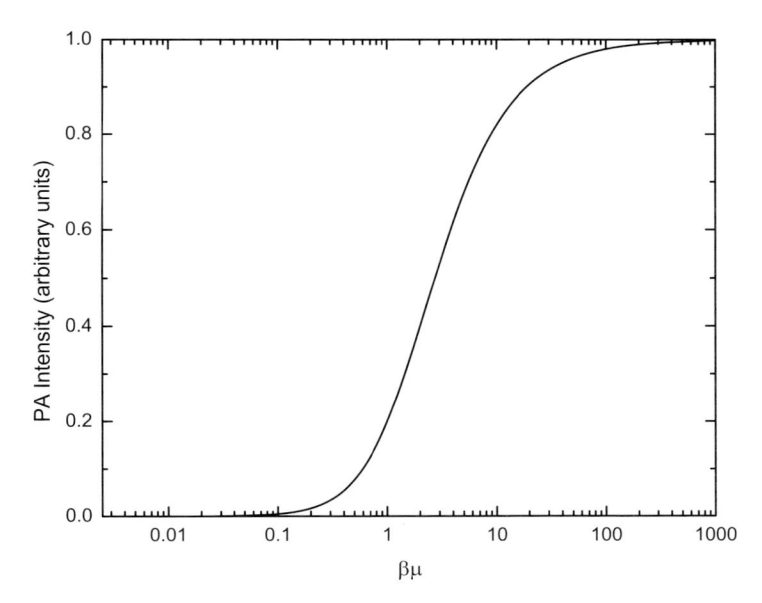

Figure 6.10 Variation of the in-quadrature (imaginary) component of the PA intensity with the product $\beta\mu$.

latter sample included a rather impressive diminution of saturation effects, as well as the elimination of the nonzero baseline (Monchalin et al., 1984) that frequently appears in PA infrared spectra (Figure 6.11). Subsequently, Marchand et al. (1997) reviewed the phase analysis method in detail, and stressed the importance of correctly identifying the onset of saturation (the point where the phase angle is equal to 67.5°). These authors pointed out that further improvements to the method and the eventual realization of quantitative PA infrared spectroscopy would require that several experimental parameters be taken into account: these include sample quantity, the position of the sample with respect to the window of the PA cell, and carrier gas pressure.

The phase analysis method has also been successfully applied to samples where surface absorption is important. For example, Bordeleau et al. (1986) calculated β^s and $\beta\mu$ spectra for EPR and cross-linked polyethylene, in which sorbed water affects dielectric properties. β^s increased concomitantly with relative humidity in these spectra. Mongeau, Rousset, and Bertrand (1986) separated surface and bulk absorption at a series of CO_2 laser wavelengths for samples of SiO deposited on CaF_2, and showed that the resolved bulk PA signals were similar to those obtained for uncoated CaF_2. Photothermal beam deflection spectroscopy (PBDS, mirage) detection was used in this experiment.

A very good example of the distinction between the surface and volume (bulk) of a sample was given in the PA infrared spectra reported by Bordeleau, Bertrand, and Sacher (1987). As shown in Figure 6.12, the $N-H_n$ band at $3350\,cm^{-1}$

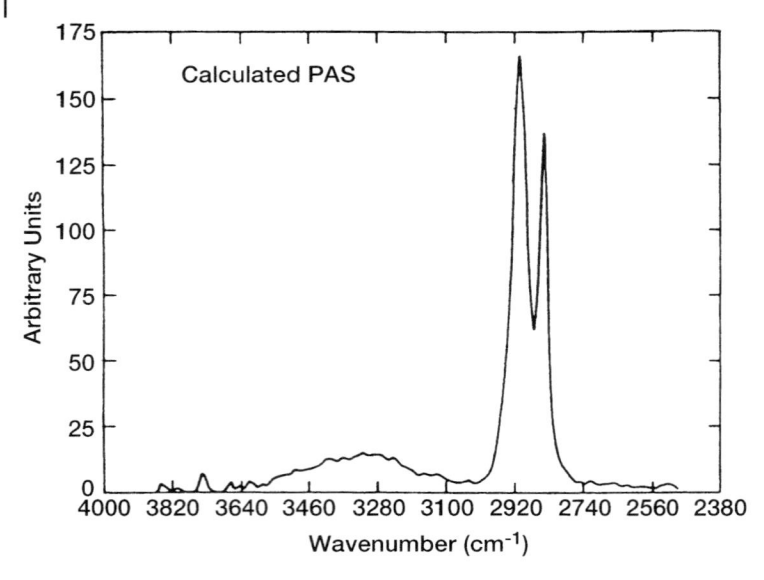

Figure 6.11 Corrected PA infrared spectrum of ethylene-propylene rubber, plotted as the dependence of $\beta\mu$ on wavenumber (Bertrand, 1988).

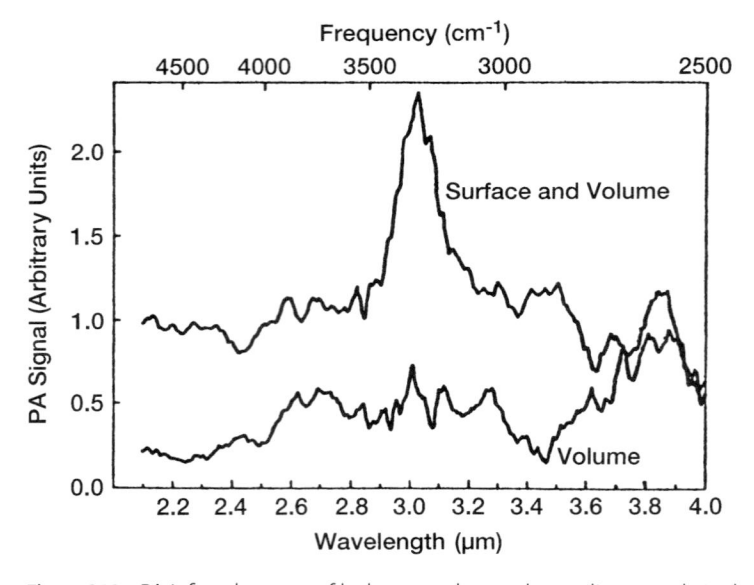

Figure 6.12 PA infrared spectra of hydrogenated amorphous silicon nitride in the N–H$_n$ stretching region. Upper curve, quadrature spectrum; lower curve, in-phase spectrum.

is prominent in the surface-sensitive (quadrature) spectrum of plasma-deposited hydrogenated amorphous silicon nitride (a-SiN$_x$:H) but virtually undetectable in the volume (bulk) spectrum. This clear distinction between the composition of the surface and bulk of the sample is consistent with the conclusion that the surface contaminants are localized in a layer having a thickness of about 50–200 Å. Similarly, Raveh et al. (1992) observed features due to OH and C=O groups in the quadrature spectrum of an amorphous carbon film grown in a microwave plasma. These oxygen-containing species were known to occur mostly near the film surface. The same bands were not detectable in the in-phase spectrum.

Finally, the phase analysis method was used in conjunction with several other techniques to study graphite-epoxy laminates by Dubois et al. (1994). The spectra in this paper differ in a fundamental way from those discussed in the previous paragraph: optical penetration depth, rather than PA intensity, is plotted as the dependent variable. The authors compared results from phase analysis with those obtained by two other techniques, specifically the use of two rapid-scan mirror velocities and transmission. Good agreement was observed between the phase analysis results and transmission measurements. This is another indication of the reliability of the phase analysis method.

References

Bertrand, L. (1988) Advantages of phase analysis in Fourier transform infrared photoacoustic spectroscopy. *Appl. Spectrosc.*, **42** (1), 134–138.

Birch, J.R. (1980) Reduction of the noise component of spectra derived from double-sided Fourier transform spectrometry. *Infrared Phys.*, **20** (5), 349–350.

Bordeleau, A., Bertrand, L., and Sacher, E. (1987) Photoacoustic infrared identification of N–H$_n$ vibrations in hydrogenated amorphous silicon nitride. *Spectrochim. Acta, Part A*, **43** (9), 1189–1190.

Bordeleau, A., Rousset, G., Bertrand, L., and Crine, J.P. (1986) Water detection in polymer dielectrics using photoacoustic spectroscopy. *Can. J. Phys.*, **64**, 1093–1097.

Bosch, R.A. and Julian, R.L. (2002) Reducing the sensitivity of Fourier transform infrared spectroscopy to line-frequency source variations. *Rev. Sci. Instrum.*, **73** (3), 1420–1422.

Burggraf, L.W. and Leyden, D.E. (1981) Quantitative photoacoustic spectroscopy of intensely light-scattering thermally thick samples. *Anal. Chem.*, **53** (6), 759–764.

Carter, R.O., III (1992) The application of linear PA/FT-IR to polymer-related problems. *Appl. Spectrosc.*, **46** (2), 219–224.

Carter, R.O., III (1997) Ultraviolet photoacoustic spectroscopy evaluation of the distribution ultraviolet absorber of paint additives as a result of processing. *Opt. Eng.*, **36** (2), 326–331.

Choquet, M., Rousset, G., and Bertrand, L. (1985) Phase analysis of infrared fourier transform photoacoustic spectra. *Proc. SPIE Int. Soc. Opt. Eng.*, **553**, 224–225.

Choquet, M., Rousset, G., and Bertrand, L. (1986) Fourier-transform photoacoustic spectroscopy: a more complete method for quantitative analysis. *Can. J. Phys.*, **64** (9), 1081–1085.

Dittmar, R.M., Palmer, R.A., and Carter, R.O., III (1994) Fourier transform photoacoustic spectroscopy of polymers. *Appl. Spectrosc. Rev.*, **29** (2), 171–231.

Dubois, M., Enguehard, F., Bertrand, L., Choquet, M., and Monchalin, J.-P. (1994) Optical penetration depth determination in a graphite-epoxy laminate by photoacoustic F.T.I.R. spectroscopy. *J. Phys. IV*, **C7** (Suppl.), 377–380.

Gajić, R. and Merkle, M.J. (1988) Signal averaging in Fourier-transform spectroscopy (two-sided interferogram). *Infrared Phys.*, **28** (5), 333–335.

Jiang, E.Y., McCarthy, W.J., and Drapcho, D.L. (1998) FT-IR PAS: a versatile tool for spectral depth profiling chemical analysis. *Spectroscopy*, **13** (2), 21–40.

McClelland, J.F. (1983) Photoacoustic spectroscopy. *Anal. Chem.*, **55** (1), 89A–105A.

McClelland, J.F., Jones, R.W., and Bajic, S.J. (2002) Photoacoustic spectroscopy, in *Handbook of Vibrational Spectroscopy* (eds J.M. Chalmers and P.R. Griffiths), John Wiley & Sons, Ltd, Chichester, pp. 1231–1251.

Marchand, H., Cournoyer, A., Enguehard, F., and Bertrand, L. (1997) Phase optimization for quantitative analysis using phase Fourier transform photoacoustic spectroscopy. *Opt. Eng.*, **36** (2), 312–320.

Mertz, L. (1965) *Transformations in Optics*, John Wiley & Sons, Inc., New York.

Michaelian, K.H. (1987) Signal averaging of photoacoustic FTIR Data. I. Computation of spectra from double-sided low resolution interferograms. *Infrared Phys.*, **27** (5), 287–296.

Michaelian, K.H. (2007) Invited article: linearization and signal recovery in photoacoustic infrared spectroscopy. *Rev. Sci. Instrum.*, **78** (5), 051301 (12 pages).

Michaelian, K.H. and Birch, J.R. (1991) Signal averaging of photoacoustic FT-IR data – II. Incoherent co-adding and interferogram symmetrization. *Infrared Phys.*, **31** (5), 527–537.

Michaelian, K.H., May, T.E., and Hyett, C. (2008) Photoacoustic infrared spectroscopy at the Canadian Light Source: commissioning experiments. *Rev. Sci. Instrum.*, **79** (1), 014903 (5 pages).

Monchalin, J.-P., Bertrand, L., Rousset, G., and Lepoutre, F. (1984) Photoacoustic spectroscopy of thick powdered or porous samples at low frequency. *J. Appl. Phys.*, **56** (1), 190–210.

Mongeau, B., Rousset, G., and Bertrand, L. (1986) Separation of surface and volume absorption in photothermal spectroscopy. *Can. J. Phys.*, **64** (9), 1056–1058.

Pichler, A. and Sowa, M.G. (2004) Using the linearization approach for synchronizing the phase of photoacoustic reference and sample data. *Appl. Spectrosc.*, **58** (10), 1228–1235.

Raveh, A., Martinu, L., Domingue, A., Wertheimer, M.R., and Bertrand, L. (1992) Fourier transform infrared photoacoustic spectroscopy of amorphous carbon films, in *Photoacoustic and Photothermal Phenomena III* (ed. D. Bićanić), Springer-Verlag, Berlin, pp. 151–154.

Sowa, M.G. and Mantsch, H.H. (1994) FT-IR step-scan photoacoustic phase analysis and depth profiling of calcified tissue. *Appl. Spectrosc.*, **48** (3), 316–319.

7
Applications

7.1
Carbons

The measurement of infrared spectra of carbons clearly presents a nontrivial challenge to the spectroscopist. Dispersion of a finely divided carbon (or, more generally, a carbonaceous solid) in an infrared-transparent diluent so as to record a transmission or diffuse reflectance spectrum may be successful; however, preparation of a well-mixed sample can sometimes be difficult or even impossible. PA spectroscopy enables the acquisition of useful data for neat carbons while eliminating the need for sample preparation. Therefore, it is not surprising to find that a significant number of research teams have successfully used this technique to study carbons during the last three decades. The results of these investigations are summarized in this section.

A general definition of carbons is adopted here. This term is taken to include – in addition to elemental carbon – samples with very high (~80–90%) carbon contents such as chars, soots, and cokes. Since several types of functional groups exist in these substances, viable PA spectra are readily obtainable. The work reviewed below confirms this statement. It should be noted that coals are not included in this discussion; PA infrared spectra of coals are described together with those of other hydrocarbon fuels in the next section of this chapter.

7.1.1
Research of Low's Group

Among the researchers who have investigated the PA infrared spectra of carbons during the last three decades, M. J. D. Low and his coworkers at New York University were the most prolific, with over 35 publications to their credit. This significant body of literature is reviewed in the first part of this section. After this, the findings of a number of other research groups are summarized.

Three papers in the extensive series published by Low's team mentioned the PA infrared spectra of carbons in the context of spectrum normalization (Low and Parodi, 1980b; Low, 1983, 1985). Several chars and soots were examined in this work. A particularly salient point (Low and Parodi, 1980b) is that the results obtained for a given carbon may be qualitatively different according to the wavelength region under consideration. For example, charring sucrose for 1 and 12 h produced carbons that yielded featureless visible and near-infrared PA spectra. However, mid-infrared PA spectra of the 1-h char exhibited bands due to C–O, C=C, C=O, CH$_2$, CH$_3$, and OH groups. The result for the 12-h char exhibited typical changes that accompany aromatization; only the C=C, CH$_2$, and CH$_3$ bands, plus strong aromatic C–H absorption, were identified. Observation of these bands shows that sucrose chars actually contain a variety of functional groups, and therefore are not suitable for normalizing PA mid-infrared spectra. In contrast, soots obtained from benzene, propane, and an ordinary candle gave essentially featureless spectra from 2000 to 4000 cm^{-1}.

The potential use of the spectrum of hexane soot as a reference for normalizing spectra of various solids was also considered by Low (1985). This work contains a rebuttal of the arguments of Rockley et al. (1984), which are discussed further below. Low concluded that the PA spectrum of hexane soot also contains structure, which may be the source of the anomalies observed by Rockley et al. in a comparison of PA and other types of mid-infrared spectra.

PA infrared spectra of carbons were mentioned in a more incidental fashion in another group of papers that dealt primarily with instrumentation and development of the PA infrared technique by Low's group. The earlier publications in this series were discussed in Chapter 2 (Low and Parodi, 1978, 1980a, 1980b, 1980c). In contrast with the perspective described in the preceding paragraph, the possible presence of functional groups in the carbons was not considered explicitly in this second group of articles. Instead, charcoal was used simply to obtain a background spectrum that could be utilized to correct the PA spectra of other samples. Low and Lacroix (1982) also utilized charcoal as a reference in photothermal beam deflection spectroscopy (PBDS) experiments, assuming that the resulting spectrum was a realistic representation of instrument throughput. Similarly, charcoal was mixed with KBr and pressed into a pellet for use as a reference in PBDS experiments on solids submerged in liquids (Varlashkin and Low, 1986). The PA spectrum of the layer of black paint on an infrared detector was found to differ from the usual 'empty instrument' spectrum obtained using the detector itself (Low, 1984). The dependence of the PA spectra on mirror velocity in this investigation was attributed to varying sampling depths. A similar effect was thought to have influenced the PA data of Riseman and Eyring (1981) on carbons, which are discussed below.

In addition to the work described in the preceding paragraphs, Low's group published a series of 22 related articles, variously entitled 'IR studies of carbons', 'Infrared studies of carbons', and 'Spectroscopic studies of carbons' between 1983 and 1991. The bibliographic data for these publications, and the topics discussed, are listed in Table 7.1. These articles are also included in the reference list at the end of this chapter. Unlike many of the other works on the PA spectra of carbons published by these researchers, these publications mostly involved the traditional use of infrared spectroscopy for identification of functional groups; normalization of PA spectra was hardly discussed. The PBDS technique was used to obtain the results in these articles.

Table 7.1 Topics discussed in 'Infrared Studies of Carbons' and 'Spectroscopic Studies of Carbons', a series of PBDS investigations published by Low et al. between 1983 and 1991.

Article number	Publication	Topic
1	Carbon, **21**, 275–281	Cellulose, chromatographic carbon
2	Carbon, **21**, 283–288	Pyrolysis of cellulose
3	Carbon, **22**, 5–12	Oxidation of cellulose chars
4	Carbon, **23**, 301–310	Pyrolysis of oxidized cellulose
5	Carbon, **23**, 311–316	Effect of NaCl on cellulose pyrolysis
6	Carbon, **23**, 335–341	Effect of $KHCO_3$ on cellulose pyrolysis
7	Carbon, **23**, 525–530	Pyrolysis of phenol-formaldehyde resin
8	Langmuir, **1**, 320–326	Oxidation of phenol-formaldehyde chars
9	Mater. Chem. Phys., **20**, 123–144	Pyrolysis of polyvinyl chloride
10	Structure and Reactivity of Surfaces, pp. 601–609	Medium-temperature chars
11	Mater. Chem. Phys., **23**, 499–516	Pyrolysis of polyvinyl bromide
12	Carbon, **28**, 529–538	Polycarbonate resin chars
13	Carbon, **28**, 855–865	Oxidation of polycarbonate resin chars
14	Mater. Chem. Phys., **25**, 501–521	Pyrolysis of polyvinyl fluoride
15	DOE/PC/79920–10, pp. 1–14	Pyrolysis of lignin
16	Polym. Degrad. Stab., **32**, 331–356	Carbonization of diallyl-diglycol- polycarbonate
17	Mater. Chem. Phys., **26**, 193–209	Pyrolysis of polyvinylidene fluoride
18	Mater. Chem. Phys., **26**, 117–130	Rice hull char
19	Mater. Chem. Phys., **26**, 465–481	Sucrose char
20	Mater. Chem. Phys., **27**, 155–179	Pyrolysis of polyvinylidene chloride
21	Mater. Chem. Phys., **27**, 359–374	Coconut shell char
22	Mater. Chem. Phys., **28**, 9–31	Oxidation of polyvinyl halide chars

The first work on 'IR studies of carbons' (Low and Morterra, 1983) contains a useful literature review on the absorption of infrared radiation by carbons that covers about 50 publications. Samples studied in this work included cellulose pyrolyzed in vacuum at temperatures up to 600 °C, as well as a series of chromatographic carbons. The latter are noteworthy because of their high aromaticity, indicated by the occurrence of relatively strong aromatic C–H bands near 3050 cm^{-1}. In fact, both pyrolyzed cellulose samples and the chromatographic carbons yielded

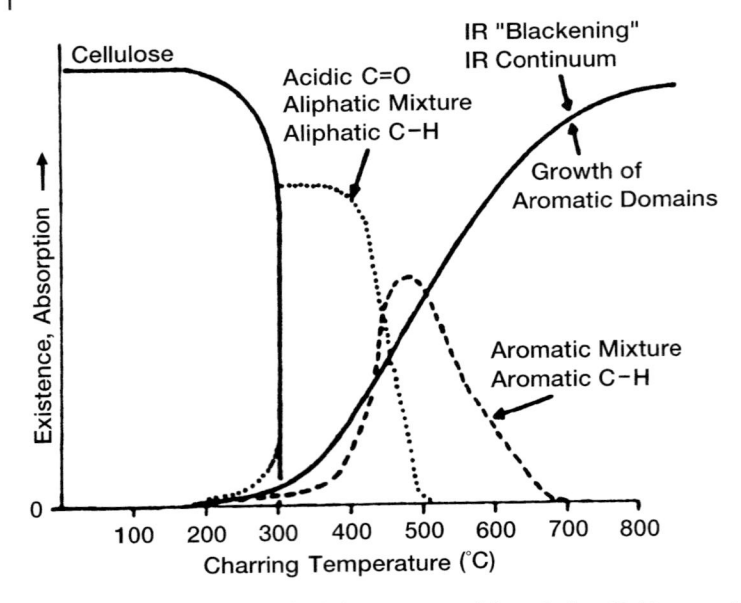

Figure 7.1 Decomposition of cellulose. (Reprinted from *Carbon* 21, Morterra, C. and Low, M.J.D., IR studies of carbons-II. The vacuum pyrolysis of cellulose, 283–288, copyright © 1983, with permission from Elsevier Science.)

spectra displaying a significant number of bands, affirming the general statement that the PA infrared spectra of carbons often contain substantial information regarding particular chemical groups.

The next four studies in this series discussed experiments that began with the pyrolysis of cellulose and continued with oxidation studies of the produced chars. In the first article, PBDS was used to follow the vacuum pyrolysis of cellulose at temperatures that ranged above 700 °C (Morterra and Low, 1983). The spectra of the pyrolysis products showed that aliphaticity is maintained up to about 500 °C, whereas aromatization is observed initially near 300 °C and then increases rapidly at higher temperatures. The increase in aromaticity is indicated by a reduction in the intensities of the aliphatic C–H stretching bands and concomitant appearance of bands due to aromatic C–H stretching and bending modes in the spectra. The thermal decomposition of cellulose is depicted schematically in Figure 7.1.

The cellulose chars prepared by pyrolysis were subsequently oxidized under conditions that led to the formation of either acidic or basic carbons (Morterra, Low, and Severdia, 1984). Acidic carbons were prepared through oxidation at temperatures up to about 530 °C, while basic carbons required higher temperatures. Interestingly, bands due to oxygen-containing functional groups did not occur in the spectra of the products of the latter process.

The next publication (Morterra and Low, 1985a) described the pyrolysis of NO_2-oxidized cellulose (NOC). PA spectra dramatically showed the successive changes that occurred at each stage of the pyrolysis (Figure 7.2); this includes aromatization

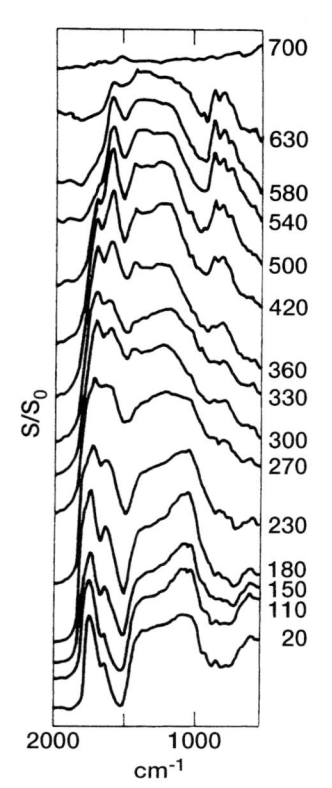

Figure 7.2 PBD spectra depicting the pyrolysis of cellulose. The numbers on the right-hand side of the figure give the pyrolysis temperatures in °C. (Reprinted from *Carbon* **23**, Morterra, C. and Low, M.J.D., IR studies of carbons-IV. The vacuum pyrolysis of oxidized cellulose and the characterization of the chars, 301–310, copyright © 1985, with permission from Elsevier Science.)

at temperatures above 300 °C, as well as the eventual disappearance of all functionality at 700 °C. The spectra obtained in this work showed that NOC degrades more readily than cellulose, yielding chars containing different functional groups at temperatures up to 500 °C.

Alkali metal compounds are known to facilitate carbonization of cellulose and oxidation of the chars. Accordingly, Low and Morterra (1985) studied the pyrolysis of cellulose containing NaCl. The resulting chars were oxidized and compared to those obtained from pure cellulose. NaCl was found to hasten decomposition and change the aliphatic/aromatic distribution in the residues. Pyrolysis at low to moderate temperatures (up to 450 °C) led to oxidation and was followed by aromatization; oxygen was eventually eliminated at higher temperatures. At 550 °C most of the infrared bands diminished in intensity, and at 650 °C only an absorption continuum remained. By contrast, aromatic groups persisted up to 600 °C in the absence of NaCl.

The presence of $KHCO_3$ during pyrolysis and oxidation also caused significant changes with respect to the reactions in pure cellulose; the temperature at which cellulose structure disappeared was reduced and a higher degree of carbonization was achieved. No aromatic C–H groups were detected when cellulose was pyrolyzed with $KHCO_3$ (Morterra and Low, 1985b).

Two articles in this series dealt with phenol-formaldehyde resins. Morterra and Low (1985c) found that resins produced in the presence of Novolac (an acid catalyst) in either vacuum or a nitrogen atmosphere yielded the same pyrolysis results; branching and cross-linking occurred near 350 °C, where diphenyl ethers were formed. Aryl-aryl ethers were detected at higher temperatures. The aliphatic bridges between the aromatic rings started to break down above 500 °C. Polyaromatic domains were formed, which subsequently exhibited oxidation behavior similar to that of chars derived from other precursors (Morterra and Low, 1985d). Oxidation of the low-temperature chars led to the formation of benzophenones and carboxylic acids.

The pyrolysis of several polymers that lack oxygen was also investigated. Morterra, O'Shea, and Low (1988) showed that the pyrolysis of polyvinyl chloride (PVC) exhibited many similarities with the processes described in their preceding studies. The fingerprint region of the spectra showed that the pyrolysis of PVC occurs in the following steps: (i) Cl elimination, (ii) formation of alkenic structures, (iii) alkane formation, and (iv) aromatization and polyaromatization. The appearance of the seemingly ubiquitous '1600-cm^{-1} band' in the spectra of samples that lack oxygen apparently confirms that this feature is due to C=C, rather than C=O, groups. This point is taken up again later in this discussion.

O'Shea, Low, and Morterra (1989) extended this work to include polyvinyl bromide (PVBr). The pyrolysis mechanism was similar to that in PVC, although the formation of polyenic structures occurred at much lower temperatures for PVBr. Next, a similar study showed that polyvinyl fluoride (PVF) is more stable than the other two polymers, with the first evidence of degradation occurring above 300 °C (O'Shea, Morterra, and Low, 1990a). The oxidation of all three polyvinyl halide chars was subsequently investigated (O'Shea, Morterra, and Low, 1991a). Low-temperature chars – where the extent of polyaromatization was quite limited – were more susceptible to oxidation, yielding carboxylic acids, anhydrides and lactones.

In addition to the polyvinyl halides, these authors studied the behavior of two polyvinylidene halides. The onset of degradation in polyvinylidene fluoride (PVDF) was near 300 °C. This process eventually yielded a pyrolytic residue containing both aliphatic and fluoroaromatic structures (O'shea, Morterra, and Low, 1990b). Polyvinylidene chloride was investigated both as a homopolymer and in the form of a copolymer with PVC (O'Shea, Morterra, and Low, 1991b). The spectra for the copolymer contained discrete bands after high-temperature pyrolysis, in contrast with the homopolymer. Because the polyvinylidene halides contain much less hydrogen than the corresponding polyvinyl halides, C–H bands disappear sooner during the pyrolysis of the former compounds.

The 1600-cm^{-1} band in the infrared spectra of many different carbons was mentioned above. Low and Glass (1989) discussed this band in detail, emphasizing the need for caution when comparing spectra of different carbons. Despite the appearance of this band in the infrared spectra of many different types of carbons, one should not assume that a common origin obtains in all cases. Indeed, both the location and the profile of the ~1600-cm^{-1} band are observed to vary in infrared spectra of various carbons.

The 1600-cm^{-1} band in PA spectra of high-temperature carbons is sometimes thought of as the infrared analog of the Raman 'G' (graphite) band. The reader may already know that the G band is accompanied by a 'D' (defect) band at about 1350 cm^{-1} in the Raman spectra of many carbonaceous solids. In their analysis of the infrared spectra of medium-temperature chars, Low and Morterra (1989) noticed that a dip (valley) near 1350 cm^{-1} gradually filled in as the pyrolysis temperature was progressively raised. The authors showed that numerical addition of a broad 1350-cm^{-1} Gaussian band to the PBDS spectrum of the 480 °C char yielded model spectra similar to those obtained for higher-temperature samples. This approach was reiterated by Wang and Low during their investigations of lignin (1989) and sucrose (1990b). This model suggests a possible link between the Raman and infrared spectra of carbons and demonstrates the complementarity of the two techniques.

Two studies in this series consider the vacuum pyrolysis of bisphenol-A polycarbonate resins and the subsequent oxidation of the chars formed in this process (Politou, Morterra, and Low, 1990a, 1990b). These thermoplastic resins are of particular interest because of their electrical and mechanical properties and chemical inertness. Pyrolysis at 420–490 °C produced the most dramatic changes, such as the formation of ester, diaryl ether, and unsaturated hydrocarbon bridges. Aromatic ring structures were formed near the upper end of this temperature range and persisted up to at least 700 °C. At still higher temperatures, the familiar continuum due to electronic absorption was observed. Oxidation of the low-temperature chars yielded products similar to those described above for the phenol-formaldehyde resins. In related work, Politou, Morterra, and Low (1991) described the spectra of pyrolyzates produced from diallyl-diglycol-polycarbonate. Low temperature oxidation of this aliphatic thermosetting resin was found to be necessary before it would char. Pyrolysis then yielded products resembling intermediate-temperature carbons.

A number of earlier articles in this series described pyrolysis of cellulose (see Table 7.1). Lignin, another important component of wood, was also studied by this group of researchers. Wang and Low (1989) showed that lignin pyrolysis occurs in stages that can be described as depolymerization, aromatic ring consolidation, and polyaromatic domain creation. These chemical changes take place more gradually during cellulose pyrolysis, where larger amounts of carbonyls are formed.

Three of the later works in this series describe the charring of food products. Wang and Low (1990a) discussed studies of rice hulls, where the products reflect the presence of both aromatics and aliphatics (derived from lignins and cellulose, respectively). Silica is the other major product. In one of the companion investigations, Wang and Low (1990b) described the charring of sucrose. Although the original sucrose spectrum contains a significant amount of detail, even the mild heating employed in the preparation of confectionery products leads to a noticeable simplification of the spectrum. The charring of sucrose proceeds in several steps: (i) conversion from aliphatic to aromatic material, (ii) dehydroxylation, (iii) elimination of aliphatic groups, (iv) loss of carbonylic species, and (v) formation of polyaromatic structures. Finally, the third article (Wang and Low, 1991) described the charring of coconut shell. At temperatures up to 300 °C, the observed changes were mainly textural. By contrast, the mixture was converted from aliphatic to aromatic character at 300–500 °C. Above 550 °C, the material became predominantly polyaromatic. The original coconut shell yielded an infrared spectrum similar to that of a hardwood, exhibiting bands due to cellulose, hemicellulose, and lignin.

Several other important works on the PA infrared spectra of carbons were published by Low and his colleagues at about the same time as the series of articles summarized above. For example, Morterra and Low (1982) studied the ~1600-cm^{-1} band that was discussed earlier. A band at this position might be thought to arise from a highly conjugated carbonyl group or, alternatively, a carbon-carbon double bond or an aromatic ring. To evaluate these possibilities, the authors analyzed a commercial carbon and a cellulose that had been charred in a vacuum. The cellulose char was subsequently oxidized at 450 and 600 °C in $^{16}O_2$ and $^{18}O_2$. The occurrence and intensity of the 1600-cm^{-1} band depended on the presence and concentration of surface oxidic species, but the location of this feature did not depend on the isotope. By contrast, a neighboring C=O band exhibited an isotope shift from 1760 to 1725 cm^{-1}, as might be expected from the difference in oxygen mass. These results led to the conclusion that the 1600-cm^{-1} band arises from aromatic C=C groups, and that an oxidized layer cross-linking aromatic chains causes the asymmetry necessary for the appearance of this Raman-active vibration in the infrared. In contradistinction with this model, the 1600-cm^{-1} band is also observed for hydrocarbon cokes that contain virtually no oxygen (see below). Moreover, this band was observed in spectra of polyvinyl halides that lack oxygen by Low's group. In these cases Morterra and Low's proposal regarding an oxidized layer cannot be invoked, and another potential reason for the occurrence of the 1600-cm^{-1} band in the PA infrared spectra must be identified.

As this research progressed, PBDS was used to study several other carbonaceous materials. Morterra and Low (1985e) reported spectra of wood charcoal, coals, and chromatographic carbons; the results for the coals are discussed in the next section of this chapter. Low and Morterra (1986) discussed charcoal, chromatographic carbon, and a phenol-formaldehyde resin. These samples were heated in vacuum and then oxidized. For charcoal and carbons, trends similar to those already described were observed; carbonization caused a progressive reduction in the aromatic C–H bands between 700 and 900 cm^{-1} and the appearance of both C=O and C–O bands. At higher temperatures the absorption continuum intensified and all bands due to specific groups eventually disappeared. This continuum may obscure the weaker absorption bands of some of the functional groups, the electronic absorption limiting the penetration of the infrared radiation into the sample.

In work related to the charring of sucrose and coconut shell described above, Low and Wang (1990) studied these processes at temperatures up to 800 °C. This investigation showed that chars heated to 450 °C exhibited similar spectra, displaying aromatic C–H bands between 700 and 900 cm^{-1}, a minimum near 1350 cm^{-1}, and bands at 1600 and 1700 cm^{-1}. Further heating to 600–650 °C caused the 1350-cm^{-1} dip to fill in, as described above. Although Low and Wang did not obtain

Raman spectra of the pyrolyzates – which might have confirmed their suggestion that the latter effect was due to infrared activation of the 1350-cm^{-1} Raman band – their proposal remains an interesting possibility that could apply to infrared spectra of other carbons.

The final works from this impressive body of research to be mentioned here describe 'unusual' bands between about 2050 and 2250 cm^{-1} in infrared spectra of chars. Most vibrational spectroscopists will realize that only a few types of functional groups produce bands in this region. Low *et al.* (1990) suggested that bands in their spectra were due to nitrogen-containing groups such as nitriles or isonitriles. The possible role of these compounds in charring could be of considerable significance. Similar bands were observed in spectra of oxidized coals and coal chars (Low, 1993).

7.1.2
Other Groups

Research on PA infrared spectra of carbons carried out by other groups is summarized here. This work began at about the same time that the first investigations in the previous section were carried out and spans a period of more than two decades.

Riseman and Eyring (1981) obtained spectra of a series of eight commercially available carbons during an investigation on the normalization of PA spectra. This research demonstrated the dependence of carbon spectra on mirror velocity in rapid-scan PA FT-IR spectroscopy. Because the spectra were saturated, bands due to specific functional groups were not identifiable in these spectra. Moreover, the different carbons tended to produce very similar spectra at each velocity. Although these spectra were essentially featureless, Riseman and Eyring emphasized the important fact that the PA signal passes through successive stages of saturation as f changes. Because f varies by a factor of 10 across the mid-infrared region in a rapid-scan experiment, the low-wavenumber region of the PA spectrum of a carbon may be fully saturated while the high-wavenumber region is only partially saturated.

Carbonaceous refractory gold ore contains organic carbon constituents that interact with the gold and make the ore non-amenable to conventional cyanidation. Nelson *et al.* (1982) studied oxidized and carbonaceous Carlin gold ores using PA infrared and EPR spectroscopies. The infrared spectra of the two ores were similar and, moreover, bore no resemblance to the spectrum of humic acid that was also obtained in this work. In fact, no hydrocarbons were detected in the ores. The organic carbon in the ore was concluded to be some type of activated carbon.

Rockley *et al.* (1984) compared PA, attenuated total reflectance (ATR), diffuse reflectance, and transmission infrared spectra of a carbon black (soot) prepared by combusting hexane. Elemental analysis yielded a composition of 94.8% carbon, 1.2% hydrogen, and 2.4% oxygen by weight; hence, spectroscopic evidence for the existence of several functional groups was anticipated. Indeed, transmission and ATR spectra contained bands due to aromatic C–H and C=C groups between 600 and 2000 cm^{-1}. However, the diffuse reflectance infrared Fourier transform (DRIFT) and PA infrared spectra displayed no reliable features, and the PA spectrum of soot deposited on an NaCl window differed from that of soot alone. A definitive explanation for the PA results was not given, and it was concluded that the PA spectrum of carbon black should be used only as a reference. This reasoning was questioned by Low (1985), whose arguments were summarized in the previous section.

Papendorf and Riepe (1989) carried out detailed surface studies of activated carbons, using PA infrared spectroscopy to characterize adsorbed chlorinated hydrocarbons. Consistent with the widespread use of carbon black to obtain background (reference) PA spectra, a spectrum of adsorbate-free carbon was employed to correct the spectra of adsorbates on carbons. Spectra were obtained for carbons coated with tri/perchloroethylene, 1,1,1-trichloroethane, 1,1,2,2-tetrachloroethane, and pentachloroethane. The concentrations of these substances on the carbons varied between 15 and 40%. Up to five bands were observed in the 700–1000 cm^{-1} region; the results resembled the spectra of the substances taken from a reference library. The characteristic C–Cl bands disappeared at concentrations below about 7%, indicating a rather high detection limit for these compounds by this technique.

As noted above, finely divided carbons are commonly used to normalize PA infrared spectra. This approach is, of course, partly based on their very strong absorption of infrared radiation. A related question concerns the maximum concentration of carbon that a sample may contain before it exhibits PA saturation.

This issue was investigated by Carter *et al.* (1989). These authors found that six carbon blacks with different particle sizes and reactivities gave essentially the same, featureless, PA spectra. These carbons were also incorporated into vulcanized natural rubber at a concentration of 8.5% by weight. PA spectra of these samples exhibited several bands due to the rubber and showed minor differences from one sample to the next. One rubber was also studied at progressively higher carbon concentrations. A significant number of bands were clearly visible in the PA spectra of samples with carbon contents up to 8%. On the other hand, only a few features remained at 15%, and a carbon concentration of 25% produced a PA spectrum that resembled a blackbody curve. The latter spectrum could therefore be used as a reference to normalize sample spectra (Carter and Wright, 1991); indeed, some PA practitioners advocate the use of carbon-filled rubber for the acquisition of reference spectra. The results of Carter *et al.* (1989) also suggest an estimate of the maximum amount of carbon that is acceptable in samples that are to be analyzed by PA infrared spectroscopy. However, as is shown below, far higher percentages of carbon can be tolerated in the study of hydrocarbon cokes.

PA infrared spectra of amorphous hydrogenated carbon films (a:C–H) were reported by Raveh *et al.* (1992). These films were grown from a methane plasma under microwave or radio-frequency discharge. The spectra were used to elucidate bonding, which is correlated with physical properties such as density and microhardness. In-phase and quadrature PA spectra (Chapter 6) showed that the bonded oxygen occurs near the film surface. PA spectra were generally superior to transmission spectra, exhibiting more intense and well-resolved bands due to C=O, CH_3, CH_2, C–C and CH groups. The microwave films were polymer-like, with lower hardness and higher hydrogen content.

Diamond is another form of carbon that has been investigated by means of PA spectroscopy. Ando et al. (1993) employed PA and diffuse reflectance spectra in a comparison of non-hydrogenated and plasma hydrogenated synthetic diamond powder. Hydrogenation produced surface CH_2 and CH_3 functionality. PA infrared spectra of small (0.5–2 μm) particles displayed several well-defined bands, but little structure was visible in the spectra of larger particles.

Yang and Simms (1993, 1995) utilized PA infrared spectroscopy to study carbon fibers derived from petroleum pitch. The production of these fibers requires several steps: (i) formation of a precursor (green) fiber; (ii) stabilization of the precursor fiber; and (iii) carbonization. PA spectroscopy offers a considerable advantage with regard to the characterization of these fibers because it eliminates the need for traditional sample preparation. For example, transmission spectra of carbon fibers had previously been obtained after they were ground and then incorporated into KBr pellets. As an alternative, a fiber was dissolved in an organic solvent, after which the solution was deposited onto an infrared-transparent window and evaporated to dryness! It need hardly be stated that the elimination of such invasive procedures is highly desirable.

The PA spectra of these fibers resemble those of high-rank coals. Specifically, the fibers exhibit features from C=O, O–H, and both aliphatic and aromatic C–H groups. The stabilization process occurs at high temperatures and involves oxidation: this is confirmed by the occurrence of bands due to ketones, aldehydes, acids, esters and anhydrides (Simms and Yang, 1994). Carbonization at progressively higher temperatures (up to 900 °C) eventually led to the elimination of all functionality and the observation of a nearly featureless spectrum.

Figure 7.3 shows typical data obtained by Yang and Simms (1995). Curve A, the PA infrared spectrum of a green fiber, clearly displays several bands attributable to the groups mentioned above. After stabilization, a much simpler spectrum (curve B) was obtained. Grinding of the stabilized powder dramatically revealed the difference between the fiber surface and bulk (curve C).

Saab et al. (1998) examined thin films of pure and derivatized C_{60}, which exhibits water-sensitive conductivity. K_3C_{60} films were exposed to water, oxygen, and radiation from a Xe lamp. PA infrared spectra indicated derivatization with hydrogen, hydroxyl, and oxygen groups. Four bands were observed in the C_{60} spectrum, the features at 526 and 580 cm^{-1} being characteristic of the functionalized material. The behavior of the fullerene material is similar to that of oxidized amorphous carbons, which also absorb water vapor.

Bouzerar et al. (2001) used PBDS to study the electronic density of states in the near-infrared and visible regions for amorphous hydrogenated carbon thin films. While no spectra were displayed in this article, variations in the absorption coefficient with photon energy were noted. The data were interpreted with regard to two density of state models, and a connection was established between the bandgap and disorder in the carbon films.

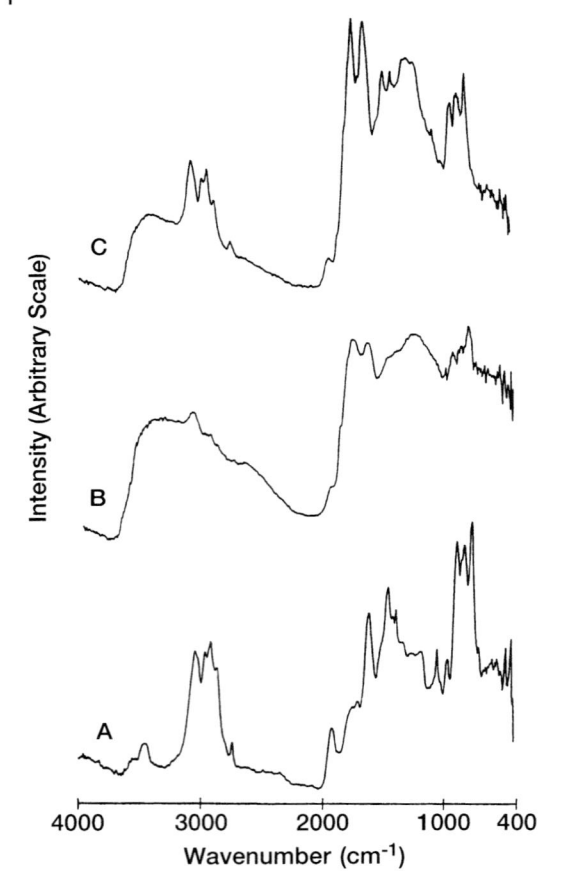

Figure 7.3 PA infrared spectra of carbon fibers. Curve A, green fiber; curve B, stabilized fiber; curve C, stabilized fiber (ground sample). (Reprinted from *Fuel* **74**, Yang, C.Q. and Simms, J.R., Comparison of photoacoustic, diffuse reflectance and transmission infrared spectroscopy for the study of carbon fibres, 543–548, copyright © 1995, with permission from Elsevier Science.)

7.1.3
Hydrocarbon Cokes

This discussion of PA infrared spectra of carbons ends with results for hydrocarbon cokes obtained in the author's laboratory. The cokes were obtained from thermal cokers operated by Syncrude Canada Ltd., and are a byproduct of the process used to create synthetic crude oil from the vast oil sand deposits in northeastern Alberta. Typically, these cokes consist of approximately 90% carbon, 4% sulfur, and residual aromatic hydrocarbons at concentrations up to approximately 1%.

Figure 7.4 shows representative PA spectra recorded for three of these cokes. Aromatic hydrocarbons are obviously detectable in some cokes, but not in others.

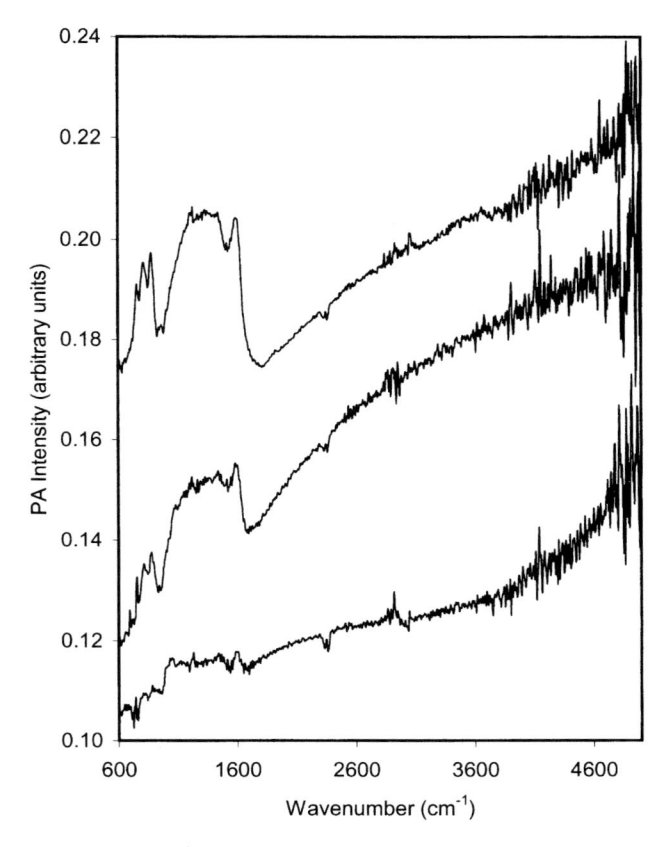

Figure 7.4 PA infrared spectra of three wall cokes. The coke spectra were ratioed against carbon black spectra obtained under similar conditions, and have been slightly offset along the vertical axis for clarity. It should be noted that the background intensity may vary due to differences in the particle sizes of the coke samples.

In the upper spectrum, prominent aromatic C–H bands occur at about 750, 810 and 870 cm^{-1}. The 1600-cm^{-1} peak is due to aromatic C=C stretching, and was discussed above in the context of work by Morterra and Low (1982) as well as Low and Glass (1989). The third indication of aromaticity in these spectra is the absorption continuum that gradually intensifies with increasing wavenumber, extending into the near-infrared: this band is due to electronic transitions. All of these features are less intense in the middle curve, and virtually absent in the lower spectrum. Thus these cokes can be classified with regard to their aromatic hydrocarbon contents by means of PA infrared spectroscopy; a large number of different cokes have been successfully studied in this way. Importantly, increased aromatic hydrocarbon contents are correlated with reduced κ and α values, as well as thermal effusivities, of the cokes (Michaelian, Hall, and Bulmer, 2002a; Michaelian, 2003). These results were interpreted in terms of a model in which the aromatic hydrocarbons occupy pore structures in the cokes, reducing the efficacy of heat propagation.

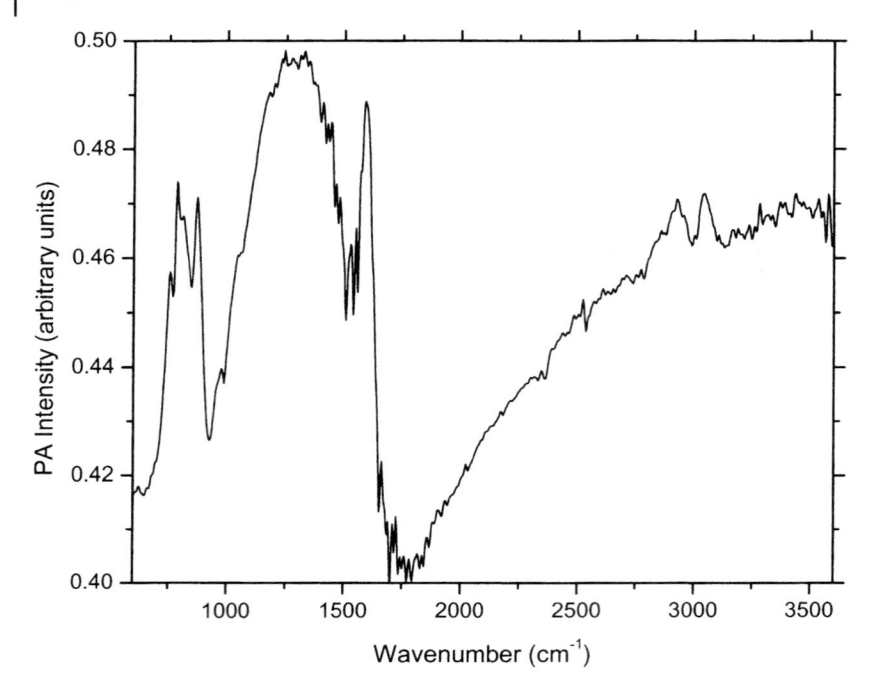

Figure 7.5 PA infrared spectrum of a different wall coke.

A more impressive PA spectrum, obtained for a different coke, appears in Figure 7.5. In addition to the features described in the previous paragraph, both aliphatic and aromatic C–H stretching bands are clearly visible in this spectrum. Moreover, the aromatic C–H region between approximately 700 and 900 cm^{-1} is observed to comprise at least four bands. These features arise from differently substituted aromatic ring structures, and provide information on the number of contiguous hydrogens in these rings.

7.2
Hydrocarbon Fuels

PA spectra of several hydrocarbon fuels are discussed in this section. Results are presented and reviewed first for coals, which have been analyzed using PA infrared spectroscopy since the emergence of the technique about 30 years ago. Second, liquid (middle distillate) fuels derived from bitumen – an energy source of increasing importance – are described. PA spectroscopy has also been used to analyze unfractionated bitumen and very high-boiling residual hydrocarbons. Asphaltenes, highly aromatic solids that cause catalyst poisoning and other deleterious effects during bitumen upgrading, are also mentioned.

7.2.1
Coals

The preceding discussion of the PA infrared spectra of carbons leads naturally to the subject of coals. This progression would seem logical from the perspective of many spectroscopists, particularly those who choose to describe coals simply as naturally abundant carbonaceous solids. Indeed, the close relationship between carbons and coals is reflected by the words used to refer to coal in some languages: typical examples occur in Spanish (*carbón*), Italian (*carbone*) and French (*charbon*). This link calls to mind the many similarities among the spectra of carbons (as defined in the previous section) and coals. PA infrared spectra of coals and carbons are discussed separately in this book because of the considerable scientific literature, as well as the role of coal as a hydrocarbon fuel.

It should be noted that a large body of literature exists on the general subject of (non-PA) infrared spectroscopy of coals. These publications span the entire period of modern infrared spectroscopy, that is, they cover more than six decades of research in coal science. No attempt is made to summarize this literature here. The interested reader may wish to consult review articles and books on coal characterization for more information. While the present discussion is restricted to the PA infrared spectra of coals, it should be mentioned that many present-day workers utilize the transmission and diffuse reflectance techniques for coal analysis with good success.

7.2.1.1 Early PA Infrared Spectra of Coals

As discussed in Chapter 2, many early successful investigations of PA infrared spectroscopy emphasized its capabilities for the analysis of problematic solid samples. Coals were included among these solids because they often require extensive grinding before they can be dispersed in alkali halides and studied by the classical transmission technique. This grinding of coals is undesirable for several reasons: it is time consuming, leads to heating and oxidation, and may liberate entrained clays and minerals. It is clearly preferable to eliminate (or at least minimize) the use of this sample preparation step when possible. The early accomplishment of this objective and the accompanying measurement of research-quality PA spectra are thus particularly relevant. The investigations of Mead *et al.* (1979), Vidrine (1980), Low and Parodi (1980d), Herres and Zachmann (1984), Larsen (1988), and Low (1986) should all be mentioned in this context. Although the spectra of coals in these articles were mostly not interpreted in detail, the data were certainly of sufficient quality to justify the subsequent use of the technique by other researchers.

Two early publications discussed PA infrared spectra of coals more thoroughly. A pioneering contribution by Rockley and Devlin (1980) compared spectra of aged and freshly cleaved surfaces of sub-bituminous, bituminous, and anthracitic coals. (These terms are arranged in order of increasing rank, or fixed carbon content, of the samples.) C–H stretching, CH_2 deformation and C=C stretching bands due to the coals, as well as several features arising from clay, were identifiable in the PA spectra. The results suggested that volatilization and oxidation occur during the aging process. Similarly, Royce, Teng, and Enns (1980) reported a PA study of coal pyrolysis, and showed that bands due to aliphatic groups diminished in intensity during this process. This behavior resembles that described for carbons in the previous section.

Krishnan (1981) observed the close similarity between the PA and diffuse reflectance spectra of a finely powdered coal sample of unspecified origin. The results in this study demonstrate the sensitivity of PA spectroscopy to the clays that frequently exist in coals, an effect that can be put down to the very small (less than ~2 µm) particle sizes of clays. In another comparison of infrared methods, Solomon and Carangelo (1982) examined PA, diffuse reflectance, and transmission spectra of a Pittsburgh seam coal with regard to their abilities to quantify hydroxyl absorbance; the region above $3500\,cm^{-1}$ in the PA spectrum was partly obscured by bands due to water vapor, hindering the study of the much broader hydroxyl band. A greatly improved spectrum of this sample was subsequently obtained by McClelland (1983).

A more detailed discussion of PA infrared spectra of coals was presented by Zerlia (1985). Semianthracite, medium- and high-volatile bituminous, and subbituminous coals were examined using standard analytical techniques in addition to PA infrared spectroscopy. As shown in Figure 7.6, detailed band assignments were given for most of the infrared features. The ability of infrared spectroscopy to provide relevant technological data on coals was also established in this work. Specifically, the content of volatile matter was shown to vary linearly with the integrated area of the $2800-3800\,cm^{-1}$ region, which includes contributions from bands due to the OH, NH, aliphatic CH_n, and aromatic C–H groups. The correlation between the spectroscopic data and the results from traditional coal analysis is particularly significant in light of the considerable time and effort required for the latter measurements.

The aromatics contents of coals can be compared using the three low-wavenumber C–H bands at approximately 750, 810, and $865\,cm^{-1}$ or, alternatively, the broad band between about 3000 and $3100\,cm^{-1}$. Gerson et al. (1984) obtained PA spectra of a series of eight coals representing different ranks, finding that higher volatile matter contents corresponded to lower aromatic/aliphatic ratios in the C–H stretching region. The same trend was reported by Zerlia (1985). Both findings were consistent with rank determinations based on reflectivity, a widely accepted parameter in coal characterization. To summarize, increased aromatics contents in coals are correlated with higher reflectivity and rank; greater aliphatics contents imply the presence of more volatile matter, and the opposite trends in the first three parameters.

PA infrared spectroscopy can also be used to analyze fly ash, the fine solid waste material produced during coal combustion. Seaverson et al. (1985) obtained spectra of four different coal fly ashes that were known to contain several minerals and water. These components were eliminated by thermal treatment at temperatures above 400 °C. In three fly ash samples, the eliminated inorganic compound was confirmed as $Ca(OH)_2$.

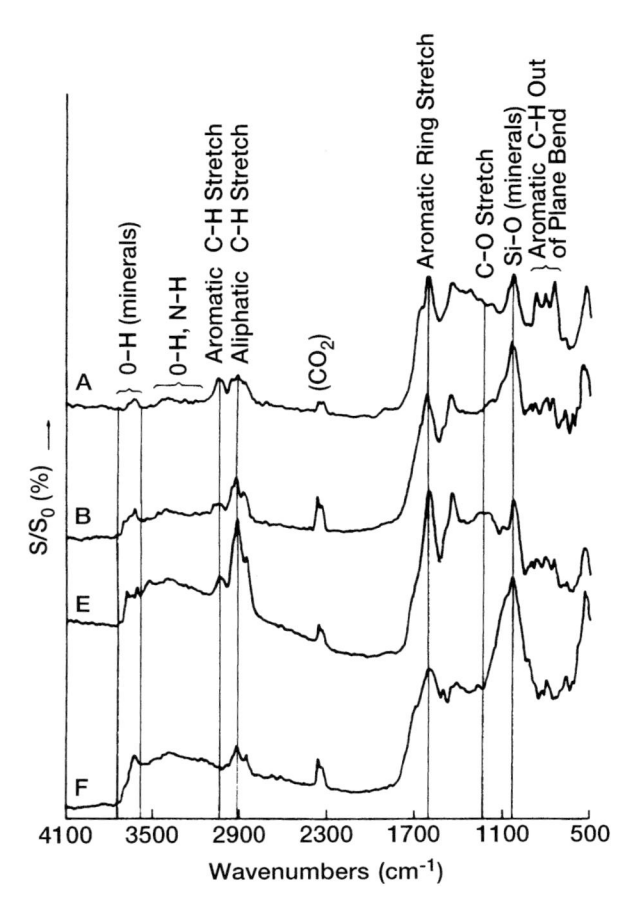

Figure 7.6 PA infrared spectra of four coals. A, semianthracite; B, medium-volatile bituminous; E, high-volatile bituminous; F, sub-bituminous. (Reprinted from *Fuel* **64**, Zerlia, T., Fourier transform infrared photoacoustic spectroscopy of raw coal, 1310–1312, copyright © 1985, with permission from Elsevier Science.)

In a rare report on polarized PA infrared spectroscopy, Cody, Larsen, and Siskin (1989) presented results for medium- and high-rank coals. The authors found little indication of a preferred orientation of the organic components, a result in apparent disaccord with optical anisotropy at visible wavelengths. Anthracite yielded an almost featureless spectrum, a result attributed to partial saturation. Dichroism was observed for coals with carbon contents above about 85%.

7.2.1.2 Near-Infrared PA Spectra of Coals

Near-infrared spectra of many coals exhibit a featureless absorption continuum that intensifies monotonically with wavenumber. This continuum arises from electronic transitions in aromatic structures and is analogous to that observed in the PA spectra of many high-temperature carbons. Moreover, electronic absorp-

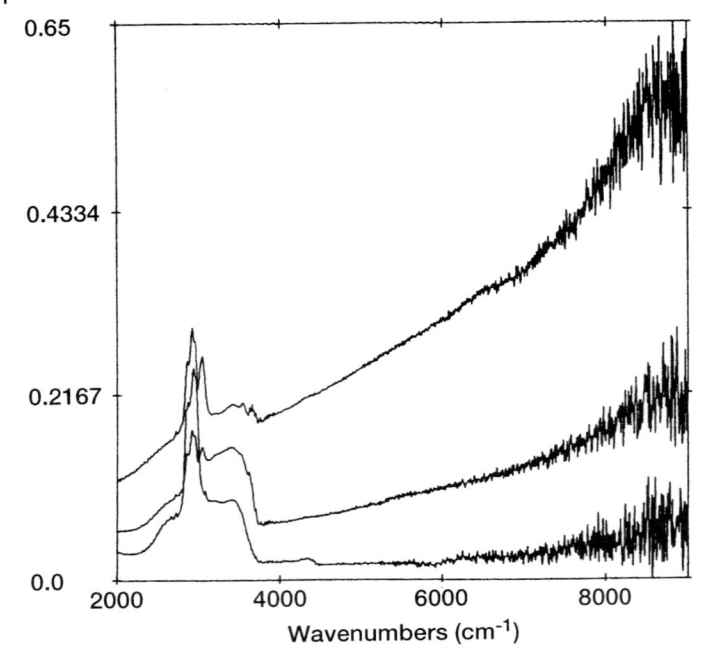

Figure 7.7 Mid- and near-infrared PA spectra of separated macerals from a sub-bituminous Alberta coal. Upper curve, fusinite; middle curve, vitrinite; lower curve, resinite. (Reprinted from *Infrared Phys.* **26**, Donini, J.C. and Michaelian, K.H., Near infrared photoacoustic FTIR spectroscopy of clay minerals and coal, 135–140, copyright © 1986, with permission from Elsevier Science.)

tion is thought to be responsible for part of the underlying intensity that is visible between the more well-defined bands in the mid-infrared spectra. The very broad electronic absorption bands of coals thus extend from the mid-infrared, throughout the near-infrared, and into the visible wavelength regions.

The near-infrared absorption continuum in coal spectra can, in principle, be observed by various means. However, its identification is sometimes complicated by the effects of scattering when techniques such as transmission and diffuse reflectance are employed. PA spectroscopy offers a distinct advantage in this situation, because results can be obtained for various particle sizes, and scattering effects can be isolated. The elimination of diluents in PA near-infrared spectroscopy also simplifies this experiment.

The assignment of the absorption continuum to electronic transitions in aromatic species is consistent with results reported for separated macerals of a sub-bituminous Alberta coal (Donini and Michaelian, 1986). As shown in Figure 7.7, the PA spectrum of fusinite displays the most intense continuum; the relative intensities of the aromatic and aliphatic C–H stretching bands confirm that this is the most highly aromatic of the three macerals. By contrast, the continuum is significantly weaker in the vitrinite spectrum, and practically unidentifiable in resinite. The C–H bands show that the latter maceral consists primarily of aliphatic material.

Three other publications also describe near-infrared PA spectra of coals. McAskill (1987) found that dispersive PA spectra of oil shales and coals of several different ranks (brown, sub-bituminous, and bituminous) all displayed the near-infrared continuum mentioned above. In fact this broad band was considered to be somewhat detrimental, since it tended to obscure narrower vibrational (combination and overtone) bands. This interference was less problematic for brown coals, and more significant for coals of higher rank. Several years later, Michaelian, Ogunsola, and Bartholomew (1995) confirmed that the near-infrared continuum intensified as the ranks of bituminous, sub-bituminous, and lignitic coals were artificially increased by thermal treatment in both oxidizing and inert atmospheres. This feature was also reported by Gentzis, Goodarzi, and McFarlane (1992).

7.2.1.3 Study of Coal Oxidation by PA Infrared Spectroscopy

The oxidation of coals results in a number of important changes to their physical and chemical characteristics. For example, the propensity of a coal for coking, its calorific value, and its floatability are all adversely affected by oxidation. A number of analytical methods – including infrared spectroscopy – have been applied to this subject with varying degrees of success. The use of PA infrared spectroscopy in the study of coal oxidation, which is naturally motivated by both the minimal sample preparation and the possibility of depth profiling, is reviewed in the following paragraphs.

The capability of infrared spectroscopy for detection of oxygen-containing functionality in coals was established decades ago. Thus it is logical that several groups investigated the feasibility of using PA infrared spectroscopy for this particular application when the technique was first gaining acceptance. The PA spectra of coals obtained in several early studies readily displayed a number of changes due to oxidation: these include the elimination of aliphatic and aromatic CH_n groups, as well as the production of anhydrides, esters, carboxylic acids, and other carbonylic species (Hamza, Michaelian, and Andersen, 1983; Chien, Markuszewski, and McClelland, 1985; Chien et al., 1985; Angle, Donini, and Hamza, 1988). It should also be noted that oxidation under laboratory conditions is not necessarily representative of the processes that occur during stockpiling (Mikula, Axelson, and Michaelian, 1985) or natural weathering.

The research of M. J. D. Low's group on the PA infrared spectra of carbons was discussed in detail in the preceding section of this chapter. Several articles published by this group mention coals or coal chars. For example, Morterra and Low (1985) proposed a model in which an anthracitic coal is comprised of a carbonaceous material and a second independent hydrocarbon phase. Oxidation of the hydrocarbon component was thought to yield compounds such as anhydrides. This investigation showed that C=O groups are formed initially in oxidation, with C–O functionality appearing later. The charring of coal yielded a product with a spectrum similar to those for charred wood, oxidized cellulose, and several other similar substances (Low and Morterra, 1989).

As noted in the previous section, Low (1993) observed bands in the 2150–2250 cm^{-1} region in PA spectra of several oxidized coals and coal chars. These bands were attributed to nitrogen-containing groups thought to be present in pyridinic or pyrrolic structures. This proposal accords with the fact that the concentration of nitrogen in the coals was on the order of two per cent. This work is significant because nitrogen-containing heterocycles also occur in other hydrocarbons such as bitumen and heavy oil.

A series of papers on the oxidation and derivatization of coals was published in the 1980s by B. M. Lynch and his collaborators at St. Francis Xavier University and several other institutions. After an evaluation of PA and diffuse reflectance spectroscopies (Lynch et al., 1983), the former technique was chosen for further work. Both naturally weathered and artificially oxidized coals were studied in this research. Carbonyl bands were examined thoroughly, with features at 1650, 1690, 1720 and 1750 cm^{-1} being of particular interest. The intensities of these bands were determined by subtracting PA spectra of fresh coals from those of the corresponding oxidized coals.

A major conclusion of this work pertains to peroxide species, which were proposed as precursors of the oxidation products. The authors showed that peroxides are present on the surfaces of virtually all coals, except for those that are freshly prepared. Carbonyl functionalities were produced by base-promoted and thermal decompositions of the peroxides (Lynch, Lancaster, and Fahey, 1986, 1987a) to yield products thought to be cyclic or open-chain esters. Importantly, total carbonyl intensity varied linearly with the bulk oxygen content (Lynch, Lancaster, and MacPhee, 1987b). Moreover, this intensity is correlated with the so-called plastic properties (Gieseler fluidity, dilatation, and melting range) of the coals (Lynch, Lancaster, and MacPhee, 1988). The findings of this research were reviewed by Lynch and MacPhee (1989). One important observation is that laboratory coal samples should be stored in sealed ampoules to prevent unwanted oxidation.

The oxidation of hand-picked resinites from Western and Arctic Canada and Australia was studied by Goodarzi and McFarlane (1991). PA infrared spectroscopy was found to be particularly well suited for the analysis of naturally exposed and weathered resinite surfaces, irrespective of their morphology. More than 20 bands were observed in the infrared spectra and assigned to various functional groups in the coal or clay that was present in the samples. Resinites are predominantly aliphatic, although oxidation at temperatures above 250 °C leads to partial aromatization (McFarlane et al., 1993). The natural oxidation products depended primarily on the rank of the coal that was the source of the resinite and the degree of weathering, which was influenced by the depositional environment. The surface sensitivity of PA infrared spectroscopy played a key role in this research.

7.2.1.4 Depth Profiling of Oxidized Coals

Oxidized coals are quite amenable to depth profiling studies, since the oxidation process is generally expected to begin at the surface and progress toward the interior of the coal particle. Moreover, differences in composition are expected to occur over dimensions on the order of a few micrometers, which are comparable to typical μ_s values in PA FT-IR spectroscopy. Thus depth profiling experiments that discriminate between the surface and bulk of a coal sample are pertinent to the study of coal oxidation.

Zerlia (1986) carried out rapid-scan depth profiling experiments on a laboratory-oxidized coal sample. Bands due to CH_n and hydroxyl groups exhibited lower intensities at higher mirror velocities, implying that these species were less abundant near the surface of the coal. Because CH_n groups are known to be consumed in oxidation, the results implied that the surface of the coal was oxidized. Different oxidation mechanisms were identified for the interior and surface regions of the relatively coarse (4-mm) coal particles in this study. There was no spectroscopic evidence for the production of several oxygen-containing functional groups (ether, carbonyl, or carboxyl) in the interior of the particles to compensate for the loss of CH_n, as expected if oxidation occurs by oxygen uptake and peroxide formation.

The most prominent change in the infrared spectrum of a coal that accompanies oxidation is often the intensification of a C=O band near 1750 cm^{-1}. As mentioned above, this peak may contain contributions from several carbonylic species, depending on the composition of the coal and the oxidation mechanism. The 1750-cm^{-1} band was investigated in amplitude-modulation depth profiling experiments on fresh and oxidized coals (Michaelian, 1989b, 1991). These studies confirmed that the oxidation of the coal was more extensive at the surface than in the interior of the coal particles. Furthermore, variation of μ_s indicated that the outer 12 µm of the coal particles was oxidized uniformly. These results were discussed in more detail in Chapter 5.

Yamada et al. (1996) performed phase-modulation depth profiling experiments on oxidized coals. At $f = 400$ Hz, the quadrature (surface sensitive) spectrum showed the intensification of the carbonyl band and diminution of C–H stretching intensity that accompany oxidation. These results are consistent with the earlier experiments described above. As the phase angle was reduced and greater depths were examined the change in the C=O band persisted, while that for the C–H bands did not. On the other hand, the change in C–H intensity remained at greater depths for $f = 100$ Hz. These results suggested an oxidation mechanism in which dehydrogenation occurs mainly in the surface layer of the coal. Alternatively, aliphatic compounds could migrate more deeply inside the coal during oxidation.

7.2.1.5 Numerical Analysis of PA Spectra of Coals

PA infrared spectra of coals have also been used to demonstrate numerical methods such as phase correction, signal averaging, curve fitting and deconvolution. The publications on these subjects generally emphasize numerical procedures rather than the analysis of the coals, which are used primarily for illustrative purposes. While these articles contribute little information on coal structure, their emphasis on the accurate calculation of coal spectra makes them relevant in the present context.

Commercial FT-IR software commonly utilizes multiplicative (Mertz) phase correction to minimize the effects of phase errors on calculated spectra. An alternative technique—interferogram symmetrization—is thought to yield more accurate spectra by some researchers, who may develop the necessary software in their own laboratories. This method was applied to PA data for separated coal macerals and other samples (Michaelian, 1989a, 1990a). The research showed that the PA spectra of a given sample calculated using either method of phase correction were generally in good agreement. It should be noted that these calculations implicitly assume that wavenumber-dependent phase information should be removed from the spectra, and do not consider the PA phase explicitly (see Chapters 4–6).

Coals sometimes yield weak PA spectra, which can make signal averaging strategies particularly important (Michaelian, 1987; 1990a). In this regard, it should be mentioned that it is often better to average a small number of 'scans' (interferograms) in FT-IR spectroscopy so as to ensure short-term instrument stability. When more averaging is required to achieve an adequate signal/noise ratio, it is better to average a series of spectra, each corresponding to a short measurement time. This situation exists because phase correction is intrinsically more accurate for shorter experiments. Signal averaging was discussed in more detail in the previous chapter.

The complexity of coals is manifested in the C–H stretching region of their PA infrared spectra. Observed spectra generally exhibit three broad, overlapping bands at about 2870, 2930 and 2960 cm^{-1} that are due to aliphatic C–H groups. In addition, many coals display an aromatic CH band centered near 3050 cm^{-1}. To increase the amount of information discernible in the C–H region of the PA spectra, numerical techniques such as second derivative computation, curve fitting and deconvolution have been used (Friesen and Michaelian, 1986, 1991; Michaelian and Friesen, 1990). These methods show that the 2800–3000 cm^{-1} region actually comprises about eight to nine bands – roughly the same number used to fit the aliphatic region in PA infrared spectra of hydrocarbon fuels (see below). These individual bands can be readily assigned to CH_n groups in environments known to exist in coals. The consistency of the results and plausibility of the band assignments both support the findings of these investigations.

An example of the results obtained by these methods is depicted in Figure 7.8. Curve fitting and deconvolution were both employed in this study (Friesen and Michaelian, 1991). The agreement between the deconvolved experimental spectrum (upper curve) and the analogous result for the fitted spectrum (lower curve) indicate that the number of bands used, as well as their parameters, are realistic.

The availability of PA infrared spectra of coal macerals enables their correlation with a number of important technological properties. For example, Gagarin et al. (1993) described an inverse procedure for the determination of the petrographic composition of a coal using the resolved spectra of separated macerals. Similarly, PA intensities at 2867 and 3019 cm^{-1} were used to predict heats of combustion (Gagarin et al., 1994, 1995a) as well as the pseudo-first-order rate constant for liquefaction in tetralin. The Roga index, a measure of the caking propensity of coals, can also be predicted from these data (Gagarin, Friesen, and Michaelian, 1995b). The ~1000–1800 cm^{-1} region in PA spectra of coals has also been analyzed by inverse methods (Michaelian et al., 1995).

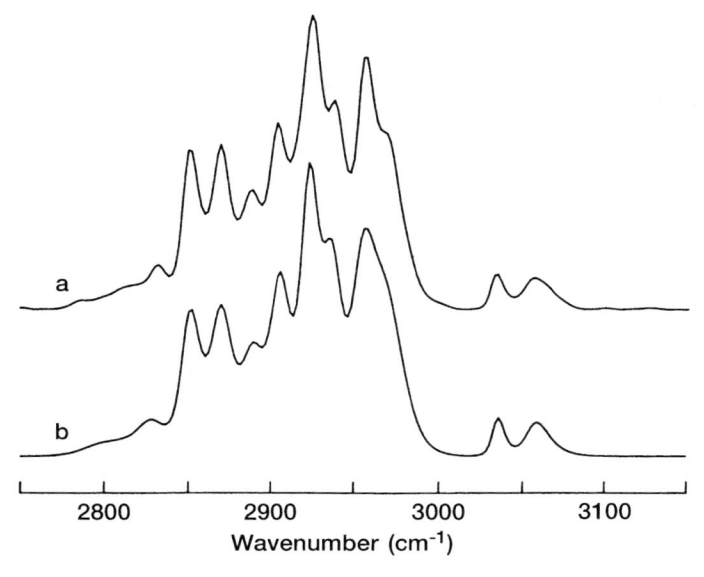

Figure 7.8 Deconvolution of measured (curve a) and fitted (curve b) PA infrared spectra of lignite. Gaussian bands were used in the curve fitting calculations. (Reproduced from Friesen, W.I. and Michaelian, K.H., *Appl. Spectrosc.* **45**: 50–56, by permission of the Society for Applied Spectroscopy; copyright © 1991.)

7.2.2
Liquid Fuels

Much of the literature discussed so far in this book refers to the investigation of samples that exist as solids at ordinary temperatures. Although there is no substantive reason why PA infrared spectroscopy cannot be successfully utilized in the study of liquid samples, one could argue that the use of the technique in this context should still be justified. As is shown below, the measurement of infrared spectra of hydrocarbon fuels – particularly highly aromatic, viscous liquids – provides one such justification.

A detailed PA infrared study of a series of distillation fractions derived from Syncrude Sweet Blend (SSB) was presented by Michaelian et al. (2001). This dark-colored liquid, alternatively referred to as synthetic crude oil or bitumen-derived crude, is obtained by extraction and upgrading of the bitumen in the vast oil sand deposits in northeastern Alberta. In this work, SSB was separated by distillation into 12 fractions, with boiling point intervals that span the range from 30 to 524 °C. Insofar as some of these fractions resemble commercial middle distillate fuels (e.g., gasoline, diesel, and jet fuel), this study demonstrates the feasibility of PA infrared analysis of hydrocarbon fuels.

This investigation revealed that much of the important information in the PA spectra of these fractions is conveyed by the C–H stretching region. Figure 7.9 shows that the intensity and shape of the band envelope between about 2800 and

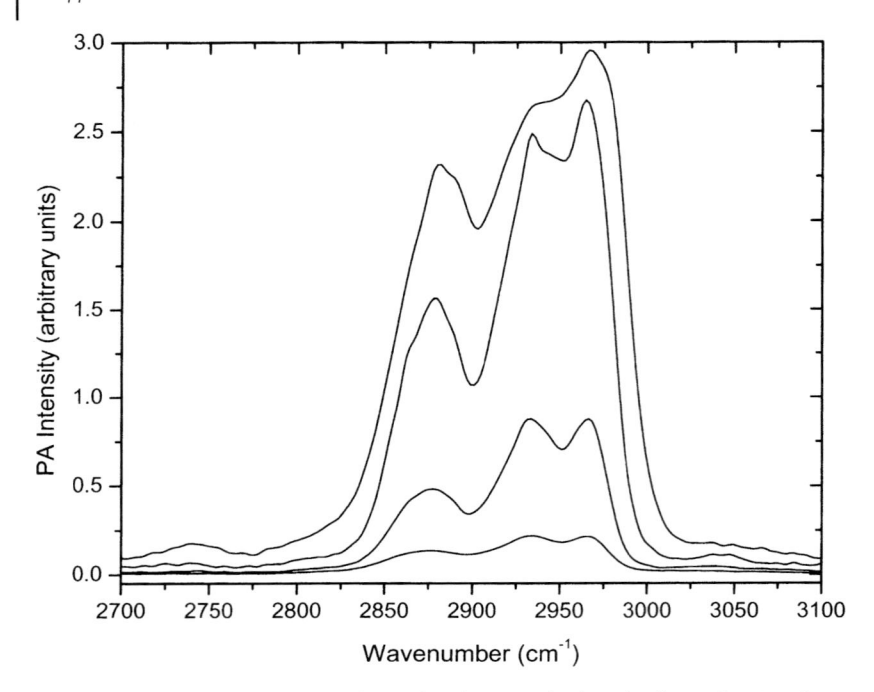

Figure 7.9 C–H stretching region of PA infrared spectra for four distillation fractions from SSB (Michaelian et al., 2001). Boiling points: top curve, 30–71 °C; second curve, 71–100 °C; third curve, 100–166 °C; bottom curve, 166–177 °C.

$3000\,cm^{-1}$ vary significantly for the first four distillation fractions; the boiling point dependence of the spectra continues less dramatically for the higher fractions. The C–H stretching bands were also analyzed by curve fitting and integration (Figure 7.10). The eight bands due to aliphatic (CH_2 and CH_3) groups systematically shift to lower wavenumbers as successively higher boiling point ranges are analyzed. Moreover, the percentage of total C–H intensity attributable to CH_2 groups increases with boiling point, while that due to CH_3 groups concomitantly decreases. Although the origin of the trend in band positions is uncertain, the increase in the CH_2/CH_3 intensity ratio is readily attributable to increasing alkyl chain length in these fractions. It is clear that the PA infrared spectra of these distillation fractions are of research quality, enabling detailed analysis and interpretation.

As might be expected, the physical characteristics of the distillation fractions differ considerably. For example, the fractions with boiling points below 100 °C are easily handled colorless liquids. PA FT-IR spectra of these fractions can be obtained with a single scan, that is, in about one second. In these cases, traditional infrared methods (e.g., ATR and transmission) and PA spectroscopy are equally convenient. By contrast, the fractions with higher boiling points (and higher aromatics contents) are darker and more viscous. These liquids are not well suited for conventional infrared sampling techniques, but are still quite amenable to PA

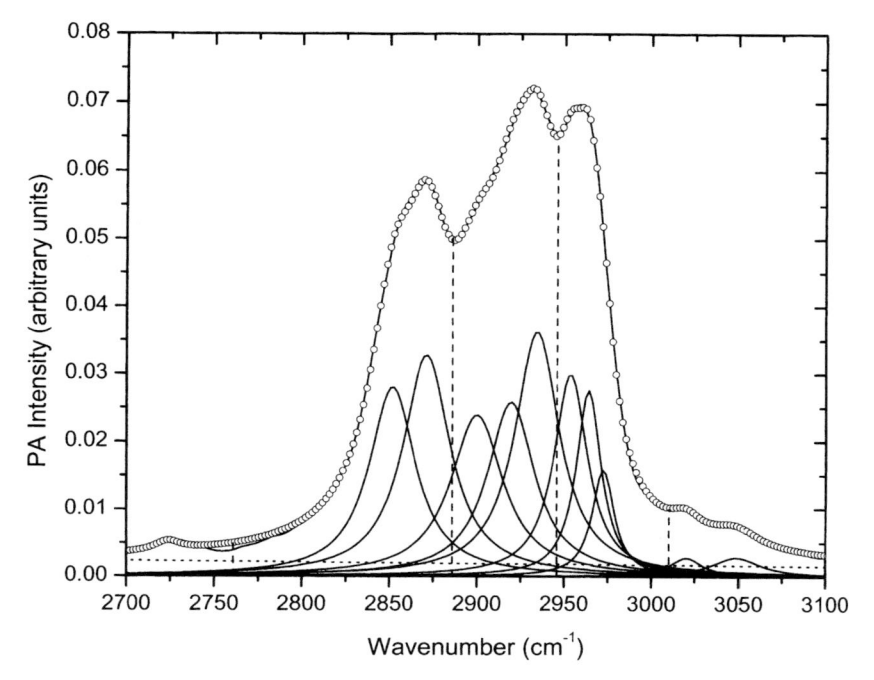

Figure 7.10 Curve fitting and integration of PA infrared spectrum of distillation fraction with boiling point range 232–288 °C. Circles represent the ordinates in the sum of the plotted component bands; dashed vertical lines show the integration limits used to calculate the areas of the asymmetric CH_2 and CH_3 stretching bands and the total area due to aliphatic C–H stretching. The dotted straight line at about 0.03 intensity units is the fitted baseline.

spectroscopy. This is another example where the elimination of sample preparation enables an analysis that would otherwise be considered nearly intractable.

The 343–524 °C fraction, referred to as heavy gas oil (HGO), is particularly important with regard to the refining of SSB into commercial fuels. PA infrared and FT-Raman spectra of a second set of distillation fractions, which comprise the entire HGO region, were obtained in a separate investigation (Michaelian, Hall, and Bulmer, 2002b). Curve fitting and integration were again used to analyze the C–H stretching region of the PA spectra. As noted above, the changes in the infrared spectra of the fractions occur mainly at lower boiling points. Thus the positions and areas of the individual bands were observed to be approximately constant for the HGO components. Although the HGO distillation fractions are quite viscous, PA infrared spectroscopy makes their analysis relatively straightforward.

PA infrared and FT-Raman spectra were also obtained for a series of narrow distillation fractions in the light gas oil (LGO) boiling range, which extends from 195 to 343 °C (Michaelian, Hall, and Bulmer, 2003). The trends in the CH_2/CH_3 intensity ratio, as well as the positions of the resolved C–H stretching bands, were consistent with those in the earlier work (Michaelian et al., 2001). Separation of these bands can be effected using integration as an alternative to the more time-consuming process of curve fitting. The lower-boiling part of the LGO region (up to 250 °C) gave similar results in another study (Michaelian, Hall, and Bulmer, 2004).

7.2.3
Other Samples

The residue ('resid') that remains after removal of the distillation fractions is a brittle solid that tends to shatter when attempts are made to grind it using a mortar and pestle. A PA spectrum obtained for several millimeter-size pieces of a typical resid is shown in Figure 7.11. The high aromatics content of this resid is indicated by the three prominent aryl C–H bands between 700 and 900 cm^{-1}, as well as the broad aromatic C=C stretching band at ~1600 cm^{-1}. These features were discussed earlier with regard to the PA spectra of both cokes and coals. In addition, aliphatic groups in the resid give rise to bands at 1377 and 1463 cm^{-1}, while the C–H stretching region of this spectrum resembles that of the higher HGO distillation fractions. The PA infrared spectrum of 'end cut' pentane-insoluble Athabasca bitumen (Bensebaa et al., 2000), which also boils above 525 °C, is quite similar to the resid spectrum in Figure 7.11.

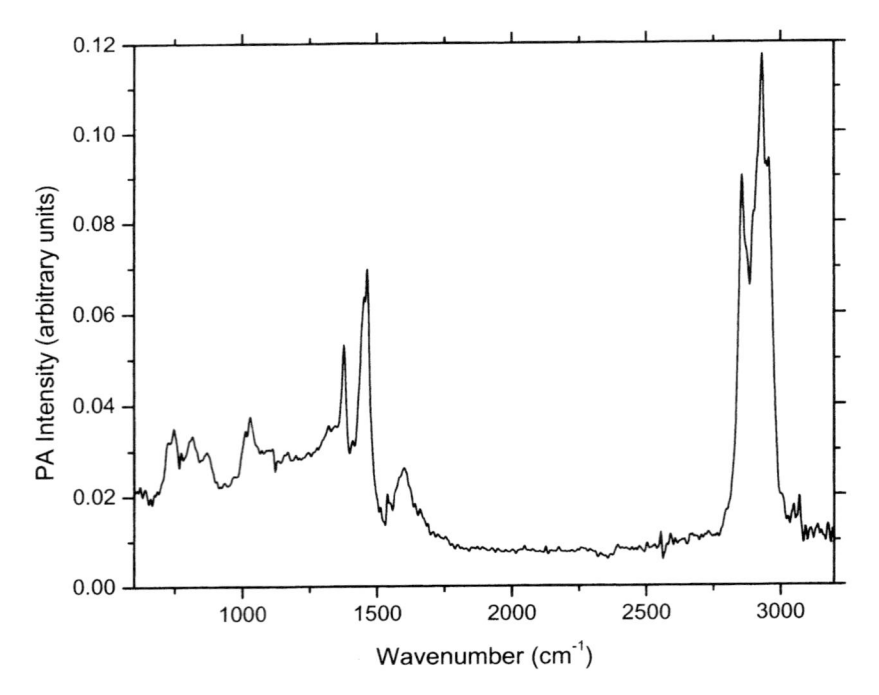

Figure 7.11 PA infrared spectrum of 'resid' from distillation; boiling point greater than 524 °C.

PA investigation of resid demonstrated another noteworthy phenomenon: as the measurement of successive spectra continued for more than one hour, a progressive diminution in overall intensity was observed. After this experiment was completed, the small particles were found to have coalesced in the sample cup, forming a disc about 5 mm in diameter and 1 mm thick. This decrease in surface area accounted for the loss of PA signal. The softening of the resid was a consequence of sample heating by the impinging radiation. This observation – which might be considered surprising in light of the high boiling range of the sample – illustrates an occasional minor disadvantage of PA infrared spectroscopy.

Spectroscopic characterization of the bitumen from which the above samples originate is obviously a nontrivial challenge. This extremely thick black liquid has a consistency similar to that of molasses, and is rarely investigated by infrared spectroscopy; its problematic physical character is one reason for this situation.

Figure 7.12 illustrates a typical PA infrared spectrum of bitumen obtained in the author's laboratory. As noted in the figure, bands due to aliphatic and aromatic CH groups are clearly identifiable. An additional peak near 1030 cm^{-1} is probably due to carbon-carbon stretching involving aliphatic groups bonded to aromatic rings. This spectrum was acquired in about 15 min.

The sample characterized in Figure 7.12 was stored in the laboratory for about four months. Because the bitumen was not removed from the sample cup, its surface was continuously exposed to the ambient environment during this period. After this interval, another PA spectrum was recorded. Figure 7.13 shows that

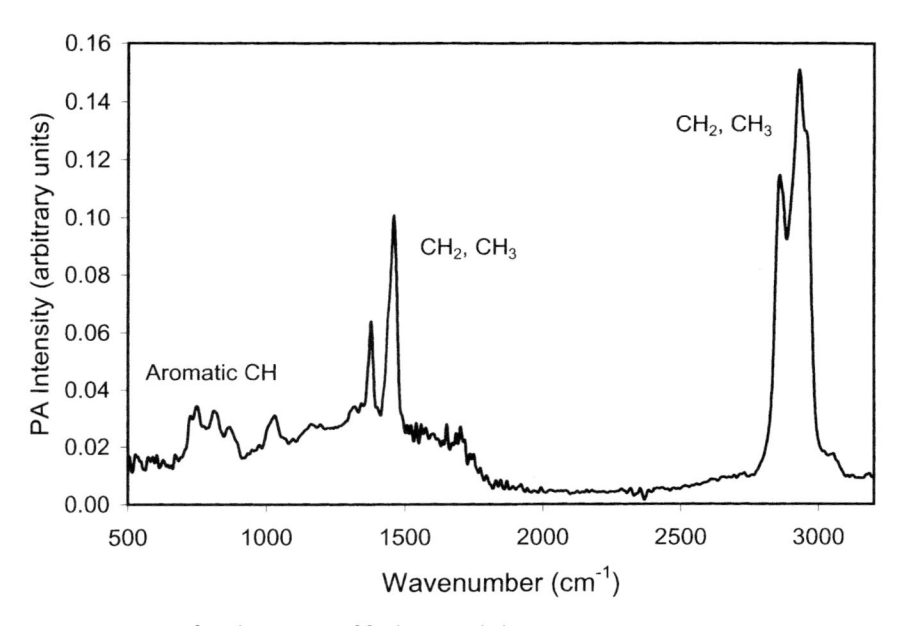

Figure 7.12 PA infrared spectrum of fresh Syncrude bitumen.

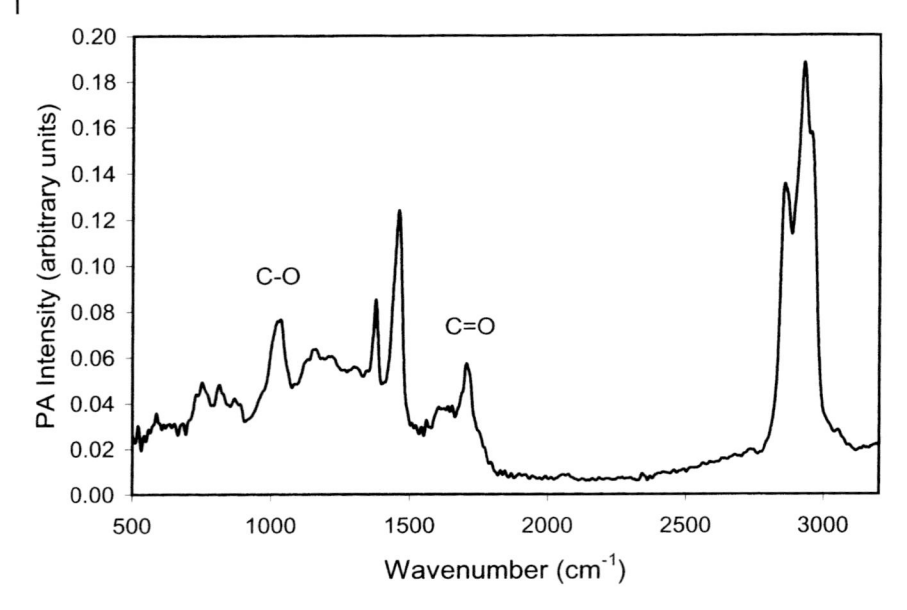

Figure 7.13 PA infrared spectrum of aged Syncrude bitumen.

prominent new bands appeared at about 1030 cm^{-1} (due to C–O or S=O groups) and 1700 cm^{-1} (C=O). Several weaker bands between 1100 and 1200 cm^{-1} are also attributable to C–O groups. Thus the aging of bitumen led to significant oxidation. However, it should be recognized that the PA spectrum is obtained from a surface layer about 10–20 μm thick: this experiment provides no information about possible bulk oxidation of the sample.

Asphaltenes, an important constituent of bitumens and other hydrocarbon fuels, have also been studied using PA infrared spectroscopy. While the composition and structure of asphaltenes – broadly defined as the alkane-insoluble, toluene-soluble component of carbonaceous fuels – is a subject of ongoing debate, spectroscopic and other analyses provide considerable useful information. Friesen et al. (2005a, 2005b) employed thermogravimetric analysis and PA infrared spectroscopy in a study of the effect of solvent/bitumen (S/B) ratio on asphaltenes isolated from Athabasca bitumen. As expected, the PA spectra exhibited bands due to clays as well as aliphatic and aromatic hydrocarbon groups. As the S/B ratio increased, it was observed that the intensities of the hydrocarbon bands increased while those of the clay bands decreased. At the same time, the CH_2/CH_3 ratio diminished slightly. The results showed that greater S/B ratio causes the precipitation of asphaltenes containing slightly larger amounts of aliphatics and that these hydrocarbons do not differ substantially. Further work on the far-, mid-, and near-infrared PA spectra of asphaltenes is under way in the author's laboratory as this book is written.

7.3
Organic Chemistry

Familiarity with the PA infrared spectra of aliphatic and aromatic hydrocarbons is essential for thorough interpretation of spectra of hydrocarbon fuels such as the coals and middle distillates described above. Accordingly, representative published near- and mid-infrared PA spectra of hydrocarbons are discussed in this section. This literature synopsis is followed by examples of PA mid- and far-infrared spectra of aromatic hydrocarbons, particularly polycyclic compounds.

Not surprisingly, examination of the scientific literature shows that many early publications on PA spectra of hydrocarbons emphasized the near-infrared region. This situation reflects the general status of PA infrared spectroscopy during its development, which began more than 30 years ago. As discussed elsewhere in this book, the ready accessibility of near-infrared sources and optics led to the initial primacy of near-infrared PA work; however, the current widespread availability of FT-IR spectrometers and commercial PA cells inevitably means that the majority of studies of hydrocarbons and other organic materials are based in the mid-infrared.

The work of Adams, Beadle, and Kirkbright (1978) is among the first near-infrared PA investigations of ordinary hydrocarbons. These authors obtained dispersive spectra between about 0.8 and 2.7 µm (12 500–3700 cm^{-1}) for n-hexane, benzene, anthracene, and benzanthracene. Several prominent bands were observed for each compound. A peak at about 2.2 µm (~4545 cm^{-1}) occurred in the spectra of all three aromatics, but was absent from the hexane spectrum. Hence, as discussed in Chapter 5, this band can be used for quantification of aromatics in aromatic-aliphatic mixtures. This suitability of PA near-infrared spectroscopy for hydrocarbon analysis was reiterated in a review by Kirkbright (1978).

Several years later, Lewis (1982) reported dispersive near-infrared (1.0–2.6 µm) PA spectra of organic and organometallic substances. The goal of this work was to provide band assignments and promote the use of PA near-infrared spectroscopy to 'fingerprint' compounds – the latter objective usually being addressed by means of mid-infrared spectroscopy. The hydrocarbons studied in this investigation included benzene, benzene-d$_6$, toluene, and cyclohexane. Although the PA spectra in this publication are displayed on a linear wavelength scale, band positions are reported in wavenumbers. For example, three bands were reported in the PA spectrum of benzene at 4100, 4675, and 6025 cm^{-1} (Adams, Beadle, and Kirkbright, 1978). The 4675-cm^{-1} peak was used for quantification of benzene in binary mixtures. Lewis (1982) assigned the near-infrared bands in the PA spectra of the organic compounds to combinations and overtones of mid-infrared transitions that were observed in separate (transmission) experiments.

Broadening the scope of this discussion to include compounds containing elements other than carbon and hydrogen leads to another PA near-infrared study of substituted benzenes. Sarma, Sastry, and Santhamma (1987) obtained dispersive spectra of three benzonitriles, five acetophenones, and three benzylbromides. About 20 bands were identified in the near-infrared spectra of these substances.

These features were readily assigned using mid- and far-infrared band positions observed in a related investigation. Because of the structural similarity of the samples, the same set of band assignments was applicable for all eleven compounds. Many bands were identified in the spectra of most or all substituted benzenes.

The final near-infrared PA study to be mentioned here was performed by Manzanares, Blunt, and Peng (1993). These authors investigated fundamental and overtone C–H stretching bands in spectra of *cis*- and *trans*-3-hexene. Piezoelectric detection was used to observe the fourth and fifth overtones, which occur in the near-infrared and visible regions, respectively. An acousto-optically modulated argon ion laser pumped a dye laser that supplied the near-infrared and visible radiation. The use of piezoelectric detection is actually rather incidental to this paper, which is primarily concerned with curve fitting the C–H bands and assignment of the spectra according to a local mode model. This implies that piezoelectric PA near-infrared spectra of alkenes are sufficiently reliable that researchers can emphasize data interpretation rather than the technique itself.

Two articles discussing mid-infrared PA spectra of hydrocarbons adsorbed on various substrates can be mentioned here. In the first, Saucy, Cabaniss, and Linton (1985) reported spectra of polynuclear aromatic compounds adsorbed on alumina and silica. The impressive detection sensitivity of PA spectroscopy, which sometimes extends to sub-monolayer coverage, was demonstrated in this work. The absence of sample preparation and capability for analysis of milligram sample quantities were noted by the authors as the most desirable attributes of PA infrared spectroscopy.

This work showed that several well-defined PA bands were detectable for a monolayer of phenanthrene on silica. Two additional analyte bands were identified after subtraction of the silica spectrum. All of these features were observed in the PA spectrum of pure phenanthrene. Moreover, the authors used PA spectroscopy to show that UV irradiation of 9-nitroanthracene adsorbed on silica led to the production of anthraquinone. Finally, comparison of spectra of intact and ground pellets demonstrated the very significant increase in PA signal that accompanies an increase in surface area.

Another example of the use of PA spectroscopy for characterization of organic compounds in the solid state was described by Gosselin *et al.* (1996). These researchers utilized PA spectroscopy to monitor the synthesis of resin compounds, which possess structures more complicated than those of the simple hydrocarbons referred to above. Transmission spectra obtained for resins in KBr pellets are complicated by light scattering and reflection. By contrast, the PA gas-microphone technique removes the need for sample grinding and preparation of pellets, while yielding superior spectra that are free from these troublesome effects. Hence Gosselin *et al.* concluded that PA infrared spectroscopy can serve as an effective analytical method in solid-phase organic chemistry.

Recently, Mistry *et al.* (2008) used PA infrared spectroscopy and several other analytical techniques to characterize proton-conducting organic-inorganic membranes. It is desirable that these membranes retain water at high temperatures. Bands due to the PTFE backbone, C–O–C, and other groups were observed. A sol-gel reaction resulted in Si–O–Si and Si–O–P formation. PA spectroscopy enabled the convenient analysis of the membranes, nicely demonstrating the versatility of the method.

7.3.1
Mid-Infrared PA Spectra of Aromatic Hydrocarbons

The PA infrared spectra of cokes and coals described in the first two sections of this chapter are generally interpreted under the reasonable assumption that these materials contain many different aromatic hydrocarbons. Aromatics are also integral components of the hydrocarbon fuels that were discussed in the preceding section. Hence it is appropriate to examine the PA spectra of pure aromatic hydrocarbons, which can be taken as models for all three types of industrial samples. Mid-infrared PA spectra of many aromatic hydrocarbons are not yet available in the primary scientific literature. Spectra of a series of 12 representative compounds were obtained in the author's laboratory and are presented here.

The results are organized according to the number of aromatic rings in each hydrocarbon studied and to the degree of alkyl substitution of the parent compound. Figure 7.14 displays PA infrared spectra of (a) 2-ethylnaphthalene, (b) 2,3,5-trimethylnaphthalene, (c) 1-methylnaphthalene, and (d) naphthalene itself. It should be noted that 2-ethylnaphthalene is a liquid at ambient temperatures. More than 20 bands occur in the fingerprint region (up to about 2000 cm^{-1}) in each spectrum, with another group of peaks appearing in the C–H stretching

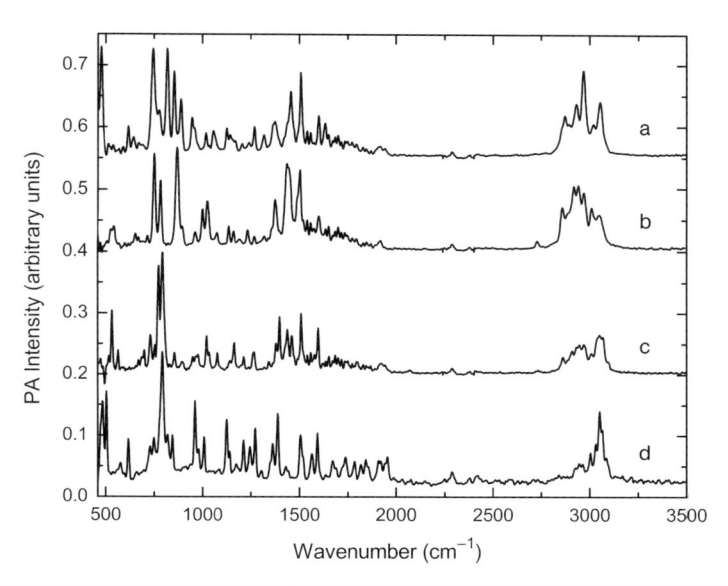

Figure 7.14 PA infrared spectra of substituted naphthalenes. The spectra have been rescaled and offset along the vertical axis for clarity. They are: (a) 2-ethylnaphthalene, (b) 2,3,5-trimethylnaphthalene, (c) 1-methylnaphthalene, and (d) naphthalene. Bands near 2350 cm^{-1} due to atmospheric CO_2 have been subtracted from some spectra.

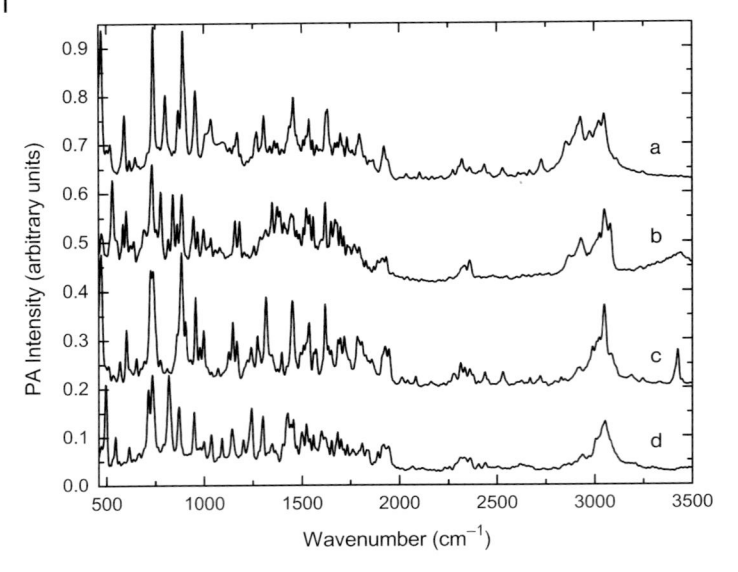

Figure 7.15 PA infrared spectra of tricyclic aromatic hydrocarbons. The spectra have been rescaled and offset along the vertical axis for clarity. They are: (a) 2-methylanthracene, (b) 9-methylanthracene, (c) anthracene, and (d) phenanthrene. Some of the features near 2350 cm^{-1} are due to atmospheric CO_2.

region. The prominent bands just below 3000 cm^{-1} in the upper three curves arise from CH_2 and CH_3 groups, while the bands at slightly higher wavenumbers are due to aromatic CH. Importantly, each spectrum displays a large number of well-defined bands – in contrast with the spectra of some of the carbonaceous materials mentioned above. Moreover, the PA spectra of these simple aromatic hydrocarbons are quite similar to infrared spectra obtained using other methods.

PA spectra of four three-ring aromatics (two substituted anthracenes, anthracene itself, and phenanthrene) are depicted in Figure 7.15. The spectra resemble those for the less complex compounds discussed in the previous paragraph, with 30 or more bands appearing in the spectrum of each tricyclic compound.

Results for four unsubstituted polycyclic aromatics: (a) 1,2:3,4-dibenzanthracene, (b) benzo[a]pyrene, (c) chrysene, and (d) pyrene are shown in Figure 7.16. The PA spectra are similar to those in the previous two figures, with the obvious exception that bands due to alkyl functionality are not present in this third group of compounds.

7.3.2
Far-Infrared PA Spectra of Aromatic Hydrocarbons

Far-infrared PA spectra of four different four- and five-ring catacondensed and pericondensed aromatic hydrocarbons have recently been reported by Michaelian

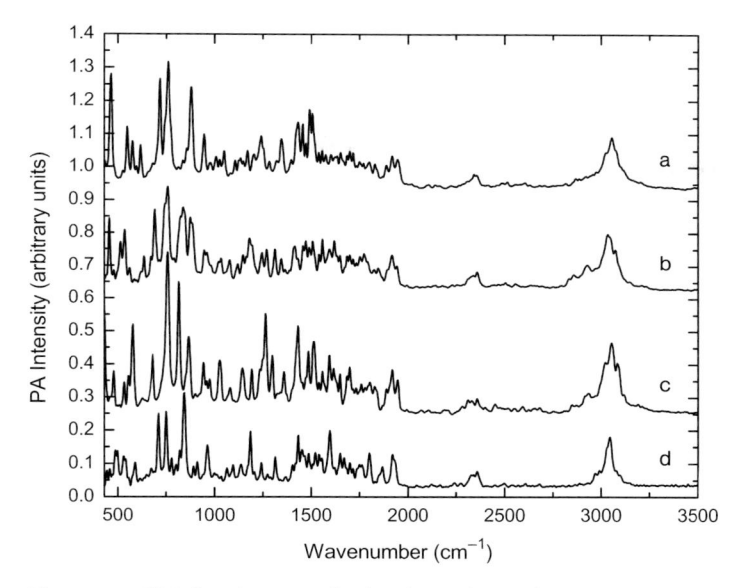

Figure 7.16 PA infrared spectra of polycyclic aromatic hydrocarbons. The bottom spectrum has been rescaled, and the other three spectra have been offset along the vertical axis for clarity. The spectra are: (a) 1,2:3,4-dibenzanthracene, (b) benzo[a]pyrene, (c) chrysene, and (d) pyrene. Some of the features near 2350 cm^{-1} are due to atmospheric CO_2.

et al. (2009).[1] These compounds can be regarded as simple models of asphaltenes, which were discussed in the previous section. Spectra were acquired using large-amplitude phase modulation and digital or lock-in demodulation. The use of large modulation amplitudes (up to 20 λ, where $\lambda = 0.6328\,\mu m$, the HeNe laser wavelength) shifts maximal available energy to about 200 cm^{-1} and enables observation of bands from about 80 to 500 cm^{-1}. PA spectroscopy facilitates the study of this region, obviating the need for solvents or mulls as well as far-infrared detectors.

Figure 7.17 shows far-infrared PA spectra of perylene ($C_{20}H_{12}$) and pyrene ($C_{16}H_{10}$). The spectra display a large number of bands, which agree with those in published spectra obtained by other techniques and with density functional theory (DFT) calculations. Lattice modes and predicted Raman-active transitions are correlated with some of the PA infrared bands.

The literature reviewed in this section, and the mid- and far-infrared results obtained for various aromatic compounds, clearly demonstrate the suitability of PA infrared spectroscopy for hydrocarbon characterization. The PA spectra of these hydrocarbons are generally quite similar to those obtained using other infrared techniques, but often do not resemble those of the naturally occurring organic substances discussed elsewhere in this book very closely.

---

1) All carbon atoms occur on the periphery of the ring system in catacondensed aromatics. Three or more rings share common carbon atoms in pericondensed compounds.

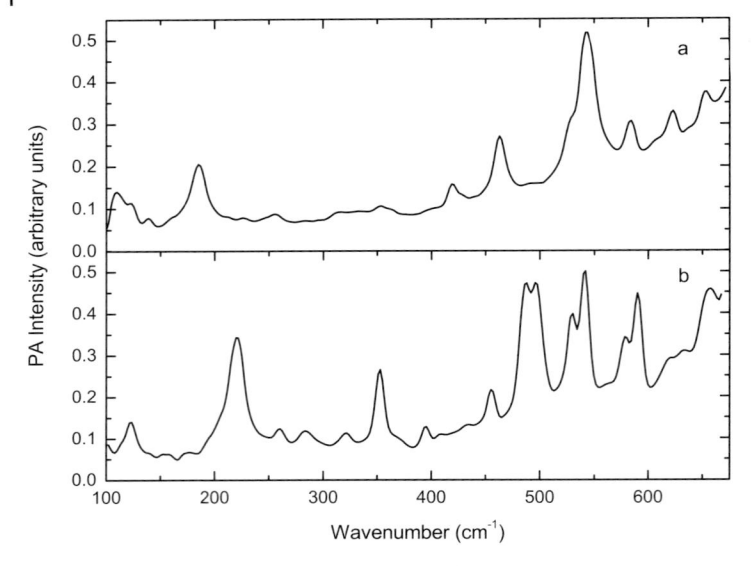

Figure 7.17 Far-infrared step-scan PA spectra of (a) perylene and (b) pyrene. Modulation frequency, 5 Hz; amplitude, 12 λ.

7.4
Inorganic Chemistry

7.4.1
Carbonyl Compounds

PA infrared spectroscopy has been successfully used for the characterization of carbonyl compounds by at least two groups of researchers. In the work reviewed here, the minimal sample preparation was cited as the principal justification for selection of the technique. Moreover, the need for nondestructive analysis of small samples also favored the choice of PA spectroscopy. Both mid- and near-infrared PA spectra of carbonyl compounds are discussed in this section.

Natale and Lewis (1982) reported dispersive near-infrared PA spectra of $Mo(CO)_6$, $Ir_4(CO)_{12}$, $[RhCl(CO)_2]_2$, and $(C_6H_6)Cr(CO)_3$. The spectral region covered in this work extended from 1.0 to 2.6 μm (10 000–3850 cm^{-1}). The metal carbonyls each displayed two or three bands near the low-wavenumber end of the near-infrared region: these peaks were attributed to overtones and combinations of the fundamental carbonyl stretching vibrations that occur near 2000 cm^{-1}. The authors found that it was necessary to utilize both Raman and infrared transmission spectra in order to fully characterize the fundamental vibration frequencies, as neither spectrum exhibited all of the bands in this region. Thus the PA near-infrared spectra of these metal carbonyls effectively combine information from two related experimental techniques in a single measurement.

The assignments of the near-infrared bands of the carbonyl compounds studied by Natale and Lewis might well be considered as tentative in view of the band overlap and noise that occurred in the PA spectra of the first two compounds. In contrast, seven well-defined near-infrared PA bands were observed for $(C_6H_6)Cr(CO)_3$; most of these features were readily assigned to overtones and combinations of the fundamental vibrations of the benzene moiety.

Both mid- and near-infrared PA spectra of a wide series of organometallic complexes were discussed in a series of papers published between 1986 and 1993 by Butler and coworkers at McGill University. In the initial PA investigation by this group (Xu, Butler, and St.-Germain, 1986) both wavelength intervals were examined for group VIB chalcocarbonyls, $M(CO)_6$ (M = Cr, Mo, W), and $Cr(CO)_5CS$. To their surprise, the authors observed gas-phase PA spectra of these metal chalcocarbonyl complexes – an observation that can be put down to sample volatility and the very high sensitivity of the PA effect in gases. Indeed, the PA spectra obtained in this work resembled published infrared absorption spectra of the gaseous complexes, and moreover differed significantly from spectra of the solid complexes in KBr disks. Thus the PA experiments yielded bands due to the vapor above the solid complexes in the gas-microphone cell. As mentioned elsewhere in this book, this phenomenon, which is not uncommon in PA infrared spectroscopy, is readily attributed to sample heating caused by absorption of the incident radiation. These researchers recognized this situation and turned it to their advantage, developing a detailed assignment of the bands in the PA spectra of the gas-phase chalcocarbonyls. Several years later, a similar PA investigation of $CH_3Mn(CO)_5$ and $CH_3Re(CO)_5$ in the same laboratory (Butler, Gilson, and Lafleur, 1992) also reported the observation of bands from vapor-phase complexes.

Butler et al. (1987a) described PA infrared spectra of 10 different solid organometallic complexes of chromium, manganese, rhenium, and iron. Near- and mid-infrared spectra were obtained in this work, and detailed band assignments were given for both regions. In the relatively narrow 3800–4600 cm^{-1} near-infrared region alone, about 15 bands were detected for $[CpMo(CO)_3]_2$, $[CpFe(CO)_2]_2$ and $Fe_2(CO)_9$ ($Cp = \eta^5\text{-}C_5H_5$). For $CpM(CO)_2(CX)$ (M = Mn, Re; X = O, S), approximately 20 near-infrared bands were identified. As illustrated in Figure 7.18, these near-infrared spectra generally displayed very good signal/noise ratios. A number of mid-infrared PA bands were also observed and assigned to fundamental vibrations of these complexes.

PA infrared spectroscopy was combined with micro-Raman and traditional infrared sampling methods in an investigation of $Ru_3(CO)_{12}$ by Butler et al. (1987b). Both ^{12}C- and ^{13}C-isotopomers were included in this study. PA and transmission (solution) infrared spectroscopies were used to examine the 3800–4200 cm^{-1} region, whereas micro-Raman and other infrared spectra were employed to characterize lower wavenumbers. The PA spectra in this work exhibited poorer signal/noise ratios than those in the previous papers by these authors. Despite this fact, a number of weak peaks that appeared in the PA spectra were assumed to be genuine and attributed to factor group splitting since they appeared only in solid-state data. The possible authenticities of some of these extra peaks would

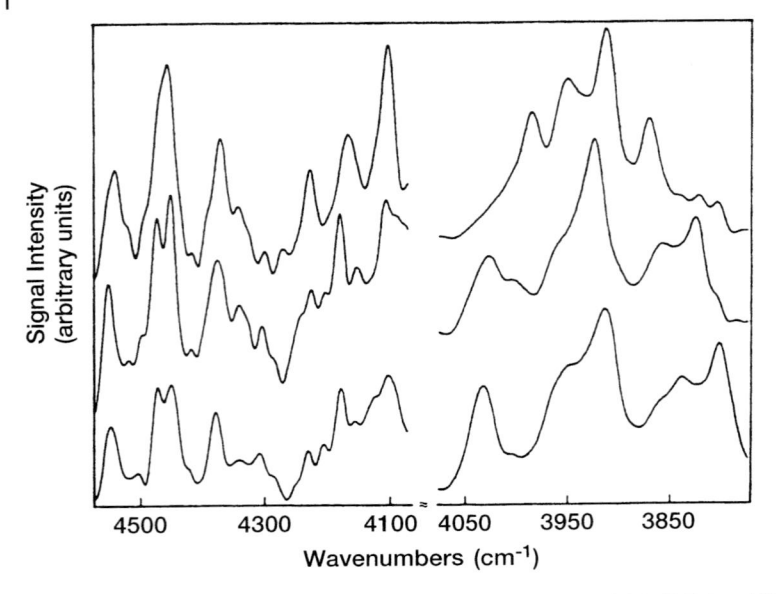

Figure 7.18 PA infrared spectra of $CpMn(CO)_2(CS)_2$, top curve; $[CpMn(CO)_3]_2$, middle curve; and $CpRe(CO)_3$, bottom curve. (Reproduced from Butler, I.S. et al., *Appl. Spectrosc.* **41**: 149–153, by permission of the Society for Applied Spectroscopy; copyright © 1987.)

appear to require confirmation, either by PA spectroscopy or another infrared method.

In a subsequent investigation that also combined PA infrared spectroscopy with other techniques, Butler, Li, and Gao (1991) obtained near-, mid- and far-infrared spectra of solid organoiron(II) carbonyl complexes $[CpFe(CO)_2R]BF_4$ (R = C_2H_4, $CH_2C(CH_3)_2$, 3-methylthiophene) and $CpFe(CO)_2I$. PA spectroscopy was found to be the most sensitive technique in the 4000–4600 cm^{-1} near-infrared region, with as many as 10 bands being identified in the data. In the mid-infrared, ATR and transmission spectra exhibited similar relative band intensities, except that the low-wavenumber bands in the ATR spectrum were stronger because of the greater depths of penetration that are characteristic of the technique at longer wavelengths. Relative intensities were somewhat different in the PA infrared spectra, where the stronger bands are broadened and appear to be partly saturated. While this saturation has a deleterious effect on the profiles of some mid-infrared bands, it also correlates with beneficial intensification of the weaker near-infrared bands and thereby facilitates their analysis.

The final two articles in this series extended the approach described above, specifically the comparison of PA infrared spectra with data obtained by other sampling methods and the use of PA near-infrared spectra to characterize the region near 4000 cm^{-1}. First, Li and Butler (1992) measured mid-infrared PA spectra of manganese(I) carbonyl halides, $Mn(CO)_5X$ and $[Mn(CO)_4X]_2$ (X = Cl, Br, I), complementing these results with infrared spectra obtained using KBr pellets and solutions of these compounds. Although the fundamental C=O stretching bands near 2000 cm^{-1} in the PA spectra of these carbonyl halides may be partly saturated, the concomitant intensification of the near-infrared bands is advantageous because it reveals features completely absent from spectra recorded for the KBr pellets. The occurrence of these near-infrared bands in the PA spectra – although fortuitous – allowed the authors to calculate the anharmonicities of the carbonyl stretching vibrations. Similarly, near-infrared PA spectra of mixed carbonyl-t-butylisocyanide complexes, $M(CO)_{6-n}(CN^tBu)_n$ (M = Cr, Mo, W; n = 1–3), were the first ever reported (Li and Butler, 1993) for this type of mixed-ligand complex. While the near-infrared features were easily detected in the PA spectra of the latter complexes, the corresponding mid-infrared PA spectra showed little indication of saturation.

7.4.2
Semiconductors

Subgap near-infrared PA spectra of semiconductors were described in a series of articles by Pilkington, Tomlinson, and coworkers at the University of Salford and other institutions. These researchers investigated $CuInSe_2$ and $Cu(In,Ga)Se_2$, which are of interest in solar cell applications. X-ray diffraction, scanning electron microscopy, and Rutherford backscattering were also employed in these studies.

The purpose-built near-infrared spectrometer used in this work was described by Zegadi, Slifkin, and Tomlinson (1994). This system included a Xe lamp and a 0.22-m monochromator, similar to that described in the section on dispersive PA spectroscopy earlier in this book. Three different types of cells were examined, and factors such as sample size and shape, as well as air leakage, were considered. Spectra were obtained for n- and p-type $CuInSe_2$ from 0.7 to 1.4 eV (~5650–11 300 cm^{-1}). Six bands in the subgap region (below 1.1 eV) were attributed to defects. The f dependence of the PA amplitude was reported for $CuInSe_2$, carbon black, and Si. This dependence was predicted for opaque materials according to Rosencwaig-Gersho (RG) theory as follows:

$$\mu_\beta < l_s, \mu_s > \mu_\beta \quad f^{-1}$$

$$\mu_\beta < l_s, \mu_s < \mu_\beta \quad f^{3/2}$$

$$\mu_\beta < l_s, \mu_s > l_s \text{ (carbon black, } \beta \approx 10^6 \text{ cm}^{-1}\text{)} \quad f^{-1}$$

In this notation, μ_β is the optical absorption length and β is the absorption coefficient; l_s is sample thickness; and μ_s is the thermal diffusion length, which varies as $f^{-1/2}$.

Two further works discussed defect states in $CuInSe_2$ single crystals. First, Zegadi et al. (1995) obtained spectra of ion-implanted $CuInSe_2$. The defect ionization energies were thought to arise from impurity-to-conduction band or valence band-to-impurity transitions. Depth profiling was accomplished by recording in-phase and quadrature spectra and calculating phase-resolved spectra at different phase angles. Similarly, Zegadi et al. (1996) examined samples with either an excess or a deficiency of Se, and studied the effect of annealing the crystals under an Se atmosphere. Non-radiative transitions that are associated with shallow and deep levels and are influenced by Se content were detected. Absorption coefficients were derived from the PA spectra.

Another group of articles discussed Cu(In,Ga)Se$_2$ compounds. Ahmed et al. (1995) reported near-infrared PA spectra of CuIn$_{0.75}$Ga$_{0.25}$Se$_2$ thin films and observed an absorption edge near 1 eV, corresponding to the band gap. Defect states were again observed in the subgap region. Spectrum intensities were reported in normalized PA amplitude units and as absorption coefficients (cm^{-1}). A plot of $(\beta h\nu)^2$ vs photon energy contains a linear region that follows the relation

$$\beta(h\nu) = \frac{A}{h\nu}[h\nu - E_g]^{1/2} \tag{7.1}$$

where A^2 and E_g are the slope and gap energy, respectively. Band gaps (obtained by extrapolation to $\beta = 0$) differed according to the composition of the film. Two broad peaks in the subgap region were observed for one film and attributed to interference, enabling calculation of film thickness. A related study (Ahmed et al., 1998) showed that the absorption edge near 1 eV is steeper in PA spectra of Se-annealed films (Figure 7.19).

Recently, Ahmed et al. (2006) described a transmission PA technique for the analysis of these thin films. This method, which could also be utilized for other sample types, is implemented as follows. A gas-microphone cell containing carbon black is fitted with a quartz window. The sample of interest (a thin film, single crystal, etc.) is placed on top of the window. Modulated light passes through the sample and is absorbed by the carbon black, giving rise to a PA signal. The sample also absorbs a fraction of the incident radiation at characteristic wavelengths,

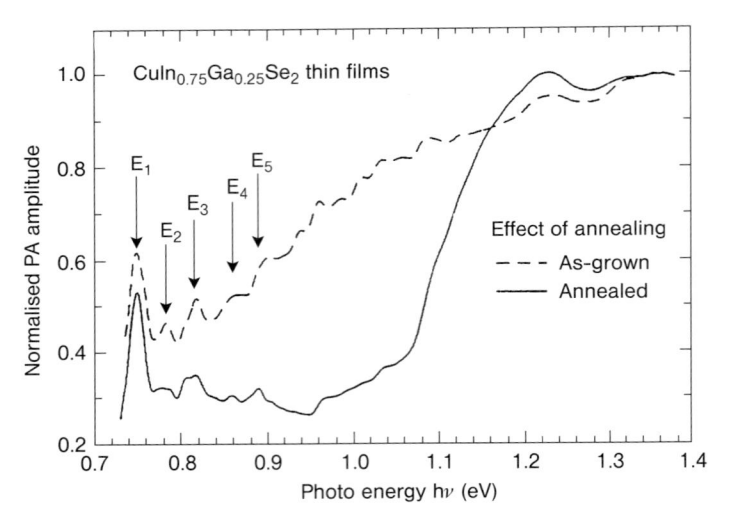

Figure 7.19 Normalized PA spectra of as-grown (dashed line) and selenium-annealed (solid line) CuIn$_{0.75}$Ga$_{0.25}$Se$_2$ thin films. (Reprinted from J. Mater. Proc. Technol. **77**, Ahmed, E., Zegadi, A., Hill, A.E., Pilkington, R.D., Tomlinson, R.D. and Ahmed, W., Thermal annealing of flash evaporated Cu(In,Ga)Se$_2$ thin films, 260–265, copyright © 1998, with permission from Elsevier Science.)

producing transmission-like features. These transmission PA bands occur at the same locations as the familiar absorptive bands in a conventional PA spectrum. The main advantage of transmission PA spectroscopy is its convenience: there is no need to re-seal the cell when samples are exchanged. The effect can also be seen to resemble the absorption of infrared radiation by CO_2 and water vapor when a PA cell is used in an unpurged FT-IR spectrometer.

Other research groups have also reported PA infrared spectra of these chalcopyrite semiconductors. Yoshino *et al.* (1999) used a disc-shaped PZT, attached to the rear surface of the sample with conducting silver paste, to acquire PA spectra of $CuInSe_2$ films between 0.6 and 1.8 eV. Data were obtained before and after quenching by illumination at a photon energy greater than 1.1 eV. Reddy *et al.* (2001) presented spectra of $CuGa_xIn_{1-x}Se_2$, calculating absorption coefficients by assuming a thermal diffusion length of 120 µm at $f = 112$ Hz. Subgap bands were attributed to defect states, while higher-energy transitions involved the crystal field and spin orbit splitting levels. These authors also described PA spectra of $CuInSe_2$ layers grown by a two-step process (Reddy, Slifkin, and Weiss, 2001).

Kuwahata, Muto, and Uehara (2000) investigated the carrier concentration dependence of visible and near-infrared PA spectra of n-type phosphorus-doped and p-type boron-doped silicon wafers. Heat generated by the absorption of light is diffused by free electrons, causing suppression of the elastic wave and a diminution of PA signal intensity. Holes give rise to a similar effect. At carrier concentrations above 10^{17} cm^{-3}, PA bands were not observed. The near-infrared PA spectrum of undoped silicon powder is shown in Figure 7.20.

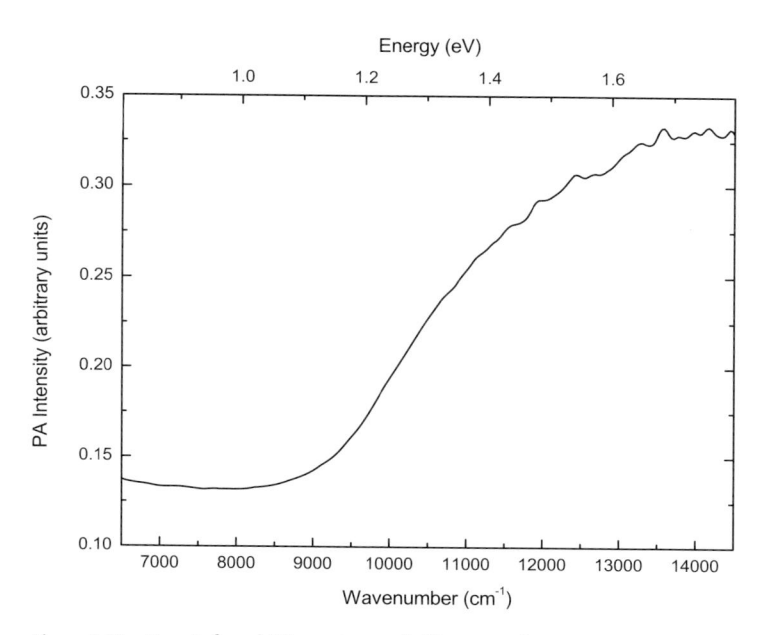

Figure 7.20 Near-infrared PA spectrum of silicon powder.

7.4.3
Superconductors

PA infrared spectroscopy was employed by Hosomi et al. (1999) in a study of superconducting phases in the Ba–Ca–Cu–O system. T_c was ≈ 126 K for an unstable phase, and ≈ 90 K for a derivative phase. The use of PA detection was somewhat incidental in this work: no spectra are shown and few experimental details are given. The 126 K phase was studied as-synthesized in the bulk form, while the 90 K phase was analyzed as a powder. Carbonate peaks in the spectrum of the 90 K sample suggested a $BaCO_3$ impurity. A broad OH band was also detected, implying the presence of water. A variety of physical and analytical methods were used to characterize these samples.

7.4.4
Other Materials

Many additional inorganic materials have also been analyzed using PA infrared spectroscopy. Three typical examples are cited here. First, Utamapanya, Klabunde, and Schlup (1991) described preparation and characterization of very-high-surface-area magnesium hydroxide and oxide. In addition to PA spectroscopy, X-ray diffraction and scanning electron microscopy were utilized in this work, which emphasized the production of ultrafine particles. Similarly, Harpness et al. (2003) employed electron microscopy, diffraction, and PA infrared spectroscopy in a study of nanosized $MoSe_2$. Kimura et al. (1996) obtained PA spectra in a wide region, extending from the ultraviolet to the mid-infrared, for neptunium (VI) compounds precipitated under CO_2. This research was motivated by the fact that hydrolysis and carbonate complexation are important for the prediction of actinide ion migration in aqueous systems. Other articles that describe PA infrared spectra of inorganic materials are listed in Appendix 2.

7.5
Biology and Biochemistry

Proteins and microorganisms such as bacteria and fungi often are not well suited to traditional infrared sample preparation techniques. Hence, the virtual elimination of sample preparation in PA spectroscopy offers an important advantage when characterization of these rather difficult samples is required. Typical examples of the use of PA infrared spectroscopy for the qualitative or quantitative analyses of these and other biological and biochemical samples are reviewed in this section.

An early application of dispersive PA spectroscopy to protein characterization occurs in the work of Sadler et al. (1984). Near-infrared spectra of proteins, in both the solid state and D_2O solution, were obtained in this investigation. PA spectroscopy was utilized because it facilitated the analysis of strongly scattering samples. Viable spectra were obtained for egg white lysozome and bovine pancreatic ribonuclease, various solid amino acids, proteins and polypeptides, and proteins in solution. The solid-state spectra generally contained much more detail than the solution spectra, enabling detailed assignments of bands to overtones and combinations of fundamentals that occur in the mid-infrared.

Work by Gordon, Greene, and their colleagues at the United States Department of Agriculture on PA spectroscopy of grains and fungi is discussed later in this chapter. Two publications by this group, which fall more naturally into the subject area of biology, biotechnology, and biochemistry, are mentioned here. In the first, PA infrared spectroscopy was used to monitor the growth of the filamentous fungus *Phanerochaete chrysosporium* on cellulose filter paper discs (Greene, Freer, and Gordon, 1988). This model system was chosen because of the possible use of waste cellulose as a substrate for fermentation. As discussed below, PA spectroscopy provides a means for the measurement of biomass formed during this process, avoiding traditional analytical methods that require extensive sample handling and sometimes lead to the ultimate destruction of the sample.

The application of PA infrared spectroscopy to this system was demonstrated using a purified protein (bovine serum albumin), a purified phospholipid (asolectin), and a carbohydrate (cellulose): a mixture of the three is an approximate representation of a reconstituted microorganism. Indeed, the PA spectrum of this mixture qualitatively resembles that of *P. chrysosporium*. Next, known amounts of the actual dry fungus were deposited on filter paper discs, and the intensities of the amide I bands were determined. For relatively small quantities, intensity increased linearly with the quantity of fungus; however, this intensity eventually reached a limiting value and did not increase further. This saturation effect was attributed to mycelial layering, a known phenomenon that is expected to limit optical penetration and thereby establish an upper limit to PA intensity. Finally, the growth of the fungus on filter paper discs was monitored by PA spectroscopy for several days and confirmed by a standard protein assay. These successful experiments clearly demonstrate the suitability of PA infrared spectroscopy for the analysis of *P. chrysosporium* on cellulose.

Gordon *et al.* (1990) continued this research, improving on the quantification of protein biomass by PA infrared spectroscopy. Bands arising from amide groups were again utilized to monitor the increase in biomass corresponding to microorganism growth. Both pure proteins and microorganisms such as a bacterium, yeast, alga, and a fungus were studied in this investigation. Whereas the cellulose band at about $3338\,\text{cm}^{-1}$ had been used to normalize intensities in the previous investigation, the authors added a fixed amount of polyacrylonitrile (PAN) as an internal standard in the later work. The nitrile band at $2243\,\text{cm}^{-1}$, which is well separated from the features due to the samples or the cellulose discs, was used to scale these spectra. Four proteins and the four microorganisms mentioned above were studied in this way. Figure 7.21 suggests saturation plateaus at high protein levels; the earlier onset of saturation for the microorganisms is a consequence of the fact that they contain about 70% non-protein biomass, which contributes to the layering mentioned above and exacerbates saturation in the spectra.

An even more impressive result appears in Figure 7.22, where the relationship between both normalized and un-normalized PA intensities and protein mass is displayed. It is obvious that the effect of saturation is ameliorated when the spectra

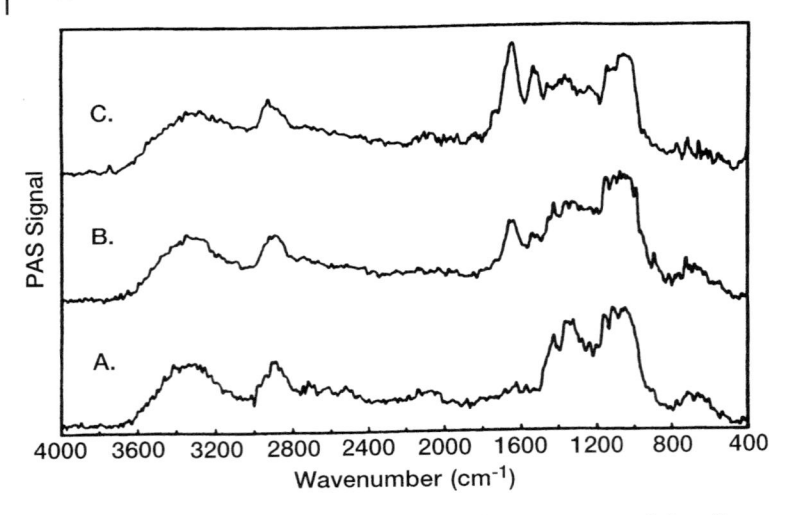

Figure 7.21 PA infrared spectra of *Phanerochaete chrysosporium* on cellulose filter paper (Gordon *et al.*, 1990). Curve A, filter paper only; curve B, 0.67 mg fungus on filter paper; curve C, 3.2 mg fungus on filter paper.

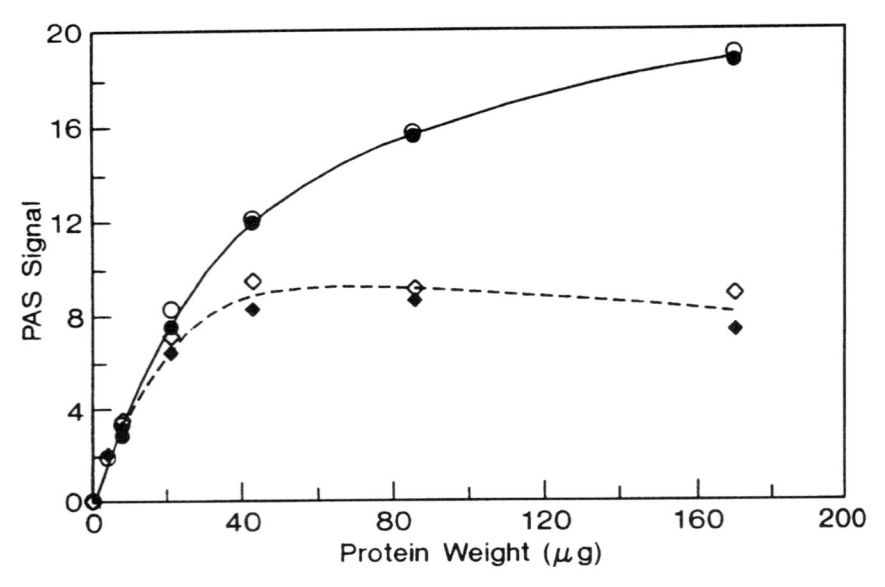

Figure 7.22 The effect of normalization on the PA intensity of the amide I (1660-cm^{-1}) band. Solid curve, data normalized using PAN intensity at 2243 cm^{-1}; dashed curve, data without normalization. The open and filled symbols refer to two different experiments. (Reprinted from *Biotech. Appl. Biochem.* **12**, Gordon, S.H. et al., Measurement of protein biomass by Fourier transform infrared-photoacoustic spectroscopy, 1–10, copyright © 1990, with permission from Elsevier Science.)

are scaled using the 2243-cm^{-1} PAN band as an internal standard. These two articles are discussed further in Chapter 5.

Ashkenazy, Gottlieb, and Yannai (1997) used PA infrared spectroscopy to characterize the biosorption of lead cations using acetone-washed biomass from beer yeast. Spectra obtained before and after lead uptake were similar, except that a carboxylate band shifted from 1405 to 1375 cm^{-1}. Chitin, a purified powder obtained from crab shells, was also studied. Biosorption by this material involves nitrogen groups, as indicated in the infrared spectra.

Other research groups also used PA infrared spectroscopy to study proteins. For example, Luo *et al.* (1994) obtained spectra of concanavalin A, hemoglobin, lysozome, and trypsin, which have different secondary structure distributions. These structures were elucidated by calculating second derivatives and then curve fitting the amide I band centered at about 1660 cm^{-1}. It was concluded that seven components are required to account for the observed lysozome spectrum. Samples were prepared as thin layers by evaporating 1–2 µL of protein solutions on Teflon membranes; in cases where saturation of the PA spectra was a problem, the thickness of the deposited layer was reduced. The authors noted the high sensitivity of PA spectroscopy, as well as the speed and ease with which the data were acquired. In fact PA spectroscopy was concluded to be superior to both infrared transmission and circular dichroism (CD) spectroscopy – the latter being commonly used to study protein structure.

This group also analyzed calmodulin (calcium-modulated protein) (Martin *et al.*, 1990, 2000) and phosphoamino acids and phosphoproteins (Graves and Luo, 1994). PA spectroscopy was well suited for the latter experiments because there was no need for crystalline materials or large sample quantities: typical amounts analyzed were 30–40 µg. Solid samples were prepared by spreading 10 µL of protein solution or phosphate ester on a 7-mm polyethylene membrane disc and evaporating to dryness. Serine phosphate was distinguished from tyrosine phosphate, the former giving rise to a band at 984 cm^{-1} and the latter yielding a 974-cm^{-1} peak. These features were attributed to dianions; by contrast, samples prepared at different pH values exhibited peaks due to monoanions. The authors also studied the interaction between Al ions and phosphate groups in phosvitin, a phosphoseryl-containing protein, and used the PA spectra to determine the corresponding pK_a value.

Huang *et al.* (1994) continued this work, using PA spectroscopy to study phosphorylase kinase and its truncated γ subunit. Metal binding sites were studied in this investigation. PA infrared spectra in the 1600–1700 cm^{-1} interval displayed a broad feature that was fitted using either three or five bands. Spectra were obtained for dialyzed enzymes, prepared as a thin layer by spreading protein solutions on Teflon disks made from disposable infrared cards. The use of PA spectroscopy was somewhat incidental to this investigation.

Lotta *et al.* (1990) demonstrated another application of PA infrared spectroscopy that is relevant in the present context. These authors investigated the interaction between the well-known electron acceptor 7,7,8,8-tetracyanoquinodimethane (TCNQ) and phospholipids, specifically diacylphosphatidylcholines and diacylphosphatidylglycerols. These interactions produce colored charge-transfer complexes, which could be studied by UV/visible absorption spectroscopy. However, PA spectroscopy was used by these researchers because sample preparation might perturb the complexes. The infrared bands of neutral TCNQ disappeared upon complexation, and new bands characteristic of the anion appeared in their place. Comparison of the spectra of the neutral and complexed species showed that a band due to phosphatidylglycerol ester carbonyl groups was modified by complex formation. PA spectroscopy was treated as an established analytical technique in this work; in other words, instrumental questions were not considered in detail.

PA infrared and ^{13}C NMR spectroscopies were used to quantify and characterize suberin, the main component of *Quercus suber* L. cork, by researchers at Universidade de Aveiro (Portugal) and Institut National Agronomique (France). The variability of cork samples acquired from different locations was examined using chemometric analysis of the spectroscopic data (Lopes *et al.*, 2001). Suberin content was closely correlated with cork quality. The minor polymeric fraction of this material was separated by solvent extraction and enzymatic treatment (Rocha *et al.*, 2001); PA spectroscopy was used to follow the progress of the isolation of polymeric suberin. Quantification of the suberin content in cork (~40–45%) by both spectroscopic methods revealed linear correlations with gravimetric data (Lopes *et al.*, 2000). Bands due to aliphatic chains, esters, and aliphatic and aromatic C=C groups were identified in the PA infrared spectra.

As noted above, the amide bands in infrared spectra of proteins are sensitive to secondary structure. van de Weert *et al.* (2001) used PA and ATR spectroscopies in a study of five different proteins, and found that differences between the spectra obtained with these techniques were not entirely due to changes in secondary structure caused by lyophilization. For example, ATR spectra were influenced by the anomalous dispersion effect and the removal of hydrating water. In the solid state, PA and KBr pellet spectra also showed differences, leading to speculation that preparation of KBr pellets might cause protein denaturation. Lysozome conformation inside polymeric matrices was discussed in a related work (van de Weert *et al.*, 2000) in which infrared microscopy and confocal laser scanning microscopy were also employed. These studies illustrate the complementary nature of various infrared techniques and show that proper interpretation of the spectra must consider the method by which they are acquired.

The nitrogen content of farmyard manure was determined by standard analytical methods and several spectroscopic techniques (Kemsley *et al.*, 2001). Near-infrared PA and reflectance spectra, as well as mid-infrared ATR data, were compared with traditional wet-chemical methods for determination of nitrogen content. The near-infrared PA measurements were made at specific wavelengths (1935, 2100, 2180, and 2230 nm) that were isolated using bandpass filters. A very low f (2 Hz) corresponded to $\mu_s \approx 100\,\mu m$. Kjeldahl (total) and ammoniacal (volatile) nitrogen were determined more accurately using mid-infrared and near-infrared reflectance spectra, with numerical analysis showing less predictive capability for the PA data.

Hydroxyapatite (HAp) $[Ca_{10}(PO_4)_6(OH)_2]$ is a mineral form of calcium apatite that occurs in bones and teeth. Coelho *et al.* (2006) used visible, near-, and mid-infrared PA spectroscopy to characterize HAp derived from the bones of Brazilian river fish. The bones were calcined and milled for different periods. The intensity and profile of a PA band at about 3165 cm^{-1} varied with calcination time, leading to the conclusion that nanostructures in HAp powder are stabilized in 8 h. This material is also mentioned in the following section on medical applications of PA spectroscopy.

7.6
Medical Applications

PA infrared spectroscopy has been utilized in a variety of medical applications, and by many research teams, during the last three decades. Examination of the relevant literature quickly reveals the diversity of this work. To illustrate this, it may be mentioned that PA spectroscopy has been employed in the following applications pertaining to human health: (i) qualitative and quantitative analysis of drugs, (ii) drug transport in membranes, (iii) the analysis of calcified tissue (teeth) and fingernails, (iv) diagnostics associated with adverse medical conditions (gallstones, cancer), (v) the identification of bacteria, and (vi) gas analysis in surgical settings. Examples of the use of PA spectroscopy in each of these areas are given in the following review.

The reader may already be aware that the spectroscopic characterization of pharmaceutical drugs is now routine. Near-infrared, terahertz, and FT-Raman spectroscopies (sometimes facilitated by the use of fiber optics) are now being used for qualitative and quantitative analysis of drugs in research laboratories and production environments with good success. At the same time, there is an obvious role for mid-infrared spectroscopy in some circumstances; one could argue that this need is best filled by PA infrared spectroscopy, since the minimal sample preparation afforded by the near-infrared and Raman techniques is retained. The advantages arising from the use of neat (undiluted) samples are obvious in many of the publications discussed below.

The suitability of PA spectroscopy for quantitative analysis of drugs was demonstrated in four early works. Krishnan (1981) described spectra of a powdered mixture of acetylsalicylic acid and phenacetin, and of the undiluted acid. Subtraction of the second spectrum from the first yielded a spectrum that was practically identical to that of pure phenacetin, a result that nicely demonstrates the additivity of PA spectra of drugs. Castleden, Kirkbright, and Long (1982) recorded dispersive near-infrared (1.3–2.5 µm, approximately 7700–4000 cm^{-1}) PA spectra of intact and manually ground tablets of propranolol, a β-andrenergic receptor blocker. This investigation was motivated by the need to quantify the analyte without the usual time-consuming physical separation of excipients. The authors found that the aromatic C–H band at 2.21 µm (4525 cm^{-1}) arising from propranolol could be used for quantification: this was accomplished by monitoring its intensity in spectra of samples containing known amounts of the drug, or by ratioing this band against the 1.65-µm (6060-cm^{-1}) band of lactose, the main excipient. Similarly, Rockley *et al.* (1983) used mid-infrared PA spectroscopy for quantitative analysis of phenobarbital/lactose mixtures. White (1985) obtained PA spectra of acetylsalicylic acid and acetoaminophen deposited on silica gel samples on TLC plates.

Several other groups used PA infrared spectroscopy for drug characterization. Belton, Saffa, and Wilson (1988a) presented infrared spectra of two drug polymorphs; these spectra are notable for their complexity and quality. Indeed, the authors observed that PA spectroscopy could be superior to diffuse reflectance in this particular application. Huvenne and Lacroix (1988) also noted that PA infrared spectroscopy could be utilized to characterize drugs without the need for sample preparation. However, these authors observed that the relative intensities in PA spectra of drugs were somewhat different from those in absorbance spectra, and concluded that PA spectra should not be used for drug identification unless they were appropriately corrected. They obtained PA infrared spectra of Flunitrazepam, Dipyriamole, and α-lactose monohydrate, and empirically determined proportionality constants that relate the PA signal intensity and the product $\beta\mu$ for a thermally thick sample. Ashizawa (1989) combined PA spectroscopy and X-ray crystallography in a study of two polymorphic forms of 2R,4S,6-fluoro-2-methyl-spiro[chroman-4,4′-imidazoline]-2′,5-dione. DSC, SEM, and PA infrared spectroscopy were employed in a study of the interaction between polyvinylpyrrolidone and trimethoprim, a bacteriostatic antibiotic (Zerlia et al., 1989). Berbenni et al. (1996) used PA infrared spectroscopy and X-ray diffraction to study three phases of (+)-fenfluramine hydrochloride, a nonstimulant anorectic agent. Untreated and annealed samples yielded spectra different from those of water-recrystallized samples, where hydrogen bonds were absent.

The determination of drug content is particularly challenging for semisolid (ointment and cream) formulations. Traditionally this problem has been approached by extraction of the drug from the semisolid dispersant, followed by the use of standard analytical methods such as HPLC for quantification. In an effort to avoid this separation step, researchers at Martin Luther University (Neubert, Collin, and Wartewig, 1997a, 1997b) utilized PA spectroscopy for quantitative analysis of dithranol and brivudin, prepared as mixtures with Vaseline. Both rapid- and step-scan PA spectra were obtained in this investigation; for equal measurement times (about 20 min), the signal/noise ratios in the step-scan spectra were about an order of magnitude better than those in the corresponding rapid-scan spectra. A series of bands between 800 and 1800 cm^{-1} in the step-scan amplitude spectra intensified as the drug concentration increased from 0.5 to 10 wt%. The integrated intensities of bands in the 1545–1770 cm^{-1} region for brivudin, or 1560–1670 cm^{-1} for dithranol, varied linearly with drug concentration as determined by capillary zone electrophoresis or HPLC. In fact, instrumental errors in the PA spectra were significantly smaller than the variations observed in the standard analytical data. These encouraging findings show that PA infrared spectroscopy is a viable means for analyzing drugs in semisolid formulations.

This work was followed by investigations of drug penetration through human strateum corneum (outer layer of the epidermis). Schendzielorz et al. (1999) and Hanh et al. (2000) studied drug penetration from semisolid formulations into

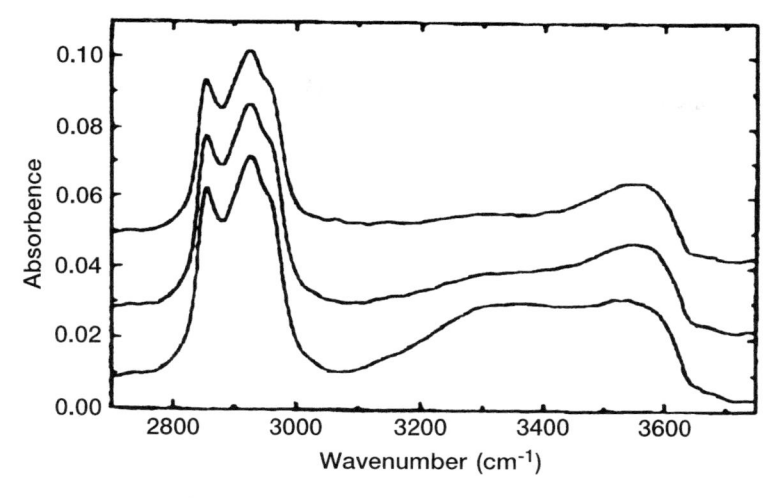

Figure 7.23 PA infrared spectra of the model membrane/clotrimazole/vaseline system recorded at various times during penetration experiments. Bottom curve, initial spectrum; middle curve, elapsed time 66 min; top curve, elapsed time 242 min. (Reproduced from Schendzielorz, A. et al., *Pharm. Res.* **16**: 42–45, by permission of Kluwer Academic/Plenum Publishers; copyright © 1999.)

dodecanol-collodion membranes, which served as models for skin. Membrane thicknesses of 31 and 17 μm were used; bands due to the drug appeared in the PA spectra of the membrane/clotrimazole/Vaseline system as penetration progressed, enabling calculation of diffusion coefficients. Results in the high-wavenumber region (Figure 7.23) show the effect of drug penetration on the hydrogen bond system of the membrane. This research was later extended to the use of actual strateum corneum samples (Hanh *et al.*, 2001). Sampling depths ranging from about 9 to 26 μm were estimated using $\alpha = 0.00029\,\text{cm}^2\,\text{s}^{-1}$. The accuracy of the PA results in these studies is noteworthy. Spectroscopic methods for measurement of skin penetration were recently reviewed by Gotter, Faubel, and Neubert (2008).

Other research groups have also used PA spectroscopy to study the diffusion of drugs through skin. Nakamura *et al.* (2001) and Mitchem, Mio, and Snook (2004) employed glass microfiber filters as skin mimetics, using PA spectroscopy to monitor the diffusion of nitroglycerin (an angina therapeutic agent) from commercial transdermal drug patches. Baby *et al.* (2007) chose shed snake skin as a model membrane; this research used PA infrared and Raman spectra to study the effects of surfactant solutions on the uptake of pharmaceuticals by skin. PA spectra are plotted to resemble transmission spectra in this article, a practice that many readers will find confusing.

Skin lipids (esters of fatty acids) have also been characterized using PA infrared spectroscopy. In an early work, Kanstad *et al.* (1981) used a CO_2 laser to analyze samples deposited as thin layers on 0.04-mm Al foils epoxied onto piezoelectric ceramic strain gage wafers. Cholesterol served as a model compound. The spectrum of a 14-μg sample contained prominent bands at 957 and 1055 cm^{-1}, although the available laser lines did not permit complete definition of either peak. Skin surface lipids obtained from a patient using a chloroform/methanol mixture yielded weaker bands, with the quantity of sample being estimated as about 5 µg. The sensitivity of PA detection is even more impressive when the results are compared with those from a transmission experiment on a KBr disk: the amount of sample required for the transmission measurement was 100 times greater than that for the PA spectrum. More recently, Neubert et al. (1998) used PA spectroscopy to study thermal degradation of ceramides, part of the lipid matrix of skin.

The obviation of sample preparation creates many potential medical and dental applications of PA infrared spectroscopy. Accordingly, several research teams have used the technique to analyze calcified tissues. This work is briefly summarized in the following paragraphs.

An important study of PA infrared spectra of extracted, intact human teeth was carried out at National Research Council Canada (Sowa and Mantsch, 1993a, 1993b, 1994a, 1994b). Rapid- and step-scan depth profiling experiments were performed in this work. Linearization was employed to ameliorate the effects of saturation; moreover, phase-modulation step-scan spectra were observed to exhibit less saturation than the corresponding rapid-scan PA spectra. Step-scan depth profiling was accomplished by changing f, or alternatively by varying the phase angle at constant f. The authors noted that the latter experiment can be carried out more efficiently using phase synthesis, a numerical method discussed in Chapter 6. Spectra are calculated at various phase angles θ using the relationship

$$I(\theta) = I(0°)\cos(\theta) + I(90°)\sin(\theta) \qquad (7.2)$$

where I denotes intensity and the angles 0° and 90° refer to in-phase and quadrature spectra, respectively. The success of phase synthesis in this experiment implies linear PA phase behavior.

The depth profiling experiments in this work showed that the protein contribution increases relative to the mineral component as the sampling depth increases, a conclusion that accords with the morphology of maturing tooth enamel. Moreover, crystallinity was found to be diminished for the subsurface apatitic structure. Although the structure of the human tooth is relatively complex, the inorganic and protein distributions observed in this investigation are reminiscent of a two-component system.

Chemical modifications of dentin (tooth) surfaces by acids and NaOCl were monitored using PA infrared spectroscopy (Di Renzo et al., 2001a, 2001b). Etching with acids (phosphoric, maleic, citric) caused a progressive increase in intensity that was attributed to greater surface area; mineral content near the surface was depleted by this treatment. Deproteination with NaOCl removed the organic phase but did not affect the carbonate and phosphate minerals. In contrast with acid etching, deproteination did not produce a front that advances into the sample. Modification of dentin surfaces is effected to improve bonding with resins, a topic that has also been investigated using PA spectroscopy (Spencer et al., 1992).

The effects of laser irradiation on tooth root structure have also been studied with PA infrared spectroscopy. Spencer, Trylovich, and Cobb (1992) found that treatment of an extracted tooth with an Nd:YAG laser led to protein breakdown and the production of NH_4^+ species. Both Nd:YAG and CO_2 lasers were used at high power levels (~10 W) in a subsequent study (Spencer et al., 1996). A band at 2015 cm^{-1} was attributed to the cyanamide ion, suggesting an organic-mineral reaction. Finally, a free electron laser (FEL) was used for ablation of bone from the tibia of rabbits (Spencer et al., 1999). Radiation at 3.0, 6.1, and 6.45 µm was absorbed by the proteins and water within bone. PA spectra confirmed the expected reduction in amide groups. The FEL caused minimal damage in the bone tissue, particularly within 10 µm of the surface.

Two additional articles that discuss dental applications can be mentioned here. Hou et al. (1988) obtained PA infrared spectra of supragingival and subgingival dental calculus (plaque), finding that the two types are distinguishable after subtraction of the hydroxyapatite spectrum. Calheiros et al. (2004) discussed the degree of conversion in restorative dental composites; no spectra are included in their publication.

Spectroscopic analysis of finger- and toe-nails is another obviously challenging problem. Sowa et al. (1995) obtained PA infrared spectra of viable and clipped nails and compared the results with ATR, transmission, and diffuse reflectance spectra. The objective of this work was to characterize the three layers that make up nails: from top to bottom, these are referred to as dorsal, intermediate, and ventral, respectively. Depth profiling was effected in step-scan phase modulation experiments. Amplitude spectra were calculated at phase angles ranging from 90° (near surface) to 180° from in-phase and quadrature spectra, as described above. These spectra showed a decrease in the methylene/methyl ratio as the sampling depth increased, consistent with the expectation that the lipid content in the intermediate layer is lower than that in the dorsal layer. Saturation affected the spectra, such that depth profiling was limited to the dorsal layer and the upper portion of the intermediate layer of the nails. Near-infrared PA spectra indicated that the ordering of the protein structure in the intermediate layer is lower than that in the dorsal layer. A mid-infrared PA spectrum of a clipped toe-nail is shown in Figure 7.24.

The infrared analysis of human hair by traditional techniques is problematic because of the requirement for sample preparation. For example, transmission measurements require dispersion of the hair sample in an infrared-transparent solid such as KBr. On the other hand, acquisition of a PA spectrum of hair is quite straightforward: only a few individual hairs, cut to lengths of about 2–3 mm, are required for this experiment. To illustrate, Figure 7.25 shows a PA spectrum of approximately 10 hairs. About a dozen bands are clearly observable, the most prominent being the amide I and II peaks at about 1660 and 1540 cm^{-1}, respectively. The PA spectrum of hair has also been obtained with a single-fiber accessory, as discussed in Chapter 5.

The suitability of PA infrared spectroscopy for the characterization of other human tissues has been demonstrated by several research groups. Wentrup-Byrne et al. (1995, 1997) used PA spectroscopy to categorize human gallstones, thereby

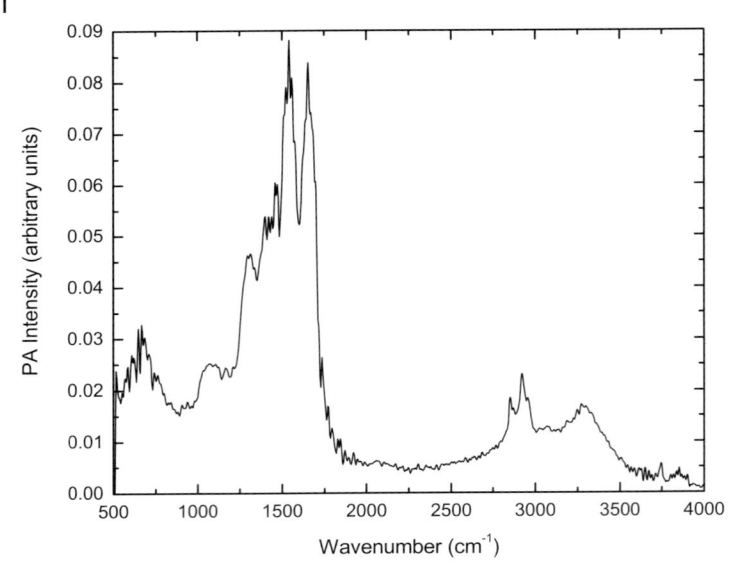

Figure 7.24 PA infrared spectrum of clipped human toe-nail. Sample was provided by the author.

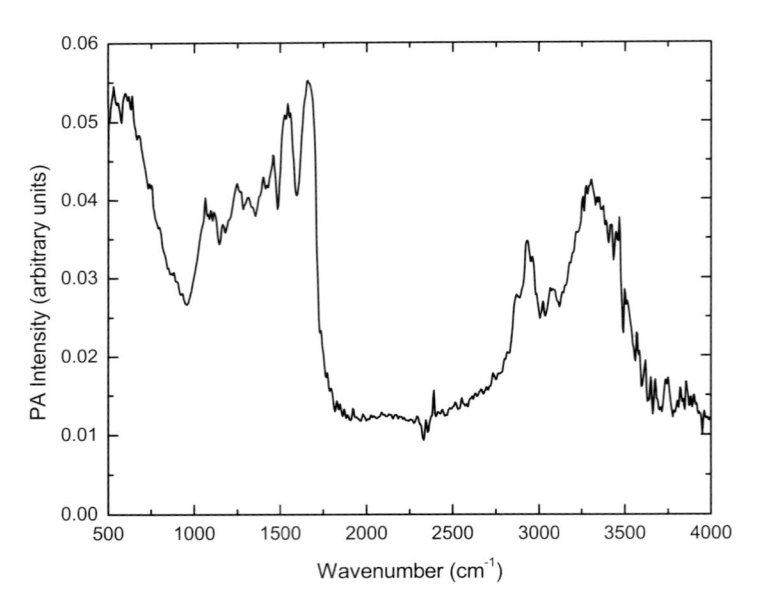

Figure 7.25 PA infrared spectrum of approximately 10 hairs (a mixture of brown and white), cut to a length of about 3 mm. Sample source was the same as in Figure 7.24.

avoiding traditional crushing, extraction, and wet chemical techniques. Because a relatively large area of the gallstone surface is examined, it is unlikely that any inhomogeneities will be overlooked. Partial least squares(PLS) analysis, a common multivariate regression algorithm, was used to model cholesterol concentration using the results from PA spectroscopy. The model predictions yielded reasonable agreement between predicted and measured (by traditional, non-spectroscopic techniques) cholesterol concentration, with discrepancies being attributed to inhomogeneities in the stones. Grinding the stones would tend to distribute these inhomogeneities more evenly, but at the expense of structural integrity.

Measurement of thermal diffusion properties provides a possible means for the determination of tissue layer thickness in cancer diagnostics. Schüle, Schmitz, and Steiner (1999a, 1999b) utilized PA infrared spectroscopy and photothermal response to 10.6-μm CO_2 laser radiation to characterize skin, liver, and muscle tissue. Step-scan PA amplitude and phase spectra of the three tissues displayed a number of features between 1000 and 1800 cm^{-1}, ensuring positive sample identification. These spectra were simulated by the finite-difference method, using data from other absorption measurements and assumed thermal properties. Both amplitude and phase spectra agreed well with the absorption spectrum, plotted as the variation of the absorption coefficient with wavenumber. The phase spectra exhibited better band definition, demonstrating the expected lower susceptibility to PA saturation.

PA measurement of glucose concentration has been implemented in several laboratories. Shen et al. (1999) carried out time-resolved experiments at visible and near-infrared wavelengths, using a wide-band hydrophone for detection. Sound velocity, absorption coefficients, and thermal properties were determined. Khalil (2004) and Sieg, Guy, and Begoña Delgado-Charro (2005) reviewed a number of analytical techniques, including near-infrared PA spectroscopy, for transdermal glucose monitoring. An increase in glucose level leads to a reduction in specific heat capacity and greater acoustic velocity, enabling quantification. At longer wavelengths, von Lilienfeld-Toal et al. (2005) utilized QCLs emitting at 1066 and 1080 cm^{-1} for non-invasive glucose monitoring. A two-chamber (measurement and reference) cell was employed in this investigation. In similar work, pulsed PA spectroscopy was used to study tissue oxygenation (Fainchtein et al., 2000) and blood oxygen saturation (Laufer et al., 2005).

PA infrared spectroscopy has also been used to study biomaterials used for bone replacement. These composite materials typically consist of HAp and polymers. Helwig et al. (2001) investigated polymerized lactones in the presence of HAp, while Bhowmik et al. (2007) employed molecular dynamics simulations to study the interaction between polyacrylic acid and HAp. The PA technique is not discussed in detail in these articles.

Traditional methods for the identification of bacteria can require up to two days to complete. Ardeleanu, Morriset, and Bertrand (1992) utilized PA infrared spectroscopy for this task in an attempt to develop a more rapid procedure. Four different bacteria were studied: *Staphylococcus aureus*, *Streptococcus pyogenes*, *Escherichia coli* and *Proteus vulgaris*. The PA spectra of these bacteria were of very good quality and resembled published infrared spectra obtained in transmission experiments. Bands due to carbohydrates were observed in the 900–1200 cm^{-1} region; between 1200 and 1500 cm^{-1}, peaks were attributed to proteins, fatty acids, and phosphates; the amide I and II maxima appeared at 1656 and 1547 cm^{-1}, respectively; an ester C=O band was observed in the 1735–1745 cm^{-1} region; and, finally, C–H features occurred between 2800 and 3000 cm^{-1}. Although the PA spectra displayed in this article did not differ greatly, the authors concluded that the technique might eventually be used to identify different types of bacteria.

The preceding discussion pertains to the PA characterization of condensed-phase samples. Medical applications of PA infrared spectroscopy involving gas samples are also found in the literature; some representative examples are briefly mentioned below.

Mitsui and coworkers at Nagoya University used PA spectroscopy for the detection of N_2O in exhaled human breath. The technique was initially compared with gas chromatography (Mitsui et al., 1997), and a linear correlation between results from the two methods was observed. A Multi-Gas Monitor, manufactured by Brüel & Kjær, was used to detect the reduction of nitrate to N_2O by microflora; the gas was measured after ingestion of lettuce or vegetable juice by volunteers (Mitsui and Kondo, 1999). No spectra are included in these articles, nor is the wavelength used for detection specified. Breath samples from Chinese and Japanese subjects revealed significant differences between the two ethnic groups, possibly indicating different metabolic pathways (Mitsui, Kato, and Kondo, 2000). The PA gas analyzer was also used to investigate the effects of stomach diseases on the concentration of N_2O in breath (Kondo et al., 2000; Mitsui and Kondo, 2004).

PA gas analysis has recently been employed in hospital settings. Schiewe-Langgartner et al. (2005) used a commercial gas analyzer to monitor the exposure of personnel to anaesthetics (sevoflurane, N_2O). Rey et al. (2008) used CO_2 laser PA spectroscopy to detect by-products emitted during surgical cautery. In related work, Hübner et al. (2008) employed a $^{13}CO_2$ laser for the PA analysis of surgical smoke produced during laparoscopic colonic surgery.

The results described in this section show that PA infrared spectroscopy is very well suited for a wide range of medical applications. These have been shown to include analysis of drugs and tissue samples; medical researchers may suggest additional problems such as the analysis of biofluids (Wang et al., 1996) or cancer research (Harvey et al., 2007) that are amenable to the PA techniques described in this book. Implementation of these suggestions will further enhance the role of PA spectroscopy in medical applications, where several additional photothermal methods are already in active use.

7.7
Polymers

The reader may already know of the existence of a large body of published literature describing the PA mid-infrared spectra of polymers. Inspection of Appendix

2 confirms that polymers comprise one of the most popular sample classes which have been studied using PA infrared spectroscopy: more than 270 articles that discuss spectra of polymers are included in this list. This popularity is partly due to the fact that polymer films and laminates are quite amenable to depth profiling experiments. Indeed, polymer layers with thicknesses on the order of a few micrometers–dimensions comparable to typical μ_s values in PA FT-IR spectroscopy–can be prepared by several well-known methods. A number of articles on the subject of depth profiling in layered polymer samples are discussed in Chapter 5; most of these publications are not mentioned in the present section.

The second reason for the widespread use of PA infrared spectroscopy in polymer characterization is, of course, the minimal sample preparation required with the technique. Numerous authors have reported PA infrared spectra of polymers that were obtained during investigations of polymerization reactions, morphological changes, interfacial phenomena, and thermal stability. The use of PA detection was partly a matter of convenience in many of these studies; accordingly, few particulars of PA spectroscopy are discussed in these works. This extensive utilization of PA detection of mid-infrared spectra by polymer scientists is obviously encouraged by the straightforward use of commercial PA gas-microphone cells and FT-IR spectrometers.

The considerable amount of scientific literature on the PA infrared spectra of polymers precludes the presentation of a comprehensive review here. Because instrumentation in PA infrared spectroscopy is a focus of this book, many references in Appendix 2 that emphasize polymer chemistry and structure are not discussed in this section. Instead, a limited number of representative articles are cited in the following paragraphs. A long series of articles by M. W. Urban and coworkers is also briefly summarized. Researchers who require more details on this subject may choose to consult the appropriate articles in the primary scientific literature or the reviews mentioned below.

Fortunately, the large extent of the scientific literature pertaining to PA infrared spectra of polymers has prompted a number of authors to write helpful review articles during the last three decades. In chronological order (and, therefore, in sequence of increasing literature coverage) these include Vidrine and Lowry (1983), Koenig (1985), Garbassi and Occhiello (1987), Urban (1987, 1989a, 1989b), Jasse (1989), and Dittmar, Palmer, and Carter (1994). This last review was also mentioned in Chapter 1 because of the considerable amount of useful information that it conveys in general with regard to PA infrared spectroscopy.

During the period that could be referred to as the demonstration phase of PA infrared spectroscopy (the late 1970s and early 1980s), some researchers chose polymers as typical solid samples that might be used to validate the technique. The resulting publications presented spectra that were mostly of good quality, although the data tended to be accompanied by little or no interpretation (Low and Parodi, 1980a; Royce, Teng, and Enns, 1980; Vidrine, 1980; Chalmers *et al.*, 1981; Zachmann, 1984). Another group of somewhat more rigorous investigations examined the differences between the PA and ATR infrared spectra of polymers, particularly with regard to sampling depth. These studies confirmed the now well-known fact that PA FT-IR spectroscopy ordinarily characterizes depths significantly greater than those analyzed in ATR spectroscopy, where the evanescent wave penetrates the sample to a distance on the order of 2 μm or less (Krishnan, 1981; Krishnan et al., 1982; Gardella et al., 1984; Vidrine, 1984; Saucy, Simko, and Linton, 1985). The requirement for little or no sample preparation in PA infrared spectroscopy was also relevant in this context, since many of the polymer samples studied were irregular in shape.

PA infrared spectra of both undoped and doped polyacetylene were described by two research groups at about the same time as the work just mentioned. The emerging PA technique was employed primarily because it allowed measurement of infrared spectra of these opaque samples without the usual requirement that they be available as thin films suitable for transmission experiments.

Two articles on this topic were published by E. M. Eyring's group. In the first, Riseman et al. (1981) obtained spectra of a mixture of cis and trans isomers of undoped polyacetylene. An n-doped sample containing the tetrabutylammonium ion was also investigated in this study. Although the spectra reported by these authors were rather noisy, several new bands were successfully identified for the doped sample. The second study by this group (Yaniger et al., 1982) showed that acetylene doped with iodine yielded a PA spectrum containing a strong band at 1400 cm^{-1} due to carbon-carbon stretching, as well as a very broad band near 950 cm^{-1} and a narrower peak at 750 cm^{-1} (Figure 7.26). These results were consistent with a model involving charge transfer into soliton levels. Metallic semiconducting and insulating regions were both concluded to exist within moderately doped bulk samples.

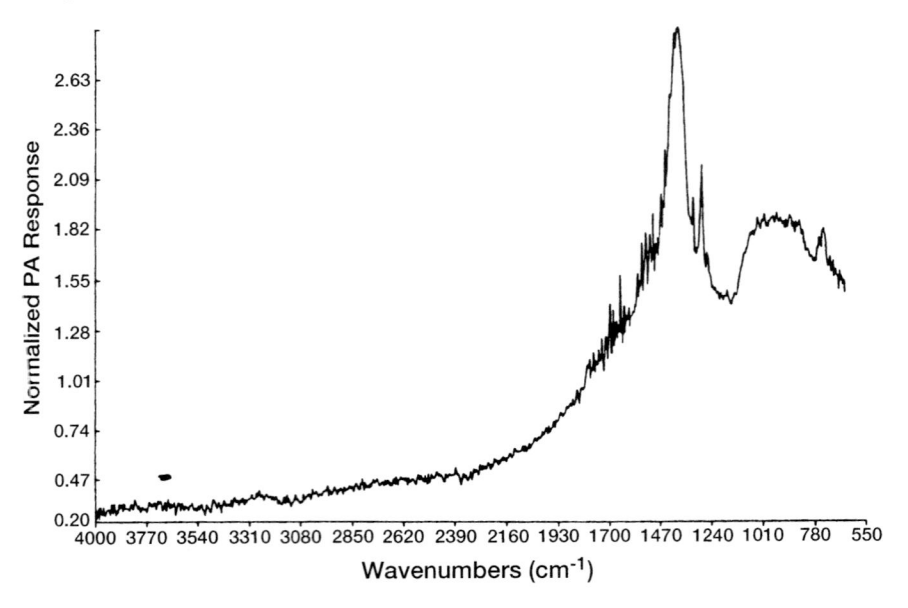

Figure 7.26 PA infrared spectrum of polyacetylene doped with iodine. (Reproduced from Yaniger, S.I. et al., *J. Chem. Phys.* **76**: 4298–4299; used with permission. Copyright © American Institute of Physics 1982.)

In subsequent work, Eckhardt and Chance (1983) reported near-infrared PA spectra of thick films of cis and trans polyacetylene. The only band in these spectra occurred at 0.74 eV (5970 cm^{-1}). Doping with oxygen led to the appearance of another band at 0.67 eV (5400 cm^{-1}) that exhibited a shape consistent with soliton absorption. The data allowed the authors to develop a model of polyacetylene fibers in the defect structure. Yaniger *et al.* (1984a, 1984b) subsequently extended their research to include doped and undoped poly(*p*-phenylene) as well as its derivatives. The mid-infrared PA spectrum of this semiconductor displayed a number of well-defined bands.

A series of articles on the PA infrared spectra of polymers was published by Teramae and Tanaka of The University of Tokyo at about the same time as the work mentioned in the previous paragraphs. This work began with two exploratory investigations of textiles and polymers that are also mentioned in the discussion on PA spectra of textiles (Teramae and Tanaka, 1981; Teramae, Hiroguchi, and Tanaka, 1982). The design of the PA cell constructed by the authors and its Helmholtz resonance were also discussed in these articles. Very satisfactory PA infrared spectra of carbon-filled rubbers and phenolic resins were obtained in these early studies.

These authors then turned their attention to the characterization of polymer films. A sample consisting of a 40-μm layer of polyethylene and a 10-μm layer of PET yielded a surprisingly complicated spectrum containing bands from both layers, even though Rosencwaig-Gersho theory predicted that a PA signal should be obtained only from the thicker top layer (Teramae and Tanaka, 1984, 1985a). This result was concluded to arise from the generation of heat at the rear surface of the sample and to depend on the magnitudes of the optical absorption coefficients of individual bands (Teramae and Tanaka, 1985b). The positioning of the polymer film in the PA cell also had a major effect on the results: better spectra were obtained when the film was either immediately behind the entrance window or in the cell cavity.

The detection of the subsurface layer of a bilayered film was also turned to advantage in a relatively simple depth profiling experiment. Teramae and Tanaka (1985a) showed that a progressive reduction of the thickness of the top layer from about 25 to 3 μm in layered samples of epoxy resin on either polypropylene or polyethyleneterephthalate led to the concomitant intensification of several bands from the lower layer (Figure 7.27). Thus depth profiling of polymer films can be accomplished even when only a single rapid-scan FT-IR mirror velocity is available to the spectroscopist. These authors reviewed their results in a subsequent publication (Teramae and Tanaka, 1987).

PA and ATR infrared spectroscopies were compared with regard to their capabilities for the characterization of polymeric coatings in more recent publications by several research groups. Carter, Paputa Peck, and Bauer (1989) used these two techniques, as well as transmission, specular reflection, and diffuse reflectance, to investigate the weathering of paints used in the automotive industry; the greater sampling depth in PA infrared spectroscopy mentioned above was again observed in this work. Factor, Tilley, and Codella (1991) investigated a photo-curable coating

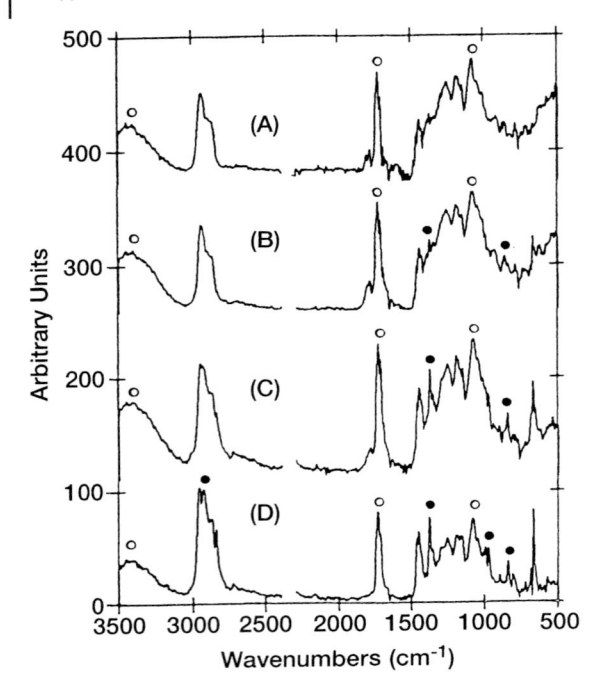

Figure 7.27 PA infrared spectra of a series of bilayered films of epoxy resin on 23-μm polypropylene. The thickness of the top layer was (A) 23 μm, (B) 9 μm, (C) 5 μm, and (D) 3 μm. The filled circles denote bands from the lower layer, whereas the open circles signify the upper layer. (Reproduced from Teramae, N. and Tanaka, S., *Appl. Spectrosc.* **39**: 797–799, by permission of the Society for Applied Spectroscopy; copyright © 1985.)

for a polycarbonate substrate and obtained infrared results that agreed with those from photo-differential scanning calorimetry. A saturated feature in the PA spectra was successfully used to normalize the intensities of two bands at 1620 and 1635 cm^{-1} that were employed to monitor curing in this work. Finally, Wetzel and Carter (1998) employed ATR microspectroscopy and PA infrared spectroscopy to study the degradation of an acrylic polymer by ultraviolet radiation. Disappearance of the acrylic bands and the corresponding appearance of new bands due to degradation products were both noted in the PA spectra.

It is perhaps appropriate to mention a few studies in which PA infrared spectroscopy was used to characterize polymers primarily because of the minimal sample preparation involved. The adsorption of polymers on gamma-iron oxides, which pertain to magnetic memory media, was investigated a number of years ago by Cook, Luo, and McClelland (1991) and Nishikawa et al. (1992). Interactions between the polymers and the iron oxides were successfully elucidated using the PA spectra obtained in both studies. In another example, Hocking et al. (1990a, 1990b, 1990c, 2000, 2001) and Klimchuk, Hocking, and Lowen (2000) employed PA infrared spectroscopy to characterize a series of newly synthesized imide-amide copolymers; the use of PA detection was rather incidental to this work.

Other polymer-based research topics that have benefited from the measurement of PA infrared spectra include organic–inorganic hybrids (Gao *et al.*, 2001, 2002), carbon fiber–polyimide composites (Moyer and Wightman, 1989), characterization of polyurethane clearcoats (van der Ven and Leuverink, 2002), and diffusion in elastomers (Paputa Peck *et al.*, 1991). The thermal diffusivity of polymers, often estimated as $\alpha_s = 0.001\,\text{cm}^2\,\text{s}^{-1}$, was measured using an 'open PA cell' by D'Almeida *et al.* (1998). The value of α_s was actually observed to vary between 0.0008 and $0.0012\,\text{cm}^2\,\text{s}^{-1}$ for a series of resin mixtures in this work.

7.7.1
Research of Urban's Group

An impressive scientific literature on PA infrared spectra of polymers, spanning a period of more than 20 years, has been published by M. W. Urban and a series of coworkers. Much of this research was carried out at North Dakota State University; the more recent investigations were performed at the University of Southern Mississippi. Table 7.2 lists 46 publications on PA infrared spectroscopy by this group, together with the main topics discussed in these articles.

Table 7.2 Topics Discussed in Publications by M. W. Urban *et al.* on PA Spectra of Polymers.

Publication	Year	Topic
Appl. Spectrosc., **40**, 994–998	1986	Depth profiling
Appl. Spectrosc., **40**, 1103–1107	1986	Phase transitions
Polymer, **27**, 1850–1854	1986	Kevlar 49 fibers
J. Coat. Technol., **59**, 29–34	1987	Coatings (review)
Macromolecules, **21**, 372–378	1988	Poly(vinylidene fluoride)
Polym. Mater. Sci. Eng., **59**, 311–315	1988	Ceramic fibers
Polym. Mater. Sci. Eng., **59**, 316–320	1988	Crosslinking
Polym. Prepr. (Am. Chem. Soc., Div. Polym. Chem.), **29**, 356–357	1988	Crosslinking
Appl. Spectrosc., **43**, 1387–1393	1989	Rheo-photoacoustics
Composites, **20**, 145–150	1989	Silane coupling agents
Composites, **20**, 585–588	1989	Ceramic fibers
Macromolecules, **22**, 1486–1487	1989	Crosslinking reactions
Polym. Mater. Sci. Eng., **60**, 739–743	1989	Rheo-photoacoustics
Polym. Mater. Sci. Eng., **60**, 875–879	1989	Crosslinking of polyester/styrene
Polym. Mater. Sci. Eng., **61**, 132–136	1989	Review
Prog. Org. Coatings **16**, 321–353	1989	Coatings (review)

Table 7.2 Continued

Publication	Year	Topic
Prog. Org. Coatings **16**, 371–386	1989	Curing of alkyd coatings
Polym. Mater. Sci. Eng., **62**, 895–899	1990	Surfaces and interfaces of latices
J. Polym. Sci., Part A: Polym. Chem., **28**, 1593–1613	1990	Curing in coatings
Polym. Commun., **31**, 279–282	1990	Surfactants
Composites, **22**, 307–318	1991	Fiber/matrix interfaces
Polym. Mater. Sci. Eng., **64**, 31–32	1991	Interfacial interactions
J. Appl. Polym. Sci., **42**, 2287–2296	1991	Surfaces and interfaces of latices
Polymer, **33**, 3343–3350	1992	Ethyl acetate in poly(vinylidene fluoride)
Polymer, **34**, 1995–2002	1993	Germanium phthalocyanine polymers
Polymer, **34**, 3376–3379	1993	Ethyl acetate in PVDF
J. Coatings Technol., **66**, 59–67	1994	Film delamination
Polymer, **35**, 5130–5137	1994	Rheo-photoacoustics
ACS Symp. Ser., **648**, 301–331	1996	Latex film formation
Polymer, **38**, 2077–2091	1997	Rheo-photoacoustics
Prog. Org. Coatings, **32**, 215–229	1997	Surfactants
J. Appl. Polym. Sci., **70**, 1321–1348	1998	Surfactants
Polymer, **39**, 5899–5912	1998	Rheo-photoacoustics of PVDF
Appl. Spectrosc., **53**, 1520–1527	1999	Stratification; depth profiling
J. Adhes. Sci. Technol., **13**, 19–34	1999	Rheo-photoacoustics of bilayers
Macromol. Symp., **141**, 15–31	1999	Stratification; depth profiling
Polymer, **40**, 4795–4803	1999	Stratification; depth profiling
Langmuir, **16**, 5382–5390	2000	Degradation processes; depth profiling
Macromolecules, **33**, 2184–2191	2000	Stratification; depth profiling
Polymer, **41**, 1597–1606	2000	Stratification; depth profiling
Prog. Org. Coatings, **40**, 195–202	2000	Stratification
Polymer, **42**, 337–344	2001	Depth profiling
Polymer, **42**, 5479–5484	2001	Stratification
Polymer, **44**, 3319–3325	2003	Depth profiling
Langmuir, **20**, 10691–10699	2004	Depth profiling
Langmuir, **24**, 1808–1813	2008	Stratification

As noted earlier in this section, the size of the complete literature pertaining to the PA infrared spectra of polymers makes a detailed review impossible in this book. Similar reasoning applies to the numerous Urban publications. Moreover, a majority of the latter work is given over to polymer structure, coatings, interfacial phenomena, and related topics. These subjects are not reviewed here. However, it can be noted that PA depth profiling was discussed in a number of the works in Table 7.2 (Urban and Koenig, 1986; Urban, 1999; Pennington, Ryntz, and Urban, 1999; Kim and Urban, 2000; Zhao and Urban, 2000; Kiland, Urban, and Ryntz, 2000, 2001; Katti and Urban, 2003; Zhang and Urban, 2004). Rheo-photoacoustic spectroscopy, a technique involving the study of polymer deformation (stretching), was also investigated extensively (McDonald, Goettler, and Urban, 1989; McDonald and Urban, 1989; Ludwig and Urban, 1992, 1994, 1997, 1998; Pennington and Urban, 1999). Several important reviews were also published by this group, the early work by Urban (1989a) being particularly noteworthy. The bibliographic particulars of this body of literature are given in the reference list at the end of this chapter and Appendix 2.

7.7.2
Other Groups

Two other research groups that have utilized PA infrared spectroscopy for the characterization of polymers can be specifically mentioned here. First, the Groupe des Couches Minces at the Université de Montréal and École Polytechnique has utilized the technique to investigate surfaces, interfaces, and thin films in a variety of materials for a number of years. Typical examples include fluoropolymer surfaces (Piyakis et al., 1995) and low permittivity insulators (Poulin et al., 2000; Yang et al., 2001; Yang and Sacher, 2001). The PA phase-analysis numerical method, discussed in Chapter 6, was utilized in the research of Piyakis and coworkers.

Structure, composition, and orientation of various polymer blends were investigated in an extensive collaboration among the Academy of Sciences of the Czech Republic, Institute of Polymer Research Dresden, and University of Valladolid. PA infrared and Raman spectroscopies were both employed in this work. For example, Schmidt et al. (1997) investigated the chemical composition of poly(ε-caprolactam)-block-polybutadiene copolymers using PA and FT-Raman spectroscopy. Polyethylene/polypropylene blends were studied by López Quintana et al. (2002) and Schmidt et al. (2001, 2002), while polycarbonate/poly(styrene-co-acrylonitrile) was characterized using polarized PA and confocal Raman imaging spectroscopy (Schmidt et al., 2000a). Polarized PA and ATR spectra were compared with respect to the analysis of oriented bulk polymers by Schmidt et al. (2000b). The two types of spectra were generally comparable, except in the 1100–1300 cm^{-1} region, where the ATR spectra exhibited noticeably better band definition. PA spectra polarized parallel or perpendicular to the direction of cold-drawing were quite similar (Schmidt et al., 2000a, 2000b), suggesting minimal chain orientation. Polarized PA spectra were also obtained for blends of poly(ether-urethanes) with polypropylenes (Eichhorn et al., 2000). Noticeable orientation effects were observed in the latter study.

7.8
Catalysts

The requirement for infrared spectra of solid-state catalysts often presents a significant challenge to the spectroscopist. This is partly due to the fact that many heterogeneous catalysts are supported on substrates such as alumina, silica, and metal oxides which strongly absorb or scatter light of various wavelengths; this characteristic complicates acquisition of their vibrational (infrared and Raman) spectra. Moreover, it is highly desirable that catalyst samples should not be ground or otherwise perturbed prior to their analysis. Fortunately, PA infrared spectroscopy affords an opportunity for the study of these systems while minimizing potentially destructive sample preparation. The fact that PA spectra characterize the outer surfaces of these samples provides additional motivation for the use of this technique, as does the capability for depth profiling.

7.8.1
Early Work

Some of the earliest demonstrations of PA infrared spectroscopy included the investigation of catalysts. A number of relevant articles were discussed earlier in this book; the reader can also consult a brief review by Spencer (1986). Examples of this work occur in the research of Low and Parodi (1978), who obtained PA spectra of a chromia-alumina catalyst above $2000\,\text{cm}^{-1}$ and were able to detect the substitution of hydroxyls by methoxy groups. The silanization of silica was also reported in this article. Several years later, this group used PBDS to study the chemisorption of CO on Ni–C and other metal–carbon catalysts (Low, Lacroix, and Morterra, 1982).

PA infrared spectra of catalysts were also described in two presentations at the 1981 conference on Fourier Transform Infrared Spectroscopy. Vidrine (1981) examined a partially poisoned catalyst, demonstrating depth profiling by varying the interferometer mirror velocity in a series of rapid-scan measurements. Mehicic, Kollar, and Grasselli (1981) briefly discussed their PA investigations on metal oxide catalysts, putting forward the useful suggestion that the metal oxide of interest could be sprinkled on an alkali halide support disc to facilitate measurement of its infrared spectrum.

Dispersive near-infrared PA spectra of catalysts were reported in two early publications by Lochmüller and Wilder (1980a, 1980b). These studies are discussed in Chapter 2. The first investigation demonstrated the suitability of PA spectroscopy for the qualitative analysis of chemically modified silica gel, while the second showed that surface coverage of silica by organic compounds could also be quantified by this technique. The authors noted that the intrinsic infrared intensity associated with a particular adsorbate functional group may be modified by its interaction with the substrate; this implies that a calibration plot used for quantification of adsorbed species must be developed using intensities obtained under similar conditions.

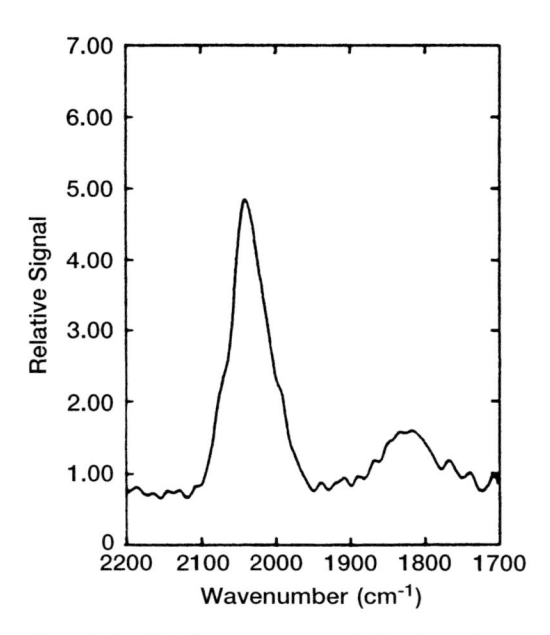

Figure 7.28 PA infrared spectrum of CO adsorbed on 5% Pt on alumina. (Reprinted with permission from Kinney, J.B. and Staley, R.H., *Anal. Chem.* **55**: 343–348. Copyright © 1983 American Chemical Society.)

Kinney and Staley (1983) designed and constructed a PA cell for use with an FT-IR spectrometer. These authors went on to investigate the infrared spectra of a variety of catalyst systems, including palmitic acid deposited on silica, CO on alumina-supported platinum, and Ag on alumina after reaction with gaseous $(CN)_2$. A typical result for the reaction of CO with Pt on alumina is shown in Figure 7.28. The authors reported that PA spectra of good quality were generally obtained within a few minutes, although scattering and absorption by the support occasionally proved to be troublesome.

Pioneering investigations by E. M. Eyring and his group on various aspects of PA infrared spectroscopy are discussed in several places in this book. During the first decade of widespread activity in this field, these researchers published a number of articles on the study of catalysts that are pertinent to the present discussion. For example, quantitative determination of surface adsorption sites in silica-alumina and γ-alumina was described by Riseman et al. (1982). Brønsted (proton donor) and Lewis (electron acceptor) acid sites were characterized by their effect on the infrared spectrum of adsorbed pyridine. Because of its emphasis on quantitative analysis, this article was also discussed in Chapter 5.

In related work, Gardella, Jiang, and Eyring (1983) investigated the adsorption of CO on dehydroxylated SiO_2. Both gaseous and adsorbed CO were quantified using PA intensities; 40% of the sites on the catalyst were found to be active. Similarly, the adsorption of CO on an Ni/SiO_2 catalyst was studied by Gardella et al. (1983). A noteworthy result in the latter work was the gradual conversion of adsorbed CO to gaseous CO_2. In fact, the authors of this investigation demonstrated that the infrared source in the FT-IR spectrometer promoted this reaction; many PA spectroscopists may recall similar results in their own work on different materials.

Two additional relevant articles were published by these researchers shortly after the initial investigations. PA infrared spectroscopy was used to study chromocene supported on silica, a catalyst for olefin polymerization (McKenna, Bandyopadhyay, and Eyring, 1984). Spectra acquired for chromocene/silica after treatment with ethylene and 2,3-dimethylbutene showed that a cyclopentadiene moiety remained attached to the catalyst after the reaction with silica. In another application, McKenna et al. (1985) investigated the photolysis of $Re_2(CO)_{10}$ adsorbed on silica and alumina. New carbonyl bands were observed in the ~1800–2500 cm^{-1} region and assigned to $Re_x(CO)_y$ entities in a number of different environments that were created by the photolysis reactions.

Several years later, Wang, Eyring, and Huai (1991) used PA, diffuse reflectance and transmission infrared spectroscopies to study the adsorption of pyridine on HZSM-5 and HY zeolites. These researchers observed PA intensity enhancement when He was used as the transducing gas and ascribed their results to signal generation by means of volumetric expansion of the interstitial gas, rather than the more familiar transfer of heat to the carrier gas above the sample (McGovern, Royce, and Benziger, 1985). Interstitial gas expansion can play an important role in powders and other highly porous samples such as zeolites. Wang, Eyring, and Huai (1991) reported intensity enhancements by a factor of at least two for adsorbed pyridine when He was used in their experiments. The effect was smaller when N_2 was utilized in the PA cell because thermal conductivity and sound velocity are lower in this gas than in He.

The research of B. S. H. Royce and coworkers at Princeton University, briefly alluded to in the previous paragraph, constituted another important contribution to the study of catalysts during the 1980s. This work began with a brief report that exposure of zeolite to moist air for 30 min prior to measurement of a PA spectrum resulted in the appearance of CO_2 and water vapor bands (Royce, Teng, and Enns, 1980). The spectrum of a fresh zeolite, by contrast, did not exhibit these bands. This experiment confirmed the suitability of PA spectroscopy for these samples and also illustrated the surface activity of zeolite.

In subsequent research, this group used PA infrared spectroscopy to characterize the methoxylation of silica, methanol adsorption on Na–Y zeolite, and the adsorption of CO on platinum black and Pt/Al_2O_3 (McGovern, Royce, and Benziger, 1984). Figure 7.29 illustrates the high quality of the spectra and confirms the adsorption of methanol on zeolite. The general absence of saturation in the PA spectra was one noteworthy result in this study. The powdered samples displayed several important optical and thermal characteristics: (i) the optical absorption coefficient was equal to the product of the absorption coefficient of the solid phase and its volume fraction; (ii) the applicable thermal conductivity was derived from the interstitial gas; and (iii) the heat capacity was that of the solid. This behavior distinguishes fine powders from coarse solids.

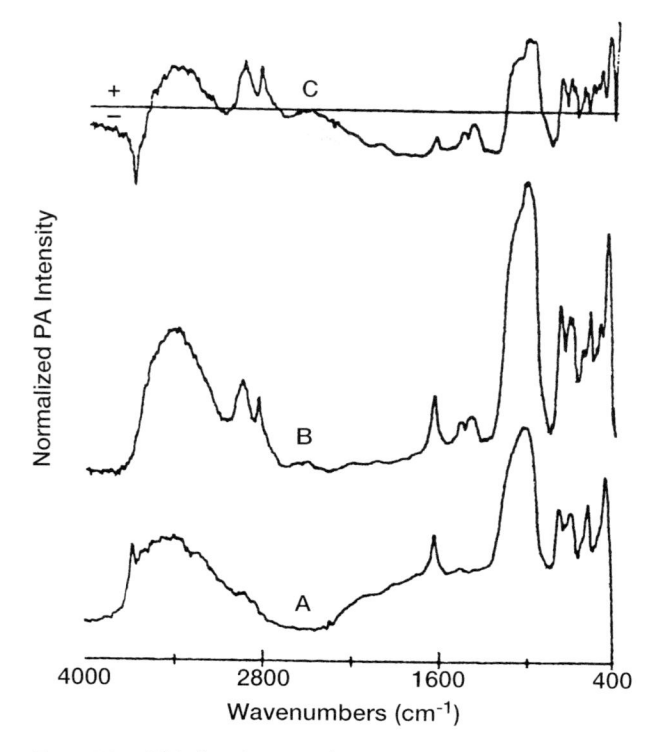

Figure 7.29 PA infrared spectra depicting methanol adsorption on Na-Y zeolite. Curve A, adsorbed sample; curve B, dried zeolite; curve C, difference, (A − B) × 2. (Reprinted from *Appl. Surf. Sci.* **18**, McGovern, S.J., Royce, B.S.H. and Benziger, J.B., Infrared photoacoustic spectroscopy of adsorption on powders, 401–413, copyright © 1984 with permission from Elsevier Science.)

PA infrared spectra of silica, methoxylated silica, aluminum oxides, and hydroxides were described in detail by Benziger, McGovern, and Royce (1985). The emphasis in this article was primarily on the information revealed by the spectra, which were of very good quality. For example, adsorbed water affected the silica lattice vibrations, whereas particle–particle interactions influenced the vibrations of the surface species. The results obtained for the aluminum-containing compounds – some of which resemble particular clays and minerals discussed later in this chapter – demonstrated the suitability of PA spectroscopy for the study of structural (phase) transformations. The PA infrared method and results obtained in earlier studies were thoroughly reviewed by Royce and Benziger (1986), who also presented results for supported metal oxide catalysts. *In situ* measurements at temperatures up to 400 °C were also described in this publication.

PA infrared spectroscopy was also used to characterize heteropoly catalysts, which are of interest with regard to the conversion of methanol to hydrocarbons. Highfield and Moffat studied the thermal stability of 12-tungstophosphoric acid, $H_3PW_{12}O_{40}$, and its interactions with NH_3 (Highfield and Moffat, 1984a), pyridine (Highfield and Moffat, 1984b), and methanol (Highfield and Moffat, 1985a). The acidic properties of the ammonium, aluminum, sodium, and pyridinium salts of $H_3PW_{12}O_{40}$ were studied by means of these interactions. Heating generally caused a loss of detail in the PA spectra of these salts. Brønsted acidity was found to be essential for catalyzing the elimination of water from methanol, a process thought to be related to hydrocarbon formation. During the investigation of NH_3 uptake, it was concluded that saturation did not affect the spectra and hence that PA spectra could be used for quantification. However, subsequent work (Highfield and Moffat, 1985b) showed that the use of PA intensities for quantification of pyridine and ammonia on $H_3PW_{12}O_{40}$ was reliable only at low concentrations, since saturation tended to limit band intensities for greater amounts of these compounds. This research was summarized in Chapter 5.

7.8.2
Surface Characterization and Reactions

Adsorption and surface reactions on silica, zirconia, alumina, and other materials were studied with the use of PA infrared spectroscopy by a number of researchers. In an early example, Nadler, Nissan, and Hollins (1988) investigated the hydrolysis of diisopropyl fluorophosphate, a toxic organophosphorus compound, on silica and alumina surfaces. The spectra indicated bonding to surfaces at the P=O site; hydrolysis rates were obtained by monitoring the time dependence of the P=O band at $1270 cm^{-1}$.

Mohamed and Vansant (1995a) examined the reduction of copper(II) oxide supported on mordenite zeolite through adsorption of carbon monoxide and H_2. In related work, PA infrared spectra were used to characterize pyridine adsorption on silica-supported copper-molybdenum and copper-silica catalysts (Mohamed, 1995; Mohamed and Vansant, 1995b). The interaction between molybdenum and silica was studied with reference to hydroxyl groups and pore structure by El Shafei and Mokhtar (1995). More recently, Mohamed (2003) combined PA infrared spectroscopy with other techniques for characterization of dealuminated mordenite zeolites, which display an enhanced capacity for adsorption of NH_3.[2]

Adsorption of pyridine on catalyst surfaces was monitored using PA infrared spectra in a series of articles by Hess, Kemnitz, and coworkers at Humboldt-Universität zu Berlin. For example aluminum oxides, hydroxyfluorides, and fluorides were considered as halogen exchange catalysts, with the nature of the acidic sites being determined from spectra of chemisorbed pyridine (Hess and Kemnitz, 1994). Activation of γ-alumina and $AlF_2(OH)$ increased catalytic activity and the strength of ammonia adsorption. A particular vibration of the pyridine ring was observed to shift by a few wavenumbers according to the choice of catalyst in this work. Similarly, PA spectra revealed the nature of the acid sites in zirconium and titanium dioxides, which were considered as possible catalysts for the alkylation or double bond isomerization of 1-butene, as well as the disproportionation of

---

2) Most PA spectra are (without comment) plotted to resemble transmission spectra in this series of articles. This practice leads to confusion and should be avoided.

CHClF$_2$ (Hess and Kemnitz, 1997). Finally, precipitated and aerogel-synthesized zirconia were compared with regard to activity and conversion rates for the benzoylation of anisole (Quaschning et al., 1998). Bands at 1450 and 1490 cm^{-1} were used to quantify the Brønsted and Lewis acid sites in samples prepared at various temperatures between 110 and 550 °C. Nitrogen adsorption, X-ray diffraction, and other characterization techniques were also utilized in this work.

Oxidative coupling of hydrocarbon radicals to form larger molecules by means of a metal-containing catalyst was investigated by Jianjun et al. (1992). Lithium-doped Mn–MgO catalysts were characterized using PA infrared spectroscopy, X-ray diffraction, and atomic absorption. The PA spectra displayed an MgO band at 410 cm^{-1}, whereas MnMg$_6$O$_8$ yielded bands at 450 and 600 cm^{-1}. Addition of Li produced LiMnMg$_6$O$_8$ and resulted in the appearance of a band at 500 cm^{-1}. This work showed that the addition of Li$^+$ improves the selectivity for C$_2$ production in methane coupling.

Morey, Davidson, and Stucky (1996) used PA infrared spectroscopy, Raman spectroscopy, and other analytical methods to characterize Ti-MCM-48, a large-pore molecular sieve containing highly dispersed titanium within the cubic MCM-48 silicate structure. This material was considered as a possible catalyst for the oxidative transformation of large organic molecules. The PA spectra suggested the isomorphous substitution of Si by Ti atoms inside the silica-based walls.

Reactions on zirconia and titania surfaces were studied by several groups. Examples include Van Der Voort et al. (2002), who utilized PA infrared, diffuse reflectance, and FT-Raman spectroscopies in an investigation of the interaction of Fe(acac)$_3$ (acac = acetyl acetonate, the ionized form of 2,4-pentanedione) with ZrO$_2$. Pawsey, Yach, and Reven (2002) described self-assembled monolayers formed by adsorption of carboxyalkylphosphonic acids on TiO$_2$ and ZrO$_2$. PA infrared and ^{31}P NMR spectra showed that the phosphonic group binds to the surface during monolayer formation. Particle sizes down to 5 nm were investigated in this work. Finally, Segura et al. (2006) described the preparation and characterization of vanadium oxide deposited on mesoporous titania. The vanadyl complex VO(acac)$_2$ served as the vanadium source in this synthesis. PA spectra were obtained for the calcined titania blank, precursor V-acac/TiO$_2$, high concentration V-acac complex, and final VO$_x$/mesoporous TiO$_2$ complex. The hydroxyl bands and acac features in the 1300–1600 cm^{-1} region differed significantly for the intermediates and final product. The materials prepared in these three studies were of interest as potential catalysts and in other applications.

Preparation of catalysts through introduction of silver into support materials was recently described in two articles. Gac (2007) modified cryptomelane-type manganese oxides with silver, finding that the produced catalysts yielded high activity during low-temperature oxidation of CO. Samples were characterized using nitrogen adsorption/desorption, X-ray diffraction, and PA infrared spectroscopy. Spectra obtained after calcination of the samples displayed a few broad bands due to hydroxyl groups and isolated water molecules. In related work, Gac et al. (2007) studied silver-doped mesoporous MCM-41 silica prepared by two different methods. The introduction of silver caused structural changes in the silica materials. Bands in the PA spectra were attributed to several different functional groups. Most features were quite broad, except for a sharp peak due to isolated silanol (Si–OH) groups. The band broadening was attributed to the amorphous nature of the samples.

7.8.3
Research of Ryczkowski's Group

The research group of J. Ryczkowski and coworkers at the University of Maria Curie-Sklodowska has used PA infrared spectroscopy extensively in catalysis research during the last two decades. An extensive series of articles has been published in the scientific literature on this work; however, only a few representative works in this literature are mentioned here and are included in the reference list in Appendix 2.

A wide variety of research topics has been investigated by this group. PA spectroscopy was found to be an essential component of this work because it enabled the analysis of opaque materials. The adsorption of ethylenediaminetetraacetic acid (EDTA), a popular chelating compound, on alumina was mentioned briefly (Ryczkowski, 1994) and later discussed in detail (Ryczkowski, 2007b). Preparation of metal-supported catalysts involves addition of a metal precursor to the support; the chelating agent can prevent migration of the metal (such as Ni) into the alumina lattice. PA infrared spectra were used to differentiate between H_4EDTA, H_2Na_2EDTA and Na_4EDTA on γ-Al_2O_3 (Figure 7.30). Nickel alumina-supported catalysts prepared from acidic solutions by classical impregnation (CIM) and double impregnation (DIM) were discussed by Pasieczna-Patkowska and Ryczkowski (2007). PA spectra suggested that catalysts prepared using DIM possessed better metal reduction than those derived from CIM. PA infrared spectra exhibited bands broader than those in corresponding transmission spectra, possibly indicating partial PA saturation.

A similar approach was employed in the preparation of alumina-supported ruthenium catalysts, which are used for conversion of higher hydrocarbons. The alumina support was modified using the disodium salt of EDTA before introduction of ruthenium red (ammoniated ruthenium oxychloride) as the metal precursor. The PA spectrum of the ruthenium red changed quickly as physical adsorption was replaced by chemisorption.

Removal of an organic template from MCM-41 silica was described by Ryczkowski et al. (2005). Temperature-programmed desorption of a surfactant molecule was monitored using mass spectrometry, differential thermal analysis, and PA infrared spectroscopy. Several organic functional groups remained on the silica surface after removal of the template; these groups were eventually eliminated by oxidation of the calcined sample.

Finally, two reviews published by this group should be mentioned. Pasieczna and Ryczkowski (2003) discussed the use of PA infrared spectroscopy for surface and interface analysis, referring to examples of pyridine chemisorption on γ-alumina and zirconia as well as two different fields (coal analysis and kinetics). PA spectroscopy was described as an 'undervalued method' in heterogeneous

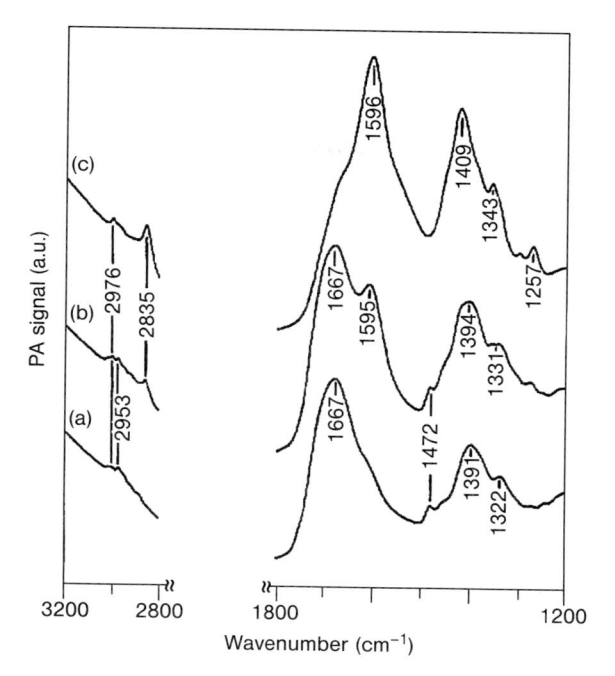

Figure 7.30 PA infrared spectra of EDTA adsorbed on γ-alumina. (a) H$_4$EDTA, (b) H$_2$Na$_2$EDTA, (c) Na$_4$EDTA. (Reprinted from *Vib. Spectrosc.* **43**, Ryczkowski, J., Spectroscopic evidences of EDTA interaction with inorganic supports during the preparation of supported metal catalysts, 203–209, copyright © 2007.)

catalysis, a comment that undoubtedly still holds true. Ryczkowski (2007a) also reviewed PA infrared spectroscopy in catalysis research, mentioning investigations on Al$_2$O$_3$, MoO$_3$, SiO$_2$, ZrO$_2$, Rh$_4$(CO)$_{12}$, and γ-Al$_2$O$_3$. Typical studies mentioned in this review discuss adsorption on surfaces, characterization of the support surfaces, and catalytic reactions. As noted above, identification of Brønsted and Lewis acidity in catalysts is of particular interest. This question pertains to calcined and sulfided catalysts.

7.9
Gases

Photoacoustic spectroscopy has achieved perhaps its greatest successes in gas analysis. The impressive sensitivity of the PA effect in gases sets a standard that is rarely attained by vibrational spectroscopists who investigate condensed-phase substances: trace gas analysis by PA spectroscopy routinely realizes detection limits at the parts-per-million (ppm) or parts-per-billion (ppb) levels.[3] Such

---

3) These levels are frequently denoted as ppmv and ppbv (parts-per-million and parts-per-billion by volume, respectively) in the scientific literature.

concentrations are, of course, orders of magnitude lower than those accessible in infrared and Raman spectra of solids and liquids in ordinary circumstances.

PA infrared spectroscopy of gases has been practiced extensively over the last four decades, predating the development of PA FT-IR spectroscopy of solids and liquids by more than 10 years. The relatively long duration of the gas-phase work has naturally led to the existence of a large body of literature. Several important early references are mentioned in the historical review in Chapter 2. A considerable number of references are discussed in this section and/or listed in Appendix 3; additional articles on PA spectra of gases are known to the present author but are not cited in this book.

Much of the relevant work on PA spectroscopy of gases completed prior to 1994 was summarized in *Air Monitoring by Spectroscopic Techniques*, edited by M. W. Sigrist and published as Volume 127 in the Wiley-Interscience Chemical Analysis series. Many additional review articles and other summaries have also been published. A comprehensive review of this literature – obviously a major undertaking – is well beyond the scope of this book. Instead, the 1994 publication by Sigrist will be taken as a demarcation point in this extensive literature: the present discussion is restricted to work published after this date. Readers who require further information may wish to consult the Chemical Analysis text or the original scientific literature on PA infrared spectra of particular gases or experimental techniques.

It should also be appreciated that a large fraction of the research on PA infrared spectra of gases has been carried out using near- and mid-infrared lasers and other tunable radiation sources: these include CO and CO_2 gas lasers, diode lasers, quantum cascade lasers, optical parametric oscillators, and other devices. By contrast, the number of articles describing PA FT-IR spectra of gases remains relatively small. Hence – recalling the definitions introduced in Chapters 1 and 2 – the majority of work on this subject can be described as discrete- (single- or multiple-) wavelength spectroscopy. This means that PA infrared spectra of gases can be discussed from a perspective different from that for most of the condensed-phase samples described in this book. The use of tunable infrared radiation sources provides the framework for organization of the following discussion.

7.9.1
CO Laser Radiation

Representative publications describing PA infrared spectra of gases, obtained using a CO laser as the infrared source and published since 1994, are listed in Table 7.3. In its most common ($\Delta v = 1$) mode of operation, this laser produces a series of emission lines between about 1250 and 2000 cm^{-1} (~5–8 μm), making it particularly well suited for the study of compounds containing C=C or C=O functionality. Ethylene and the carboxylic acids are typical examples and are included together with various other compounds in these investigations.

The research team of Persijn and coworkers at the University of Nijmegen utilized PA infrared spectroscopy to study gases associated with the ripening of fruits. Four articles published by these investigators describe elegant experiments

Table 7.3 Recent PA studies of gases: CO laser used as infrared source.

Gases	Region (cm^{-1})	Reference
$CH_3(CH_2)_nCOOH$, $n = 8, 10, 12, 14$	1730–1890	Jalink et al. (1995)
CH_4, H_2O, CO_2	1260–2000	Bijnen et al. (1996)
CH_3COOH, CD_3COOD, C_2H_5COOH	1600–1950	Kästle and Sigrist (1996a)
CD_3COOD	1700–1840	Kästle and Sigrist (1996b)
CH_4	1250–2040	Ageev, Ponomarev, and Sapozhnikova (1998)
C_2H_4, CH_3CHO, CO_2, C_2H_5OH, H_2O	1250–2080	Bijnen et al. (1998)
C_2H_5OH, CH_3CHO	1300–2000	Oomens et al. (1998)
HCOOH	1500–2000	Merker et al. (1999)
C_2H_5OH, CH_3CHO, C_2H_4, CO_2, H_2O	1500–1700	Persijn et al. (1999)
C_2H_4, C_2H_6, C_5H_{12}	2630–3600	Santosa et al. (1999)
C_5H_8	2820–3175	Dahnke et al. (2000)
C_2H_5OH, CH_3CHO, C_2H_4, CO_2, H_2O	1300–2000	Persijn et al. (2000)
CH_4, CO_2, C_2H_6	1250–2000, 2400–3600	Persijn, Santosa, and Harren (2002)

in which three CO-laser intracavity PA cells were employed for simultaneous detection of trace gases. In the first two studies, the laser was cooled to 77 K to achieve single-line operation over the 1250–2080 cm^{-1} tuning range; C_2H_4, acetaldehyde, CO_2, C_2H_5OH, and H_2O vapor were studied during aerobic and anaerobic gas emission by cherry tomatoes (Bijnen et al., 1998) and the fermentation of red bell peppers and apples (Oomens et al., 1998). Absorption coefficients were obtained for each of these gases using a series of lines between approximately 1350 and 1950 cm^{-1}. C_2H_5OH and acetaldehyde were successfully detected at ppb levels in multicomponent gas mixtures.

Trace gases emitted by pears were discussed in two subsequent publications by this group. Persijn et al. (1999) established detection limits ranging from 0.3 ppb to 10 ppm for the five gases mentioned above during a study of the effect of CO_2 on fermentation. The other work (Persijn et al., 2000) reported a detailed list of absorption coefficients for these gases, obtained by scaling intensity data to the HITRAN database.[4] Low-resolution PA spectra between 1300 and 2000 cm^{-1}, plotted as the variation of absorption coefficient with infrared wavenumber, were also reported in this work.

---

4) HITRAN is the acronym for High Resolution Transmission, a compilation of spectroscopic parameters used to predict transmission and emission spectra of selected small molecules.

Intracavity CO laser PA spectra of CH_4, H_2O, and CO_2 obtained with the use of a single PA cell were described by Bijnen et al. (1996). Detection of CH_4 in air and H_2O in the presence of O_2 was carried out in this work. In a very unusual application of PA infrared spectroscopy, these authors measured the emission of CH_4 and H_2O by cockroaches and scarab beetles! Motivation for this experiment is not as abstruse as might be anticipated: arthropods are believed to be responsible for as much as 25% of the CH_4 in the atmosphere. Hence this study has both biological and environmental implications. Trace gas concentrations at ppb levels were measured in this work.

A significant modification to this equipment was described by Persijn, Santosa, and Harren (2002). A newly designed CO laser, consisting of three concentric tubes, allowed cooling of the gas discharge mixture (He, CO, N_2, air). This laser exhibited increased gain and a wider emission range: both fundamental ($\Delta v = 1$; 1250–2000 cm^{-1}) and overtone ($\Delta v = 2$; 2400–3600 cm^{-1}) lines were available. Three intracavity cells were again used in this work. The resonance frequencies of these cells must agree within a few Hz. Where necessary, differences were compensated by cell temperature increases of a few degrees. To demonstrate the sensitivity of their PA gas detection system, the authors monitored C_2H_6 emission in a herbicide-treated maize leaf, as well as CH_4 and CO_2 release by a cockroach. Detection of sub-ppb concentrations was possible with this equipment.

Another application to biological systems was described by Ageev, Ponomarev, and Sapozhnikova (1998). These authors utilized PA spectroscopy to study the effects of C_2H_4 and other pollutants on pea, wheat, barley, and pine seedlings. CO_2, CO, and HeNe lasers were used in this work, the latter two being employed for CH_4 quantification. The kinetics of CO_2 evolution was examined, with concentrations being on the order of hundreds to thousands of ppm.

Long-chain saturated fatty acids, which occur as solids at ambient temperatures, were studied in the gas phase by Jalink et al. (1995). These authors utilized a heat pipe cell and CO laser excitation to measure PA spectra at elevated temperatures for the even-carbon-numbered capric, lauric, myristic, and palmitic acids (C_{10}, C_{12}, C_{14}, and C_{16} acids, respectively). About 25 different laser lines were available in the 1730–1850 cm^{-1} region, permitting observation of the ~1780-cm^{-1} C=O stretching band for each compound. PA intensities were measured as a function of temperature for capric acid, allowing confirmation of the validity of the Clausius-Clapeyron equation.

PA spectra of two lower carboxylic acids were obtained with the use of a CO laser by Kästle and Sigrist (1996a, 1996b). These authors studied acetic acid -h_4 and -d_4, as well as propionic (propanoic) acid, in the C=O stretching region (1700–1840 cm^{-1}). The PA gas cell was immersed in a temperature-controlled water bath so as to maintain its temperature between about 280 and 350 K in these experiments. Bands due to the acid monomers (~1780 cm^{-1}) and dimers (~1730 cm^{-1}) were observed, the former being partially resolved into P and R branches. The temperature dependence of the spectra enabled calculation of the thermodynamic quantities ΔH and ΔS associated with dimerization. In addition, absolute integrated absorbances were determined from transmission measurements performed in parallel with the PA experiments.

PA spectra of the simplest carboxylic acid (formic) were studied in an innovative experiment that utilized a CO laser (Merker et al., 1999). High-resolution spectra of HCOOH vapor were recorded by mixing CO laser lines with microwave radiation, which permitted tuning the emission over the 1500–2000 cm^{-1} region with a spectral coverage of about 50%. Despite the existence of gaps among the available laser lines, the authors were able to obtain the first rotationally resolved spectrum of the (HCOOH)$_2$ dimer. Definite assignments of some of the newly observed features in this PA spectrum were not immediately available – a typical situation when pioneering spectroscopic data are first obtained.

The overtone CO laser, which was mentioned above, has been employed in additional PA experiments. Santosa et al. (1999) discussed the use of the $\Delta v = 2$ CO laser, which potentially produces hundreds of lines in the 2.6–4.0 μm (2500–3850 cm^{-1}) region, and noted that the lines are well placed for observation of C–H, O–H, and N–H bands. Ethylene, ethane, and pentane were monitored using an intracavity PA cell in this work. The experimental apparatus was designed to analyze gases produced during the storage of fruit and to detect ethane and pentane in human breath or, alternatively, plant tissue emissions. The latter two gases are generated during the course of lipid peroxidation in both situations. Excellent detection limits at or below the ppb level were established in this investigation. Similarly, Dahnke et al. (2000) utilized a CO overtone laser and intracavity PA cell for measurement of isoprene (C$_5$H$_8$) emission by eucalyptus leaves under varying light conditions. The authors noted that other volatile organic compounds (VOCs) also give rise to bands in the ~3–4 μm region and found that the use of 13 laser lines was adequate to reliably determine isoprene concentration in the presence of other VOCs. A detection limit of 1 ppb was established for isoprene in this work.

The results discussed in the preceding paragraphs confirm that about half of the mid-infrared region (1250–2000 and 2500–3850 cm^{-1}) is amenable to gas-phase PA spectroscopy when fundamental and overtone CO laser radiation is employed as the infrared source. Longer wavelengths (lower wavenumbers) are accessible using CO$_2$ lasers, as discussed in the next section.

7.9.2
CO$_2$ Laser Radiation

The CO$_2$ laser has been widely used as an infrared radiation source for the measurement of PA spectra for many years. A selection of articles published between 1994 and 2009 that describe PA spectra of gases acquired using ^{12}CO$_2$ and ^{13}CO$_2$ lasers appears in Table 7.4. Tunable radiation between approximately 900 and 1100 cm^{-1} (~9–11 μm) is produced by these lasers, which are a logical choice for measurement of PA spectra of oxygen-containing compounds, NH$_3$, and various other small molecules. Typical results from such experiments are discussed in the following paragraphs.

The research of M. W. Sigrist and coworkers in the Institute of Quantum Electronics at the ETH in Zürich, alluded to earlier, has made a major contribution to the PA spectroscopy of gases for many years. An important part of these

Table 7.4 Recent PA studies of gases: CO_2 laser used as infrared source.

Gases	Region (cm^{-1})	Reference
C_2H_4, CH_3OH, C_2H_5OH, C_7H_8	1040–1057	Repond and Sigrist (1994)
CO_2, NH_3, O_3, C_2H_4, CH_3OH, C_2H_5OH, toluene	932–1088	Repond and Sigrist (1996)
NH_3, NH_2D, NHD_2, ND_3	930–1085	Petkovska and Miljanić (1997)
C_2H_4, SO_2	930–960	Gondal (1997)
C_2H_4, NH_3	944 + others	Radak et al. (1998)
C_2H_4	949, 953	Schäfer et al. (1998b)
O_3	1064	Zeninari et al. (1998)
C_2H_4	925–1085	Calasso and Sigrist (1999)
C_2H_4, N_2	1069	Zeninari et al. (1999)
C_2H_4, CH_3OH, C_2H_5OH, C_6H_6, CO_2, H_2O	949–1079	Nägele and Sigrist (2000)
CH_3OH, C_2H_5OH, C_2H_4, CO_2, $(CH_3)_2O$	930–1090	Sigrist et al. (2000a)
C_2H_4	949	Sigrist et al. (2000b)
C_7H_8, C_8H_{10}	975–1055	Zelinger et al. (2000)
O_3, O_2, N_2, He, Ne, Ar, Kr, Xe	1064	Zeninari et al. (2000)
NH_3	1085	Pushkarsky et al. (2002)
NH_3, $(CH_3)_2CO$, CH_3CHO, CH_3Br, $(CH_3)_2O$, C_2H_5OH, C_2H_4	920–1100[a]	Romann and Sigrist (2002)
AsH_3, CO, CO_2, COS, NO, H_2O	1350–2130[b]	
NH_3, C_2H_4	949, 978	Marinov and Sigrist (2003)
NH_3	927	Pushkarsky, Webber, and Patel (2003)
N_2O, NO_2, SO_2	920–1090	Schramm et al. (2003a)
N_2O, NO_2, SO_2	920–1090	Schramm et al. (2003b)
C_6H_6, C_2HCl_3, $C_4H_8O_2$	930–990	Radak et al. (2004)
NH_3	900–1100	Schilt et al. (2004a)
CH_3OH; various air pollutants	~1050	Zelinger et al. (2004)
'Chemical warfare agents'	927	Pushkarsky, Webber, and Patel (2005)
NH_3	925–1085	Zhang, Wu, and Yu (2007)
CH_3OH, C_2H_5OH, NH_3, H_2O, CO_2	900–1090	Rey et al. (2008)
n-Pentylacetate	900–1100	Herecová et al. (2009)
CH_3OH, O_3	900–1100	Skřínský et al. (2009)
C_2H_5OH, SF_6, O_3	900–1090	Zelinger et al. (2009)

a) Fundamental radiation.
b) Frequency-doubled radiation.

investigations involved the use of CO_2 lasers as infrared sources. For example, Repond and Sigrist (1994, 1996) used a high-pressure CO_2 laser, tunable from ~1039–1057 cm^{-1}, and a nonresonant cylindrical stainless steel cell to obtain PA spectra of C_2H_4, CH_3OH, C_2H_5OH, and toluene. C_2H_4 was detected at concentrations as low as 50 ppb in this investigation. The capabilities of the technique for quantification were clearly demonstrated by two important results: first, the PA signal varied linearly with concentration over a wide range (four orders of magnitude), and second, the spectrum of a multicomponent mixture was shown to be equal to the sum of the spectra of the individual gases. The later use of a resonant multipass PA cell (Nägele and Sigrist, 2000) capable of flow-mode operation yielded a reduction in the C_2H_4 detection limit to an even more impressive 70 parts per trillion (ppt).

A number of additional articles written by this research group describe many important instrumental developments that pertain to the PA infrared spectroscopy of gases. Calasso and Sigrist (1999) discussed in detail the characteristics of the two types of microphones (condenser and electret) used in nonresonant gas phase PA experiments. Two useful reviews on the detection of trace gases (Sigrist *et al.*, 2000a, 2000b) mentioned high-pressure (continuously tunable) and line-tunable CO_2 lasers, an optical parametric oscillator (OPO)-based difference-frequency laser, and a diode-based difference-frequency laser. A mobile CO_2 laser PA system was constructed for field measurements, while the laboratory-based difference-frequency lasers allowed access to the C–H stretching region (2860–3180 cm^{-1}) for a mixture of benzene, toluene, and xylene (BTX). Detection limits at or below the ppb level were achieved for many gases with this equipment. In addition to this high sensitivity, very large dynamic ranges (up to seven orders of magnitude) were demonstrated in this excellent work. The mobile PA system was subsequently used to monitor NH_3 and C_2H_4 in road-traffic emissions (Marinov and Sigrist, 2003); typical concentrations were about 600 ppb and 400 ppb, respectively.

Another significant advance in instrumentation was described by Romann and Sigrist (2002), who constructed a PA spectrometer based on a continuously tunable CO_2 laser emitting both fundamental and frequency-doubled radiation. This laser provided access to the ~10-μm (900–1100 cm^{-1}) and ~5-μm (1900–2100 cm^{-1}) regions. A large number of gases, listed in Table 7.4, were analyzed using this system. The dependence of the PA signal on gas pressure was investigated for NO, and a detection limit of 493 ppm was established for this gas. The availability of both fundamental and second-harmonic radiation makes this experimental arrangement an attractive option for the acquisition of PA infrared spectra of gases.

The research interests of the Sigrist group have recently widened to include the use of PA spectroscopy for the detection of trace amounts of gases released during surgery. Rey *et al.* (2008) used a CO_2 laser to obtain spectra of by-products emitted during surgical cautery of porcine liver. The gases identified in this study included CO_2 (the major component), CH_3OH, C_2H_5OH, NH_3, and H_2O. A total of 55 laser lines between about 935 and 1080 cm^{-1} were available in this study.

As mentioned above, NH_3 is one of the important molecules that absorbs in the region accessible to CO_2 laser radiation. In the first of two related articles, Petkovska and Miljanić (1997) described PA infrared spectra of NH_3 and its various deuterated forms. PA intensities were found to diminish with increasing deuterium content: NH_3 absorbs at more laser emission wavelengths than the other isotopomers and also displays larger absorption coefficients. A companion publication (Radak et al., 1998) examined the coincidences between the CO_2 laser lines and the absorption bands of both NH_3 and C_2H_4. As might be expected, pressure broadening had a significant effect on the overlap of the absorption bands with the narrower laser lines. PA signals also depended on the pressure of the absorbing gas. More recently, Schilt et al. (2004a) used a commercial gas analyzer that incorporates a CO_2 laser and resonant PA cell to detect NH_3 at sub-ppb levels. Longitudinal, azimuthal, and radial eigenmodes of cylindrical acoustic cavities were discussed in detail. NH_3 detection at very low concentrations is affected by atmospheric CO_2 and H_2O, which also absorb radiation in the 10-µm region. The laser wavelength(s) utilized by the gas analyzer were not specified in this article.

Radak et al. (2004) extended their work on the coincidences of CO_2 laser lines with analyte absorption bands near 10 µm to include gas-phase trichloroethylene (TCE), benzene, and dioxan – common industrial solvents of known toxicity. PA data were presented as signal intensities ($mV W^{-1}$) at about 25 line positions between 930 and 990 cm^{-1}. A detection limit of 10 ppm was established for TCE and C_6H_6; the corresponding value for dioxan was 20 ppm.

Gondal (1997) designed and constructed a resonant CO_2 laser PA spectrometer for detection of specific air pollutants. C_2H_4 and SO_2 were analyzed with this equipment; the minimum detectable concentrations were 50 ppt and 50 ppb, respectively. Cell resonance was studied in detail for 1 ppm mixtures of C_2H_4 in N_2, Ar, and He. Three or four resonances were identified at frequencies that ranged up to 8 kHz. The longitudinal resonance frequency did not vary with buffer gas pressure for N_2 and Ar, but increased with pressure when He was used. In gas-phase PA infrared spectroscopy, these acoustic resonances should be fully characterized so that the response of the apparatus (and the Q factor) can be maximized.

Trace gas monitoring was also discussed by Schäfer et al. (1998b), who noted the advantages of the use of a pulsed CO_2 laser for PA detection of C_2H_4. These authors observed a linear relationship between PA signal intensity and analyte concentration, which implies that it is possible to determine the optical absorption coefficient if the concentration is known. Optical saturation can also be studied using this technique.

A collaboration between scientists at the Université de Reims (France) and the Russian Academy of Sciences on the PA spectroscopy of ozone and other gases led to several publications relevant to this discussion. Zeninari et al. (1998) utilized a CO_2 laser to selectively excite the v_3 vibrational level of ozone. A lock-in amplifier was used to measure the phase shift between the PA signal and the incident radiation. The observed phase lag was consistent with a model based on a two-step de-excitation process rather than a simple single-step mechanism. A subsequent investigation (Zeninari et al., 2000) confirmed the suitability of this three-level model and reported rate constants for the collision of O_3 with various noble gases (He, Ne, Ar, Kr and Xe). These authors also described a differential resonant Helmholtz cell in considerable detail (Zeninari et al., 1999), the low-pressure case being of greatest relevance to their research. Ethylene and nitrogen were studied in this work.

A series of three articles by Pushkarsky, Patel, and coworkers described the use of CO_2 lasers for the detection and quantification of NH_3 and other gases. First, Pushkarsky et al. (2002) developed an NH_3 sensor based on use of the CO_2 laser line at 9.22 μm (1085 cm^{-1}), estimating a detection threshold of 220 ppt with an integration time of 30 s. Flow-through measurement cells were used in this work (a common practice in the detection of trace amounts of NH_3) because of the well-known tendency for the polar NH_3 gas to adhere to surfaces in the gas-handling system and the PA cells. Longitudinal cell resonance at 1915 Hz was characterized in detail; the microphone output at resonance was confirmed to be a coherent sine wave.

Two further works by these researchers were based on the use of $^{13}CO_2$ lasers. Pushkarsky, Webber, and Patel (2003) utilized a line at 10.784 μm (927 cm^{-1}) for detection of NH_3 in a simple transmission-type geometry (Figure 7.31). Background signals, which can arise from H_2O, CO_2, and window absorption as well as ambient noise, were measured by tuning the laser to neighboring lines where NH_3 does not absorb. The minimum detectable absorption coefficient was estimated as an impressive 9.6×10^{-10} cm^{-1}, equivalent to 32 ppt. Subsequently, Pushkarsky, Webber, and Patel (2005) discussed the feasibility of $^{13}CO_2$ laser-based PA detection of chemical warfare agents (CWAs).[5] Diisopropyl methylphosphonate (DIMP), a surrogate for the nerve agent sarin, was studied in synthetic clean air, typical ambient air, and other synthetic gas mixtures. Detection at a concentration of 4 ppb was concluded to be achievable. Although the lowest possible noise is determined by the transducers (power meter, microphone), ambient acoustical and vibrational noise sources often give rise to larger interfering signals.

PA infrared spectra of gases emitted by diesel engines were reported by Schramm et al. (2003a, 2003b). The contaminant gases detected and relevant wavenumbers

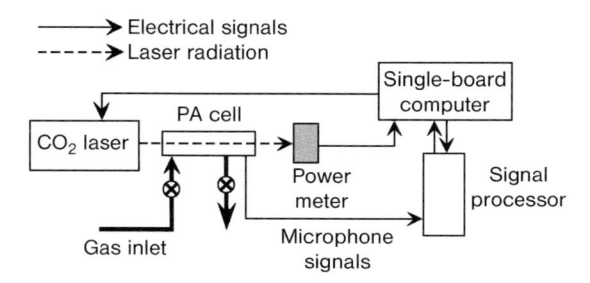

Figure 7.31 CO_2 laser-based PA sensor. (Reproduced from Pushkarsky, M.B., Webber, M.E. and Patel, C.K.N., Appl. Phys. B **77**: 381–385, Fig. 1, by permission of Springer; copyright © 2003.)

5) This work was motivated by the perceived rise in terrorism 'all over the world'.

were as follows: N_2O (949.5 cm^{-1}), NO_2 (982.1 cm^{-1}), and SO_2 (1090 cm^{-1}). PA signals for these species, measured with purpose-built equipment, ranged from 1 to 2 mV W^{-1} and corresponded to concentrations between 1 and 16 ppm. These results are of considerable interest because of the increasing use of diesel fuel for power generation and transportation.

The final series of articles on CO_2 laser PA spectroscopy to be mentioned in this section describes work carried out by Z. Zelinger, S. Civiš, and a number of coworkers at the J. Heyrovský Institute of Physical Chemistry and other institutions in the Czech Republic and abroad. Quantitative analysis of gas mixtures and air pollutants has been a central research topic for this group during the last decade. For example, a study of toluene and the three isomers of xylene (Zelinger et al., 2000) compared results obtained using different numbers of laser lines; selectivity was improved through the use of as many as six lines in a calculation of the lowest component entropy, whereas utilization of only four lines gave optimal results for the least-squares method. Detection limits between 1 and 3.6 ppm were established for the four gases studied in this investigation.

Dispersion of air pollutants was studied extensively by this group. Zelinger et al. (2004) described wind tunnel-based NO_2 measurements using PA infrared spectroscopy and DIAL (differential absorption LIDAR, LIDAR = Light Detection and Ranging). The PA system, which included a 38-cm brass tube fitted with KBr windows, was also capable of detecting CH_3OH, C_2H_5OH, freons and other gases. The linear dynamic range of PA spectroscopy reached six orders of magnitude in this work. More recently, the distribution of pollutants in urban scale models was investigated using CO_2 laser PA spectroscopy and computational fluid dynamics (CFD) modeling (Zelinger et al., 2009). Combination of spectroscopy and modeling facilitated investigation of both light (CH_3OH, C_2H_5OH, O_3) and heavy (SF_6, 1,2-dichloroethane) pollutants. Other recent works by this group discuss the PA detection of n-pentylacetate, another simulant for sarin (Herecová et al., 2009), and the use of the Allan variance to determine optimal signal-averaging times (Skřínský et al., 2009). The latter method could also prove to be important in PA FT-IR spectroscopy and other techniques discussed in this book.

7.9.3
Other Radiation Sources

As noted above, a variety of other mid- and near-infrared radiation sources have also been utilized for PA investigations of gases. A number of articles published between 1994 and 2009 on this subject are listed in Table 7.5 and reviewed below. The mid-infrared region is discussed first.

Research of Sigrist and coworkers has already been discussed with regard to the use of both CO and CO_2 lasers. In addition to those studies, this group has investigated the use of a number of other mid-infrared sources in the last 15 years. For example, Bohren and Sigrist (1997) employed a nanosecond pulsed OPO to obtain PA spectra of gases in the 2.5–4.5-µm (2200–4000-cm^{-1}) region. C_6H_6, toluene, CH_3OH, C_2H_5OH, CH_4, and isopentane vapors were examined at concentrations

Table 7.5 Recent PA studies of gases: various lasers and infrared radiation sources[a].

Source	Gases	Wavelength (μm)	Region (cm^{-1})	Reference
DFB diode laser	NH_3	1.53	6529	Fehér et al. (1994)
Color center laser	C_2H_2	1.56–1.59	6300–6400	Hornberger et al. (1995)
Raman-shifted dye laser	D_2O	1.25–1.29	7765–8030	Cohen, Bar, and Rosenwaks (1996)
DFB diode laser	NH_3	1.55	6450	Miklós and Fehér (1996)
OPO	CH_3OH, CH_3CH_2OH, CH_4, C_5H_{12}, C_6H_6, C_7H_8	2.5–4.5	2220–4000	Bohren and Sigrist (1997)
OPO	CH_4	3.31–3.32	3012–3019	Baxter, Barth, and Orr (1998)
diode laser	H_2O, toluene	0.91, 1.31	7634, 10989	Beenen and Niessner (1998)
OPO	C_2H_6	3.34	2990	Kühnemann et al. (1998)
DFB diode laser	CH_4	1.653–1.657	6035–6050	Schäfer et al. (1998a)
Diode laser	C_6H_6, C_7H_8, C_8H_{10}, H_2O	1.31, 1.67	5990, 7630	Beenen and Niessner (1999a)
Diode laser	C_6H_6, C_7H_8, C_8H_{10}, H_2O	1.31, 1.67	5990, 7630	Beenen and Niessner (1999b)
Raman-shifted dye laser	HNCO	0.62–1.5	6750–16200	Coffey et al. (1999)
Ti:sapphire ring laser	ClCN	0.77–0.97	10300–13000	Lecoutre and Hadj Bachir (1999)
DFB diode laser	NH_3	1.53	6529	Miklós et al. (1999a)
EC diode laser	CH_4	1.32–1.34	7450–7590	Miklós et al. (1999b)
DFB QCL	NH_3, H_2O	8.49–8.52	1173–1178	Paldus et al. (1999)
Ti:sapphire ring laser	OCS	0.77–0.83	12000–13000	Tranchart et al. (1999)
Ti:sapphire ring laser	$^{12}CO_2$	0.70–0.85	11700–14200	Yang and Noda (1999)
Ti:sapphire ring laser	AsH_3, H_2Se	0.85–0.87, 0.77–0.79	11500–11650, 12600–12925	Hao et al. (2000)
F-center laser	N_2–DCCH, OC–DCCH	3	3334–3337	Hünig, Oudejans, and Miller (2000)
Ti:sapphire ring laser	HCN	0.77–0.88	11390–13020	Lecoutre et al. (2000a)
Ti:sapphire ring laser	$HSiF_3$	0.91–0.92, 0.77–0.78	10900–10960, 12875–12925	Lecoutre et al. (2000b)
OPO	CH_4	1.645–1.67	5988–6079	Liang et al. (2000)

Table 7.5 Continued

Source	Gases	Wavelength (μm)	Region (cm^{-1})	Reference
OPO	CH_4	3.31–3.32	3011–3020	Oomens et al. (2000)
Difference-frequency generation	HCHO	3.53	2833	Seiter and Sigrist (2000)
DFB diode laser	HF	1.3	7665	Wolff and Harde (2000)
Ti:sapphire ring laser	H_3SiD	0.81–0.85	11797–12278	Bürger et al. (2001)
Difference-frequency generation	CH_4	3.2–3.7	2703–3125	Fischer et al. (2001)
DFB QCL	CO_2, CH_3OH, NH_3	10.33–10.36, 10.16–10.19	965–968, 981–984	Hofstetter et al. (2001)
QCL	C_2H_5OH	9.4	1064	Barbieri et al. (2002)
DFB QCL	C_2H_4	10.3	970	Schilt et al. (2002)
Diode laser	NH_3	1.53	6536	Schmohl, Miklós, and Hess (2002)
OPO	C_2H_6	3.0–3.8	2632–3333	van Herpen et al. (2002a)
OPO	C_2H_6	3.0–3.8	2632–3333	van Herpen et al. (2002b)
OPO	C_2H_6	3.0–3.8	2632–3333	van Herpen et al. (2002c)
QCL	N_2O	8	1248	Danworaphong et al. (2003)
OPO	C_2H_4, C_2H_6	3.34–3.35	2982–2990	Müller, Popp, and Kühnemann (2003)
OPO	CO_2	3.7–4.7	2128–2703	van Herpen, Bisson, and Harren (2003)
Diode laser	CO_2	1.43	6988	Veres et al. (2003)
Diode laser	CH_4	2.372	4215	Schilt et al. (2004b)
Diode laser	H_2O	1.37	7299	Szakáll et al. (2004)
QCL	NO, HMDS[b]	5.34, 8.44	1185, 1871	Elia et al. (2006a)
Fabry-Pérot QCL	HMDS[b]	8.44	1185	Elia et al. (2006b)
DFB QCL	NH_3	9.51–9.57	1045–1052	Filho et al. (2006)
Diode laser; QCL	CH_4	1.65;7.86–7.89	6060; 1267–1272	Grossel et al. (2006)
DFB QCL	NO_2, N_2O	6.2; 7.96–7.98	1253–1256; 1613	Lima et al. (2006)
EC QCL	NO_2	6.15–6.45	1550–1625	Pushkarsky et al. (2006a)
EC QCL	2,4,6-Trinitrotoluene	7.1–7.45	1345–1408	Pushkarsky et al.(2006b)
	CH_4	1.65	6060	Schilt, Besson, and Thévenaz (2006)

Table 7.5 Continued

Source	Gases	Wavelength (μm)	Region (cm^{-1})	Reference
Diode laser	NH_3, CH_4, C_2H_4	1.63	6115	Scotoni et al. (2006)
Diode laser	H_2S	1.57	6351	Varga et al. (2006)
OPO	CO_2	4.23	2364	van Herpen et al. (2006)
EC QCL	TATP[c]	7.18–7.40	1350–1393	Dunayevskiy et al. (2007)
DFB QCL	CH_4, N_2O	7.86–7.91, 7.79–7.86	1265–1273, 1272–1283	Grossel et al. (2007a)
DFB QCL	NO	5.37–5.39	1856–1862	Grossel et al. (2007b)
EC QCL	NH_3, DMMP[d]	9.425–9.65	1035–1060	Mukherjee et al. (2008a)
EC QCL	NH_3, NO_2, SO_2, DMMP[d], $(CH_3)_2CO$, ethylene glycol	6.3, 6.4, 7.3, 9.5, 10.5	952, 1053, 1370, 1563, 1587	Mukherjee et al. (2008b)
Diode laser	H_2CO	2.29–2.30	4350–4361	Cihelka, Matulková, and Civiš (2009)
Fiber laser	NH_3	1.53	6536	Peng et al. (2009)
OPO	CH_4, N_2O	3.816–3.820, 3.86–3.88, 4.021–4.032	2618–2621, 2577–2591, 2480–2487	Berrou et al. (2010)

a) DFB = distributed feedback; OPO = optical parametric oscillator; EC = external cavity; QCL = quantum cascade laser.
b) HMDS = hexamethyldisilazane.
c) TATP = triacetone triperoxide.
d) DMMP = dimethyl methyl phosphonate.

of 100 ppm in synthetic air in this work. The detection limits for most of these gases with this system were on the order of a few ppm, although sub-ppm sensitivity was achieved for CH_4 and isopentane.

Seiter and Sigrist (2000) utilized a pulsed laser based on difference-frequency mixing of the outputs from a cw external-cavity diode laser and a Q-switched Nd:YAG laser in periodically poled $LiNbO_3$ (PPLN, a nonlinear optical material) for trace gas analysis in the 3–4-μm region. This source was used together with a 36-m multipass cell to detect formaldehyde at concentrations below 8 ppb. In a related investigation (Fischer et al., 2001) a single-pass resonant cell was used to obtain PA spectra of CH_4. A detection limit of 1.5 ppm, roughly equal to the typical concentration of this substance in air, was established in this work.

Pulsed and cw OPOs have been employed as infrared radiation sources in PA spectrometers by a number of other research groups. Baxter, Barth, and Orr (1998) used a pulsed OPO, pumped at 1.064 μm and injection-seeded by a tunable diode laser, for PA and coherent anti-Stokes Raman spectroscopy (CARS). A high-resolution PA spectrum of CH_4 was obtained in the ~3012–3019-cm^{-1} interval in this work. Kühnemann et al. (1998) utilized a cw OPO for measurements on C_2H_6, establishing a detection limit of less than one ppb with their equipment; this result compared favorably with data from CO laser-based systems. The OPO used in this work was tunable from 2.3 to 4 µm, enabling the acquisition of PA spectra of many other gases. These authors subsequently developed a transportable dual-cavity singly-resonant cw OPO system capable of providing radiation between 3.1 and 3.9 µm (2565–3225 cm^{-1}) and reported an impressive detection limit of 110 ppt for C_2H_6 (Müller, Popp, and Kühnemann, 2003). In similar work, Oomens et al. (2000) described a cw OPO source that produced about 300 mW of radiation near 3000 cm^{-1}; the lower alkanes were detected at concentrations near one ppb in this study. Van Herpen et al. used OPOs that emit at 3.0–3.8 µm (2002a–c), 3.7–4.7 µm (2003), and 4.23 µm (2006) to detect C_2H_6 and CO_2. Very recently, Berrou et al. (2010) developed an entangled cavity doubly resonant OPO that provided nanosecond pulsed radiation in the 3.8–4.3-µm (2325–2630-cm^{-1}) region. High-resolution PA spectra of N_2O and CH_4 were obtained using this apparatus, which was also employed for direct absorption spectroscopy.

The quantum-cascade laser (QCL) is a relatively new mid-infrared source that was discussed in Chapter 3 with regard to quartz-enhanced PA spectra of gases, as well as PA spectra of solid substances. A number of research groups have also used QCLs together with microphone transducers to obtain PA spectra of gases in the last decade. These lasers can be configured in several ways; the distributed feedback (DFB), external-cavity (EC), and Fabry-Pérot QCLs have become popular choices among PA spectroscopists. A few examples of QCL PA spectra of gases are mentioned in the following paragraphs.

As might be expected, initial QCL-based work on PA infrared spectra of gases emphasized the analysis of small molecules. For example, Paldus et al. (1999) obtained spectra of NH_3 and H_2O vapor in one of the first successful QCL PA experiments. This pioneering study demonstrated a detection limit of 100 ppb for NH_3. Hofstetter et al. (2001) analyzed CO_2, CH_3OH, and NH_3 using a QCL and a Herriott multipass arrangement around a PA cell equipped with a 16-microphone array. NH_3 was monitored in the 966–968-cm^{-1} region at a concentration of 300 ppb with this apparatus. Barbieri et al. (2002) constructed a PA gas sensor that included a Peltier-cooled Fabry-Pérot InGaAs-based QCL emitting at 9.4 µm and a purpose-built PA cell; a detection limit of about 1 ppm was obtained for C_2H_5OH with this equipment. Danworaphong et al. (2003) used an 8-µm (1248-cm^{-1}) QCL, mounted in the center of an open-ended cylindrical chamber, to study N_2O. The theoretical detection limit, which was determined by the intrinsic noise of the microphone and electronics, was on the order of $10^{-9}\,W\,cm^{-1}$. Specification of the limit in these units provides a value that is normalized with respect to laser power.

The use of QCLs for PA detection of trace gases was briefly reviewed recently by Elia et al. (2006a). Applications in which high-sensitivity PA detection of trace gases plays a key role include pollution monitoring, toxic-gas detection, medical diagnostics, and industrial process control. The authors discussed other publications on QCL PA spectra of gases as well as their own results for NO and hexamethyldisilazane (HMDS, [(CH$_3$)$_3$Si]$_2$NH), a VOC used in organic synthesis and semiconductor manufacturing processes. A related investigation on HMDS (Elia et al., 2006b) utilized a Fabry-Pérot QCL emitting at 8.4 μm (1183 cm^{-1}). The PA spectrometer constructed for this work incorporated a resonant PA cell and four electret microphones. A detection limit of 200 ppb was obtained for HMDS in this work.

Three articles by the Groupe de Spectrométrie Moléculaire et Atmosphérique in Reims described QCL PA infrared spectra of CH$_4$ and two different nitrogen oxides. Grossel et al. (2006) first compared a 1.65-μm diode laser with a 7.9-μm QCL for monitoring CH$_4$, obtaining detection limits of 0.15 ppm and 3 ppb respectively. The PA cell used in this work included two equal-volume cylinders, each fitted with a microphone, linked by two capillaries. The acoustic waves were opposite in phase at resonance, making it possible to eliminate most contributions to the PA signal that arise from ambient noise. Grossel et al. (2007a) next used this cell with two DFB QCLs to observe CH$_4$ (1265–1273 cm^{-1}) and N$_2$O (1272–1283 cm^{-1}). Two modulation schemes were described: (i) in amplitude modulation a current ramp tunes the laser emission across the band of interest while the beam is modulated at the resonance frequency; (ii) to achieve wavelength modulation, a sinusoidal modulation is added to the current ramp and the harmonic (derivative) signal is recorded. Application of the current ramp to the laser produces a wavelength scan in both cases. The detection limits obtained using flowing gases in this work were 34 ppb (CH$_4$) and 14 ppb (N$_2$O). Similarly, a DFB QCL with a tuning range from 1856 to 1862 cm^{-1} was used to identify NO gas (Grossel et al., 2007b). Four NO bands, as well as four interfering H$_2$O bands, fall within this relatively narrow region. A detection limit of 20 ppb was achieved for NO with the laser power at 3 mW.

A series of articles by Patel and coworkers on CO$_2$ laser-based PA spectroscopy was mentioned earlier in this section. These authors also used QCLs as radiation sources in recent work; both light and heavy molecules were studied in this research. For example, Pushkarsky et al. (2006a) used an EC-QCL that was tunable from about 1550 to 1625 cm^{-1} to observe trace amounts of NO$_2$. A triplet at 1599.8, 1599.95, and 1600.07 cm^{-1} was found to be free from interfering H$_2$O bands and selected for quantification of NO$_2$. In related work, NH$_3$ was studied in the 9.425–9.65-μm region using an EC-QCL (Mukherjee et al., 2008a). These researchers also described a more complex system in which five EC-QCLs were multiplexed in a single sensor, with a galvanometer mirror being used to select radiation from the laser of interest (Mukherjee et al., 2008b). Nominal QCL wavelengths in this system were 6.3, 6.4, 7.3, 9.5, and 10.5 μm. NH$_3$, NO$_2$, and SO$_2$ were observed with this apparatus. This research group also used QCL-based PA spectroscopy to detect CWA simulants (Pushkarsky et al., 2006b; Dunayevskiy et al., 2007; Mukherjee et al., 2008a, 2008b).

As shown in Table 7.5, several different types of near-infrared lasers have been used since 1994 to obtain PA spectra of gases. Among these, diode lasers with emission wavelengths greater than 1 μm have been utilized by a large number of researchers. Publications describing this work are summarized next, after which spectra obtained using other near-infrared lasers are considered.

Ammonia has been the subject of several PA investigations that utilized DFB near-infrared lasers. Fehér et al. (1994) constructed a system using a laser diode that emitted at about 1.53 μm (6529 cm^{-1}). Frequency modulation was achieved over a very narrow range by adjusting the laser drive current. The authors detected NH_3 at pressures down to 5.3×10^{-5} Torr using this apparatus and estimated the detection limit of the PA system as 8 ppb. In related subsequent work, two of these authors (Miklós and Fehér, 1996) confirmed this limit and showed that it was approximately three orders of magnitude lower than that obtainable by absorption measurements using a one-meter cell.

As noted above, PA analysis of NH_3 is complicated by the tendency of this gas to adsorb on surfaces such as the walls of the detector and tubing of the gas transfer system. This factor prompted Miklós et al. (1999a) to develop a differential flow-through system based on the same 1.53-μm 5-mW DFB laser. The gas flowed through two tubes, one of which was irradiated collinearly, in this improved arrangement. The second tube facilitated subtraction of the background signal from the flowing gas. PA signals from NH_3 varied linearly with concentration down to 1 ppm with this apparatus. The sensitivity of such a PA system can, of course, be further improved through the use of a more powerful laser.

Other DFB diode lasers have also been used to obtain PA near-infrared spectra of gases. For example, Schäfer et al. (1998a) used a 1.65-μm DFB laser to detect CH_4 at concentrations down to 60 ppm. These researchers found that absorption spectroscopy actually yielded significantly better sensitivity than their PA measurements, which were limited by low laser power. In contrast, Wolff and Harde (2000), who employed a 1.30-μm DFB laser to detect gaseous HF, attained much higher sensitivity using PA infrared spectroscopy: the detection limit in this experiment was about 80 ppb. Recently, Cihelka, Matulková, and Civiš (2009) used near- and mid-infrared laser diodes to obtain PA spectra of formaldehyde, achieving detection limits of < 1 ppm and < 10 ppb in these two respective regions.

Beenen and Niessner (1998, 1999a, 1999b) utilized NIR laser diodes and a resonant PA cell to detect C_6H_6, toluene, xylene, and H_2O vapor. The long-term objective of this work was the development of portable gas detection apparatus. In this laboratory-based study, the laser output was coupled to a flow-through cell using optical fibers. Emission wavelengths of 0.908, 0.911 and ~1.67 μm were used for the hydrocarbons, whereas water was monitored at 1.31 μm. These wavelengths can be tuned over narrow ranges by varying the temperature of the diode, a capability that was utilized to distinguish between the different compounds. For example, C_6H_6 was monitored at 1.67 μm (5988 cm^{-1}), while toluene was detected at the slightly different wavelength of 1.68 μm (5952 cm^{-1}). The sensitivities achieved in this work (70 μg L^{-1}, C_6H_6; 100 μg L^{-1}, toluene; 160 μg L^{-1}, xylene) are sufficient for environmental monitoring.

The literature described in the preceding paragraphs shows that PA infrared spectroscopy is indeed capable of extremely sensitive gas analysis in the near-infrared region. However, it should be kept in mind that the wavelength ranges covered by the diode lasers used in these studies are quite narrow; obviously, PA detection of trace gases is feasible only when the available laser lines coincide with the absorption bands of the gases of interest.

Table 7.5 lists a number of publications in which a titanium:sapphire ring laser was used to obtain near-infrared PA spectra of gases. Five articles resulted from collaborations between Lecoutre, Huet, and collaborators at l'Université de Sciences et Technologies de Lille and at several other institutions. These investigations involved the measurement–and more particularly the interpretation–of high-resolution spectra of ClCN (Lecoutre and Hadj Bachir, 1999), OCS (Tranchart et al., 1999), HCN (Lecoutre et al., 2000a), $HSiF_3$ (Lecoutre et al., 2000b) and H_3SiD (Bürger et al., 2001). The high sensitivity of the PA apparatus allowed the authors to study very weak bands using only small volumes of gas, and to obtain research-quality results.

Similar systems have been used by other researchers. Yang and Noda (1999) reported the first observation of combination bands for $^{12}CO_2$ between 11 700 and 14 200 cm^{-1} by use of PA spectroscopy. Hao et al. (2000) obtained Doppler-limited spectra of AsH_3 and H_2Se. A minimum detectable absorption coefficient of 6.35×10^{-9} cm^{-1} was calculated for water vapor in this work, which examined noise sources and cell resonance in considerable detail.

Other near-infrared lasers have also been used in the measurement of PA spectra of gases. Hornberger et al. (1995) employed a 1.5-μm color center laser and a multipass cell to record high-resolution spectra of acetylene. The minimum detectable absorption coefficient with this apparatus is on the order of 10^{-10} cm^{-1}. Miklós et al. (1999b) utilized an external-cavity diode laser to obtain PA spectra of CH_4 in the 7450–7590-cm^{-1} range. Close agreement between PA and absorption spectra was demonstrated for the Q branch of the combination band centered near 7510 cm^{-1}. Cohen, Bar, and Rosenwaks (1996) used a Raman-shifted dye laser to obtain PA spectra of D_2O, while Coffey et al. (1999) utilized a similar laser for measurement of overtone spectra of HNCO. The latter two articles are mainly concerned with interpretation of spectra rather than PA instrumentation.

7.9.4
FT-IR PA Spectra of Gases

The preceding discussion of the recent literature on the PA infrared spectroscopy of gases reviewed work in which various mid- and near-infrared radiation sources were used for excitation. The large number of references in Tables 7.3–7.5 show that a considerable amount of research has been carried out in this field within the last 15 years. By contrast, only a few articles describing the measurement of PA spectra of gases with an FT-IR spectrometer are known to exist. This literature, which also spans a longer period of time, is briefly summarized in this section.

PA infrared spectra of solid carbonyl compounds were discussed earlier in this chapter. Investigations by I. S. Butler and his group comprise most of the published work in this area. These researchers also observed gas-phase spectra during their studies. For example, experiments on the group VIB metal chalcocarbonyl complexes $M(CO)_6$ (M = Cr, Mo and W) and $Cr(CO)_5CS$ by Xu, Butler, and

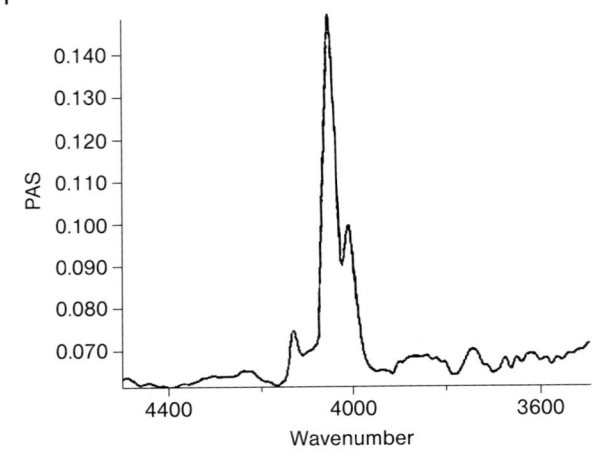

Figure 7.32 PA infrared spectrum of gaseous $CH_3Mn(CO)_5$. (Reproduced from Butler, I.S. et al., *Appl. Spectrosc.* **46**: 1605–1607, by permission of the Society for Applied Spectroscopy; copyright © 1992.)

St.-Germain (1986) yielded PA spectra of the corresponding gases rather than the expected spectra of the solid compounds. Indeed, the PA spectra agreed with published spectra of the gaseous complexes and differed significantly from absorption spectra of the same complexes in KBr disks. Observation of gas-phase spectra in this work was put down to heating of the samples by the infrared beam.

Several years later, Butler, Gilson, and Lafleur (1992) reported gas-phase PA infrared spectra of $CH_3Mn(CO)_5$ and $CH_3Re(CO)_5$. The spectrum of the former compound is depicted in Figure 7.32. This result shows typical examples of the near-infrared bands that are clearly visible in these spectra; similar results were obtained in the previous study and in the investigations of the solid complexes described below. The authors assigned the bands near $4000\,cm^{-1}$ to overtones and combinations of the fundamental transitions that were identified at lower frequencies.

Olafsson et al. (1999) described a multipass cell optimized for the detection of gases in PA FT-IR spectroscopy. Two factors ensured that the signal arising from the cell wall was negligible, even though reflection losses were relatively high. First, the thermal time constant of the cell wall was much less than the width of the interferogram centerburst; and second, the heat capacity of the cell wall was much greater than that of the gas. An impressive detection limit of about 0.1 ppm was achieved in this experiment, corresponding to an absorption coefficient of $10^{-5}\,cm^{-1}$. These values compare favorably with those mentioned earlier with regard to laser-excited PA spectra of gases.

Two additional articles that discuss PA FT-IR spectra of gases can be mentioned. Liu et al. (1999) described the determination of total inorganic carbonate content in solids (rock, soil, sediment) and water; CO_2 was detected and quantified in this work. Similarly, Booth and Sowa (2001) measured CO_2 evolution in bitterbrush

(*Purshia tridentata* DC., a member of the rose family) seeds. PA spectra are not displayed in the latter publication.

As stated at the beginning of this section, some of the most important achievements in PA spectroscopy have occurred with respect to gas detection. The high sensitivity afforded by PA infrared spectroscopy is particularly impressive, and has led to the development of many successful applications and the publication of an extensive body of literature. This field of research is expected to continue to prosper in the future.

7.10
Wood and Paper

The measurement of research-quality infrared spectra of wood, paper, and wood products is obviously a challenging task, since these materials are often not amenable to traditional sample preparation methods. Wood products and paper are insoluble in many common solvents, and may not be readily dispersed in infrared-transparent solids. Fortunately, PA infrared spectroscopy proves to be an entirely viable means for their characterization. Representative published results for wood-derived samples are summarized in this section. Some of the early literature is also discussed in a review by Workman (2001).

A group based at the Pulp and Paper Research Institute of Canada and McGill University was among the first to utilize PA spectroscopy in research on wood products. St.-Germain and Gray (1987) used the technique in a study of mechanical pulp brightening, primarily because of the minimal sample preparation required. PA infrared spectra were obtained for pulp sheets and ground pulp in this work. About a dozen bands were observed, the most prominent occurring at 1056, ~2900, and ~3400 cm^{-1}. The authors noted that intensities varied with sample packing in the PA cell, and therefore scaled the PA spectra based on the observed intensities of either the broad 3400-cm^{-1} band, the 1056-cm^{-1} (C–O) band, or an aromatic (1510-cm^{-1}) band. This use of internal standards improved reproducibility to ±5% or better. Brightening of pulp with peroxide caused diminution of the 1740- and 1650-cm^{-1} bands, implying deacetylation and the removal of conjugated carbonyl structures respectively. The spectra were also consistent with quinone formation during brightening. An absorption spectrum of the pulp in a KBr pellet was found to resemble the corresponding PA spectrum.

At least 20 bands are known to occur in the infrared spectra of various types of wood. In an exploratory study carried out at about the same time as the above-mentioned work, Kuo *et al*. (1988) convincingly demonstrated the suitability of PA infrared spectroscopy for the characterization of several different kinds of wood. In fact, these authors found PA spectroscopy to be superior to traditional infrared techniques (transmission, ATR, diffuse reflectance) with regard to this analytical problem. In their PA experiments, they examined 6-mm circular samples cut from microtomed sections that were 400 μm thick. The thermal diffusivity of wood is approximately 0.002 cm^2 s^{-1}, implying sampling depths ranging from ~18 to 56 μm at the modulation frequencies used to obtain PA FT-IR spectra in this investigation.

Kuo *et al.* (1988) obtained spectra for transverse and oblique sections of ponderosa pine; cellulose bands near 1050 and 3350 cm^{-1} were stronger in the spectrum of the oblique section, consistent with the fact that the cellulose chains were oriented perpendicularly to the transverse section normal. This observation of differences due to microfibrillar orientation would, of course, not have been possible in a sample that was ground prior to a transmission or diffuse reflectance measurement. In a related experiment, these authors examined eastern cottonwood samples with various degrees of decay by the brown-rot fungus *Gleoephyllum trabeum*, and found that the intensities of several carbohydrate bands diminished as decay progressed; by contrast, lignin features intensified. Using only small sample quantities, the authors also demonstrated very significant differences between the PA infrared spectra of eastern cottonwood kraft and groundwood fibers or vessel elements (Figure 7.33). The relative lignin contents of these samples were also deduced from these spectra.

This successful analysis of wood by PA spectroscopy naturally led to further work. Comparison with other infrared methods was common to several studies. For example, Pandey and Theagarajan (1997) obtained diffuse reflectance and PA infrared spectra of hardwood species, noting insensitivity of the PA spectra to particle size and concentration. Transmission spectra of 50-μm wood sections served as benchmark data. Bajic *et al.* (1998) investigated both PA and transient

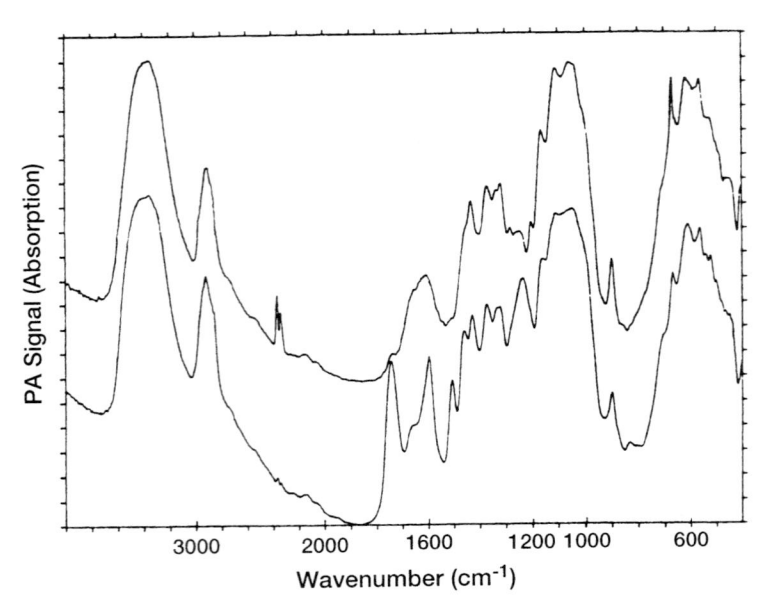

Figure 7.33 PA infrared spectra of eastern cottonwood kraft (KAPPA No. 25, top curve) and groundwood fibers (bottom curve). (Reproduced from Kuo, M.-L. et al., *Wood Fiber Sci.* **20**: 132–145, by permission of the Society of Wood Science and Technology; copyright © 1988).

infrared spectroscopy of wood in the context of the need for rapid analysis. The latter technique, which involves measurement of emission spectra of moving samples, is mentioned below in the discussion of PA spectra of food products.

Bajic et al. (1998) found that PA infrared spectra of milled wood specimens displayed bands that characterized the samples as hardwoods or softwoods, and could be attributed to particular functional groups. Principal component analysis was then employed to determine whether wood species could be distinguished on the basis of their PA spectra. Indeed, the hardwoods and softwoods clustered separately, showing the feasability of this approach. American Society for Testing and Materials (ASTM) methods were also used to analyze the wood samples; the PA results agreed with the ASTM data within 1% with correlation coefficients ranging from 0.85 to 0.94. These encouraging results show that PA infrared spectra can be used for feedstock identification and analysis of the chemical composition of wood before it is processed. Moreover, quantitative analysis of wood extracts, lignin and carbohydrates by PA infrared spectroscopy is certainly possible. Similar work in the present author's laboratory has shown that added wood preservatives are readily detected using this technique.

Several research groups have incorporated PA infrared spectroscopy in studies of photodegradation or weathering of wood. Ohkoshi (2002) characterized the changes in acetylated and polyethylene glycol-impregnated wood caused by irradiation with visible light. Bands due to carbonyl groups and lignin were monitored in this work. Pandey and Vuorinen (2008) combined PA, Raman, and ultraviolet/visible absorption spectroscopies in an investigation of the degradation caused by an ultraviolet laser or an Xe lamp. PA infrared spectra showed a reduction of the lignin aromatic structure (bands at 1462, 1506 and 1596 cm^{-1}). Polysaccharide features were less affected. Depth profiling experiments led to the reasonable conclusion that photodegradation is a surface phenomenon.

The weathering of tropical wood has also been studied using PA infrared spectroscopy. Sudiyani et al. (2003) observed changes in the spectra that were attributed to the loss of lignin and xylan. Cellulose was not affected by weathering. The estimated thermal diffusivity, 0.002 $cm^2 s^{-1}$, corresponded to μ_s values of 14 and 22 μm at 2000 and 800 cm^{-1}, respectively. By contrast, μ_α ranged from about 5 to 12 μm in the same wavenumber region; in other words, the wood samples were optically opaque and thermally thick under the conditions employed in this study. In a related study, Yamauchi et al. (2004) found that weathering led to diminution of the intensities of aromatic bands in the spectra of near-surface layers. The cellulose polymer was again observed to be minimally affected by weathering.

The use of PA infrared spectroscopy for characterization of wood is continuing. A recent article by Shaw, Karunakaran, and Tabil (2009) describes spectra of steam-exploded poplar wood. These workers investigated the quality of pellets prepared from this wood and wheat straw, employing several chemical and physical methods of analysis. Steam pretreatment was found to reduce a carboxyl band due to hemicellulose, implying that this component was hydrolyzed or removed. Lignin content was higher in the pretreated feedstocks. These results show the viability of PA spectroscopy for identifying chemical changes in feedstocks.

The early investigation of pulp mentioned above has been followed by work at other institutions. Bjarnestad and Dahlman (2002) used PA infrared spectroscopy and partial least-squares analysis to predict the compositions of hardwood and softwood pulps. Carbohydrate and lignin contents were deduced from the spectra. Insensitivity to sample morphology and minimal sample preparation, two attributes of PA spectroscopy, were both beneficial in this work.

Researchers at the Australian Pulp and Paper Institute, Monash University, have utilized PA infrared spectroscopy in several recent investigations. Bhardwaj, Dang, and Nguyen (2006) studied the carboxyl content of kraft pulps. PLS analysis predicted this content accurately, obviating the usual sample preparation and titration. An attempt at depth profiling yielded inconclusive results. Similarly Bhardwaj and Nguyen (2007) examined bleached de-inked pulps, for which brightness is of particular interest. Bands due to C=C and C=O groups were affected by bleaching, the most significant spectral changes occurring in the 1750–1775 cm^{-1} interval. Depth profiling by use of several FT-IR mirror velocities showed that the composition of the pulp surface was quite different from that of the bulk. Thermal diffusivities of 0.0085 and 0.0132 $cm^2 s^{-1}$ were reported for de-inked and peroxide-bleached pulp, respectively. Dang et al. (2007) obtained spectra of kraft pulps with varying lignin content (Kappa number), also using depth profiling and PLS analysis in their work. Quantification of residual lignin in chemical pulps was the objective of this study. The 1200–1650 cm^{-1} region of the PA spectra was found to be correlated with the Kappa number of the pulps.

Sawdust, another familiar wood product, can be analyzed after dispersion in an alkali halide by means of standard transmission or diffuse reflectance techniques. Because it is finely divided, undiluted sawdust can be studied even more readily using PA infrared spectroscopy. As an example, Figure 7.34 shows a PA infrared spectrum of hemlock sawdust in which approximately 20 bands are visible. Sawdust obtained from fir yielded a similar result. Not surprisingly, these PA spectra closely resemble those measured for the original wood samples.

The results discussed so far clearly show that PA infrared spectroscopy is very well suited for the analysis of wood, pulp and sawdust. Paper is another related product that is yet to be considered. In addition to cellulose, paper may contain lignin, hemicellulose, sizing components such as starch and clay, and polymer coatings. The heterogeneity of some papers thus provides a second motivation for the use of PA infrared spectroscopy: in addition to minimal sample preparation, the capability of depth profiling makes it possible to characterize the layers in coated papers. Both attributes of PA infrared spectroscopy were utilized in the research discussed in the following paragraphs.

The suitability of PA infrared spectroscopy for the analysis of paper was briefly illustrated in an early survey (Krishnan, 1981). PA spectra were recorded for a coated paper and the corresponding uncoated paper. Subtraction of the second spectrum from the first canceled the bands due to the paper and revealed the spectrum of the coating; several important bands due to aromatic and aliphatic C–H groups, carbonyl functionality, and other species were identified in this way.

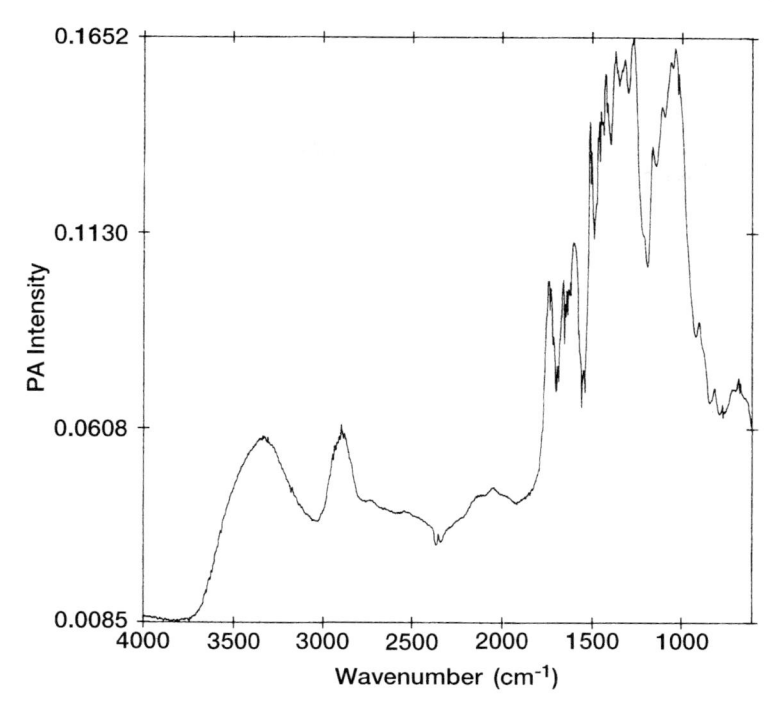

Figure 7.34 PA infrared spectrum of hemlock sawdust. The sample fraction with particle sizes less than 250 μm was analyzed.

The experiment was then repeated using the specular reflectance technique. Subtraction was less successful, because the bands arising from the paper were too strong to permit reliable cancellation. This investigation was among the first to show that PA infrared spectroscopy is particularly well suited for the analysis of paper.

An early demonstration of the suitability of PA near-infrared spectroscopy for the analysis of paper occurs in the work of Jin, Kirkbright, and Spillane (1982). These authors observed three bands in the 1.3–2.2 μm region; features arising from cellulose occurred at 1.55 and 2.05 μm (6450 and 4880 cm^{-1}, respectively). A third band at 1.9 μm (5265 cm^{-1}), due to water, varied systematically with the moisture content of the paper. Interestingly, even prolonged drying at 105 °C did not result in its complete disappearance. In the mid-infrared, Moore et al. (1990) studied embrittlement of several types of paper using HCl, H_2O_2, and oxalic acid. Carbonyl bands appeared in the spectra of several samples, suggesting the production of esters or aldehydes. Forsskåhl, Kenttä Kyyrönen, and Sundström (1995) compared PA, ATR, diffuse reflectance, and transmission spectroscopies in an investigation of light-irradiated paper. Thermal diffusivity was assumed to fall in the 10^{-4}–10^{-3} cm^2 s^{-1} range, significantly lower than the values used by other groups. This corresponded to thermal diffusion lengths of ~3–10 μm at $f = 330$ Hz.

However, the porous nature of the paper led to a suggestion that the acoustic signal might be generated both from the external surface and the internal fiber/gas surfaces.

As an alternative to the common gas-microphone PA method, PBDS can also be utilized to study paper. For example, Low, Morterra, and Severdia (1984) obtained PBDS infrared spectra of paper, observing that treatment of the sample with ethylene oxide resulted in diminution of the broad band at $3400\,cm^{-1}$ and intensification of C–H stretching bands near $3000\,cm^{-1}$. The open geometry and nondestructive sampling features of PBDS were put to particularly good use in this experiment, in which a piece of paper manufactured in 1577 was also analyzed; a blue discoloration on one side of this paper was concluded to be a type of mold, and found to yield a spectrum distinct from that of the paper itself.

Analytical chemists might suggest using PA spectroscopy to characterize solid samples deposited on filter paper. To be successful, such an experiment would normally require the spectrum of filter paper itself to be obtained, so that its contribution to the spectrum of the filtered solid could be removed numerically. In this context, it is relevant to note that PA spectra of filter paper were described in the literature a number of years ago. Harbour *et al.* (1985) obtained near-infrared PA spectra for Whatman filter paper, observing six well-defined bands between about 3800 and $8000\,cm^{-1}$. Exposure of the filter paper to atmospheres with varying relative humidities showed that bands at 3816, 5155, and $6944\,cm^{-1}$ intensified as humidity increased. It was proposed that the $I(5155)/I(4670)$ intensity ratio be used as a measure of water content relative to that of cellulose in paper and related materials.

The mid-infrared PA spectrum of Whatman filter paper was also obtained by Harbour *et al.*, and found to contain bands due to cellulose at about 700, 1100, 1350, 1450, 2900 and $3300\,cm^{-1}$. A similar PA infrared spectrum of a commercial filter paper, extending from 500 to $4500\,cm^{-1}$, is depicted in Figure 7.35. This spectrum can reliably be subtracted from the PA spectrum of a solid or oil deposited on a similar sheet of paper so as to reveal the bands of the analyte: in fact the strategy mentioned above is readily implemented.

PA infrared spectra of bleached kraft papers were described by Gurnagul, St.-Germain, and Gray (1986). This investigation was carried out at about the same time as the study of mechanical pulp brightening by this group that was mentioned above. This group examined the effects of beating the pulp for various periods of time, and of exposing the paper samples to a moist atmosphere. PA intensities were lower for the more highly beaten sample than for the unbeaten sample, which had a higher specific surface area and surface roughness. These results accord with PA studies of finely divided solids, where intensity generally increases as particle size decreases. Another significant finding from this investigation was the fact that PA intensities were lower for sheets of kraft paper with high moisture contents. The authors reasoned that this might have been due to a lowered efficiency of heat transfer between the moist cellulose surface and the carrier gas; PA spectroscopists may well recall their own experiences in which moist or oily samples yielded weaker-than-expected spectra. In the case under

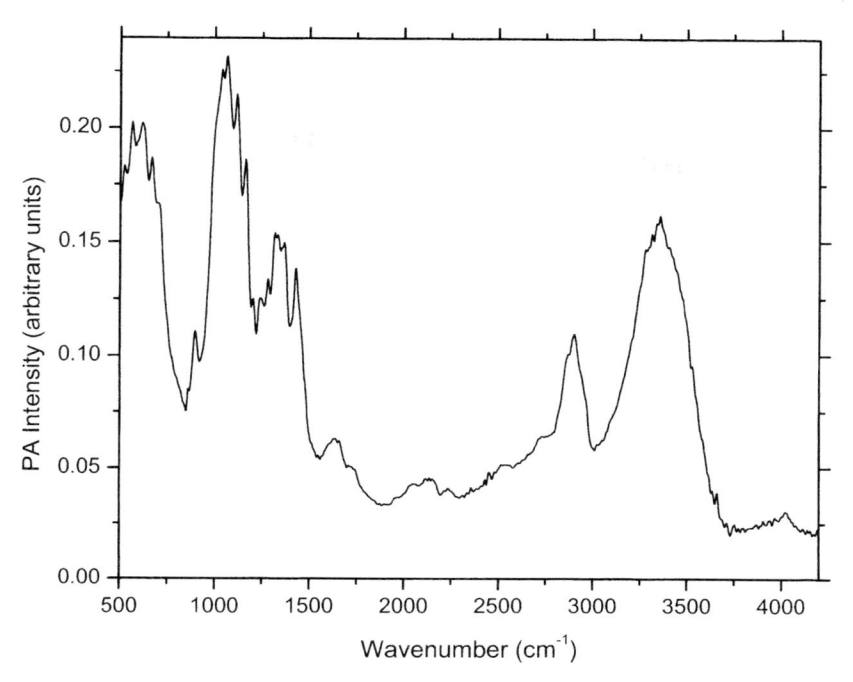

Figure 7.35 PA infrared spectrum of ordinary filter paper.

discussion here, the moist paper samples had lower surface areas, which would have also given rise to reduced PA intensities.

Photothermal radiometry (PTR) and PA infrared spectroscopy were used in a combined study of specialty papers with various cotton contents (Garcia et al., 1998, 1999). Two important thermophysical properties (thermal diffusivity, α_s; thermal conductivity, κ_s) were calculated from the PTR data, utilizing a one-dimensional photothermal model of a sheet of paper in air. α_s ranged from $1.3 \times 10^{-7}\,\mathrm{m^2 s^{-1}}$ to $1.9 \times 10^{-7}\,\mathrm{m^2 s^{-1}}$ for seven different types of paper, while κ_s assumed values between 1.00 and $3.60\,\mathrm{W\,m^{-1}\,K^{-1}}$. PTR measurements were effected using both transmission and backscattering geometries; both signals depended on the infrared emission coefficient β_{IR}, which is also equal to the average absorption/extinction coefficient across the bandwidth of the infrared detector used in this experiment. β_{IR} was calculated as 1.79–$3.10 \times 10^{-4}\,\mathrm{m^{-1}}$, approximately equal to β_s, the analogous quantity in the visible region. It should be noted that β_{IR} cannot be correlated with specific infrared absorption bands because the PTR measurements were carried out with a broad-band infrared detector, that is, no wavelength (wavenumber) discrimination was incorporated in these experiments.

The relationship between the PA infrared spectra of these papers and the PTR results is partly illustrated in Figures 7.36 and 7.37. Figure 7.36 shows the PA spectra obtained for the front and back surfaces of a sheet of paper for which the corresponding β_{IR} values were determined as $2.85 \times 10^{-4}\,\mathrm{m^{-1}}$ and $1.79 \times 10^{-4}\,\mathrm{m^{-1}}$,

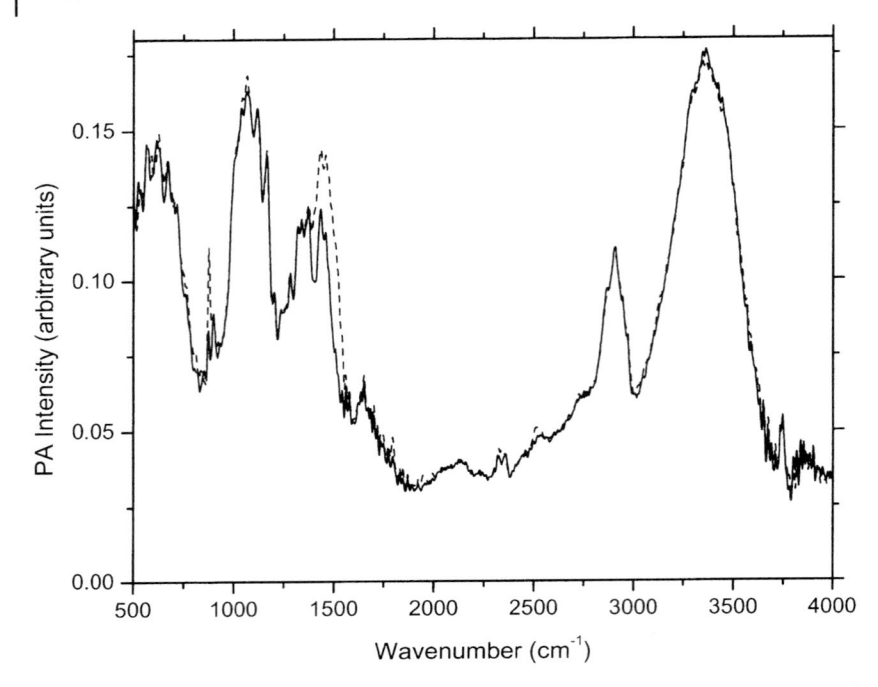

Figure 7.36 PA infrared spectra of Krypton Parchment paper (high cotton content) provided by Center Innovation DOMTAR. The solid and dashed curves represent the spectra obtained for the two sides of a single sheet.

respectively. Indeed, the differences in the infrared spectra in the 1300–1700 cm^{-1} region suggest varying depth profiles in both optical and thermophysical properties. In Figure 7.37, the difference between the PA infrared spectra of the two surfaces is more pronounced, and corresponds to an even larger differential in the thermophysical properties obtained from the transmission and backscattering PTR experiments. In general, a decrease in cotton content results in higher α_s values for these papers.

Two publications that deal with coated papers can be mentioned. In the first, Halttunen et al. (1999) described depth profiling experiments on paper coated with sodium oleate. The base paper thickness was 78 μm, with coating thicknesses of 18–58 μm. PA magnitude spectra obtained at higher f exhibited more intense bands due to the oleate, consistent with the fact that it was present on the surface of the paper. Concomitantly, a $CaCO_3$ band from the base paper diminished in intensity.

In the carbonyl stretching region, the frequency of the asymmetric –COO$^-$ band shifts from 1563 cm^{-1} in pure sodium oleate to 1573 cm^{-1} when the compound is mixed with starch to form the coating. Attachment of the coating to paper produced bands at 1539 and 1573 cm^{-1}. The asynchronous two-dimensional correlation spectrum of the coated paper showed that the band due to pure oleate (shifted

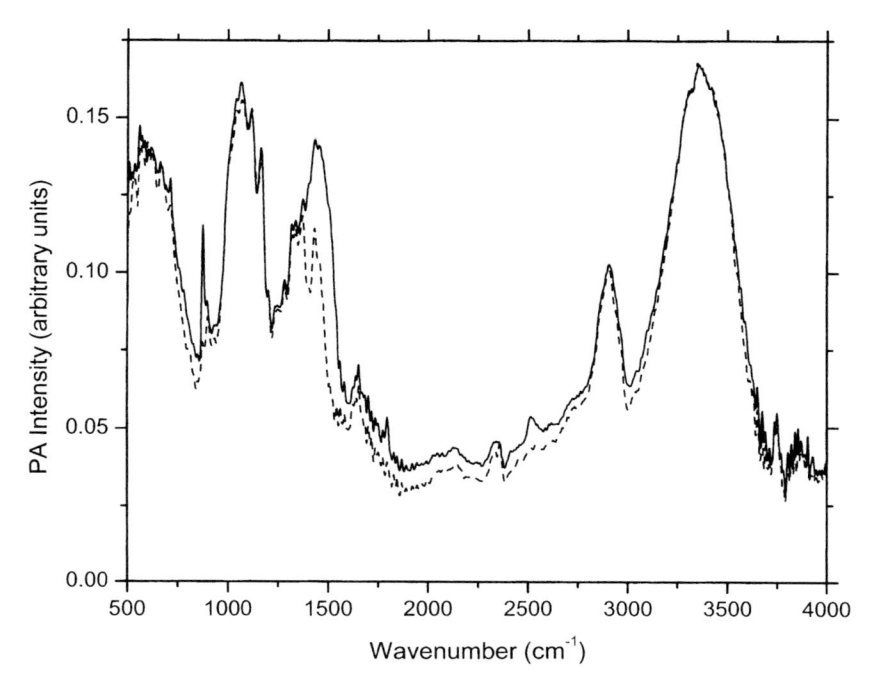

Figure 7.37 PA infrared spectra of Belfast Bond paper (intermediate cotton content). Details are the same as in Figure 7.36.

slightly to 1560 cm^{-1}) was of shallower origin, with the 1573 cm^{-1} feature arising from the intermediate layer and the 1539 cm^{-1} band originating from the deepest region. Hence the concentration of sodium oleate was increased on the surface of the coating layer. Phase spectra further confirmed that the 1539 cm^{-1} band arose from the sodium oleate at the bottom of the coating layer. These impressive results illustrate the considerable detail that can be derived from depth profiling experiments on coated paper by PA infrared spectroscopy and suggest further research on this topic.

Wahls, Kenttä, and Leyte (2000) also reported depth profiling experiments for coated paper, using the difference between the spectra of the coated and uncoated (base) papers to distinguish the spectrum of the bulk from that of the coating. Step-scan magnitude spectra, obtained using a lock-in amplifier, were calculated as $[I^2 + Q^2]^{1/2}$ where I and Q are the in-phase and quadrature spectra, respectively. The spectra of the coated and uncoated papers were scaled according to their intensities in the 2200–2300 cm^{-1} region, where no absorption bands were observed. As shown in Figure 7.38, this strategy yielded positive and negative differences, rather than nulling the bands of one component. The differences between the PA spectra of the base paper and the coated paper arose from the fact that less base paper was sampled in the second case; the difference bands varied with f because of the corresponding changes in μ_s. This behavior was modeled using synthetic triangular bands, assuming typical values for the absorption coefficients.

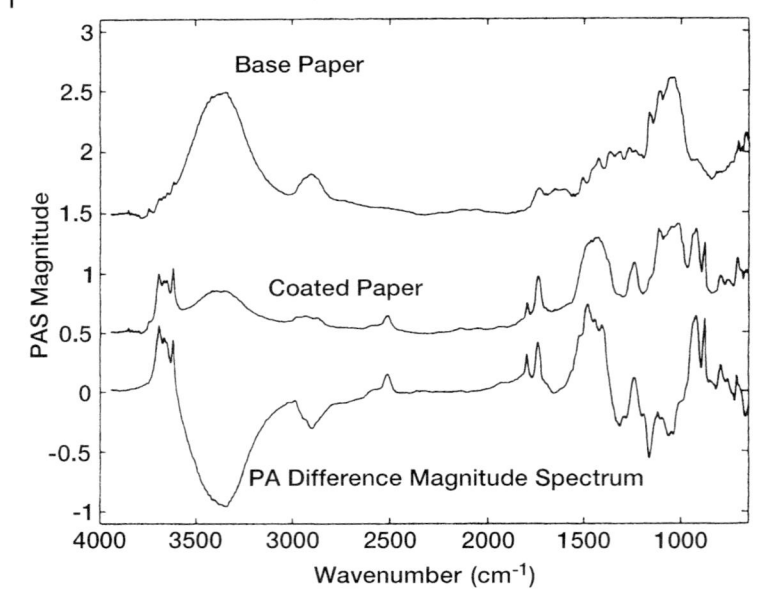

Figure 7.38 Normalized and scaled PA infrared magnitude spectra obtained at a modulation frequency of 800 Hz for base paper and coated paper. The bottom curve is the difference. Spectra have been offset along the y axis for clarity. (Reproduced from Wahls, M.W.C. et al., *Appl. Spectrosc.* **54**: 214–220, by permission of the Society for Applied Spectroscopy; copyright © 2000).

Absorption by the coating yielded positive difference bands, while bands from the bulk gave negative differences; these differences were greater when μ_s was approximately equal to the coating thickness. As pointed out by Wahls, Kenttä, and Leyte (2000), this method could also be used to study laminates.

Recently, Chia *et al.* (2009) discussed PA and ATR spectra of magnetic paper. The objective of this work was to use infrared spectroscopy to determine the degree of loading of magnetite (Fe_3O_4) particles. Magnetic properties increased approximately linearly with this loading. Reduced PA intensities at higher FT-IR mirror velocities (higher f) were put down to reduced thermal diffusion lengths. Intensity increased with loading, prompting the suggestion that introduction of the magnetic particles led to higher thermal diffusivity. Baseline-corrected spectra were divided into calibration and test sets; PLS analysis of the data yielded good agreement between the predicted and actual degrees of loading.

The results presented in this section show that PA infrared spectroscopy is particularly well suited to the study of wood, pulp, paper, and other wood products. Both major attributes of PA spectroscopy – its requirement for little sample preparation and the capability of depth profiling – provide a strong justification for the use of the technique in this context. The future analysis of coated papers by PA infrared spectroscopy appears to be particularly promising.

7.11
Food Products

Food products generally consist of proteins, fats, carbohydrates, and water. To the spectroscopist, it is obvious that many foods are not amenable to traditional infrared sample preparation techniques. Fortunately, near- and mid-infrared PA spectroscopy provide a welcome opportunity for analyzing foods using both qualitative and quantitative methods that otherwise might not exist.

Several research groups have successfully studied the PA infrared spectra of foods, food products, and packaging materials. Some of the earlier literature in this subject area was reviewed by McQueen, Wilson, and Kinnunen (1995), who compared near- and mid-infrared PA techniques with both ATR spectroscopy and wet chemical methods. In general, published analyses of food products by PA spectroscopy have addressed the following problems: (i) the near-infrared determination of moisture content, (ii) quantification of fats and edible oils, (iii) elucidation of molecular structures in foods through the assignment of mid-infrared bands to particular functional groups, and (iv) characterization of coatings and microorganisms on foods. The existing literature on these topics is briefly reviewed below. Additional examples of PA infrared spectra of typical food products are also illustrated.

7.11.1
Early PA Studies of Foods

The capability of PA spectroscopy for the quantification of moisture in single-cell protein was demonstrated by Castleden, Kirkbright, and Menon (1980). These authors obtained dispersive near-infrared PA spectra (1.3–2.3 µm, approximately 7700–4350 cm^{-1}) of protein samples with a range of particle sizes, and found that the combination band due to water at about 1.9 µm (5265 cm^{-1}) could be used to quantify moisture at levels from 0 to 12%. Bands due to NH and CH groups were also identified in the spectra. However, quantitative analysis was not feasible until the samples were separated according to particle size, since PA intensities are normally greater for smaller particles. Castleden, Kirkbright, and Menon (1980) fractionated their samples and were able to carry out successful moisture determinations for specific particle size distributions. These experiments were continued by Jin, Kirkbright, and Spillane (1982), who obtained PA near-infrared spectra of protein samples and milk substitutes. PA intensities of bands due to water were again found to vary linearly with concentration in the ~0–10% range in this study.

During the 1980s, P. S. Belton and his colleagues at the Institute of Food Research and the University of East Anglia in Norwich, UK, published a series of articles describing near- and mid-infrared PA spectra of various food products. A major goal of this work was the use of PA spectroscopy for quantitative analysis of specific constituents in foods. During this period Belton (1984) presented a brief review of PA spectroscopy that mentions possible applications to the characterization of foods.

An early investigation by this research group involved the use of dispersive PA near-infrared (1–2.5 µm, or 10 000–4000 cm^{-1}) spectra for determination of the moisture content of potato starch (Belton and Tanner, 1983). Instrumentation included an Xe lamp, a monochromator, a chopper, and a lock-in amplifier. It was observed that the PA intensity due to water absorbed on starch did not increase linearly with the amount of water present at higher concentrations. This finding disagreed with the above-mentioned results of Castleden, Kirkbright, and Menon (1980) and Jin, Kirkbright, and Spillane (1982), who reported a linear variation of near-infrared PA intensity with water concentration; however, a wider concentration range was investigated by Belton and Tanner (1983). In fact, the nonlinear variation of PA intensity with analyte concentration led the latter authors to employ a model that predicts proportionality between the reciprocal of PA intensity and the reciprocal of analyte concentration. Their experimental results agreed well with this model, which is discussed in more detail in Chapter 5.

Following their initial near-infrared study, this research group published a series of articles on mid-infrared PA spectra of foods. Belton, Wilson, and Saffa (1987) compared PA spectra with results obtained using ATR and diffuse reflectance spectroscopies. The objective of this investigation was the quantitative analysis of protein-starch mixtures, prepared using either wheat gluten or casein together with potato starch. The authors found that PA and ATR spectra gave the best results, while the diffuse reflectance spectra were less satisfactory. Problems with sample mixing and loading were encountered with both the ATR and diffuse reflectance methods. Therefore PA spectroscopy was concluded to be the most suitable technique for the analysis of these mixtures.

In a related investigation, Belton, Saffa, and Wilson (1987) described a detailed PA study of sucrose, sucrose/KBr mixtures, and carbon black. The main goal of this work was to elucidate the influence of particle size on quantitative analysis in PA infrared spectroscopy. Spectra of sucrose powders with different particle sizes were shown but not discussed in detail. This work is also mentioned in the discussion of particle size effects in PA spectroscopy (Chapter 3). Sucrose and its chars were described earlier in this chapter with respect to the PA spectra of carbons.

The final work in this series (Belton, Saffa, and Wilson, 1988b) compares PA and ATR infrared spectroscopies for the analysis of chocolate and cocoa liquors. These methods were selected because the samples were unsuitable for ordinary infrared transmission measurements. Both types of spectra shown in this work are of good quality, although the bands in the PA spectrum are rather broad because it was acquired at comparatively low resolution (16 cm^{-1}). Despite this limitation, about a dozen bands are clearly identifiable in this spectrum. In fact, the authors concluded that PA spectroscopy was the best method for the quantitative analysis of fat in chocolate. This article is also mentioned in connection with quantitative analysis in Chapter 5.

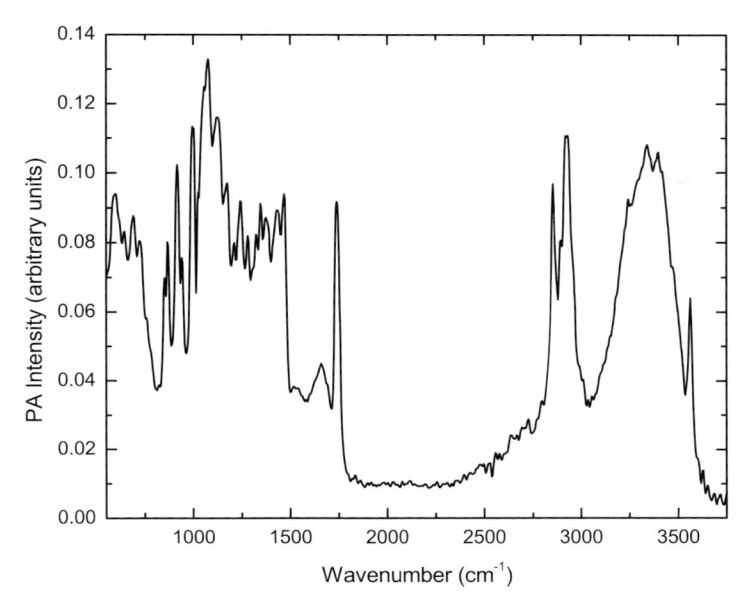

Figure 7.39 PA infrared spectrum of milk chocolate, obtained using a Bruker IFS 113v FT-IR and a Princeton Applied Research Corporation 6003 PA cell.

7.11.2
A Typical Example: Milk Chocolate

The capability of PA infrared spectroscopy for the characterization of chocolate is further demonstrated by a result for ordinary milk chocolate, depicted in Figure 7.39. This PA spectrum, recorded in the author's laboratory in approximately 25 min, displays about 20 well-defined bands. Indeed, food scientists would probably agree that this spectrum conveys considerable useful information regarding the composition of this chocolate sample. More generally, the reader can appreciate the capability for measurement of a research-quality spectrum of a soft material without any sample preparation. Of course this observation is relevant to samples other than food products.

7.11.3
Agricultural Grains

PA infrared spectroscopy can be used to characterize agricultural grains as well as prepared foods. For example, Greene *et al.* (1992) obtained PA and diffuse reflectance infrared spectra of corn in the course of a study of fungal contamination, which can potentially be a serious health hazard. These spectroscopic techniques were used because of their convenience. In this study, PA infrared spectra of corn infected with *Fusarium moniliforme* or *Aspergillus flavus* were observed to be greatly different from the spectrum of uninfected corn: major differences were observed in the amide I and II (1650- and 1550-cm^{-1}) bands, implying a significant increase in protein or acetylated sugar content as a consequence of the infection. Differences between diffuse reflectance and PA spectra were put down to the sample grinding required in the former technique. However, a minor drawback to the use of PA detection was noted: only one kernel of corn could be analyzed in each experiment! The examination of a large number of samples by this technique would therefore undoubtedly require significant time and effort. In fact, these investigators proposed an alternative to PA spectroscopy several years later; emission spectra of corn kernels on a moving conveyor belt were obtained using a technique referred to as transient infrared spectroscopy (Gordon et al., 1999). This effectively increased the amount of corn that could be analyzed in a limited period of time.

A related study by this group (Gordon et al., 1997) compared results from PA infrared spectroscopy with those from bright greenish-yellow fluorescence (BGYF), an accepted analytical method for the detection of pathogenic fungi on corn kernels. The authors noted ten specific changes that occur in the PA spectra of corn as a consequence of infection; these arise from phenomena such as an increase in the number of COOH groups, greater CO_2 evolution, and a decrease in carbohydrates, all of which are caused by the fungi. The observation of these changes in the PA spectra of a series of blind samples agreed completely with positive results in separate BGYF tests. Hence these experiments validated the use of PA infrared spectroscopy for the detection of fungal infection in corn. Because of the high information content in the PA infrared spectra of corn, the authors proposed that the PA results be combined with knowledge-based pattern recognition techniques in an expert system.

The number of features in the PA spectra that can be used to discriminate between healthy corn and samples infected with mycotoxigenic fungi was eventually increased to 12 (Gordon et al., 1998). In this later work, an artificial neural network was trained to distinguish contaminated from uncontaminated corn by pattern recognition of the infrared spectra. Although the use of individual bands was generally not adequate for discriminating between the two types of corn, the authors found that the simultaneous examination of eight or more features in the PA spectra led to a high success rate for the neural network. The question as to whether PA spectroscopy is capable of distinguishing among infections caused by different fungal species was not resolved in this work.

7.11.4
The PA Infrared Spectrum of Flour

The use of dispersive and FT near-infrared spectroscopies for the determination of the moisture and protein contents in wheat and other grains is now well established. In the context of the present discussion, it is relevant to ask whether PA infrared spectroscopy might usefully be employed for the analysis of these agricultural grains and products derived from them. To partly address this question,

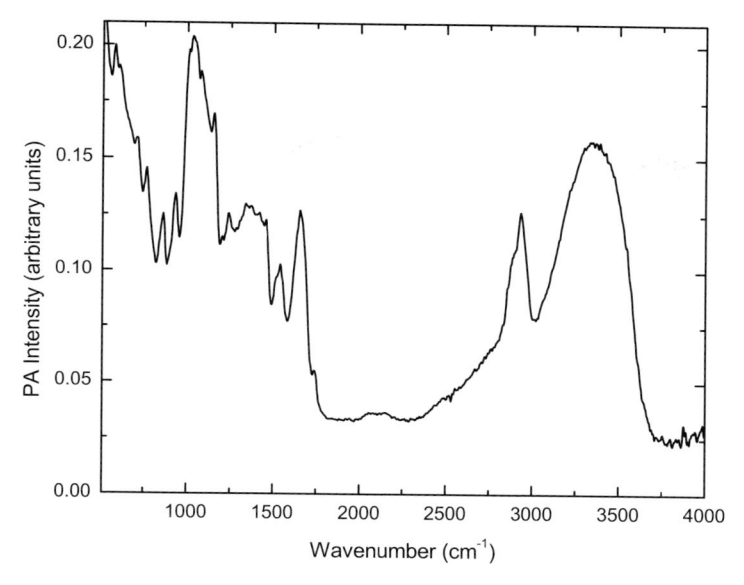

Figure 7.40 PA infrared spectrum of whole wheat flour, recorded under conditions similar to those for Figure 7.39.

a PA infrared spectrum was recorded for commercially available whole wheat flour (Figure 7.40). It should be noted that this spectrum closely resembles the result obtained for ordinary white flour under similar conditions. In the fingerprint region, the well-known amide I and II bands (see above) are clearly visible, as are a number of features arising from various other functional groups. The broad band near $3400\,cm^{-1}$ can readily be attributed to hydrogen bonding. Thus the information derived from the mid-infrared spectrum of flour is analogous to that from the corresponding near-infrared spectrum. The ease with which these spectra are recorded, as well as the intrinsically greater information content in the mid-infrared region, suggest that PA mid-infrared spectroscopy could be used successfully instead of established near-infrared techniques for the analysis of wheat, other grains, and products such as flour.

7.11.5
Further PA Studies of Food Products

McQueen *et al.* (1995) utilized optothermal near-infrared spectroscopy for the analysis of an extensive series of cheese samples. This technique is described in Chapter 3. Measurements were carried out at three wavelengths (1740, 1935, and 2180 nm; 5750, 5170, and $4590\,cm^{-1}$, respectively) that were selected by broad-bandpass optical filters. Protein, fat and moisture contents were also obtained using standard analytical techniques. Correlation coefficients ranging from 0.93 to 0.96 were calculated for the spectroscopic and wet chemical data. These results confirm the viability of the optothermal method for cheese analysis. Because data were recorded at only three wavelengths, spectra were not plotted.

Rapid quantitative analysis of the main components of pea seeds (starch, proteins and lipids) was investigated by Letzelter *et al.* (1995). Both PA infrared spectroscopy and established wet chemical methods were employed in this work. To assess the dependence of PA band intensities on concentration, the authors analyzed mixtures of starch in KBr and protein in starch. In the first case, the broad carbohydrate band at about $1000\,\text{cm}^{-1}$ exhibited partial saturation at concentrations of about 50%; by contrast, the strong amide I protein band varied linearly with concentration in the second experiment. Thus factors other than band intensities can determine whether PA saturation affects the spectra of these samples. Proceeding to the PA analysis of single pea seeds, the authors found that partial least squares (PLS), a common multivariate regression algorithm, gave reliable results with regard to sample composition. They concluded that the PA data adequately predicted the contents of the three components mentioned above, which were required for breeding purposes. This work is also mentioned in Chapter 5 with regard to quantitative analysis.

Near-infrared optothermal spectroscopy was mentioned above. This technique can also be implemented at mid-infrared wavelengths by use of a suitable laser or thermal radiation source. For example, Favier *et al.* (1996) used the 966-cm^{-1} CO_2 laser line in optothermal experiments to detect trans fatty acids (TFAs) in margarines. This wavenumber is diagnostic for the carbon-carbon bonds in these acids. The accuracy of the optothermal results was confirmed by agreement with data obtained from two different types of chromatography and transmission infrared spectroscopy.

Favier *et al.* (1998) next employed a tunable CO_2 laser in optothermal window experiments carried out to detect contaminants in extra-virgin olive oil. Laser lines at 931, 953, 966, 1041, and $1079\,\text{cm}^{-1}$ were used in this work. Olive oil does not absorb significantly in this region. On the other hand, common adulterants such as the safflower and sunflower oils utilized in this study absorb radiation at one or more of these locations, making optothermal window (OW) determinations of these contaminants feasible. Favier *et al.* (1998) showed that the observed OW signals were proportional to the amount of safflower or sunflower oil that was present in mixtures with olive oil. The limit of detection for either of these contaminants was approximately 5%.

A brief overview of infrared photothermal detection schemes – including several hyphenated techniques – that are suitable for quantification of TFAs in margarines and edible oils was presented by Bićanić *et al.* (1999). No spectra are included in this article. The methods discussed include dual beam thermal lensing (DBTL) spectrometry, the OW technique, gas chromatography combined with either PA or photopyroelectric (PPE) detection, and high-performance liquid chromatography (HPLC)/DBTL. The latter method does not appear to have been extended to infrared wavelengths.

The DBTL and OW measurements were based on the above-mentioned absorption of 966-cm^{-1} CO_2 laser radiation by a trans carbon-carbon double bond. The sensitivity of DBTL was found to be about two orders of magnitude better than that in conventional infrared spectroscopy, with an impressive detection limit of 0.002%. The OW method was used at higher concentrations and displayed an operating range of 4–60% TFA.

The GC methods were developed to quantify TFAs in the vapor state. Initial experiments with a flow-through, high-temperature PA cell using CO_2 laser radiation achieved limited success because of problems with temperature instability and varying TFA vapor pressure. Photopyroelectric (PPE) detection was more satisfactory, although its 2-µg sensitivity was still inferior to that in conventional GC.

7.11.6
Research of Irudayaraj's Group

An important series of articles describing applications of PA infrared spectroscopy to the characterization of foods was published by J. Irudayaraj and his colleagues at The Pennsylvania State University. These authors compared PA spectroscopy with other techniques in several studies, beginning with an investigation of particle size effects in PA and diffuse reflectance spectra of sucrose powder (Yang and Irudayaraj, 1999). The salient points in their work are discussed in this section.

The versatility of PA spectroscopy was nicely demonstrated in the analysis of potato chips (Sivakesava and Irudayaraj, 2000). The spectra exhibited bands from sugars, fats, and three different oils used in frying. Depth profiling of the chips was also performed. Sampling depths of 7–20 µm were calculated using $\alpha = 0.0013\,\text{cm}^2\,\text{s}^{-1}$. It was observed that fat content decreased with depth, while moisture and protein increased.

Yang and Irudayaraj (2000a) compared PA and ATR spectroscopies in the analysis of semisolid fat and edible oils, choosing butter, soybean oil, and lard as representative materials. The PA and ATR methods were selected because they require little sample preparation, obviously an important consideration in view of the rather problematic physical characteristics of fats and oils. Indeed, both sampling techniques yielded satisfactory spectra. Locations and assignments of a total of 18 bands observed in the PA infrared spectra of soybean oil and lard were also given by these authors.

An important limitation of PA infrared spectroscopy was noted during this investigation. Heat generated by the absorption of infrared radiation tended to evaporate the water contained in butter, leading to the appearance of rotation-vibration bands due to water vapor in the 1450–1650 and 3100–3700 cm^{-1} regions. These sharp features tended to obscure the broader, more relevant bands arising from the condensed-phase samples. This phenomenon is well known to many researchers who use PA infrared spectroscopy, since it occurs frequently during the acquisition of PA spectra of samples that contain significant amounts of water.

Irudayaraj et al. (2001) continued their exploratory comparison of the PA and ATR techniques through measurement of spectra of lard, peanut butter, mayonnaise, and whipped topping. These common foods were analyzed as purchased, and again after heating at either 60 or 90 °C for periods as long as 32 days. Heating caused the elimination of water from the latter two samples as well as the oxidation of the lipids in lard and peanut butter. The authors again concluded that ATR was preferable to PA infrared spectroscopy with regard to high-moisture samples. Approximately 10 to 20 bands were identifiable in both types of spectra for these four food products.

Four articles by this group (Irudayaraj and Yang, 2000, 2001; Yang and Irudayaraj, 2000b) reported depth profiling studies of cheese slices and their associated polymer-based packaging materials. The use of PA infrared spectroscopy in this research was prompted by the fact that cheese is not very amenable to most infrared sample preparation methods, and also by the possibility of successfully depth profiling both the wrapper and the cheese.

In-phase and quadrature step-scan data were used to calculate magnitude and phase PA spectra of both the packaging material and its contents. The use of widely separated phase modulation frequencies, ranging from 50 to 900 Hz, allowed the retrieval of spectra corresponding to greatly different sample depths. In this way, it was confirmed that two cheese components (fat and protein) had diffused into the surface layer of the wrapper during storage. On the other hand, bands arising from the package material mostly originated from deeper within the wrapper. Depth profiling of dried cheese confirmed the rather intuitive expectation that water had been eliminated from its surface, whereas moisture beneath the surface was less prone to evaporation.

Two of these articles described the use of generalized two-dimensional (G2D) spectral correlation analysis. This method, discussed in Chapter 5, can be used to confirm the ordering of a multi-layered system. In favorable cases it may also assist in the identification of overlapping bands, because spectral information is spread across an additional dimension. Indeed, Yang and Irudayaraj (2000b) found that G2D confirmed the diffusion of the amide I and II components from the cheese into the packaging. Moreover a 1461-cm^{-1} C–H band, not detectable in one-dimensional spectra because of the existence of several neighboring bands, was clearly revealed in the G2D correlation spectrum. Depth profiling of a model three-layer protein/starch/polyethylene system was also demonstrated using both G2D and PA phase spectra (Irudayaraj and Yang, 2001).

This group also used PA spectroscopy to study edible coatings and microorganisms on fruit surfaces. Yang et al. (2001) carried out depth profiling experiments on these systems, using either rapid- or step-scan PA infrared spectroscopy. Edible coatings were modeled using two-layer (protein/apple) and three-layer (starch/protein/apple, protein/starch/apple) samples, prepared by treating an apple skin with aqueous solutions of protein and starch. For the two-layer sample, a magnitude PA spectrum exhibited bands from both layers, even though the sampling depth was estimated to be only 6 µm. However, the corresponding phase angle spectrum facilitated assignments of the bands to the first or second layers. G2D spectral correlation analysis confirmed the known fact that the apple skin was below the layer of protein film. For the more complicated three-layer samples, G2D and phase angle spectra were used to confirm the ordering of the layers.

Microorganisms on fruit contain proteins, which may be identified using the well-known amide I and II bands. Yang *et al.* (2001) examined apple and honeydew melon surfaces covered with several different microbes in this work. Principal component and canonical variate analyzes were then used to discriminate between spectra of samples containing microorganisms and those of untreated samples. Magnitude spectra were used for depth profiling; plausibly, the amide I and II bands were stronger at higher modulation frequencies because the microbes were on the surface of the fruit. The authors suggested that the combination of PA spectroscopy and multivariate analysis might eventually become an accepted method for the determination of microorganisms on the surfaces of food products. Indeed, Irudayaraj, Yang, and Sakhamuri (2002) subsequently showed that apple skin surfaces contaminated with different microorganisms could be distinguished using PA infrared spectroscopy and suitable numerical techniques. Pathogenic and nonpathogenic *Escherichia coli* were also distinguished in these experiments.

The contamination of extra-virgin olive oil with other oils was discussed above with regard to the optothermal technique (Favier *et al.*, 1998). Yang and Irudayaraj (2001b) also studied this problem, employing near-infrared, FT-Raman, as well as mid-infrared (PA and ATR) spectroscopies. The adulterant in the more recent study was olive pomace oil, a fully refined product. FT-Raman spectra exhibited the highest correlation with the concentration of the contaminant. The PA and ATR spectra, which were practically identical, confirm the statement by Favier *et al.* (1998) that extra-virgin olive oil absorbs very weakly throughout the $931–1079\,cm^{-1}$ region where the CO_2 laser lines occur.

The final publication by this group to be included in this discussion presents infrared spectra for different types of meats. Yang and Irudayaraj (2001b) compared ATR and PA infrared spectroscopies with regard to their capabilities for the characterization of both beef and pork. The use of these particular infrared techniques obviously presents a significant advantage with regard to earlier infrared methods for the analysis of meats, in which protein was solvated and fat was either emulsified or extracted and refined. Both the ATR and the PA spectra of the beef samples obtained in this work were of very good quality. PA depth profiling experiments confirmed that the surface of each sample had a lower moisture content than the interior (Figure 7.41). Moreover, differences among different types of meats (beef, pork, chicken, turkey) were thought to be observable using PA infrared spectroscopy.

7.11.7
Food Production

PA infrared spectroscopy has recently been used by several research groups in applications related to food production. Mohamed *et al.* (2004) obtained PA spectra of wheat kernels in a study of heat damage. Three cultivars exhibited differences in sensitivities to heat in the polyester and amide I regions, suggesting morphological and secondary structural changes. Depth profiling showed that these

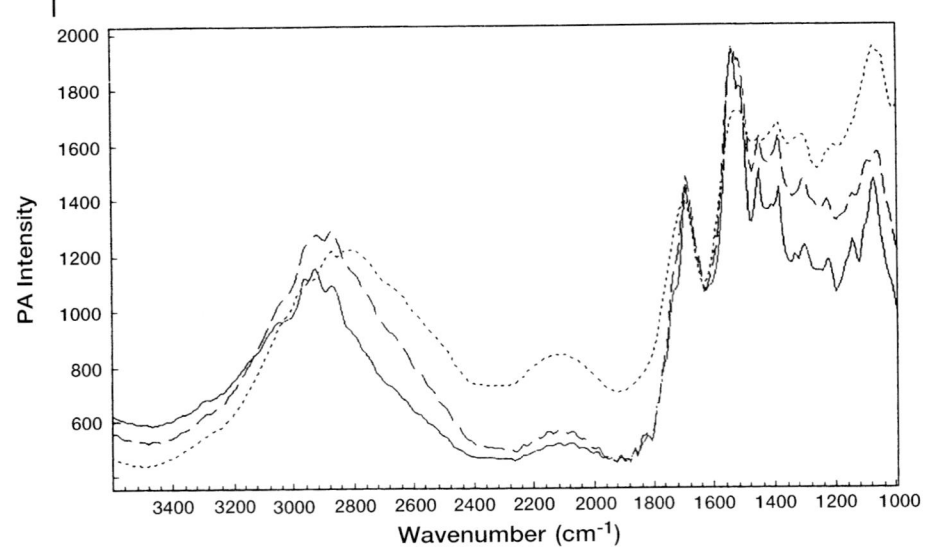

Figure 7.41 Depth profiling results for fresh beef. Phase modulation frequencies were as follows: solid curve, 900 Hz; dashed curve, 100 Hz; dotted curve, 10 Hz. (Reprinted from Lebensm.-Wiss. u.-Technol. **34**, Yang, H. and Irudayaraj, J., Characterization of beef and pork using Fourier transform infrared photoacoustic spectroscopy, 402–409, copyright © 2001, with permission from Elsevier Science.)

changes were minor near the surface of the heated kernels. Armenta et al. (2006) utilized PA spectroscopy to quantify Mancozeb, a pesticide used to protect crops against fungal diseases, in commercial formulations. PLS analysis of the spectra revealed that the results were comparable with the more time-consuming HPLC reference method. PA spectroscopy also confers an environmental advantage by reducing the requirement for reagents in HPLC. Finally, PA infrared spectra of transgenic (genetically modified) and conventional soybean seeds were compared by Caires et al. (2008). The spectra of the two types of seeds were quite similar, leading the authors to use canonical discriminant analysis to interpret the data. It was concluded that bands due to unsaturated and neutral lipids, proteins, and carbohydrates were most important for discrimination.

The considerable body of research described above clearly demonstrates that PA infrared spectroscopy is well suited for the analysis of food products. Indeed, the two principal advantages of PA spectroscopy – minimal sample preparation and the capability for depth profiling of layered samples – play an important role in much of this work. It might be anticipated that the use of PA infrared spectroscopy for the characterization of food products will eventually become sufficiently common that it no longer merits special comment. For example, the authors of a publication on sodium caseinate/glycerol and sodium caseinate/polyethylene glycol edible coatings (Siew et al., 1999) utilized PA infrared spectroscopy and a number of other analytical techniques, but did not find it necessary to display any infrared spectra.

7.12
Clays and Minerals

The reader may already be aware of the extensive literature on the infrared spectroscopy of clays and minerals: numerous texts, review articles, and original contributions have been dedicated to the subject. Indeed, some clay scientists might question the use of PA detection in the measurement of infrared spectra of these industrially important materials. A few exploratory PA studies utilized clays and minerals simply as examples of common solids; these circumstances could hardly be considered auspicious, and did not provide a particularly strong justification for the PA analysis of these substances.

Investigations carried out by a number of research groups during the last 30 years have clearly revealed more compelling reasons for the acquisition of PA infrared spectra of clays and minerals. Most importantly, the minimal sample preparation in this technique means that these solids generally require little grinding prior to measurement of their spectra. Elimination of grinding is highly desirable, since this procedure can bring about delamination and phase changes. Similarly, it is important to avoid grinding clays in situations where observation of both longitudinal and transverse optic bands is possible; this is illustrated below.

As discussed elsewhere in this book, the preparation of pellets is unnecessary in PA infrared spectroscopy. This fact is particularly relevant with regard to the study of clays and minerals, where the pressing of pellets containing an alkali halide diluent can result in unwanted ion exchange reactions. This undesirable effect was reported by Pelletier *et al.* (1999), who observed the transformation of Na-saponites into K-saponites during the preparation of KBr pellets. Fortunately, this situation can be completely avoided through measurement of PA infrared spectra.

As noted above, a number of studies presenting PA spectra of clays and minerals were published when the technique was first gaining widespread acceptance. The work of Kirkbright (1978), already mentioned in several other contexts, was among these publications. This article includes a dispersive near-infrared PA spectrum of kaolinite, a common phyllosilicate with the empirical composition $Al_4Si_4O_{10}(OH)_8$. Five bands were observed in the 1.2–2.6-μm (8330–3850-cm^{-1}) region. Features near 1.4 and 1.9 μm (7140 and 5260 cm^{-1}) were attributed to free water, while bands at slightly longer wavelengths were assigned to combination transitions involving hydroxyl groups. A number of years later, the near-infrared PA spectrum of kaolinite was obtained in the laboratory of the present author using an FT-IR spectrometer. The results of that experiment are described below.

Two publications pertaining to this discussion arose from the research of Rockley's group, which is mentioned elsewhere in this book. In the first article, Rockley, Richardson, and Davis (1982) reported the PA infrared spectrum of asbestos (chrysotile). A well-defined O–H stretching doublet at 3655 and 3692 cm^{-1} showed the expected 1:3 intensity ratio. The authors also noted the strong absorption at

950 cm^{-1} and suggested that a CO_2 laser might be a suitable radiation source for measurement of PA spectra in this region. This proposal might not seem viable to present-day researchers with access to FT-IR instruments, but was entirely reasonable at a time when the use of CO_2 lasers in PA experiments was quite common. The PA spectrum of asbestos has been studied by other groups and is mentioned again below; it is also described in Chapter 6.

Several years later, Rockley and Rockley (1987) compared PA and transmission infrared spectra for interlamellar cation-exchanged montmorillonites. Dimethyl methylphosphonate (DMMP), a phosphonate pesticide model compound, was adsorbed on the clays. Adsorption caused some DMMP bands to shift by a few wavenumbers, which suggested that the interaction between DMMP and montmorillonite occurs through P=O bonds. The PA spectra obtained in this work contain a few weak bands that were not detected in transmission spectra obtained for KBr pellets; the increased relative intensities of these bands imply that the PA spectra may have been partly saturated.

The authors unsuccessfully attempted to measure the PA spectrum of pure DMMP. The liquid was heated by the infrared beam and eventually evaporated, leading to the observation of gas-phase PA spectra. Practicing PA spectroscopists may be very familiar with this phenomenon, which occurs quite readily when attempts are made to record PA spectra of liquids with moderate to high vapor pressures.

Low and his colleagues at New York University studied a number of minerals during their development of photothermal beam deflection, reviewed in Chapter 3. An early publication on this subject (Low, Lacroix, and Morterra, 1982) presented a PBDS spectrum of calcite ($CaCO_3$). This spectrum contains about 10 narrow peaks; in addition, a broad, strong band near 1400 cm^{-1} arises from the carbonate group. A more extensive subsequent study (Low and Tascon, 1985) reported spectra of calcite-type minerals (magnesite, calcite, rhodocrosite, siderite, smithsonite), aragonites (aragonite, strontianite, witherite, cerussite), and lead minerals (crocoite, endlichite, vanadinite, mimetite, wulfenite). The spectra of the aragonites are shown in Figure 7.42. The authors observed that slight grinding of the minerals increased band intensities and reduced specular reflection, which sometimes produced severely distorted bands in spectra of unground samples. For example, PBDS spectra of a calcite single crystal showed strong orientation effects but were not easily interpreted because they displayed unusual band shapes. Spectra of ground calcite were free of this problem.

The near-infrared spectrum of kaolinite mentioned above (Kirkbright, 1978) was one of the first successful PA results for clays. Several years later, Donini and Michaelian (1985, 1986) obtained mid- and near-infrared PA spectra (~2000–10 000 cm^{-1}) of kaolinite and other clays using an FT-IR spectrometer (Figure 7.43). The kaolinite spectrum in the latter work displayed bands at 4526 and 4620 cm^{-1} that are due to combinations of the OH-stretching bands at 3620 and 3695 cm^{-1} with the Al–OH vibration at 915 cm^{-1}. At higher wavenumbers the overtones of the stretching bands occurred at 7065 and 7160 cm^{-1}, respectively. It should be noted that the latter assignment conflicts with that of Kirkbright (1978), who

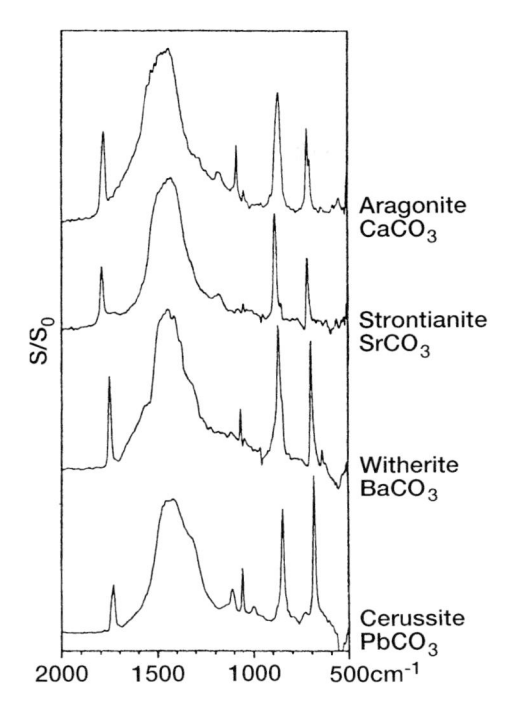

Figure 7.42 PBD spectra of aragonites. (Reproduced from Low, M.J.D. and Tascon, J.M.D., *Phys. Chem. Minerals* **12**: 19–22, Fig. 7, by permission of Springer-Verlag; copyright © 1985.)

attributed the ~7160-cm^{-1} band to adsorbed water rather than structural hydroxyl groups. Moreover, observation of the two high-wavenumber bands in the more recent work permitted calculation of anharmonicities for the 3620-cm^{-1} ('inner' hydroxyl) and 3695-cm^{-1} ('inner surface' hydroxyl) vibrations. PA near-infrared spectra of some cation-exchanged bentonites were also obtained in this investigation, and several bands were assigned by analogy with the kaolinite results.

After this work was completed, far-, mid-, and near-infrared (250–10 000 cm^{-1}) PA spectra were obtained for a kaolinite sample that was about 20% deuterated (Michaelian, Bukka, and Permann, 1987). Deuteration shifted the OH-stretching bands at 3620, 3651, 3669, 3684, and 3695 cm^{-1} to 2675, 2691, 2706, 2718, and 2725 cm^{-1}, respectively. The $\nu(OH)/\nu(OD)$ ratios for these bands are all approximately equal to 1.355. These data are significant in that the 3684-cm^{-1} band had previously been identified only in a few Raman spectra of natural kaolinite; observation of the 2718-cm^{-1} band was one of the first indications that this fifth hydroxyl stretching band is also infrared active. Subsequent investigations of infrared (both PA and transmission) and Raman spectra of kaolinites obtained from different sources confirmed this point. Those results are discussed below.

Far-infrared PA spectra of clays and minerals were also discussed in a paper by Donini and Michaelian (1988). The weakness of these spectra demanded the systematic elimination of noise sources that had deleterious effects on the data. In

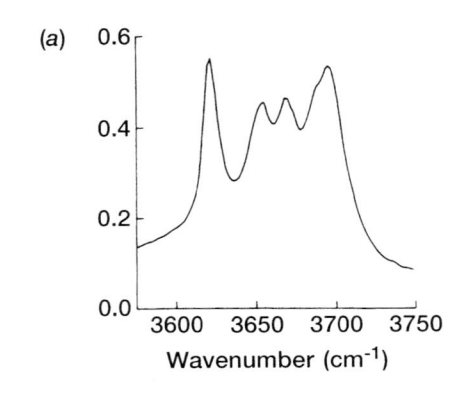

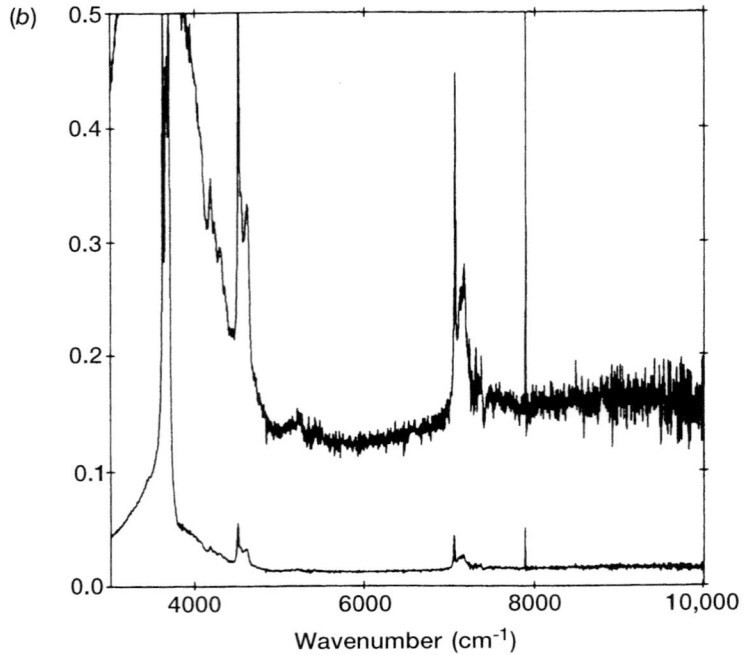

Figure 7.43 PA mid- and near-infrared spectra of international kaolinite. (a) OH-stretching region; (b) near-infrared region. The upper curve in (b) is scale-expanded by a factor of 10. The spike at 7900 cm^{-1} is a sub-multiple of the HeNe laser line. (Reprinted from *Infrared Phys.* **26**, Donini, J.C. and Michaelian, K.H., Near infrared photoacoustic FTIR spectroscopy of clay minerals and coal, 135–140, copyright © 1986, with permission from Elsevier Science.)

addition to obvious sources of noise (Donini and Michaelian, 1985), the sequence of data averaging was considered in this work; it was found that the average of several spectra (S_s) exhibited less noise than the spectrum calculated from the average interferogram (S_i). These averaging strategies are discussed in Chapter 6. In the present case, it turns out that phase correction of the individual interfero-

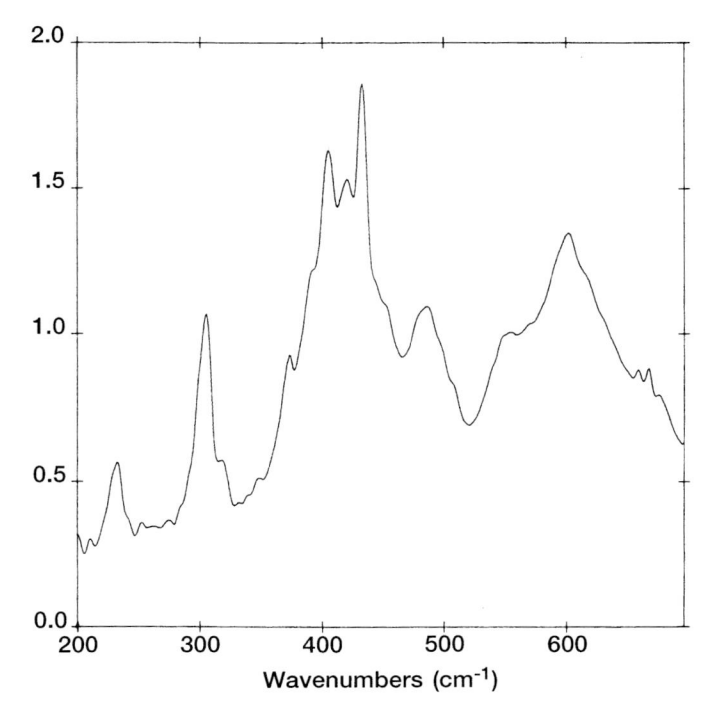

Figure 7.44 Far-infrared PA spectrum of chrysotile asbestos. (Reproduced from Donini, J.C. and Michaelian, K.H., *Appl. Spectrosc.* **42**: 289–292, by permission of the Society for Applied Spectroscopy; copyright © 1988.)

grams is preferable to that for the average interferogram. The far-infrared PA spectrum of asbestos (a hydrated magnesium silicate) obtained in this investigation is depicted in Figure 7.44.

The development of commercial step-scan FT-IR spectrometers in the late 1980s made PA saturation and phase investigations much more feasible than had been the case previously. In an early implementation of amplitude-modulation step-scan PA infrared spectroscopy (Michaelian, 1990b), mid- and near-infrared spectra were obtained for kaolinite. Quadrature spectra were observed to be superior to the corresponding in-phase data, with less saturation, a lower background, and better band definition. An important result in this work was the appearance of the ~3686-cm^{-1} hydroxyl stretching band in the quadrature spectrum – a direct observation of this band, which was mentioned above with regard to deuterated kaolinite. This result is shown in Figure 7.45.

The combination of transmission infrared, PA, and Raman spectra has revealed a total of five hydroxyl stretching bands in kaolinites. Transmission infrared spectroscopy first identified the four bands at 3620, 3651, 3669 and 3695 cm^{-1}; the fifth OH band was observed at about 3684 cm^{-1} in Raman spectra of several kaolinites more than 20 years ago, and subsequently identified in rapid- and step-scan PA infrared spectra (Friesen and Michaelian, 1986). The interpretation of this fifth band was the subject of further research. Shoval, Yariv, and Michaelian (1999)

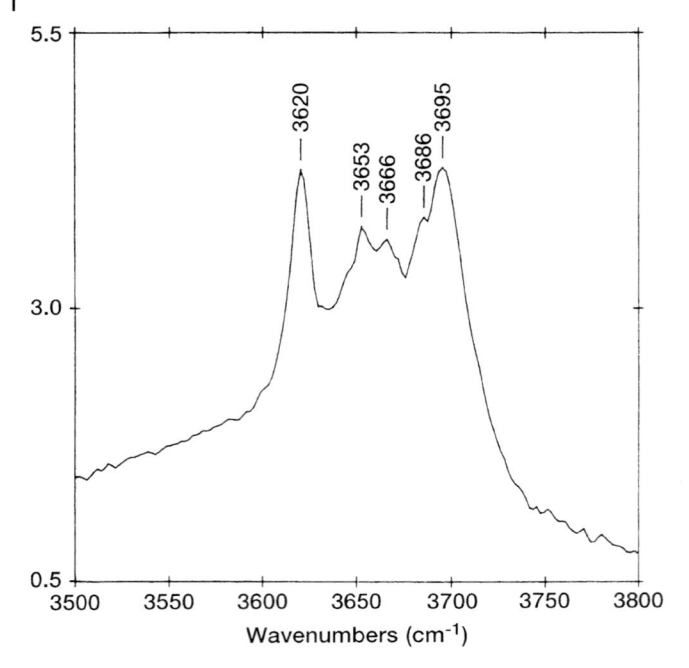

Figure 7.45 The OH-stretching region in the quadrature step-scan spectrum of kaolinite, measured at a resolution of 4 cm^{-1} and a modulation frequency of 40 Hz. (Reprinted from *Infrared Phys.*, **30**, Michaelian, K.H., Step-scan photoacoustic infrared spectra of kaolinite, 181–186, copyright © 1990, with permission from Elsevier Science.)

showed that its intensity was proportionately greater in Raman and PA infrared spectra of highly crystallized kaolinites, which have large coherent domains. A comparison of spectra for kaolinites with different particle sizes led the authors to conclude that the bands at 3695 and 3686 cm^{-1} are the longitudinal and transverse optic (LO and TO) modes of the high-wavenumber inner-surface OH-stretching band. These modes were discussed in detail by Farmer (1998, 2000).

Curve fitting results for the PA spectrum of a highly crystallized sample are shown in Figure 7.46. It should be noted that the relative intensity of the 3686-cm^{-1} band in this PA spectrum may be enhanced because of partial saturation (Michaelian, 1990b). This band has now also been identified in transmission infrared spectra of hydrothermal and authigenic kaolinites which have a high degree of crystallinity (Shoval et al., 1999). The continued grinding of a highly crystallized kaolinite causes the destruction of these domains and the disappearance of the 3686-cm^{-1} band (sometimes labeled 'band Z'); hence, transmission spectra differ from the PA spectrum by an increasing amount (Shoval et al., 2002a).

Kaolinite is usually referred to as a nonexpanding clay. In fact kaolinite can be intercalated by small polar organic molecules or alkali halides, particularly in the presence of small amounts of water. Formation of mechanochemical complexes between kaolinite and alkali halides has been investigated by PA infrared spectros-

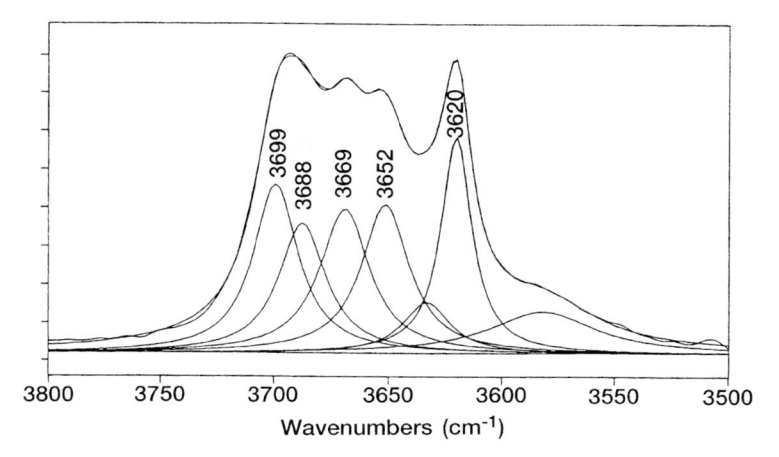

Figure 7.46 Curve fitting results for the OH-stretching region of the PA infrared spectrum of highly crystallized MS-6 kaolinite, obtained from Makhtesh-Ramon, Israel. (Reproduced with the permission of the Mineralogical Society of Great Britain & Ireland from the paper by S. Shoval et al.: *Clay Minerals* **34**, p. 561; copyright © 1999.)

copy; the motivation for this work was the critical need to avoid grinding the complexes after they are formed. Mid- and far-infrared PA spectra of kaolinite complexes with cesium and potassium halides were obtained in this research.

PA and diffuse reflectance infrared spectroscopies were used to study the reactions between CsBr and kaolinite that occur during shaking, grinding at ambient atmosphere, and grinding in the presence of a few drops of water (Michaelian, Yariv, and Nasser, 1991). An intercalation complex was formed by wet grinding, but not by the other two techniques. It is important to note that this grinding was an integral step in the intercalation process; once the CsBr-kaolinite complex was formed, it was not re-ground before its PA spectrum was obtained. Figure 7.47 shows the PA spectra of untreated kaolinite and the CsBr-kaolinite mixtures; the spectrum of the dry ground mixture (b) was similar to that of kaolinite (d), implying that no complex was formed. On the other hand, the spectrum of the wet ground sample (a) exhibited a number of changes indicating the intercalation of both CsBr and H_2O. In particular, the Al–OH deformation bands near 900 cm^{-1}, the Al–O band at 552 cm^{-1}, and the Si–O bands in the spectrum of kaolinite were all shifted to different locations; moreover, the PA spectrum of the wet ground sample also displayed the water v_2 band at 1630 cm^{-1}. These observations led to the development of a model in which a water molecule coordinated to the Cs$^+$ ion donates one proton to a siloxane oxygen and the other to the halide, while accepting a proton from a kaolinite inner surface hydroxyl group. A similar structure exists for the kaolinite/CsCl complex, which has been prepared by several different procedures and characterized by thermal analysis and PA infrared spectroscopy (Yariv *et al.*, 1994). The kaolinite/CsI complex cannot be obtained mechanochemically, although it has been produced by an indirect technique (Michaelian *et al.*, 1998).

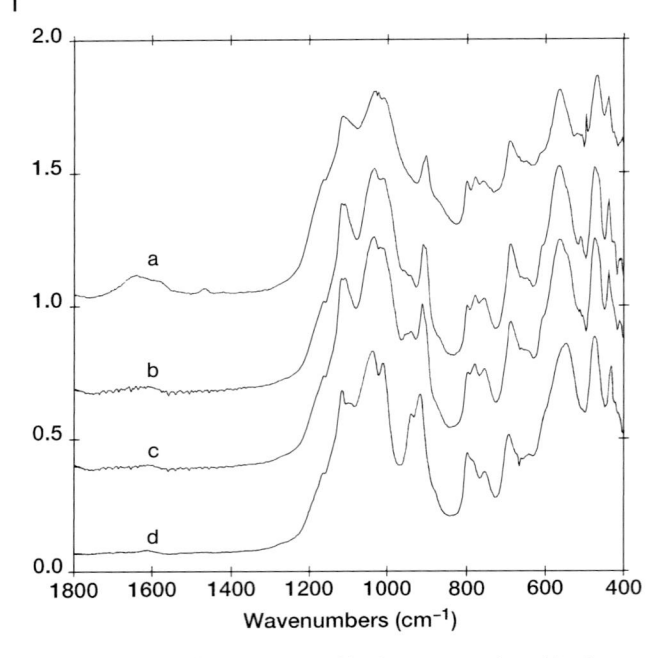

Figure 7.47 PA infrared spectra of kaolinite (curve d) and kaolinite/CsBr mixtures: dry mixed (curve c), dry ground (curve b), and wet ground (curve a).

It has already been noted that the literature on PA far-infrared spectroscopy is rather limited. Nevertheless, far-infrared PA and diffuse reflectance spectra were obtained for the kaolinite/alkali halide complexes mentioned in the previous paragraph and for complexes involving potassium halides (Michaelian et al., 1997). PA far-infrared spectra of these complexes were quite weak, partly because of the excess alkali halide in each sample. Another limitation exists in the low-wavenumber (long-wavelength) cutoff of the salts, which restricted the accessible region in some cases. Despite these restrictions, useful low-wavenumber PA spectra were obtained (Figure 7.48). Intercalation caused kaolinite bands due to mixed Si–O deformations and octahedral sheet vibrations to become narrower or shift by as much as $13\,cm^{-1}$.

Dickite is another, less common, member of the kaolin sub-group of dioctahedral nonexpanding clays. It has the same composition as kaolinite, $Al_4Si_4O_{10}(OH)_8$, differing only in the stacking of its layers. Shoval et al. (2001) showed that LO and TO crystal modes also occur in the hydroxyl stretching region of the Raman and infrared spectra of dickite; the assignments of the six identifiable OH bands to three LO/TO pairs were determined from polarized micro-Raman spectra of single-crystal dickite and PA infrared spectra of coarse non-oriented crystals. All six bands were used to fit the PA spectra, where the bands are much broader than their counterparts in the Raman spectra of this clay.

Thermal treatment of dickite, which is pertinent to the ceramic industry, was studied by Shoval et al. (2002b). PA and transmission infrared, and micro-Raman,

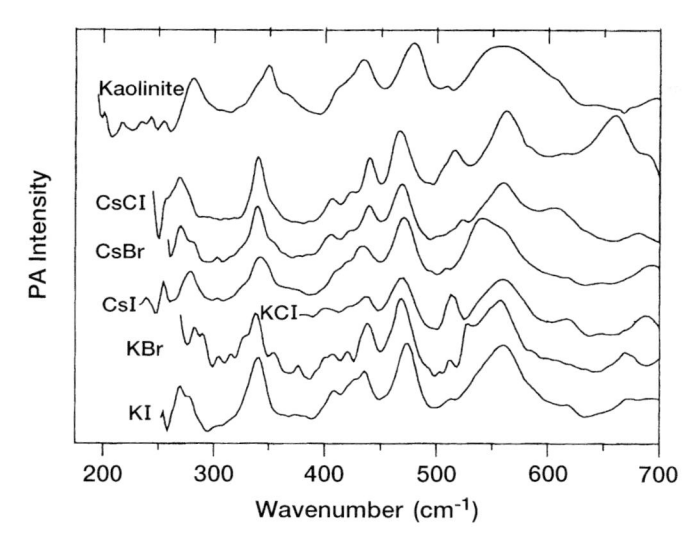

Figure 7.48 Far-infrared PA spectra of kaolinite and kaolinite/alkali halide/water complexes. (Reproduced from Michaelian, K.H. et al., *Mikrochim. Acta* **14** (Suppl.):211–212, by permission of Springer-Verlag, copyright © 1997.)

spectra were used to identify products of the treatment. The PA spectra were curve-fitted with a series of bands – many of which are quite broad – that prove the existence of mullite, Al-spinel, corundum and amorphous silica. Treatment at 700–1000 °C results in the production of amorphous meta-dickite, whereas still higher temperatures (1000–1300 °C) cause recrystallization of the minerals.

Mid- and far-infrared PA spectra of asbestos were described above. Returning briefly to this topic, a series of publications by Bertrand and co-workers at École Polytechnique de Montréal should be mentioned. Several instrumental methods were employed in this research. Monchalin *et al.* (1979) used chopped radiation from a cw HF laser to obtain spectra in the ~3400–3800 cm^{-1} region. PA intensity was greater when the laser polarization was perpendicular to the fiber axis, and lower with parallel polarization. Bertrand, Monchalin, and Lepoutre (1982) subsequently compared magnitude and phase PA infrared spectra of asbestos obtained with a dispersive spectrometer, and found the phase curve to be the more accurate representation of the absorption spectrum (obtained in a transmission experiment). Specifically, the intensity of the broad band near 3400 cm^{-1} (Rockley, Richardson, and Davis, 1982) due to adsorbed water is reduced in the phase spectrum. Choquet, Rousset, and Bertrand (1985) continued this work was with an FT-IR spectrometer. Real and imaginary spectra calculated from double-sided interferograms yielded a spectrum more accurate than that obtained using the conventional FT-IR phase correction algorithm. In addition to diminution of the broad water band, the high-wavenumber region of the PA spectrum was preferentially intensified in these calculations. This work was put on a more theoretical basis in two subsequent articles (Choquet, Rousset, and Bertrand, 1986; Bertrand,

1988). The numerical methods used by these authors are described in detail in Chapter 6.

PA infrared spectroscopy can be used to analyze soils, which are composed of clays, minerals, water, and organic matter. Quantification of these components and the study of their interactions are of primary interest. For example, Hofmann, Faubel, and Ache (1996) described the analysis of PCBs on sand, observing a linear relationship between band intensities and PCB concentration.[6] Dóka et al. (1998) used CO laser radiation at $1801\,\mathrm{cm}^{-1}$ to measure carbonate contents in soils. More recently, a collaboration between laboratories in Israel and China utilized PA spectroscopy in an extensive study of soils. Changwen, Linker, and Shaviv (2007) compared PA, ATR, and transmission spectroscopies, noting the minimal sample preparation in PA spectroscopy as an important advantage. Quantitative analysis of soil components yielded errors at per cent levels. Numerical methods (principal component analysis, probabilistic neural networks, partial least squares) were used to differentiate series of spectra, some of which were quite similar, in subsequent work (Du, Linker, and Shaviv, 2008; Du et al., 2009). Linker (2008) used PA spectroscopy to analyze nitrates on ion exchange membranes, also employing numerical methods for quantification.

Clays and minerals affect the extraction of bitumen from oil sand, as well as the upgrading of this complex hydrocarbon mixture to middle distillate fuels. PA infrared spectroscopy was used to characterize these inorganic components in a collaboration between National Research Council Canada and Syncrude Canada Ltd. Spectra of 'end cuts' (residue boiling above $525\,°\mathrm{C}$) displayed prominent kaolinite bands (Bensebaa et al., 2000). Solids in fine tailings (fluid waste) produced during extraction were also studied, and a structural model was based on depth profiling results (Bensebaa, Majid, and Delsandes, 2001). Fine solids were concluded to consist of amorphous minerals, while coarser solids were dominated by kaolinite (Majid et al., 2003). These researchers also used PA spectroscopy to characterize the solids in coker gas oils (Xu et al., 2005). Post-extraction oil sand, a residue obtained by distillation extraction of bitumen with an aromatic solvent, contains kaolinite, quartz, silica and siderite (Michaelian, Hall, and Kenny, 2006, Michaelian, 2007). PA spectra confirmed that clays are rejected together with asphaltenes when short-chain alkanes are added to bitumen froth (Friesen et al., 2005a, 2005b). PA spectra of these hydrocarbons were discussed earlier in this chapter.

As a final comment to this discussion, it can be noted that the minimal sample preparation required in PA infrared spectroscopy has provided the motivation for much of the published work on clays and minerals. The advantages offered by the other major capability of PA spectroscopy – depth profiling – have not been utilized as extensively. The study of clay- and mineral-organic interactions (Bowen, Compton, and Blanche, 1989; Brienne et al., 1996; Mendelovici, Frost, and Kloprogge, 2001), which are important in a variety of industrial applications, benefits from both aspects. PA spectroscopy has also been used recently to characterize polymer-clay nanocomposites (Sikdar, Katti, and Katti, 2006, 2008).

---

6) Several band assignments in this article are questionable.

7.13
Textiles

The analysis of textiles by infrared spectroscopy presents yet another situation in which traditional sample preparation techniques tend to be problematic. PA infrared spectroscopy nicely obviates this difficulty; moreover, because textile fibers are commonly coated to improve their performance and appearance, the depth profiling capability of PA spectroscopy is also relevant to this application. Both facts have been recognized by at least a half-dozen groups during the last three decades. The principal results of their investigations are summarized in the following discussion.

PA infrared spectroscopy was first used to characterize textiles about 30 years ago, at a time when the technique was still evolving. Consequently, some early reports on PA spectra of textiles discussed issues regarding instrumentation that are now resolved. For example, Teramae and Tanaka (1981) described a purpose-built PA cell and then proceeded to compare PA infrared spectra of carbon black with the more familiar 'empty instrument' spectrum obtained with a triglycine sulfate (TGS) detector. The Helmholtz resonance of the PA cell was also studied. To demonstrate the capabilities of their equipment, these authors reported PA spectra of cotton and nylon cloths; cellulose bands at about 1100, 2900, and 3300 cm^{-1} were observed in the former case, whereas nylon displayed characteristic polyamide bands at 1550, 1650, 2900, and 3350 cm^{-1}. These results were subsequently discussed in more detail in an article on PA infrared spectra of polymers, rubbers and resins (Teramae, Hiroguchi, and Tanaka, 1982).

Exploratory PA studies of textiles were also carried out at shorter wavelengths. Davidson and King (1983) reported near-infrared (1.2–2.4 μm, or about 8300–4200 cm^{-1}) PA spectra of wool and polyester-fiber fabrics and wool-polyester blends. The PA spectra of wool and polyester differed, with each containing about a dozen bands. As might be expected, several features due to OH, NH, and CH groups were observed in these spectra. The PA near-infrared spectrum of cotton fabric was also measured and found to resemble that of wool. The often-mentioned insensitivity of PA spectra to sample morphology was observed in this work: various types of wool fabric (challis, serge, flannel, and knitted) yielded very similar PA spectra. Moreover, the application of a fluorescent whitening agent and subsequent prolonged irradiation did not produce changes in the near-infrared spectra, although this treatment caused the samples to turn yellow.

This near-infrared investigation was followed by a related PA mid-infrared study of wool, polyester, and nylon (Davidson and Fraser, 1984). As shown in Figure 7.49, the PA infrared spectra of wool and nylon yarns are quite different, with the nylon spectrum containing the larger number of bands. The authors demonstrated the additivity of the PA spectra of textiles by examining results for blends of different yarns. For example, the polyester spectrum was successfully retrieved by subtracting the spectrum of a wool fabric from that of a wool/polyester mixture. Similarly, a suitably weighted coaddition of wool and nylon PA spectra closely resembled the spectrum of a known mixture. Poorer agreement was obtained when this experiment was extended to ternary mixtures, primarily because of band

(a)

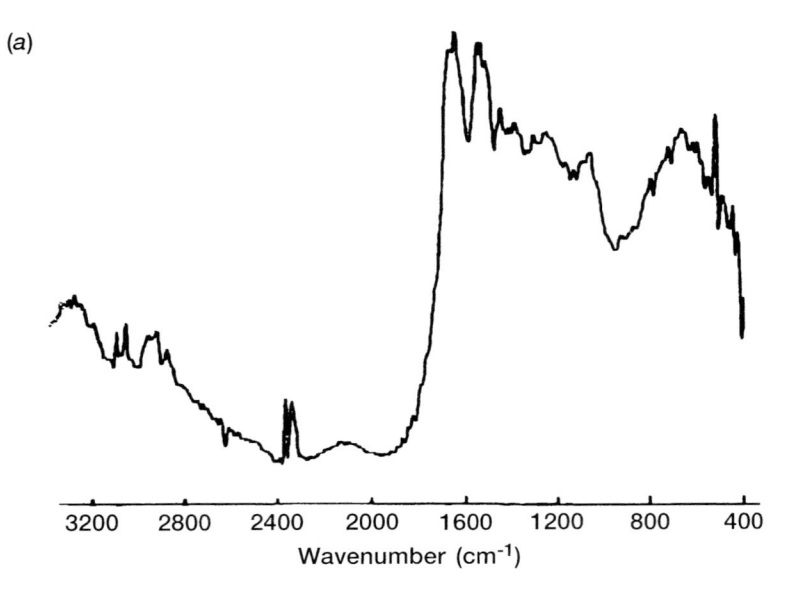

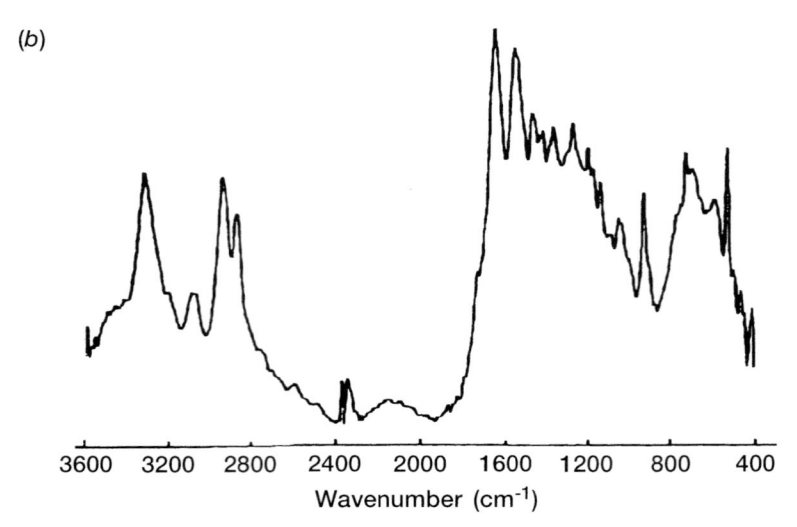

(b)

Figure 7.49 PA infrared spectra of (a) wool and (b) nylon yarns. (Davidson, R.S. and Fraser, G.V. Copyright © 1984. *J. Soc. Dyers Colour.* **100**: 167–170; reproduced by permission of the Society of Dyers and Colourists.)

overlap. Nevertheless, this research clearly showed the capabilities of PA FT-IR spectroscopy for both qualitative and quantitative analysis of textile mixtures.

Wool oxidation was studied by McKenna *et al.* (1985), who employed PA infrared spectroscopy primarily because of the minimal sample preparation required. The PA infrared spectrum of untreated wool displayed a number of amide bands (labeled A, B, I, II and III; see Figure 7.50) in addition to well-known features near

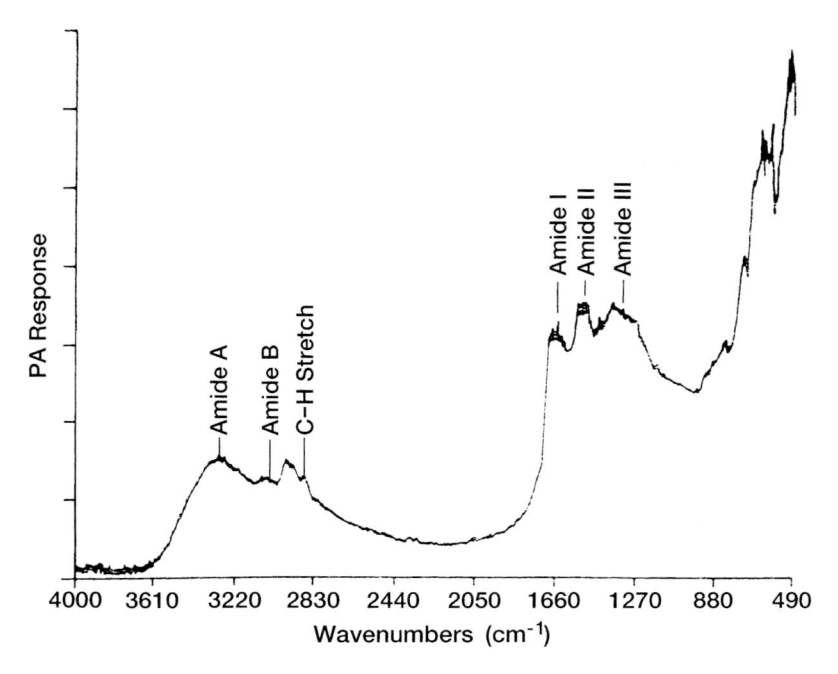

Figure 7.50 PA infrared spectrum of untreated wool. (Reprinted from McKenna, W.P. et al., *Spectrosc. Lett.* **18**: 115–122, by permission of Marcel Dekker, Inc.; copyright © 1985.)

2900 cm^{-1} due to C–H stretching. Oxidation with dichloroisocyanuric acid resulted in the appearance of an additional band at 1022 cm^{-1}, which the authors attributed to a sulfoxide or sulfinic acid (RSOOH). By contrast, oxidation in a corona discharge did not cause any observable changes in the infrared spectrum, suggesting a low concentration of oxidation products.

Keratins are fibrous proteins that occur in hair, feathers, hooves, and horns. PA infrared spectra of the surface (cuticle) and interior (cortex) of intact keratin fibers, namely three types of wool and human hair, were investigated by Jurdana et al. (1994). Diffuse reflectance and ATR infrared spectroscopies were also utilized by these authors, who were particularly interested in surface characterization. Analysis of hair is also discussed above in the section on medical applications of PA infrared spectroscopy.

All of the bands observed by McKenna et al. (1985) in the PA spectrum of wool were also detected in the spectra of wool and human hair by Jurdana et al. (1994); the amide I and II bands were studied in detail in an attempt to distinguish between the surface and interior of the fibers. μ_s was calculated using $\kappa = 9.05 \times 10^{-5}$ cal °C^{-1}s^{-1}cm^{-1} and known values for the density and heat capacity. Sampling depths ranged from 0.8 to 22.9 µm, depending on the rapid-scan mirror velocities and wavenumber region under consideration. This made it possible to distinguish the cuticle sheath of the wool from the cortex, since the former has a thickness between 1 and 10 µm.

This investigation also showed that the wavenumber separation between the amide I and II bands varied with the sampling depth. This effect is due to differences between the chemical compositions of the cuticle and cortex and is analogous to previously published ATR and diffuse reflectance findings. Some PA spectra measured at the lower velocities were partly saturated, making it impossible to obtain reliable intensities for the amide I and II bands. The diffuse reflectance spectra displayed relatively broad bands that were attributed to the heterogeneity of the fiber interiors. By contrast, the narrower bands in the ATR spectra were due to the cuticle.

These researchers extended their studies in a subsequent publication (Jurdana et al., 1995). Both wool and hair were again examined, with an emphasis on comparison of the rapid- and step-scan data. The origins of the amide I and II bands were determined in step-scan phase modulation experiments by changing the phase angle of the lock-in amplifier in small increments. Plots of the amide I band position and the amide I/II intensity ratio as a function of phase lag, which can ideally be used to locate the interface between cuticle and cortex, showed discontinuities at phase angles between 126° and 153°. Subtraction of a phase error of 45° corrected these angles to 81° and 108°, yielding μ_s values between 5.2 and 9 μm. The transition from cuticle to cortex was thus determined rather elegantly in this work. In general, saturation was more problematic in the rapid-scan spectra of wool and hair than in the step-scan spectra.

Church and Evans (1995) used both PA and ATR infrared spectroscopies to study the surface treatment of wool with fluorocarbon polymer finishes. Quantitative analysis was an objective of this investigation. The authors observed that the PA spectrum of a thin free-standing film of the polymer closely resembles the absorption spectrum obtained in a transmission experiment; the ATR spectrum, after correction for the standard wavelength dependence of the sampling depth, displayed similar relative intensities. Although the PA spectra of the treated wool were partly saturated, the integrated intensities of the C–F stretching bands were proportional to the known concentration of the polymer on the wool. The ATR detection limit was lower than that in PA spectroscopy, but the ATR response deviated from linearity at higher concentrations. In this sense the PA results could be concluded to be as good as, or better than, those obtained by ATR.

Comparison of PA and ATR infrared spectroscopies with respect to the characterization of surface-treated wool was continued by Carter, Fredericks, and Church (1996). Samples that had been chlorinated, or chlorinated and neutralized with bisulfite for shrinkproofing, were studied in this second investigation. Wool treated with a fluoropolymer was also examined in this study. Rapid-scan depth-profiling experiments confirmed that chlorination led to surface oxidation, specifically the conversion of cystine to cysteic acid. PA spectra recorded at higher velocities qualitatively resembled the ATR spectrum, although ATR yielded better signal/noise ratios. Depth profiling experiments of wool samples treated with the fluoropolymer showed that the polymer collected on the surface and did not penetrate into the fiber.

An extensive series of studies on the PA infrared spectra of fabrics, fibers, and related polymeric materials was published between 1985 and 1994 by C. Q. Yang

Table 7.6 Topics discussed in publications by C. Q. Yang et al. on PA spectra of textiles.

Publication	Year	Topic
Polym. Mater. Sci. Eng., **53**, 169–175	1985	Depth profiling of sized cotton yarn
J. Coated Fabrics, **17**, 110–128	1987	Polymeric sizing of cotton yarns
Appl. Spectrosc., **41**, 889–896	1987	Depth profiling of sized cotton yarn
Anal. Chim Acta, **194**, 303–309	1987	Weathering of cotton fabric
Text. Res. J., **59**, 562–568	1989	Foam finished fabrics
Polym. Mater. Sci. Eng., **62**, 903–906	1990	Surface treatment of cotton and PPE
Appl. Spectrosc., **44**, 1035–1039	1990	Finishing of PET fibers
Appl. Spectrosc., **45**, 102–108	1991	Polycarboxylic acids as crosslinking agents for cotton fabrics
Appl. Spectrosc., **45**, 1695–1698	1991	Photooxidation of cotton cellulose
J. Appl. Polym. Sci., **43**, 1609–1616	1991	Polycarboxylic acids as crosslinking agents for cotton fabrics
Text. Res. J., **61**, 298–305	1991	Ester crosslinkage in cotton cellulose
Polym. Mater. Sci. Eng., **64**, 33–35	1991	Oxidation of cotton fabrics
Polym. Mater. Sci. Eng., **64**, 372–374	1991	Polycarboxylic acids as crosslinking agents for cotton fabrics
Ind. Eng. Chem. Res., **31**, 617–621	1992	Oxidation and degradation in various textiles
J. Environ. Polym. Degrad, **2**, 153–160	1994	Photooxidation of PE fabric
J. Appl. Polym. Sci., **51**, 389–397	1994	Photo- and thermal oxidation of polypropylene fabric

and several collaborators. This research was initiated at Kansas State University and continued at Marshall University. Publication details are given in Table 7.6. Major findings are discussed below.

The intial articles in this series emphasized depth profiling and surface sensitivity in rapid-scan PA infrared spectra. A clear example of both effects with regard to sized (stiffened) cotton fibers was given by Yang et al. (1985); the 1735-cm^{-1} band due to the polyurethane sizing agent was significantly stronger in the PA spectrum than in the transmission spectrum, the latter having been obtained for a ground sample. Similarly, the intensity of this band was diminished in the PA spectrum of a powdered sample. Grinding obviously mixes the surface and bulk of the fibers, thereby decreasing the relative intensities of the bands due to the sizing agent on the surface. Depth profiling was also clearly demonstrated by measuring PA spectra of an intact sample at different mirror velocities: the characteristic 1735-cm^{-1} band became proportionately stronger as μ_s was reduced, consistent with the presence of polyurethane on the fiber surface and a diminished contribution of the bulk fiber to the spectrum.

Surface sensitivity and depth profiling were again observed for cotton fabrics treated by means of conventional padding and foam finishing techniques, where the distribution of finishing agents is correlated with wrinkle recovery properties (Yang and Fateley, 1987; Yang, Perenich, and Fateley, 1989). The wavenumber dependence of μ_s in rapid-scan PA spectroscopy was quite noticeable in these experiments. In a model system comprising a glass fiber coated with poly(vinyl acetate), the C–H stretching band was absent in PA spectra measured at higher velocities (smaller μ_s) while the 1735-cm^{-1} band persisted; this was due to the greater μ_s at the lower wavenumber and the penetration of the coating into the fiber (Yang, Bresee, and Fateley, 1987).

The concentration of polymeric sizing agents near the surface of cotton yarns was further examined by Yang and Bresee (1987). PA infrared spectra showed that desizing with boiling NaOH removed the sizing agent from the surface, but tended to leave it in the bulk. Although the PA spectra of cotton yarn sized with polyurethane and unsized yarn appeared to be nearly the same, subtraction dramatically recovered the spectrum of the sizing agent from that of the sized yarn (Figure 7.51). It is clear that the sizing agent was present mainly on the surface of the yarn. On the other hand, X-ray photoelectron spectroscopy (XPS) revealed that a copolymer finish on poly(ethylene terephthalate) fibers was inhomogeneous (Yang, Bresee, and Fateley, 1990), a conclusion supported by scanning electron microscopy (SEM). The apparent discrepancy between these later results and those from PA infrared spectroscopy probably arises from the differences in spatial resolution among the three techniques.

Oxidation of cotton cellulose by ultraviolet radiation was studied by Yang and Freeman (1991a). After only 4 h of irradiation at 254 nm, a band due to carboxylic

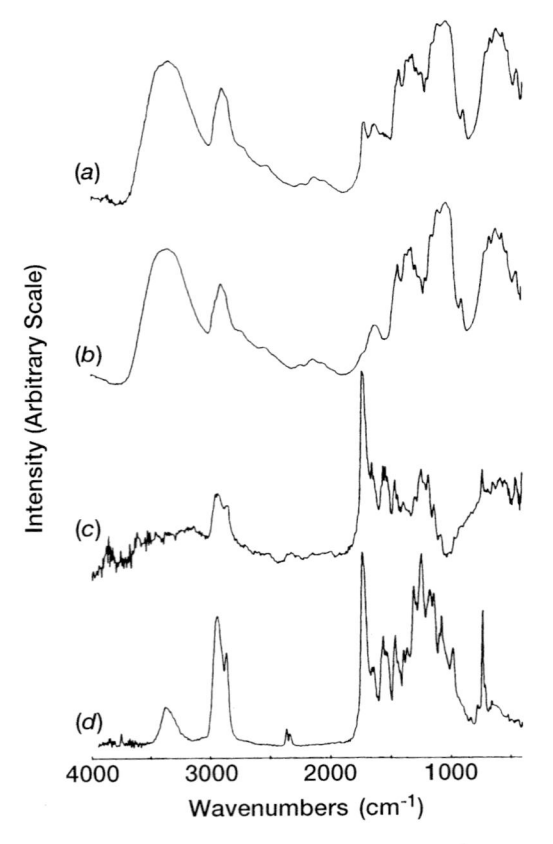

Figure 7.51 PA infrared spectra of (a) cotton yarn sized with polyurethane; (b) unsized cotton yarn; (c) difference between the spectra in (a) and (b); (d) polyurethane. (Reproduced from Yang, C.Q. and Bresee, R.R., *J. Coated Fabrics* **17**: 110–128, by permission of Sage Publications Ltd., copyright © 1987.)

acids appeared at 1735 cm^{-1} in the PA infrared spectrum. This band intensified during prolonged (118-h) exposure to UV light. Treatment with NaOH converted the acid to a carboxylate salt, causing the 1735-cm^{-1} peak to be replaced by another feature at 1618 cm^{-1}. Finally, acidification with HCl caused the reappearance of the carboxylic acid. Thermal oxidation at 180 °C (Yang and Freeman, 1991b) produced aldehydes, ketones, carboxylic acids, esters, and anhydrides. These compounds were evenly distributed between the surface and bulk.

The effectiveness of polycarboxylic acids as cross-linking agents for cotton fabrics was examined in three related studies (Yang, 1991a, 1991b; Yang and Kottes Andrews, 1991). It was concluded that polycarboxylic acids probably form anhydrides during curing, and that these anhydrides esterify the cotton cellulose. An ester carbonyl band was monitored, allowing semiquantitative comparison of cross-linkages formed under different experimental conditions. Several different polycarboxylic acids were included in these experiments.

Yang (1990, 1991c) presented detailed comparisons of PA, transmission, and diffuse reflectance spectra for fabrics, fibers, and films. The PA spectra of fabrics displayed the surface sensitivity that was already described in detail, the data being interpreted under the assumption that the PA signal originates entirely within a single thermal diffusion length. Diffuse reflectance spectra were obtained for a treated cotton fabric before and after grinding. Band definition was similar to that in the transmission spectra, and better than that in the PA spectra; this result could, of course, imply partial saturation of the PA bands. Diffuse reflectance spectroscopy was concluded to be less surface-sensitive than PA infrared spectroscopy. A comparison of PA and diffuse reflectance spectra obtained for a polypropylene fiber treated with a finishing solution supported this conclusion.

Later papers in this series examined degradation of cotton, silk, and polypropylene (Yang, 1992); polyethylene fabric (Martin and Yang, 1994); and polypropylene fabric (Yang and Martin, 1994). The surface sensitivity of PA spectroscopy was confirmed and utilized in these studies. As expected, photooxidation occurred on the surface of a fabric facing the radiation source, with less oxidation being observed on the rear surface. By contrast, the thermal oxidation products of polypropylene were homogeneously distributed between the surface and bulk.

These articles amply illustrate the suitability of PA infrared spectroscopy for textile analysis. Minimal sample preparation and the capability for depth profiling – the two most important features of PA spectroscopy – are an essential aspect of this work. Further studies are also expected to be successful.

References

Adams, M.J., Beadle, B.C., and Kirkbright, G.F. (1978) Optoacoustic spectrometry in the near-infrared region. *Anal. Chem.*, **50** (9), 1371–1374.

Ageev, B.G., Ponomarev, Y.N., and Sapozhnikova, V.A. (1998) Laser photoacoustic spectroscopy of biosystems gas exchange with the atmosphere. *Appl. Phys. B*, **67** (4), 467–473.

Ahmed, E., Zegadi, A., Hill, A.E., Pilkington, R.D., and Tomlinson, R.D. (1995) Optical properties of flash-evaporated

$CuIn_{0.75}Ga_{0.25}Se_2$ thin films by photoacoustic spectroscopy. *Thin Solid Films*, **268**, 144–151.

Ahmed, E., Zegadi, A., Hill, A.E., Pilkington, R.D., Tomlinson, R.D., and Ahmed, W. (1998) Thermal annealing of flash evaporated $Cu(In, Ga)Se_2$ thin films. *J. Mater. Process. Technol.*, **77** (1–3), 260–265.

Ahmed, E., Pilkington, R.D., Hill, A.E., Ali, N., Ahmed, W., and Hassan, I.U. (2006) Transmission photoacoustic spectroscopy of flash-evaporated $CuIn_{0.75}Ga_{0.25}Se_2$ thin films. *Thin Solid Films*, **515**, 239–244.

Ando, T., Inoue, S., Ishii, M., Kamo, M., Sato, Y., Yamada, O., and Nakano, T. (1993) Fourier-transform infrared photoacoustic studies of hydrogenated diamond surfaces. *J. Chem. Soc. Faraday Trans.*, **89** (4), 749–751.

Angle, C.W., Donini, J.C., and Hamza, H.A. (1988) The effect of ultrasonication on the surface properties, ionic composition and electrophoretic mobility of an aqueous coal suspension. *Colloids Surf. B Biointerfaces*, **30** (3–4), 373–385.

Ardeleanu, M., Morriset, R., and Bertrand, L. (1992) Fourier transform infrared photoacoustic spectra of bacteria, in *Photoacoustic and Photothermal Phenomena III* (ed. D. Bićanić), Springer, Berlin, pp. 81–84.

Armenta, S., Moros, J., Garrigues, S., and de la Guardia, M. (2006) Direct determination of Mancozeb by photoacoustic spectrometry. *Anal. Chim. Acta*, **567** (2), 255–261.

Ashizawa, K. (1989) Polymorphism and crystal structure of 2R,4S,6-fluoro-2-methyl-spiro(chroman-4,4'-imidazoline]-2',5-dione (M79175). *J. Pharm. Sci.*, **78** (3), 256–260.

Ashkenazy, R., Gottlieb, L., and Yannai, S. (1997) Characterization of acetone-washed yeast biomass functional groups involved in lead biosorption. *Biotechnol. Bioeng.*, **55** (1), 1–10.

Baby, A.R., Lacerda, A.C.L., Kawano, Y., Velasco, M.V.R., and Kaneko, T.M. (2007) PAS-FTIR and FT-Raman qualitative characterization of sodium dodecyl sulfate interaction with an alternative stratum corneum model membrane. *Pharmazie*, **62** (10), 727–731.

Bajic, S.J., Jones, R.W., McClelland, J.F., Hames, B.R., and Meglen, R.R. (1998) Rapid analysis of wood using transient infrared spectroscopy and photoacoustic spectroscopy with PLS regression. *AIP Conf. Proc.*, **430**, 466–469.

Barbieri, S., Pellaux, J.-P., Studemann, E., and Rosset, D. (2002) Gas detection with quantum cascade lasers: an adapted photoacoustic sensor based on Helmholtz resonance. *Rev. Sci. Instrum.*, **73** (6), 2458–2461.

Baxter, G.W., Barth, H.-D., and Orr, B.J. (1998) Laser spectroscopy with a pulsed, narrowband infrared optical parametric oscillator system: a practical, modular approach. *Appl. Phys. B*, **66** (5), 653–657.

Beenen, A. and Niessner, R. (1998) Development of a photoacoustic gas sensor for *in situ* and on-line measurement of gaseous water and toluene. *Analyst*, **123** (4), 543–545.

Beenen, A. and Niessner, R. (1999a) Development of a photoacoustic trace gas sensor based on fiber-optically coupled NIR laser diodes. *Appl. Spectrosc.*, **53** (9), 1040–1044.

Beenen, A. and Niessner, R. (1999b) Trace gas analysis by photoacoustic spectroscopy with NIR laser diodes. *AIP Conf. Proc.*, **463**, 211–213.

Belton, P.S. (1984) Photoacoustic spectroscopy (PAS), in *Biophysical Methods in Food Research* (ed. H.W.-S. Chan), Blackwell Scientific Publications, Oxford, pp. 123–135.

Belton, P.S. and Tanner, S.F. (1983) Determination of the moisture content of starch using near infrared photoacoustic spectroscopy. *Analyst*, **108** (1286), 591–596.

Belton, P.S., Wilson, R.H., and Saffa, A.M. (1987) Effects of particle size on quantitative photoacoustic spectroscopy using a gas-microphone cell. *Anal. Chem.*, **59** (19), 2378–2382.

Belton, P.S., Saffa, A.M., and Wilson, R.H. (1987) Use of Fourier transform infrared spectroscopy for quantitative analysis: a comparative study of different detection methods. *Analyst*, **112** (8), 1117–1120.

Belton, P.S., Saffa, A.M., and Wilson, R.H. (1988a) Photoacoustic infrared spectroscopy, in *Analytical Applications of Spectroscopy* (eds C.S. Creaser and A.M.C. Davies), The Royal Society of Chemistry, Cambridge, pp. 245–250.

Belton, P.S., Saffa, A.M., and Wilson, R.H. (1988b) The potential of Fourier transform infrared spectroscopy for the analysis of confectionery products. *Food Chem.*, **28** (1), 53–61.

Belton, P.S., Saffa, A.M., and Wilson, R.H. (1988c) Quantitative analysis by Fourier transform infrared photoacoustic spectroscopy. *Proc. SPIE Int. Soc. Opt. Eng.*, **917**, 72–77.

Bensebaa, F., Kotlyar, L., Pleizier, G., Sparks, B., Deslandes, Y., and Chung, K. (2000) Surface chemistry of end cuts from Athabasca bitumen. *Surf. Interface Anal.*, **30** (1), 207–211.

Bensebaa, F., Majid, A., and Delsandes, Y. (2001) Step-scan photoacoustic Fourier transform and X-ray photoelectron spectroscopy of oil sands fine tailings: new structural insights. *Spectrochim. Acta A*, **57** (13), 2695–2702.

Benziger, J.B., McGovern, S.J., and Royce, B.S.H. (1985) IR photoacoustic spectroscopy of silica and aluminum oxide, in *Catalyst Characterization Science* (ed. M.L. Deviney, J.L. Glard), American Chemical Society, pp. 449–463.

Berbenni, V., Marini, A., Bruni, G., Bini, M., Magnone-Grato, A., and Villa, M. (1996) Spectroscopic and thermoanalytical characterizatioin of (+)-Fenfluramine hydrochloride. *Appl. Spectrosc.*, **50** (7), 871–879.

Berrou, A., Raybaut, M., Godard, A., and Lefebvre, M. (2010) High-resolution photoacoustic and direct absorption spectroscopy of main greenhouse gases by use of a pulsed entangled cavity doubly resonant OPO. *Appl. Phys. B*, **98** (1), 217–230.

Bertrand, L. (1988) Advantages of phase analysis in Fourier transform infrared photoacoustic spectroscopy. *Appl. Spectrosc.*, **42** (1), 134–138.

Bertrand, L., Monchalin, J.-P., and Lepoutre, F. (1982) Magnitude and phase photoacoustic spectra of chrysotile asbestos, a powdered sample. *Appl. Opt.*, **21** (2), 248–252.

Bhardwaj, N.K. and Nguyen, K.L. (2007) Photoacoustic Fourier transform infrared spectroscopic study of hydrogen peroxide bleached de-inked pulps. *Colloids Surf. A*, **301**, 323–328.

Bhardwaj, N.K., Dang, V.Q., and Nguyen, K.L. (2006) Determination of carboxyl content in high-yield kraft pulps using photoacoustic rapid-scan Fourier transform infrared spectroscopy. *Anal. Chem.*, **78** (19), 6818–6825.

Bhowmik, R., Katti, K.S., Verma, D., and Katti, D.R. (2007) Probing moecular interactions in bone biomaterials: through molecular dynamics and Fourier transform infrared spectroscopy. *Mater. Sci. Eng. C*, **27** (3), 352–371.

Bićanić, D., Fink, T., Franko, M., Moćnik, G., van de Bovenkamp, P., van Veldhuizen, B., and Gerkema, E. (1999) Infrared photothermal spectroscopy in the science of human nutrition. *AIP Conf. Proc.*, **463**, 637–639.

Bijnen, F.G.C., Harren, F.J.M., Hackstein, J.H.P., and Reuss, J. (1996) Intracavity CO laser photoacoustic trace gas detection: cyclic CH_4, H_2O and CO_2 emission by cockroaches and scarab beetles. *Appl. Opt.*, **35** (27), 5357–5368.

Bijnen, F.G.C., Zuckermann, H., Harren, F.J.M., and Reuss, J. (1998) Multicomponent trace-gas analysis by three intracavity photoacoustic cells in a CO laser: observation of anaerobic and postanaerobic emission of acetaldehyde and ethanol in cherry tomatoes. *Appl. Opt.*, **37** (15), 3345–3353.

Bjarnestad, S. and Dahlman, O. (2002) Chemical compositions of hardwood and softwood pulps employing photoacoustic Fourier transform infrared spectroscopy in combination with partial least-squares analysis. *Anal. Chem.*, **74** (22), 5851–5858.

Bohren, A. and Sigrist, M.W. (1997) Optical parametric oscillator based difference frequency laser source for photoacoustic trace gas spectroscopy in the 3 μm mid-IR range. *Infrared Phys. Technol.*, **38** (7), 423–435.

Booth, D.T. and Sowa, S. (2001) Respiration in dormant and non-dormant bitterbrush seeds. *J. Arid Environ.*, **48** (1), 35–39.

Bouzerar, R., Amory, C., Zeinert, A., Benlahsen, M., Racine, B., Durand-Drouhin, O., and Clin, M. (2001) Optical properties of amorphous hydrogenated carbon thin films. *J. Non-Cryst. Solids*, **281** (1–3), 171–180.

Bowen, J.M., Compton, S.V., and Blanche, M.S. (1989) Comparison of sample preparation methods for the Fourier transform infrared analysis of an organo-clay mineral sorption mechanism. *Anal. Chem.*, **61** (18), 2047–2050.

Brienne, S.H.R., Volkan Bozkurt, M.M., Rao, S.R., Xu, Z., Butler, I.S., and Finch, J.A. (1996) *Appl. Spectrosc.*, **50** (4), 521–527.

Bürger, H., Lecoutre, M., Huet, T.R., Breidung, J., Thiel, W., Hänninen, V., and Jalonen, L. (2001) The (n00), n = 3, 4, and 6, local mode states of H_3SiD: Fourier transform infrared and laser photoacoustic spectra and ab initio calculations of spectroscopic parameters. *J. Chem. Phys.*, **114** (20), 8844–8854.

Butler, I.S., Xu, Z.H., Werbowyj, R.S., and St.-Germain, F. (1987a) FT-IR photoacoustic spectra of some solid organometallic complexes of chromium, manganese, rhenium, and iron. *Appl. Spectrosc.*, **41** (1), 149–153.

Butler, I.S., Xu, Z.H., Darensbourg, D.J., and Pala, M. (1987b) FT-IR, photoacoustic and micro-Raman spectra of the dodecacarbonyltriruthenium(0) complexes $Ru_3(^{13}CO)_{12}$ and $Ru_3(CO)_{12}$. *J. Raman Spectrosc.*, **18**, 357–363.

Butler, I.S., Li, H., and Gao, J.P. (1991) Comparison of photoacoustic, attenuated total reflection, and transmission infrared spectra of crystalline organoiron(II) carbonyl complexes. *Appl. Spectrosc.*, **45** (2), 223–226.

Butler, I.S., Gilson, D.F.R., and Lafleur, D. (1992) Infrared photoacoustic spectra of gaseous pentacarbonyl(methyl)manganese(I) and pentacarbonyl(methyl)rhenium(I). *Appl. Spectrosc.*, **46** (11), 1605–1607.

Caires, A.R.L., Teixeira, M.R.O., Suarez, Y.R., Andrade, L.H.C., and Lima, S.M. (2008) Discrimination of transgenic and conventional soybean seeds by Fourier transform infrared photoacoustic spectroscopy. *Appl. Spectrosc.*, **62** (9), 1044–1047.

Calasso, I.G. and Sigrist, M.W. (1999) Selection criteria for microphones used in pulsed nonresonant gas-phase photoacoustics. *Rev. Sci. Instrum.*, **70** (12), 4569–4578.

Calheiros, F.C., Braga, R.R., Kawano, Y., and Ballester, R.Y. (2004) Relationship between contraction stress and degree of conversion in restorative composites. *Dent. Mater.*, **20** (10), 939–946.

Carter, E.A., Fredericks, P.M., and Church, J.S. (1996) Fourier transform infrared photoacoustic spectroscopy of surface-treated wool. *Text. Res. J.*, **66** (12), 787–794.

Carter, R.O. and Wright, S.L. (1991) Evaluation of the appropriate sample position in a PAS/FT-IR experiment. *Appl. Spectrosc.*, **45** (7), 1101–1103.

Carter, R.O., Paputa Peck, M.C., Samus, M.A., and Killgoar, P.C. (1989) Infrared photoacoustic spectroscopy of carbon black filled rubber: concentration limits for samples and background. *Appl. Spectrosc.*, **43** (8), 1350–1354.

Carter, R.O., Paputa Peck, M.C., and Bauer, D.R. (1989) The characterization of polymer surfaces by photoacoustic Fourier transform infrared spectroscopy. *Polym. Degrad. Stab.*, **23** (2), 121–134.

Castleden, S.L., Kirkbright, G.F., and Menon, K.R. (1980) Determination of moisture in single-cell protein utilising photoacoustic spectroscopy in the near-infrared region. *Analyst*, **105** (1256), 1076–1081.

Castleden, S.L., Kirkbright, G.F., and Long, S.E. (1982) Quantitative assay of propanolol by photoacoustic spectroscopy. *Can. J. Spectrosc.*, **27** (1), 244–248.

Chalmers, J.M., Stay, B.J., Kirkbright, G.F., Spillane, D.E.M., and Beadle, B.C. (1981) Some observations on the capabilities of photoacoustic Fourier transform infrared spectroscopy (PAFTIR). *Analyst*, **106** (1268), 1179–1186.

Changwen, D., Linker, R., and Shaviv, A. (2007) Characterization of soils using photoacoustic mid-infrared spectroscopy. *Appl. Spectrosc.*, **61** (10), 1063–1067.

Chia, C.H., Zakaria, S., Nguyen, K.L., Dang, V.Q., and Duong, T.D. (2009) Characterization of magnetic paper using Fourier transform infrared spectroscopy. *Mater. Chem. Phys.*, **113**, 768–772.

Chien, P.-L., Markuszewski, R., and McClelland, J.F. (1985) Comparison of Fourier transform infrared-photoacoustic spectroscopy (FTIR-PAS) and conventional methods for analysis of coal oxidation. *Prepr. Pap. Am. Chem. Soc. Div. Fuel Chem.*, **30** (1), 13–20.

Chien, P.-L., Markuszewski, R., Araghi, H.G., and McClelland, J.F. (1985) Study of coal oxidation kinetics by Fourier transform infrared-photoacoustic spectroscopy (FTIR-PAS), in *Proceedings, 1985 International Conference on Coal Science*, Pergamon Press, Sydney, pp. 818–821.

Choquet, M., Rousset, G., and Bertrand, L. (1985) Phase analysis of infrared Fourier transform photoacoustic spectra. *Proc. SPIE Int. Soc. Opt. Eng.*, **553**, 224–225.

Choquet, M., Rousset, G., and Bertrand, L. (1986) Fourier-transform photoacoustic spectroscopy: a more complete method for quantitative analysis. *Can. J. Phys.*, **64** (9), 1081–1085.

Church, J.S. and Evans, D.J. (1995) The quantitative analysis of fluorocarbon polymer finishes on wool by FT-IR spectroscopy. *J. Appl. Polym. Sci.*, **57** (13), 1585–1594.

Cihelka, J., Matulková, I., and Civiš, S. (2009) Laser diode photoacoustic and FTIR laser spectroscopy of formaldehyde in the 2.3 μm and 3.5 μm spectral range. *J. Mol. Spectrosc.*, **256** (1), 68–74.

Cody, G.D., Larsen, J.W., and Siskin, M. (1989) Investigation of linear dichroism in coals using polarized photoacoustic FTIR. *Energy Fuels*, **3** (5), 544–551.

Coelho, T.M., Nogueira, E.S., Steimacher, A., Medina, A.N., Weinand, W.R., Lima, W.M., Baesso, M.L., and Bento, A.C. (2006) Characterization of natural nanostructured hydroxyapatite obtained from the bones of Brazilian river fish. *J. Appl. Phys.*, **100** (9), 094312-1–6.

Coffey, M.J., Berghout, H.L., Woods, E., and Crim, F.F. (1999) Vibrational spectroscopy and intramolecular energy transfer in isocyanic acid (HNCO). *J. Chem. Phys.*, **110** (22), 10850–10862.

Cohen, Y., Bar, I., and Rosenwaks, S. (1996) Spectroscopy of D_2O (2,0,1). *J. Mol. Spectrosc.*, **180** (2), 298–304.

Cook, L.E., Luo, S.Q., and McClelland, J.F. (1991) Fourier transform infrared photoacoustic spectroscopy of polymers adsorbed from solution by gamma-iron oxide. *Appl. Spectrosc.*, **45** (1), 124–126.

Dahnke, H., Kahl, J., Schüler, G., Boland, W., Urban, W., and Kühnemann, F. (2000) On-line monitoring of biogenic isoprene emissions using photoacoustic spectroscopy. *Appl. Phys. B*, **70** (2), 275–280.

D'Almeida, J.R.M., Cella, N., Monteiro, S.N., and Miranda, L.C.M. (1998) Thermal diffusivity of an epoxy system as a function of the hardener content. *J. Appl. Polym. Sci.*, **69** (7), 1335–1341.

Dang, V.Q., Bharwaj, N.K., Hoang, V., and Nguyen, K.L. (2007) Determination of lignin content in high-yield kraft pulps using photoacoustic rapid scan Fourier transform infrared spectroscopy. *Carbohydr. Polym.*, **68**, 489–494.

Danworaphong, S., Calasso, I.G., Beveridge, A., Diebold, G.J., Gmachl, C., Capasso, F., Sivco, D.L., and Cho, A.Y. (2003) Internally excited acoustic resonator for photoacoustic trace detection. *Appl. Opt.*, **27** (42), 5561–5565.

Davidson, R.S. and Fraser, G.V. (1984) The analysis of textile mixtures using Fourier transform infrared photoacoustic spectroscopy. *J. Soc. Dyers Colour.*, **100** (5–6), 167–170.

Davidson, R.S. and King, D. (1983) A new method of distinguishing wool from polyester-fibre and cotton fabrics. *J. Textile Inst.*, **74** (6), 382–384.

Di Renzo, M., Ellis, T.H., Sacher, E., and Stangel, I. (2001a) A photoacoustic FTIRS study of the chemical modifications of human dentin surfaces: I. Demineralization. *Biomaterials*, **22** (8), 787–792.

Di Renzo, M., Ellis, T.H., Sacher, E., and Stangel, I. (2001b) A photoacoustic FTIRS study of the chemical modifications of human dentin surfaces: II. Deproteination. *Biomaterials*, **22** (8), 793–797.

Dittmar, R.M., Palmer, R.A., and Carter, R.O. (1994) Fourier transform photoacoustic spectroscopy of polymers. *Appl. Spectrosc. Rev.*, **29** (2), 171–231.

Dóka, O., Bicanic, D., Szücs, M., and Lubbers, M. (1998) Direct measurement of carbonate content in soil samples by means of CO laser infrared photoacoustic spectroscopy. *Appl. Spectrosc.*, **52** (12), 1526–1529.

Donini, J.C. and Michaelian, K.H. (1985) Near-, mid-, and far-infrared photoacoustic spectroscopy. *Proc. SPIE Int. Soc. Opt. Eng.*, **553**, 344–345.

Donini, J.C. and Michaelian, K.H. (1986) Near infrared photoacoustic FTIR spectroscopy of clay minerals and coal. *Infrared Phys.*, **26** (3), 135–140.

Donini, J.C. and Michaelian, K.H. (1988) Low frequency photoacoustic spectroscopy of solids. *Appl. Spectrosc.*, **42** (2), 289–292.

Du, C., Linker, R., and Shaviv, A. (2008) Identification of agricultural Mediterranean soils using mid-infrared photoacoustic spectroscopy. *Geoderma*, **143** (1–2), 85–90.

Du, C., Zhou, J., Wang, H., Chen, X., Zhu, A., and Zhang, J. (2009) Determination of soil properties using Fourier transform mid-infrared photoacoustic spectroscopy. *Vib. Spectrosc.*, **49** (1), 32–37.

Dunayevskiy, I., Tsekoun, A., Prasanna, M., Go, R., and Patel, C.K.N. (2007) High-sensitivity detection of triacetone triperoxide (TATP) and its precursor acetone. *Appl. Opt.*, **46** (25), 6397–6404.

Eckhardt, H. and Chance, R.R. (1983) Defect states in polyacetylene: a photoacoustic and diffuse reflectance study. *J. Chem. Phys.*, **79** (11), 5698–5704.

Eichhorn, K.-J., Hopfe, I., Pötschke, P., and Schmidt, P. (2000) Polarized FTIR photoacoustic spectroscopy on blends of thermoplastic poly(ether-urethanes) with modified polypropylenes. *J. Appl. Polym. Sci.*, **75** (9), 1194–1204.

Elia, A., Di Franco, C., Lugarà, P.M., and Scamarcio, G. (2006a) Photoacoustic spectroscopy with quantum cascade lasers for trace gas detection. *Sensors*, **6** (10), 1411–1419.

Elia, A., Rizzi, F., Di Franco, C., Lugarà, P.M., and Scamarcio, G. (2006b) Quantum cascade laser-based photoacoustic spectroscopy of volatile chemicals: application to hexamethyldisilazane. *Spectrochim. Acta A*, **64** (2), 426–429.

El Shafei, G.M.S.M. and Mokhtar, M. (1995) Interaction between molybdena and silica: FT-IR/PA studies of surface hydroxyl groups and pore structure assessment. *Colloids Surf. A*, **94** (2–3), 267–277.

Factor, A., Tilley, M.G., and Codella, P.J. (1991) Determination of residual unsaturation in photo-cured acrylate formulations using photoacoustic FT-IR spectroscopy. *Appl. Spectrosc.*, **45** (1), 135–138.

Fainchtein, R., Stoyanov, B.J., Murphy, J.C., Wilson, D.A., and Hanley, D.F. (2000) Local determination of hemoglobin concentration and degree of oxygenation in tissue by pulsed photoacoustic spectroscopy. *Proc. SPIE Int. Soc. Opt. Eng.*, **3916**, 19–33.

Farmer, V.C. (1998) Differing effects of particle size and shape in the infrared and Raman spectra of kaolinite. *Clay Miner.*, **33** (4), 601–604.

Farmer, V.C. (2000) Transverse and longitudinal crystal modes associated with OH stretching vibrations in single crystals of kaolinite and dickite. *Spectrochim. Acta A*, **56** (5), 927–930.

Favier, J.P., Bićanić, D., van de Bovenkamp, P., Chirtoc, M., and Helander, P. (1996) Detection of total trans fatty acids content in margarine: an intercomparison study of GLC, GLC + TLC, FT-IR, and optothermal window (open photoacoustic cell). *Anal. Chem.*, **68** (5), 729–733.

Favier, J.P., Bićanić, D., Cozijnsen, J., van Veldhuizen, B., and Helander, P. (1998) CO_2 laser infrared optothermal spectroscopy for quantitative adulteration studies in binary mixtures of extra-virgin olive oil. *J. Am. Oil Chem. Soc.*, **75** (3), 359–362.

Fehér, M., Jiang, Y., Maier, J.P., and Miklós, A. (1994) Optoacoustic trace-gas monitoring with near-infrared diode lasers. *Appl. Opt.*, **33** (9), 1655–1658.

Filho, M.B., da Silva, M.G., Sthel, M.S., Schramm, D.U., Vargas, H., Miklós, A., and Hess, P. (2006) Ammonia detection by using quantum-cascade laser photoacoustic spectroscopy. *Appl. Opt.*, **45** (20), 4966–4971.

Fischer, C., Sigrist, M.W., Yu, Q., and Seiter, M. (2001) Photoacoustic monitoring of trace gases by use of a diode-based difference frequency laser source. *Opt. Lett.*, **26** (20), 1609–1611.

Forsskåhl, I., Kenttä Kyyrönen, P., and Sundström, O. (1995) Depth profiling of a photochemically yellowed paper. Part I: FT-IR techniques. *Appl. Spectrosc.*, **49** (2), 163–170.

Friesen, W.I. and Michaelian, K.H. (1986) Fourier deconvolution of photoacoustic FTIR spectra. *Infrared Phys.*, **26** (4), 235–242.

Friesen, W.I. and Michaelian, K.H. (1991) Deconvolution and curve fitting in the analysis of complex spectra. The CH stretching region in infrared spectra of coal. *Appl. Spectrosc.*, **45** (1), 50–56.

Friesen, W.I., Michaelian, K.H., Long, Y., and Dabros, T. (2005a) Effect of solvent-to-bitumen ratio on the pyrolysis properties of precipitated Athabasca asphaltenes. *Energy Fuels*, **19** (3), 1109–1115.

Friesen, W.I., Michaelian, K.H., Long, Y., and Dabros, T. (2005b) Thermogravimetric and infrared characterization of asphaltenes precipitated from bitumen. *Prepr. Pap. Am. Chem. Soc. Div. Fuel Chem.*, **50** (1), 228–229.

Gac, W. (2007) The influence of silver on the structural, redox and catalytic properties of the cryptomelane-type manganese oxides in the low-temperature CO oxidation reaction. *Appl. Catal. B*, **75** (1–2), 107–117.

Gac, W., Derylo-Marczewska, A., Pasieczna-Patkowska, S., Popivnyak, N., and Zukocinski, G. (2007) The influence of the preparation methods and pretreatment conditions on the properties of Ag-MCM-41 catalysts. *J. Mol. Catal. A Chem.*, **268** (1–2), 15–23.

Gagarin, S.G., Gladun, T.G., Friesen, W.I., and Michaelian, K.H. (1993) Simulating infrared spectra of macerals and estimation of petrographic composition of coals by spectra of fractions with various density. *Coke Chem.*, (4), 9–15.

Gagarin, S.G., Friesen, W.I., Michaelian, K.H., and Gladun, T.G. (1994) Reactivity of coal based on photoacoustic ir spectroscopic investigations. *Solid Fuel Chem.*, **28** (3), 35–42.

Gagarin, S.G., Friesen, W.I., and Michaelian, K.H. (1995a) Estimation of the heat of combustion of coal by infrared spectroscopy. *Coke Chem.*, (4), 6–16.

Gagarin, S.G., Friesen, W.I., and Michaelian, K.H. (1995b) Prediction of the roga index of coal blends based on data of IR-spectroscopy. *Coke Chem.*, (8), 23–28.

Gao, Y., Choudhury, N.R., Dutta, N., Matisons, J., Reading, M., and Delmotte, L. (2001) Organic-inorganic hybrid from ionomer via sol-gel reaction. *Chem. Mater.*, **13** (10), 3644–3652.

Gao, Y., Choudhury, N.R., Matisons, J., Schubert, U., and Moraru, B. (2002) Part 2: inorganic-organic hybrid polymers by polymerization of methacrylate-substituted oxotitanium clusters with methyl methacrylate: thermomechanical and morphological properties. *Chem. Mater.*, **14** (11), 4522–4529.

Garbassi, F. and Occhiello, E. (1987) Spectroscopic techniques for the analysis of polymer surfaces and interfaces. *Anal. Chim. Acta*, **197** (C), 1–42.

Garcia, J.A., Mandelis, A., Marinova, M., Michaelian, K.H., and Afrashtehfar, S. (1998) Quantitative photothermal radiometric and FT-IR photoacoustic measurements of specialty papers. *Appl. Spectrosc.*, **52** (9), 1222–1229.

Garcia, J.A., Mandelis, A., Marinova, M., Michaelian, K.H., and Afrashtehfar, S. (1999) Quantitative photothermal radiometric and FTIR photoacoustic measurements of specialty papers. *AIP Conf. Proc.*, **463**, 395–397.

Gardella, J.A., Jiang, D.-Z., and Eyring, E.M. (1983) Quantitative determination of catalytic surface adsorption sites by

Fourier transform infrared photoacoustic spectroscopy. *Appl. Spectrosc.*, **37** (2), 131–133.

Gardella, J.A., Jiang, D.-Z., McKenna, W.P., and Eyring, E.M. (1983) Applications of Fourier transform infrared photoacoustic spectroscopy (FT-IR/PAS) to surface and corrosion phenomena. *Appl. Surf. Sci.*, **15** (1–4), 36–49.

Gardella, J.A., Grobe, G.L., Hopson, W.L., and Eyring, E.M. (1984) Comparison of attenuated total reflectance and photoacoustic sampling for surface analysis of polymer mixtures by Fourier transform infrared spectroscopy. *Anal. Chem.*, **56** (7), 1169–1177.

Gentzis, T., Goodarzi, F., and McFarlane, R.A. (1992) Molecular structure of reactive coals during oxidation, carbonization and hydrogenation – an infrared photoacoustic spectrosopic and optical microscopic study. *Org. Geochem.*, **18** (3), 249–258.

Gerson, D.J., McClelland, J.F., Veysey, S., and Markuszewski, R. (1984) Characterization of coal using Fourier transform infrared photoacoustic spectroscopy. *Appl. Spectrosc.*, **38** (6), 902–904.

Gondal, M.A. (1997) Laser photoacoustic spectrometer for remote monitoring of atmospheric pollutants. *Appl. Opt.*, **36** (15), 3195–3201.

Goodarzi, F. and McFarlane, R.A. (1991) Chemistry of fresh and weathered resinites – An infrared photoacoustic spectroscopic study. *Int. J. Coal Geol.*, **19** (1–4), 283–301.

Gordon, S.H., Greene, R.V., Freer, S.N., and James, C. (1990) Measurement of protein biomass by Fourier transform infrared-photoacoustic spectroscopy. *Biotechnol. Appl. Biochem.*, **12**, 1–10.

Gordon, S.H., Schudy, R.B., Wheeler, B.C., Wicklow, D.T., and Greene, R.V. (1997) Identification of Fourier tranform infrared photoacoustic spectral features for detection of *Aspergillus flavus* infection in corn. *Int. J. Food Microbiol.*, **35** (2), 179–186.

Gordon, S.H., Wheeler, B.C., Schudy, R.B., Wicklow, D.T., and Greene, R.V. (1998) Neural network pattern recognition of photoacoustic FTIR spectra and knowledge-based techniques for detection of mycotoxigenic fungi in food grains. *J. Food Prot.*, **61** (2), 221–230.

Gordon, S.H., Jones, R.W., McClelland, J.F., Wicklow, D.T., and Greene, R.V. (1999) Transient infrared spectroscopy for detection of toxigenic fungi in corn: potential for on-line evaluation. *J. Agric. Food Chem.*, **47** (12), 5267–5272.

Gosselin, F., Di Renzo, M., Ellis, T.H., and Lubell, W.D. (1996) Photoacoustic FTIR spectroscopy, a nondestructive method for sensitive analysis of solid-phase organic chemistry. *J. Org. Chem.*, **61** (23), 7980–7981.

Gotter, B., Faubel, W., and Neubert, R.H.H. (2008) Optical methods for measurements of skin penetration. *Skin Pharmacol. Physiol.*, **21** (3), 156–165.

Graves, D.J. and Luo, S. (1994) Use of photoacoustic Fourier-transform infrared spectroscopy to study phosphates in proteins. *Biochem. Biophys. Res. Commun.*, **205** (1), 618–624.

Greene, R.V., Freer, S.N., and Gordon, S.H. (1988) Determination of solid-state fungal growth by Fourier transform infrared-photoacoustic spectroscopy. *FEMS Microbiol. Lett.*, **52**, 73–78.

Greene, R.V., Gordon, S.H., Jackson, M.A., Bennett, G.A., McClelland, J.F., and Jones, R.W. (1992) Detection of fungal contamination in corn: potential of FTIR-PAS and -DRS. *J. Agric. Food Chem.*, **40** (7), 1144–1149.

Grossel, A., Zeninari, V., Joly, L., Parvitte, B., Courtois, D., and Durry, G. (2006) New improvements in methane detection using a Helmholtz resonant photoacoustic laser sensor: a comparison between near-IR diode lasers and mid-IR quantum cascade lasers. *Spectrochim. Acta A*, **63** (5), 1021–1028.

Grossel, A., Zéninari, V., Parvitte, B., Joly, L., and Courtois, D. (2007a) Optimization of a compact photoacoustic quantum cascade laser spectrometer for atmospheric flux measurements: application to the detection of methane and nitrous oxide. *Appl. Phys. B*, **88** (3), 483–492.

Grossel, A., Zéninari, V., Joly, L., Parvitte, B., Durry, G., and Courtois, D. (2007b) Photoacoustic detection of nitric oxide with a Helmholtz resonant quantum cascade laser sensor. *Infrared Phys. Technol.*, **51** (2), 95–101.

Gurnagul, N., St.-Germain, F.G.T., and Gray, D.G. (1986) Photoacoustic Fourier transform infrared measurements on paper. *J. Pulp Pap. Sci.*, **12** (5), J156–J159.

Halttunen, M., Tenhunen, J., Saarinen, T., and Stenius, P. (1999) Applicability of FTIR/PAS depth profiling for the study of coated papers. *Vib. Spectrosc.*, **19** (2), 261–269.

Hamza, H.A., Michaelian, K.H., and Andersen, N.E. (1983) A fundamental approach to beneficiation of fine oxidized coals, in Proceedings, 1983 International Conference on Coal Science, Center for Conference Management, Pittsburgh, pp. 248–251.

Hanh, B.D., Neubert, R.H.H., Wartewig, S., Christ, A., and Hentzsch, C. (2000) Drug penetration as studied by noninvasive methods: Fourier transform infrared-attenuated total reflection, Fourier transform infrared, and ultraviolet photoacoustic spectroscopy. *J. Pharm. Sci.*, **89** (9), 1106–1113.

Hanh, B.D., Neubert, R.H.H., Wartewig, S., and Lasch, J. (2001) Penetration of compounds through human stratum corneum as studied by Fourier transform infrared photoacoustic spectroscopy. *J. Control. Release*, **70** (3), 393–398.

Hao, L.-Y., Han, J.-X., Shi, Q., Zhang, J.-H., Zheng, J.-J., and Zhu, Q.-S. (2000) A highly sensitive photoacoustic spectrometer for near infrared overtone. *Rev. Sci. Instrum.*, **71** (5), 1975–1980.

Harbour, J.R., Hopper, M.A., Marchessault, R.H., Dobbin, C.J., and Anczurowski, E. (1985) Photoacoustic spectroscopy of cellulose, paper and wood. *J. Pulp Pap. Sci.*, **11** (2), J42–J47.

Harpness, R., Gedanken, A., Weiss, A.M., and Slifkin, M.A. (2003) Microwave-assisted synthesis of nanosized $MoSe_2$. *J. Mater. Chem.*, **13** (10), 2603–2606.

Harvey, T.J., Henderson, A., Gazi, E., Clarke, N.W., Brown, M., Correia Faria, E., Snook, R.D., and Gardner, P. (2007) Discriminaton of prostate cancer cells by reflection mode FTIR photoacoustic spectroscopy. *Analyst*, **132** (4), 292–295.

Helwig, E., Sandner, B., Gopp, U., Vogt, F., Wartewig, S., and Henning, S. (2001) *Biomaterials*, **22** (19), 2695–2702.

Herecová, L., Hejzlar, T., Pavlovský, J., Míček, D., Zelinger, Z., Kubát, P., Janečková, R., Nevrlý, V., Bitala, P., Střižík, M., Klouda, K., and Civiš, S. (2009) CO_2-laser photoacoustic detection of gaseous *n*-pentylacetate. *J. Mol. Spectrosc.*, **256** (1), 109–110.

Herres, W. and Zachmann, G. (1984) FT-IR photoakustische Spektroskopie in der Feststoffanalytik. *LaborPraxis*, 632–638.

Hess, A. and Kemnitz, E. (1994) Characterization of catalytically active sites on aluminum oxides, hydroxyfluorides, and fluorides in correlation with their catalytic behavior. *J. Catal.*, **149** (2), 449–457.

Hess, A. and Kemnitz, E. (1997) Surface acidity and catalytic behavior of modified zirconium and titanium dioxides. *Appl. Catal. A*, **149** (2), 373–389.

Highfield, J.G. and Moffat, J.B. (1984a) Characterization of 12-tungstophosphoric acid and related salts using photoacoustic spectroscopy in the infrared region. I. Thermal stability and interactions with ammonia. *J. Catal.*, **88** (1), 177–187.

Highfield, J.G. and Moffat, J.B. (1984b) Characterization of 12-tungstophosphoric acid and related salts using photoacoustic spectroscopy in the infrared region. II. Interactions with pyridine. *J. Catal.*, **89** (2), 185–195.

Highfield, J.G. and Moffat, J.B. (1985a) Elucidation of the mechanism of dehydration of methanol over 12-tungstophosphoric acid using infrared photoacoustic spectroscopy. *J. Catal.*, **95** (1), 108–119.

Highfield, J.G. and Moffat, J.B. (1985b) The influence of experimental conditions in quantitative analysis of powdered samples by Fourier transform infrared photoacoustic spectroscopy. *Appl. Spectrosc.*, **39** (3), 550–552.

Hocking, M.B., Syme, D.T., Axelson, D.E., and Michaelian, K.H. (1990a) Water-soluble imide-amide copolymers. I. Preparation and characterization of poly [acrylamide-co-sodium *N*-(4-sulfophenyl) maleimide]. *J. Polym. Sci. A1*, **28** (11), 2949–2968.

Hocking, M.B., Syme, D.T., Axelson, D.E., and Michaelian, K.H. (1990b) Water-soluble imide-amide copolymers. II. Preparation and characterization of poly (acrylamide-co-*p*-maleimidobenzoic acid). *J. Polym. Sci A1*, **28** (11), 2969–2982.

Hocking, M.B., Syme, D.T., Axelson, D.E., and Michaelian, K.H. (1990c) Water-soluble imide-amide copolymers. III. Preparation and characterization of poly (acrylamide-co-*N,N*-diallylaniline) and poly (acrylamide-co-sodium *N,N*-diallylsulfanilate). *J. Polym. Sci. A1*, **28** (11), 2983–2996.

Hocking, M.B., Klimchuk, K.A., and Lowen, S. (2000) Water-soluble acrylamide copolymers. VI. Preparation and characterization of poly(*N,N*-dimethylacrylamide-*co*-acrylamide) and control polyacrylamides. *J. Polym. Sci. A1*, **38** (17), 3128–3145.

Hocking, M.B., Klimchuk, K.A., and Lowen, S. (2001) Water-soluble acrylamide copolymers. VIII. Preparation and characterization of polyacrylamide-co-*t*-butylacrylamide. *J. Polym. Sci. A1*, **39** (12), 1960–1977.

Hofmann, G.R., Faubel, W., and Ache, H.J. (1996) Fourier transform infrared photoacoustic spectroscopy of PCB contaminated soil. *Prog. Nat. Sci.*, **6** (Suppl.), S630–S633.

Hofstetter, D., Beck, M., Faist, J., Nägele, M., and Sigrist, M.W. (2001) Photoacoustic spectroscopy with quantum cascade distributed-feedback lasers. *Opt. Lett.*, **26** (12), 887–889.

Hornberger, Ch., König, M., Rai, S.B., and Demtröder, W. (1995) Sensitive photoacoustic overtone spectroscopy of acetylene with a multipass photoacoustic cell and a colour centre laser at 1.5 μm. *Chem. Phys.*, **190** (2–3), 171–177.

Hosomi, T., Suematsu, H., Fjellvåg, H., Karppinen, M., and Yamauchi, H. (1999) Identification of superconducting phases in the Ba–Ca–Cu–O system: an unstable phase with $T_c \approx 90$ K. *J. Mater. Chem.*, **9** (5), 1141–1148.

Hou, R.-Z., Wu, J.-G., Soloway, R.D., Guo, H., Zhang, Y.-F., Du, Y.-C., Liu, F., and Xu, G.-X. (1988) Fourier transform infrared photoacoustic spectroscopy of dental calculus. *Mikrochim. Acta*, **II**, 133–136.

Huang, C.-Y.F., Yuan, C.-J., Luo, S., and Graves, D.J. (1994) Mutational analyses of the metal ion and substrate binding sites of phosphorylase kinase γ subunit. *Biochemistry*, **33** (19), 5877–5883.

Hübner, M., Sigrist, M.W., Demartines, N., Gianella, M., Clavien, P.A., and Hahnloser, D. (2008) Gas emission during laparoscopic colorectal surgery using a bipolar vessel sealing device: a pilot study on four patients. *Patient Saf. Surg.*, **2**, (5 pages).

Hünig, I., Oudejans, L., and Miller, R.E. (2000) Infrared optothermal spectroscopy of N_2– and OC–DCCH: the C–H stretching region. *J. Mol. Spectrosc.*, **204** (1), 148–152.

Huvenne, J.P. and Lacroix, B. (1988) Spectres d'absorption détectés par effet photoacoustique en infrarouge par transformée de Fourier – applications à quelques médicaments. *Spectrochim. Acta A*, **44** (1), 109–113.

Irudayaraj, J. and Yang, H. (2000) Analysis of cheese using step-scan Fourier transform infrared photoacoustic spectroscopy. *Appl. Spectrosc.*, **54** (4), 595–600.

Irudayaraj, J. and Yang, H. (2001) Depth profiling of a heterogeneous food-packaging model using step-scan Fourier transform infrared photoacoustic spectroscopy. *J. Food Eng.*, **55** (1), 25–33.

Irudayaraj, J., Sivakesava, S., Kamath, S., and Yang, H. (2001) Monitoring chemical changes in some foods using Fourier transform photoacoustic spectroscopy. *J. Food Sci.*, **66** (9), 1416–1421.

Irudayaraj, J., Yang, H., and Sakhamuri, S. (2002) Differentiation and detection of microorganisms using Fourier transform infrared photoacoustic spectroscopy. *J. Mol. Struct.*, **606** (1–3), 181–188.

Jalink, H., Bićanić, D., Franko, M., and Bozóki, Z. (1995) Vapor-phase spectra and the pressure-temperature dependence of long-chain carboxylic acids studied by a CO laser and the photoacoustic heat-pipe detector. *Appl. Spectrosc.*, **49** (7), 994–999.

Jasse, B. (1989) Fourier-transform infrared photoacoustic spectroscopy of synthetic polymers. *J. Macromol. Sci. Part A Pure Appl. Chem.*, **26** (1), 43–67.

Jianjun, Z., Xiangguong, Y., Yingli, B., and Kaiji, Z. (1992) Role of Li ion in Li-Mn-MgO catalyst for oxidative coupling of methane. *Catal. Today*, **13** (4), 555–558.

Jin, Q., Kirkbright, G.G., and Spillane, D.E.M. (1982) The determination of moisture in some solid materials by near infrared photoacoustic spectroscopy. *Appl. Spectrosc.*, **36** (2), 120–124.

Jurdana, L.E., Ghiggino, K.P., Leaver, I.H., Barraclough, C.G., and Cole-Clarke, P. (1994) Depth profile analysis of keratin fibers by FT-IR photoacoustic spectroscopy. *Appl. Spectrosc.*, **48** (1), 44–49.

Jurdana, L.E., Ghiggino, K.P., Leaver, I.H., and Cole-Clarke, P. (1995) Application of FT-IR step-scan photoacoustic phase modulation methods to keratin fibers. *Appl. Spectrosc.*, **49** (3), 361–366.

Kanstad, S.O., Nordal, P.-E., Hellgren, L., and Vincent, J. (1981) Infrared photoacoustic spectroscopy of skin lipids. *Naturwissenschaften*, **68**, 47–48.

Kästle, R. and Sigrist, M.W. (1996a) CO laser photoacoustic spectroscopy of acetic, deuterated acetic and propionic acid molecules. *Spectrochim. Acta A*, **52** (10), 1221–1228.

Kästle, R. and Sigrist, M.W. (1996b) Temperature-dependent photoacoustic spectroscopy with a Helmholtz resonator. *Appl. Phys. B*, **63** (4), 389–397.

Katti, K.S. and Urban, M.W. (2003) Conductivity model and photoacoustic FT-IR surface depth profiling of heterogeneous polymers. *Polymer*, **44** (11), 3319–3325.

Kemsley, E.K., Tapp, H.S., Scarlett, A.J., Miles, S.J., Hammond, R., and Wilson, R.H. (2001) Comparison of spectroscopic techniques for the determination of Kjeldahl and ammoniacal nitrogen content of farmyard manure. *J. Agric. Food Chem.*, **49** (2), 603–609.

Khalil, O.S. (2004) Non-invasive glucose measurement technologies: an update from 1999 to the dawn of the new millennium. *Diabetes Technol. Ther.*, **6** (5), 660–697.

Kiland, B.R., Urban, M.W., and Ryntz, R.A. (2000) Distribution of individual components in thermoplastic olefins: step-scan FT-IR photoacoustic phase analysis. *Polymer*, **41** (4), 1597–1606.

Kiland, B.R., Urban, M.W., and Ryntz, R.A. (2001) Surface depth-profiling of polymer compounds using step-scan photoacoustic spectroscopy (S^2 PAS). *Polymer*, **42** (1), 337–344.

Kim, H. and Urban, M.W. (2000) Molecular level chain scission mechanisms of epoxy and urethane polymeric films exposed to UV/H_2O. Multidimensional spectroscopic studies. *Langmuir*, **16** (12), 5382–5390.

Kimura, T., Kato, Y., Yoshida, Z., and Nitani, N. (1996) Speciation of neptunium(VI) solid phases formed in aqueous carbonate systems. *J. Nucl. Sci. Technol.*, **33** (6), 519–521.

Kinney, J.B. and Staley, R.H. (1983) Photoacoustic cell for Fourier transform infrared spectrometry of surface species. *Anal. Chem.*, **55** (2), 343–348.

Kirkbright, G.F. (1978) Analytical optoacoustic spectrometry. *Optica Pura y Aplicada*, **11**, 125–136.

Klimchuk, K.A., Hocking, M.B., and Lowen, S. (2000) Water-soluble acrylamide copolymers. VII. Preparation and characterization of poly(methylacrylamide-co-acrylamide). *J. Polym. Sci. A1*, **38** (17), 3146–3160.

Koenig, J.L. (1985) Recent advances in FT-IR (Fourier transform infrared spectroscopy) of polymers. *Pure Appl. Chem.*, **57** (7), 971–976.

Kondo, T., Mitsui, T., Kitagawa, M., and Nakae, Y. (2000) Association of fasting breath nitrous oxide concentration with gastric juice nitrate and nitrite concentrations and *Helicobacter pylori* infection. *Dig. Dis. Sci.*, **45** (10), 2054–2057.

Krishnan, K. (1981) Some applications of Fourier transform infrared photoacoustic spectroscopy. *Appl. Spectrosc.*, **35** (6), 549–557.

Krishnan, K., Hill, S., Hobbs, J.P., and Sung, C.S.P. (1982) Orientation measurements from polymer surfaces using Fourier transform infrared photoacoustic spectroscopy. *Appl. Spectrosc.*, **36** (3), 257–259.

Kühnemann, F., Schneider, K., Hecker, A., Martis, A.A.E., Urban, W., Schiller, S., and Mlynek, J. (1998) Photoacoustic trace-gas detection using a cw single-frequency parametric oscillator. *Appl. Phys. B*, **66** (6), 741–745.

Kuo, M.-L., McClelland, J.F., Luo, S., Chien, P.-L., Walker, R.D., and Hse, C.-Y. (1988) Applications of infrared photoacoustic spectroscopy for wood samples. *Wood Fiber Sci.*, **20** (1), 132–145.

Kuwahata, H., Muto, N., and Uehara, F. (2000) Carrier concentration dependence of photoacoustic spectra of silicon by a piezoelectric transducer method. *Jpn. J. Appl. Phys. Part 1*, **39** (5B), 3169–3171.

Larsen, J.W. (1988) The macromolecular structure of bituminous coals: macromolecular anisotropy, aromatic-aromatic interactions, and other complexities. *Prepr. Am. Chem. Soc. Div. Fuel Chem.*, **33**, 400–406.

Laufer, J., Elwell, C., Delpy, D., and Beard, P. (2005) *In vitro* measurements of absolute blood oxygen saturation using pulsed near-infrared photoacoustic spectroscopy: accuracy and resolution. *Phys. Med. Biol.*, **50** (18), 4409–4428.

Lecoutre, M. and Hadj Bachir, I. (1999) Laser photoacoustic detection of high overtone transitions of ClCN in the near infrared. *J. Mol. Spectrosc.*, **197** (1), 64–67.

Lecoutre, M., Huet, T.R., Mkadmi, E.B., and Bürger, H. (2000a) High-resolution laser photoacoustic spectroscopy of $HSiF_3$: the 5_{v1} and 6_{v1} overtone bands. *J. Mol. Spectrosc.*, **202** (2), 207–212.

Lecoutre, M., Rohart, F., Huet, T.R., and Maki, A.G. (2000b) Photoacoustic detection of new bands of HCN between 11 390 and 13 020 cm^{-1}. *J. Mol. Spectrosc.*, **203** (1), 158–164.

Letzelter, N.S., Wilson, R.H., Jones, A.D., and Sinnaeve, G. (1995) Quantitative determination of the composition of individual pea seeds by Fourier transform infrared photoacoustic spectroscopy. *J. Sci. Food Agric.*, **67** (2), 239–245.

Lewis, L.N. (1982) The analysis of the near IR of some organic and organo-metallic compounds by photoacoustic spectroscopy. *J. Organomet. Chem.*, **234** (3), 355–365.

Li, H. and Butler, I.S. (1992) Infrared photoacoustic spectra of the solid manganese(I) carbonyl halides, $Mn(CO)_5X$ and $[Mn(CO)_4X]_2$ (X = Cl, Br, I). *Appl. Spectrosc.*, **46** (12), 1785–1789.

Li, H. and Butler, I.S. (1993) Solid-state infrared photoacoustic spectra of group 6B metal mixed carbonyl-*t*-butylisocyanide complexes, $M(CO)_{6-n}(CN^tBu)_n$ (M = Cr, Mo, W; n = 1–3). *Appl. Spectrosc.*, **47** (2), 218–221.

Liang, G.-C., Liu, H.-H., Kung, A.H., Mohacsi, A., Miklos, A., and Hess, P. (2000) Photoacoustic trace detection of methane using compact solid-state lasers. *J. Phys. Chem. A*, **104** (45), 10179–10183.

Lima, J.P., Vargas, H., Miklós, A., Angelmahr, M., and Hess, P. (2006) Photoacoustic detection of NO_2 and N_2O using quantum cascade lasers. *Appl. Phys. B*, **85** (2–3), 279–284.

Linker, R. (2008) Determination of nitrate concentration in soil via photoacoustic spectroscopy analysis of ion exchange membranes. *Appl. Spectrosc.*, **62** (2), 248–250.

Liu, W., Sun, Z., Ranheimer, M., Forsling, W., and Tang, H. (1999) A flexible method of carbonate determination using an automatic gas analyzer equipped with an FTIR photoacoustic measurement chamber. *Analyst*, **124** (3), 361–365.

Lochmüller, C.H. and Wilder, D.R. (1980a) Quantitative examination of chemically-modified silica surfaces by near-infrared photoacoustic spectroscopy. *Anal. Chim. Acta*, **116** (1), 19–24.

Lochmüller, C.H. and Wilder, D.R. (1980b) Quantitative photoacoustic spectroscopy of chemically modified silica surfaces. *Anal. Chim. Acta*, **118** (1), 101–108.

Lopes, M.H., Pascoal Neto, C., Barros, A.S., Rutledge, D., Delgadillo, I., and Gil, A.M. (2000) Quantitation of aliphatic suberin in *Quercus suber* L. cork by FTIR spectroscopy and solid-state ^{13}C-NMR spectroscopy. *Biopolymers (Biospectroscopy)*, **57** (6), 344–351.

Lopes, M.H., Barros, A.S., Pascoal Neto, C., Rutledge, D., Delgadillo, I., and Gil, A.M. (2001) Variability of cork from Portuguese *Quercus suber* studied by solid-state ^{13}C-NMR and FTIR spectroscopies. *Biopolymers (Biospectroscopy)*, **62** (5), 268–277.

López Quintana, S., Schmidt, P., Dybal, J., Kratochvíl, J., Pastor, J.M., and Merino, J.C. (2002) Microdomain structure and chain orientation in polypropylene/polyethylene blends investigated by micro-Raman confocal imaging spectroscopy. *Polymer*, **43** (19), 5187–5195.

Lotta, T.I., Tulkki, A.P., Virtanen, J.A., and Kinnunen, P.K.J. (1990) Interaction of 7,7,8,8-tetracyanoquinodimethane with diacylphosphatidylcholines and -phosphatidylglycerols. A photoacoustic Fourier transform infrared study. *Chem. Phys. Lipids*, **52**, 11–27.

Low, M.J.D. (1983) Normalizing infrared photoacoustic spectra of solids. *Spectrosc. Lett.*, **16**, 913–922.

Low, M.J.D. (1984) Infrared spectra of infrared detectors. *Spectrosc. Lett.*, **17** (8), 455–461.

Low, M.J.D. (1985) Hexane soot as standard for normalizing infrared photothermal spectra of solids. *Spectrosc. Lett.*, **18** (8), 619–625.

Low, M.J.D. (1986) Some practical aspects of FT-IR/PBDS. Part I: vibrational noise. *Appl. Spectrosc.*, **40** (7), 1011–1019.

Low, M.J.D. (1993) Unusual bands in the infrared spectra of some oxidized coals and coal chars. *Spectrosc. Lett.*, **26** (3), 453–459.

Low, M.J.D. and Glass, A.S. (1989) The assignment of the 1600 cm^{-1} mystery band of carbons. *Spectrosc. Lett.*, **22** (4), 417–429.

Low, M.J.D. and Lacroix, M. (1982) An infrared photothermal beam deflection Fourier transform spectrometer. *Infrared Phys.*, **22** (3), 139–147.

Low, M.J.D. and Morterra, C. (1983) IR studies of carbons – I. IR photothermal beam deflection spectroscopy. *Carbon*, **21** (3), 275–281.

Low, M.J.D. and Morterra, C. (1985) IR studies of carbons – V. Effects of NaCl on cellulose pyrolysis and char oxidation. *Carbon*, **23** (3), 311–316.

Low, M.J.D. and Morterra, C. (1986) The infrared examination of carbons with beam deflection spectroscopy. *IEEE Trans. Ultrason. Ferroelectr. Freq. Control*, **33** (5), 585–589.

Low, M.J.D. and Morterra, C. (1989) IR studies of carbons. X. The spectral profile of medium-temperature chars, in *Structure and Reactivity of Surfaces* (eds C. Morterra, A. Zecchina, and G. Costa), Elsevier, Amsterdam, pp. 601–609.

Low, M.J.D. and Parodi, G.A. (1978) Infrared photoacoustic spectra of surface species in the 4000–2000 cm^{-1} region using a broad band source. *Spectrosc. Lett.*, **11** (8), 581–588.

Low, M.J.D. and Parodi, G.A. (1980a) An infrared photoacoustic spectrometer. *Infrared Phys.*, **20** (5), 333–340.

Low, M.J.D. and Parodi, G.A. (1980b) Carbon as reference for normalizing infrared photoacoustic spectra. *Spectrosc. Lett.*, **13** (9), 663–669.

Low, M.J.D. and Parodi, G.A. (1980c) Infrared photoacoustic spectra of solids. *Spectrosc. Lett.*, **13** (2–3), 151–158.

Low, M.J.D. and Parodi, G.A. (1980d) Infrared photoacoustic spectroscopy of solids and surface species. *Appl. Spectrosc.*, **34** (1), 76–80.

Low, M.J.D. and Tascon, J.M.D. (1985) An approach to the study of minerals using infrared photothermal beam deflection spectroscopy. *Phys. Chem. Miner.*, **12**, 19–22.

Low, M.J.D. and Wang, N. (1990) A relation between the spectral profile and the infrared continuum of spectra of medium-temperature carbons. *Spectrosc. Lett.*, **23** (8), 983–990.

Low, M.J.D., Lacroix, M., and Morterra, C. (1982) Infrared photothermal beam deflection Fourier transform spectroscopy of solids. *Appl. Spectrosc.*, **36** (5), 582–584.

Low, M.J.D., Morterra, C., and Severdia, A.G. (1982) Infrared photothermal deflection spectroscopy of carbon-supported metal catalysts. *Spectrosc. Lett.*, **15** (5), 415–421.

Low, M.J.D., Morterra, C., and Severdia, A.G. (1984) An approach to the infrared study of materials by photothermal beam deflection spectroscopy. *Mater. Chem. Phys.*, **10**, 519–528.

Low, M.J.D., Politou, A.S., Varlashkin, P.G., and Wang, N. (1990) Unusual bands in the infrared spectra of some chars. *Spectrosc. Lett.*, **23** (4), 527–531.

Ludwig, B.W. and Urban, M.W. (1992) The mobility of ethyl acetate in poly(vinylidene fluoride): rheo-photoacoustic Fourier transform infra-red studies. *Polymer*, **33** (16), 3343–3350.

Ludwig, B.W. and Urban, M.W. (1994) Negative Poisson ratio in poly(tetrafluoroethylene) monitored spectroscopically. *Polymer*, **35** (23), 5130–5137.

Ludwig, B.W. and Urban, M.W. (1997) Rheo-photoacoustic *FT*i.r. studies of thermal history-strain dependence in poly(vinylidene fluoride). *Polymer*, **38** (9), 2077–2091.

Ludwig, B.W. and Urban, M.W. (1998) Rheo-photoacoustic *FT*i.r. and morphology studies of molecular weight-strain dependence in poly(vinylidene fluoride). *Polymer*, **39** (24), 5899–5912.

Luo, S., Huang, C.-Y.F., McClelland, J.F., and Graves, D.J. (1994) A study of protein secondary structure by Fourier transform infrared/photoacoustic spectroscopy and its application for recombinant proteins. *Anal. Biochem.*, **216**, 67–76.

Lynch, B.M. and MacPhee, J.A. (1989) Photoacoustic FTIR spectroscopy in detection of coal oxidation, in *Chemistry of Coal Weathering* (ed. C.R. Nelson), Elsevier, Amsterdam, pp. 83–106.

Lynch, B.M., MacEachern, A.M., MacPhee, J.A., Nandi, B.N., Hamza, H., and Michaelian, K.H. (1983) Photoacoustic infrared Fourier transform (PAIFT) and diffuse reflectance infrared Fourier transform (DRIFT) spectroscopies in studies of oxidation, conversion, and derivatization reactions of bituminous and sub-bituminous coals, in Proceedings, 1983 International Conference on Coal Science, Center for Conference Management, Pittsburgh, pp. 653–654.

Lynch, B.M., Lancaster, L., and Fahey, J.T. (1986) Detection by photoacoustic infrared Fourier transform spectroscopy of surface peroxide species in chemically and thermally modified coals. *Prepr. Pap. Am. Chem. Soc. Div. Fuel Chem.*, **31** (1), 43–48.

Lynch, B.M., Lancaster, L.-I., and MacPhee, J.A. (1987a) Characterization of surface functionality of coals by photoacoustic FTIR (PAIFT) spectroscopy, reflectance infrared microspectrometry, and X-ray photoelectron spectroscopy (XPS). *Prepr. Pap. Am. Chem. Soc. Div. Fuel Chem.*, **32** (1), 138–145.

Lynch, B.M., Lancaster, L.-I., and MacPhee, J.A. (1987b) Carbonyl groups from chemically and thermally promoted decomposition of peroxides on coal surfaces. Detection of specific types using photoacoustic infrared Fourier transform spectroscopy. *Fuel*, **66** (7), 979–983.

Lynch, B.M., Lancaster, L., and MacPhee, J.A. (1988) Detection of carbonyl functionality of oxidized, vacuum-dried coals by photoacoustic infrared Fourier transform (PAIFT) spectroscopy: correlations with added oxygen and with plastic properties. *Energy Fuels*, **2** (1), 13–17.

McAskill, N.A. (1987) Near-infrared photoacoustic spectroscopy of coals and shales. *Appl. Spectrosc.*, **41**, 313–317.

McClelland, J.F. (1983) Photoacoustic spectroscopy. *Anal. Chem.*, **55** (1), 89A–105A.

McDonald, W.F. and Urban, M.W. (1989) A novel rheo-photoacoustic FT-IT technique to monitor deformations in polymers. *Polym. Mater. Sci. Eng.*, **60**, 739–743.

McDonald, W.F., Goettler, H., and Urban, M.W. (1989) A novel approach to photoacoustic FT-IR spectroscopy: rheo-photoacoustic measurements. *Appl. Spectrosc.*, **43** (8), 1387–1393.

McFarlane, R.A., Gentzis, T., Goodarzi, F., Hanna, J.V., and Vassallo, A.M. (1993) Evolution of the chemical structure of Hat Creek resinite during oxidation; a combined FT-IR photoacoustic, NMR and optical microscopic study. *Int. J. Coal Geol.*, **22** (2), 119–147.

McGovern, S.J., Royce, B.S.H., and Benziger, J.B. (1984) Infrared photoacoustic spectroscopy of adsorption on powders. *Appl. Surf. Sci.*, **18** (4), 401–413.

McGovern, S.J., Royce, B.S.H., and Benziger, J.B. (1985) The importance of interstitial gas expansion in infrared photoacoustic spectroscopy of powders. *J. Appl. Phys.*, **57** (5), 1710–1718.

McKenna, W.P., Bandyopadhyay, S., and Eyring, E.M. (1984) FT-IR/PAS investigation of chromocene supported on silica. *Appl. Spectrosc.*, **38** (6), 834–837.

McKenna, W.P., Gale, D.J., Rivett, D.E., and Eyring, E.M. (1985) FT-IR photoacoustic spectroscopy investigation of oxidized wool. *Spectrosc. Lett.*, **18** (2), 115–122.

McQueen, D.H., Wilson, R., and Kinnunen, A. (1995) Near and mid-infrared photoacoustic analysis of principal components of foodstuffs. *Trends Analyt. Chem.*, **14** (10), 482–492.

McQueen, D.H., Wilson, R., Kinnunen, A., and Jensen, E.P. (1995) Comparison of two infrared spectroscopic methods for cheese analysis. *Talanta*, **42** (12), 2007–2015.

Majid, A., Argue, S., Boyko, V., Pleizier, G., L'Ecuyer, P., Tunney, J., and Lang, S. (2003) Characterization of sol–gel-derived nano-particles separated from oil sands fine tailings. *Colloids Surf. A*, **224** (1–3), 33–44.

Manzanares, C., Blunt, V.M., and Peng, J. (1993) Vibrational spectroscopy of nonequivalent C–H bonds in liquid *cis*- and *trans*-3-hexene. *Spectrochim. Acta A*, **49** (8), 1139–1152.

Marinov, D. and Sigrist, M.W. (2003) Monitoring of road-traffic emissions with a mobile photoacoustic system. *Photochem. Photobiol. Sci.*, **2** (7), 774–778.

Martin, B.L., Wu, D., Tabatabai, L., and Graves, D.J. (1990) Formation of cyclic imide-like structures upon the treatment of Calmodulin and a Calmodulin peptide with heat. *Arch. Biochem. Biophys.*, **276** (1), 94–101.

Martin, B.L., Li, B., Liao, C., and Rhode, D.J. (2000) Differences between Mg^{2+} and transition metal ions in the activation of calcineurin. *Arch. Biochem. Biophys.*, **380** (1), 71–77.

Martin, L.K. and Yang, C.Q. (1994) Infrared spectroscopy studies of the photooxidation of a polyethylene nonwoven fabric. *J. Environ. Polym. Degrad.*, **2** (2), 153–160.

Mead, D.G., Lowry, S.R., Vidrine, D.W., and Mattson, D.R. (1979) Infrared spectroscopy using a photoacoustic cell, in Fourth international conference on infrared and millimeter waves and their applications, p. 231.

Mehicic, M., Kollar, R.G., and Grasselli, J.G. (1981) Analytical applications of photoacoustic spectroscopy using Fourier transform infrared (FTIR). *Proc. SPIE Int. Soc. Opt. Eng.*, **289**, 99–101.

Mendelovici, E., Frost, R.L., and Kloprogge, J.T. (2001) Modification of chrysotile surface by organosilanes: an IR-photoacoustic spectroscopy study. *J. Colloid Interface Sci.*, **238** (2), 273–278.

Merker, U., Engels, P., Madeja, F., Havenith, M., and Urban, W. (1999) High-resolution CO-laser sideband spectrometer for molecular-beam optothermal spectroscopy in the 5–6.6 μm wavelength region. *Rev. Sci. Instrum.*, **70** (4), 1933–1938.

Michaelian, K.H. (1987) Signal averaging of photoacoustic FTIR data. I. Computation of spectra from double-sided low resolution interferograms. *Infrared Phys.*, **27**, 287–296.

Michaelian, K.H. (1989a) Interferogram symmetrization and multiplicative phase correction of rapid-scan and step-scan photoacoustic FT-IR data. *Infrared Phys.*, **29** (1), 87–100.

Michaelian, K.H. (1989b) Depth profiling and signal saturation in photoacoustic FT-IR spectra measured with a step-scan interferometer. *Appl. Spectrosc.*, **43** (2), 185–190.

Michaelian, K.H. (1990a) Data treatment in photoacoustic FT-IR spectroscopy, in *Vibrational Spectra and Structure*, vol. 18 (ed. J.R. Durig), Elsevier, Amsterdam, pp. 81–126.

Michaelian, K.H. (1990b) Step-scan photoacoustic infrared spectra of kaolinite. *Infrared Phys.*, **30** (2), 181–186.

Michaelian, K.H. (1991) Depth profiling of oxidized coal by step-scan photoacoustic FT-IR spectroscopy. *Appl. Spectrosc.*, **45** (2), 302–304.

Michaelian, K.H. (2003) Photoacoustic infrared spectroscopy of industrial materials (invited). *Rev. Sci. Instrum.*, **74** (1), 659–662.

Michaelian, K.H. (2007) Invited article: linearization and signal recovery in photoacoustic infrared spectroscopy. *Rev. Sci. Instrum.*, **78** (5), 051301, (12 pages).

Michaelian, K.H. and Friesen, W.I. (1990) Photoacoustic FT-IR spectra of separated western Canadian coal macerals. Analysis of the CH stretching region by curve fitting and deconvolution. *Fuel*, **69** (10), 1271–1275.

Michaelian, K.H., Bukka, K., and Permann, D.N.S. (1987) Photoacoustic infrared spectra (250–10000 cm^{-1}) of partially deuterated kaolinite #9. *Can. J. Chem.*, **65** (6), 1420–1423.

Michaelian, K.H., Yariv, S., and Nasser, A. (1991) Study of the interactions between caesium bromide and kaolinite by photoacoustic and diffuse reflectance infrared spectroscopy. *Can. J. Chem.*, **69** (4), 749–754.

Michaelian, K.H., Ogunsola, O.I., and Bartholomew, R.J. (1995) Infrared spectroscopy of thermally treated low-rank coals. II. Mid- and near-infrared photoacoustic spectra of coals heated in oxidizing and inert atmospheres. *Can. J. Appl. Spectrosc.*, **40** (4), 94–99.

Michaelian, K.H., Friesen, W.I., Zhang, S.L., Gentzis, T., Crelling, J.C., and Gagarin, S.G. (1995) FT-IR spectroscopy of western Canadian coals and macerals, in *Coal Science* (eds J.A. Pajares and J.M.D. Tascón), Elsevier, Amsterdam, pp. 255–258.

Michaelian, K.H., Akers, K.L., Zhang, S.L., Yariv, S., and Lapides, I. (1997) Far-infrared spectra of kaolinite/alkali metal halide complexes. *Mikrochim. Acta*, **14** (Suppl.), 211–212.

Michaelian, K.H., Lapides, I., Lahav, N., Yariv, S., and Brodsky, I. (1998) Infrared study of the intercalation of kaolinite by caesium bromide and caesium iodide. *J. Colloid Interface Sci.*, **204** (2), 389–393.

Michaelian, K.H., Zhang, S.L., Hall, R.H., and Bulmer, J.T. (2001) Photoacoustic infrared spectroscopy of distillation fractions from Syncrude Sweet Blend.

Curve fitting and integration of C–H stretching bands. *Can. J. Anal. Sci. Spectrosc.*, **46** (1), 10–22.

Michaelian, K.H., Hall, R.H., and Bulmer, J.T. (2002a) Photoacoustic infrared spectroscopy and thermophysical properties of Syncrude cokes. *J. Therm. Anal. Calorim.*, **69** (1), 135–147.

Michaelian, K.H., Hall, R.H., and Bulmer, J.T. (2002b) FT-Raman and photoacoustic infrared spectroscopy of Syncrude heavy gas oil distillation fractions. *Spectrochim. Acta A*, **59** (4), 811–824.

Michaelian, K.H., Hall, R.H., and Bulmer, J.T. (2003) FT-Raman and photoacoustic infrared spectroscopy of syncrude light gas oil distillation fractions. *Spectrochim. Acta A*, **59** (13), 2971–2984.

Michaelian, K.H., Hall, R.H., and Bulmer, J.T. (2004) Raman, photoacoustic infrared and absorption spectroscopy of Syncrude sweet blend distillation fractions. *Can. J. Anal. Sci. Spectrosc.*, **49** (1), 43–52.

Michaelian, K.H., Hall, R.H., and Kenny, K.I. (2006) Photoacoustic infrared spectroscopy of Syncrude post-extraction oil sand. *Spectrochim. Acta A*, **64** (3), 703–710.

Michaelian, K.H., Billinghurst, B.E., Shaw, J.M., and Lastovka, V. (2009) Far-infrared photoacoustic spectra of tetracene, pentacene, perylene and pyrene. *Vib. Spectrosc.*, **49** (1), 28–31.

Miklós, A. and Fehér, M. (1996) Optoacoustic detection with near-infrared diode lasers: trace gases and short-lived molecules. *Infrared Phys. Technol.*, **37** (1), 21–27.

Miklós, A., Hess, P., Mohácsi, Á., Sneider, J., Kamm, S., and Schäfer, S. (1999a) Improved photoacoustic detector for monitoring polar molecules such as ammonia with a 1.53 μm DFB diode laser. *AIP Conf. Proc.*, **463**, 126–128.

Miklós, A., Hess, P., Romolini, A., Spada, C., Lancia, A., Kamm, S., and Schäfer, S. (1999b) Measurement of the ($v_2 + 2v_3$) band of methane by photoacoustic and long path absorption spectroscopy. *AIP Conf. Proc.*, **463**, 217–219.

Mikula, R.J., Axelson, D.E., and Michaelian, K.H. (1985) Oxidation and weathering of stockpiled western Canadian coals, in *Proceedings, 1985 International Conference* on *Coal Science*, Pergamon Press, Sydney, pp. 495–498.

Mistry, M.K., Choudhury, N.R., Dutta, N.K., Knott, R., Shi, Z., and Holdcroft, S. (2008) Novel organic–inorganic hybrids with increased water retention for elevated temperature proton exchange membrane application. *Chem. Mater.*, **20** (21), 6857–6870.

Mitchem, L., Mio, C., and Snook, R.D. (2004) Diffusion of transdermally delivered nitroglycerin through skin mimetics using photoacoustic and attenuated total reflectance spectrometry. *Anal. Chim. Acta*, **511** (2), 281–288.

Mitsui, T. and Kondo, T. (1999) Vegetables, high nitrate foods, increased breath nitrous oxide. *Dig. Dis. Sci.*, **44** (6), 1216–1219.

Mitsui, T. and Kondo, T. (2004) Increased breath nitrous oxide after ingesting nitrate in patients with atrophic gastritis and partial gastrectomy. *Clin. Chim. Acta*, **345** (1–2), 129–133.

Mitsui, T., Miyamura, M., Matsunami, A., Kitagawa, K., and Arai, N. (1997) Measuring nitrous oxide in exhaled air by gas chromatography and infrared photoacoustic spectrometry. *Clin. Chem.*, **43** (10), 1993–1995.

Mitsui, T., Kato, N., and Kondo, T. (2000) Estimate of nitrate metabolism in intestinal tract by measuring breath nitrous oxide concentration in Chinese and Japanese. *Dig. Dis. Sci.*, **45** (5), 1002–1005.

Mohamed, A., Rayas-Duarte, P., Gordon, S.H., and Xu, J. (2004) Estimation of HRW wheat heat damage by DSC, capillary zone electrophoresis, photoacoustic spectroscopy and rheometry. *Food Chem.*, **87** (2), 195–203.

Mohamed, M.M. (1995) Fourier-transform infrared/photoacoustic study of pyridine adsorbed on silica supported copper-molybdenum catalysts. *Spectrochim. Acta A*, **51** (1), 1–9.

Mohamed, M.M. (2003) Structural and acidic characteristics of Cu–Ni-modified acid-leached mordenites. *J. Colloid Interface Sci.*, **265** (1), 106–114.

Mohamed, M.M. and Vansant, E.F. (1995a) Redox behaviour of copper mordenite zeolite. *J. Mater. Sci.*, **30** (19), 4834–4838.

Mohamed, M.M. and Vansant, E.F. (1995b) Structural and acidic properties of copper-silica catalysts. 1. A differential scanning calorimetry and Fourier transform-infrared/photoacoustic study. *Colloids Surf. A*, **96** (3), 253–260.

Monchalin, J.-P., Gagné, J.-M., Parpal, J.-L., and Bertrand, L. (1979) Photoacoustic spectroscopy of chrysotile asbestos using a cw HF laser. *Appl. Phys. Lett.*, **35** (5), 360–363.

Moore, P., Poslusny, M., Daugherty, K.E., Venables, B.J., and Okuda, T. (1990) Detection of cellulose embrittlement by infrared spectroscopy. *Appl. Spectrosc.*, **44** (2), 326–328.

Morey, M., Davidson, A., and Stucky, G. (1996) A new step toward transition metal incorporation in cubic microporous materials: preparation and characterization of Ti-MCM-48. *Microporous Mater.*, **6** (2), 99–104.

Morterra, C. and Low, M.J.D. (1982) The nature of the $1600\,cm^{-1}$ band of carbons. *Spectrosc. Lett.*, **15** (9), 689–697.

Morterra, C. and Low, M.J.D. (1983) IR studies of carbons–II. The vacuum pyrolysis of cellulose. *Carbon*, **21** (3), 283–288.

Morterra, C. and Low, M.J.D. (1985a) IR studies of carbons–IV. The vacuum pyrolysis of oxidized cellulose and the characterization of the chars. *Carbon*, **23** (3), 301–310.

Morterra, C. and Low, M.J.D. (1985b) IR studies of carbons–VI. The effects of $KHCO_3$ on cellulose pyrolysis and char oxidation. *Carbon*, **23** (4), 335–341.

Morterra, C. and Low, M.J.D. (1985c) I.R. studies of carbons–VII. The pyrolysis of a phenol-formaldehyde resin. *Carbon*, **23** (5), 525–530.

Morterra, C. and Low, M.J.D. (1985d) Infrared studies of carbons. 8. The oxidation of phenol-formaldehyde chars. *Langmuir*, **1** (3), 320–326.

Morterra, C. and Low, M.J.D. (1985e) An infrared spectroscopic study of some carbonaceous materials. *Mater. Chem. Phys.*, **12** (3), 207–233.

Morterra, C., Low, M.J.D., and Severdia, A.G. (1984) IR studies of carbons–III. The oxidation of cellulose chars. *Carbon*, **22** (1), 5–12.

Morterra, C., O'Shea, M.L., and Low, M.J.D. (1988) Infrared studies of carbons. IX. The vacuum pyrolysis of non-oxygen-containing materials: PVC. *Mater. Chem. Phys.*, **20** (2), 123–144.

Moyer, D.J.D. and Wightman, J.P. (1989) Characterization of surface pretreatments of carbon fiber-polyimide matrix composites. *Surf. Interface Anal.*, **14** (9), 496–504.

Mukherjee, A., Dunayevskiy, I., Prasanna, M., Go, R., Tsekoun, A., Wang, X., Fan, J., and Patel, C.K.N. (2008a) Sub-parts-per-billion level detection of dimethyl methyl phosphonate (DMMP) by quantum cascade laser photoacoustic spectroscopy. *Appl. Opt.*, **47** (10), 1543–1548.

Mukherjee, A., Prasanna, M., Lane, M., Go, R., Dunayevskiy, I., Tsekoun, A., and Patel, C.K.N. (2008b) Optically multiplexed multi-gas detection using quantum cascade laser photoacoustic spectroscopy. *Appl. Opt.*, **47** (27), 4884–4887.

Müller, F., Popp, A., and Kühnemann, F. (2003) Transportable, highly sensitive photoacoustic spectrometer based on a continuous-wave dual-cavity optical parametric oscillator. *Opt. Express*, **11** (22), 2820–2825.

Nadler, M.P., Nissan, R.A., and Hollins, R.A. (1988) FT-IR and FT-NMR study of organophosphorus surface reactions. *Appl. Spectrosc.*, **42** (2), 634–642.

Nägele, M. and Sigrist, M.W. (2000) Mobile laser spectrometer with novel resonant multipass photoacoustic cell for trace-gas sensing. *Appl. Phys. B*, **70** (2–6), 895–901.

Nakamura, O., Lowe, R.D., Mitchem, L., and Snook, R.D. (2001) Diffusion of nitroglycerine from drug delivery patches through micro-fibre filters using Fourier transform infrared photoacoustic spectrometry. *Anal. Chim. Acta*, **427** (1), 63–73.

Natale, M. and Lewis, L.N. (1982) Applications of PAS for the investigation of overtones and combinations in the near IR. *Appl. Spectrosc.*, **36**, 410–413.

Nelson, J.H., MacDougall, J.J., Baglin, F.G., Freeman, D.W., Nadler, M., and Hendrix, J.L. (1982) Characterization of Carlin-type gold ore by photoacoustic, Raman, and EPR spectroscopy. *Appl. Spectrosc.*, **36** (5), 574–576.

Neubert, R., Collin, B., and Wartewig, S. (1997a) Direct determination of drug content in semisolid formulations using step-scan FT-IR photoacoustic spectroscopy. *Pharm. Res.*, **14** (7), 946–948.

Neubert, R., Collin, B., and Wartewig, S. (1997b) Quantitative analysis of drug content in semisolid formulations using step-scan FT-IR photoacoustic spectroscopy. *Vib. Spectrosc.*, **13** (2), 241–244.

Neubert, R., Raith, K., Raudenkolb, S., and Wartewig, S. (1998) Thermal degradation of ceramides as studied by mass spectrometry and vibrational spectroscopy. *Anal. Commun.*, **35** (5), 161–164.

Nishikawa, Y., Kimura, K., Matsuda, A., and Kenpo, T. (1992) Surface characterization of gamma-iron oxides for magnetic memory media using Fourier transform infrared photoacoustic spectroscopy. *Appl. Spectrosc.*, **46** (11), 1695–1698.

Ohkoshi, M. (2002) FTIR-PAS study of light-induced changes in the surface of acetylated or polyethylene glycol-impregnated wood. *J. Wood Sci.*, **48** (5), 394–401.

Olafsson, A., Hansen, G.I., Loftsdottir, A.S., and Jakobsson, S. (1999) FTIR photoacoustic trace gas detection. *AIP Conf. Proc.*, **463**, 208–210.

Oomens, J., Zuckermann, H., Persijn, S., Parker, D.H., and Harren, F.J.M. (1998) CO-laser-based photoacoustic trace-gas detection: applications in postharvest physiology. *Appl. Phys. B*, **67** (4), 459–466.

Oomens, J., Bisson, S., Harting, M., Kulp, T., and Harren, F.J.M. (2000) New laser sources for photoacoustic trace gas detection with applications in biomedical science. *Proc. SPIE Int. Soc. Opt. Eng.*, **3916**, 295–300.

O'Shea, M.L., Low, M.J.D., and Morterra, C. (1989) Spectroscopic studies of carbons. XI. The vacuum pyrolysis of non-oxygen-containing materials: PVBr. *Mater. Chem. Phys.*, **23** (5), 499–516.

O'Shea, M.L., Morterra, C., and Low, M.J.D. (1990a) Spectroscopic studies of carbons. XIV. The vacuum pyrolysis of non-oxygen containing materials: PVF. *Mater. Chem. Phys.*, **25** (5), 501–521.

O'Shea, M.L., Morterra, C., and Low, M.J.D. (1990b) Spectroscopic studies of carbons. XVII. Pyrolysis of polyvinylidene fluoride. *Mater. Chem. Phys.*, **26** (2), 193–209.

O'Shea, M.L., Morterra, C., and Low, M.J.D. (1991a) Spectroscopic studies of carbon. XX. The pyrolysis of polyvinylidene chloride and of Saran. *Mater. Chem. Phys.*, **27** (2), 155–179.

O'Shea, M.L., Morterra, C., and Low, M.J.D. (1991b) Spectroscopic studies of carbons. XXII. The oxidation of polyvinyl halide chars. *Mater. Chem. Phys.*, **28** (1), 9–31.

Paldus, B.A., Spence, T.G., Zare, R.N., Oomens, J., Harren, F.J.M., Parker, D.H., Gmachl, C., Cappasso, F., Sivco, D.L., Baillargeon, J.N., Hutchinson, A.L., and Cho, A.Y. (1999) Photoacoustic spectroscopy using quantum-cascade lasers. *Opt. Lett.*, **24** (3), 178–180.

Pandey, K.K. and Theagarajan, K.S. (1997) Analysis of wood surfaces and ground wood by diffuse reflectance (DRIFT) and photoacoustic (PAS) Fourier transform infrared spectroscopic techniques. *Holz Roh Werkst.*, **55** (6), 383–390.

Pandey, K.K. and Vuorinen, T. (2008) Comparative study of photodegradation of wood by a UV laser and a xenon light source. *Polym. Degrad. Stab.*, **93** (12), 2138–2146.

Papendorf, U. and Riepe, W. (1989) Identification of adsorbed species on activated carbon by photoacoustic spectroscopy, in Proceedings, 1989 International Conference on Coal Science, Tokyo, vol. II, pp. 1111–1113.

Paputa Peck, M.C., Samus, M.A., Killgoar, P.C., and Carter, R.O. (1991) Examination of diffusing materials on rubber surfaces by photoacoustic infrared spectroscopy. *Rubber Chem. Technol.*, **64** (4), 610–621.

Pasieczna, S. and Ryczkowski, J. (2003) Infrared photoacoustic spectroscopy – Advantages and disadvantages in surface science and catalysis research. *J. Phys. IV*, **109**, 65–71.

Pasieczna-Patkowska, S. and Ryczkowski, J. (2007) Spectroscopic studies of alumina-supported nickel catalysts precursors Part I. Catalysts prepared from acidic solutions. *Appl. Surf. Sci.*, **253** (13), 5910–5913.

Pawsey, S., Yach, K., and Reven, L. (2002) Self-assembly of carboxyalkylphosphonic acids on metal oxide powders. *Langmuir*, **18** (13), 5205–5212.

Pelletier, M., Michot, L.J., Barrès, O., Humbert, B., Petit, S., and Robert, J.-L. (1999) Influence of KBr conditioning on the infrared hydroxyl-stretching region of saponites. *Clay Miner.*, **34** (3), 439–445.

Peng, Y., Zhang, W., Li, L., and Yu, Q. (2009) Tunable fiber laser and fiber amplifier based photoacoustic spectrometer for trace gas detection. *Spectrochim. Acta A*, **74** (4), 924–927.

Pennington, B.D. and Urban, M.W. (1999) Rheo-photoacoustic (RPA) FT-IR spectroscopic studies of the adhesion of thermosets to polyolefins. *J. Adhes. Sci. Technol.*, **13** (1), 19–34.

Pennington, B.D., Ryntz, R.A., and Urban, M.W. (1999) Stratification in thermoplastic olefins (TPO); photoacoustic FT-IR depth profiling studies. *Polymer*, **40** (17), 4795–4803.

Persijn, S.T., Veltman, R.H., Oomens, J., Harren, F.J.M., and Parker, D.H. (1999) A CO laser based photoacoustic system applied to the detection of trace gases emitted by conference pears stored at high CO_2 and low O_2 levels. *AIP Conf. Proc.*, **463**, 609–611.

Persijn, S.T., Veltman, R.H., Oomens, J., Harren, F.J.M., and Parker, D.H. (2000) CO laser absorption coefficients for gases of biological relevance: H_2O, CO_2, ethanol, acetaldehyde, and ethylene. *Appl. Spectrosc.*, **54** (1), 62–71.

Persijn, S.T., Santosa, E., and Harren, F.J.M. (2002) A versatile photoacoustic spectrometer for sensitive trace-gas analysis in the mid-infrared wavelength region (5.1–8.0 and 2.8–4.1 µm). *Appl. Phys. B*, **75** (2–3), 335–342.

Petkovska, L.T. and Miljanić, Š.S. (1997) CO_2-laser photoacoustic spectroscopy of deuterated ammonia. *Infrared Phys. Technol.*, **38** (6), 331–336.

Piyakis, K., Sacher, E., Domingue, A., Pireaux, J.-J., Leclerc, G., Bertrand, P., and Lhoest, J.B. (1995) A multitechnique analysis of the outermost layers of the Teflon PFA surface. *Appl. Surf. Sci.*, **84** (3), 227–235.

Politou, A.S., Morterra, C., and Low, M.J.D. (1990a) Infrared studies of carbons. XII. The formation of chars from a polycarbonate. *Carbon*, **28** (4), 529–538.

Politou, A.S., Morterra, C., and Low, M.J.D. (1990b) Infrared studies of carbons. XIII. The oxidation of polycarbonate chars. *Carbon*, **28** (6), 855–865.

Politou, A.S., Morterra, C., and Low, M.J.D. (1991) Infrared studies of carbons. XVI. The carbonization of an aliphatic allyl polycarbonate. *Polym. Degrad. Stab.*, **32**, 331–356.

Poulin, S., Yang, D.Q., Sacher, E., Hyett, C., and Ellis, T.H. (2000) The surface structure of Dow Cyclotene 3022, as determined by photoacoustic FTIR, confocal Raman and photoelectron spectroscopies. *Appl. Surf. Sci.*, **165** (1), 15–22.

Pushkarsky, M.B., Webber, M.E., Baghdassarian, O., Narasimhan, L.R., and Patel, C.K.N. (2002) Laser-based photoacoustic ammonia sensors for industrial applications. *Appl. Phys. B*, **75** (2–3), 391–396.

Pushkarsky, M.B., Webber, M.E., and Patel, C.K.N. (2003) Ultra-sensitive ambient ammonia detection using CO_2-laser-based photoacoustic spectroscopy. *Appl. Phys. B*, **77** (4), 381–385.

Pushkarsky, M., Webber, M., and Patel, C.K.N. (2005) High sensitivity, high selectivity detection of CWAs and TICs using tunable laser photoacoustic spectroscopy. *Proc. SPIE Int. Soc. Opt. Eng.*, **5732**, 93–107.

Pushkarsky, M., Tsekoun, A., Dunayevskiy, I.G., Go, R., and Patel, C.K.N. (2006a) Sub-parts-per-billion level detection of NO_2 using room-temperature quantum cascade lasers. *Proc. Natl. Acad. Sci. U.S.A.*, **103** (29), 10846–10849.

Pushkarsky, M.B., Dunayevskiy, I.G., Prasanna, M., Tsekoun, A.G., Go, R., and Patel, C.K.N. (2006b) High-sensitivity detection of TNT. *Proc. Natl. Acad. Sci. U.S.A.*, **103** (52), 19630–19634.

Quaschning, V., Deutsch, J., Druska, P., Niclas, H.-J., and Kemnitz, E. (1998) Properties of modified zirconia used as Friedel-Crafts-acylation catalysts. *J. Catal.*, **177** (2), 164–174.

Radak, B.B., Pastirk, I., Ristić, G.S., and Petkovska, L.T. (1998) Pressure effects on CO_2-laser coincidences with ethene and ammonia investigated by photoacoustic detection. *Infrared Phys. Technol.*, **39** (1), 7–13.

Radak, B.B., Petkovska, M.T., Trtica, M.S., Miljanic, S.S., and Petkovska, L.T. (2004) Photoacoustic study of CO_2-laser coincidences with absorption of some organic solvent vapours. *Anal. Chim. Acta*, **505** (1), 67–71.

Raveh, A., Martinu, L., Domingue, A., Wertheimer, M.R., and Bertrand, L. (1992) Fourier transform infrared photoacoustic spectroscopy of amorphous carbon films, in *Photoacoustic and Photothermal Phenomena III* (ed. D. Bicanic), Springer, Berlin, pp. 151–154.

Reddy, K.T.R., Chalapathy, R.B.V., Slifkin, M.A., Weiss, A.W., and Miles, R.W. (2001) Photoacoustic spectroscopy of sprayed $CuGa_xIn_{1-x}Se_2$ thin films. *Thin Solid Films*, **387**, 205–207.

Reddy, K.T.R., Slifkin, M.A., and Weiss, A.M. (2001) Characterization of inorganic materials with photoacoustic spectrophotometry. *Opt. Mater.*, **16**, 87–91.

Repond, P. and Sigrist, M.W. (1994) Photoacoustic spectroscopy on gases with high pressure continuously tunable CO_2 laser. *J. Phys. IV Colloq. C7*, **4**, C7-523–526.

Repond, P. and Sigrist, M.W. (1996) Photoacoustic spectroscopy on trace gases with continuously tunable CO_2 laser. *Appl. Opt.*, **35** (21), 4065–4085.

Rey, J.M., Schramm, D., Hahnloser, D., Marinov, D., and Sigrist, M.W. (2008) Spectroscopic investigation of volatile compounds produced during thermal and radiofrequency bipolar cautery on porcine liver. *Meas. Sci. Technol.*, **19**, 075602, (5 pages).

Riseman, S.M. and Eyring, E.M. (1981) Normalizing infrared FT photoacoustic spectra of solids. *Spectrosc. Lett.*, **14** (3), 163–185.

Riseman, S.M., Yaniger, S.I., Eyring, E.M., MacInnes, D., Macdiarmid, A.G., and Heeger, A.J. (1981) Infrared photoacoustic spectroscopy of conducting polymers. I. Undoped and *n*-doped polyacetylene. *Appl. Spectrosc.*, **35** (6), 557–559.

Riseman, S.M., Massoth, F.E., Dhar, G.M., and Eyring, E.M. (1982) Fourier transform infrared photoacoustic spectroscopy of pyridine adsorbed on silica-alumina and γ-alumina. *J. Phys. Chem.*, **86** (10), 1760–1763.

Rocha, S.M., Goodfellow, B.J., Delgadillo, I., Neto, C.P., and Gil, A.M. (2001) Enzymatic isolation and structural characterisation of polymeric suberic of cork from Quercus suber L. Int. J. Biol. Macromol., 28 (2), 107–119.

Rockley, M.G. and Devlin, J.P. (1980) Photoacoustic infrared spectra (IR-PAS) of aged and fresh-cleaved coal surfaces. Appl. Spectrosc., 34 (4), 407–408.

Rockley, M.G., Richardson, H.H., and Davis, D.M. (1982) The Fourier-transformed infrared photoacoustic spectroscopy of asbestos fiber. J. Photoacoustics, 1 (1), 145–149.

Rockley, M.G., Woodard, M., Richardson, H.H., Davis, D.M., Purdle, N., and Bowen, J.M. (1983) Determination of phenycyclidine and phenobarbital in complex mixtures by Fourier-transformed infrared photoacoustic spectroscopy. Anal. Chem., 55 (1), 32–34.

Rockley, M.G., Ratcliffe, A.E., Davis, D.M., and Woodard, M.K. (1984) Examination of a C-black by various FT-IR spectroscopic methods. Appl. Spectrosc., 38 (4), 553–556.

Rockley, N.L. and Rockley, M.G. (1987) The FT-IR analysis by PAS and KBr pellet of cation-exchanged clay mineral and phosphonate complexes. Appl. Spectrosc., 41 (3), 471–475.

Romann, A. and Sigrist, M.W. (2002) Photoacoustic gas sensing employing fundamental and frequency-doubled radiation of a continuously tunable high-pressure CO_2 laser. Appl. Phys. B, 75 (2–3), 377–383.

Royce, B.S.H. and Benziger, J.B. (1986) Fourier transform photoacoustic spectroscopy of solids. IEEE Trans. Ultrason. Ferroelectr. Freq. Control, 33 (5), 561–572.

Royce, B.S.H., Teng, Y.C., and Enns, J. (1980) Fourier transform infrared photoacoustic spectroscopy of solids, in 1980 Ultrasonics Symposium Proceedings, pp. 652–657.

Ryczkowski, J. (1994) FTIR and FTIR PAS applications in quantitative analysis of EDTA adsorption on alumina surface. Proc. SPIE Int. Soc. Opt. Eng., 2089, 182–183.

Ryczkowski, J. (2007a) Application of infrared photoacoustic spectroscopy in catalysis. Catal. Today, 124 (1–2), 11–20.

Ryczkowski, J. (2007b) Spectroscopic evidences of EDTA interaction with inorganic supports during the preparation of supported metal catalysts. Vib. Spectrosc., 43 (1), 203–209.

Ryczkowski, J., Goworek, J., Gac, W., Pasieczna, S., and Borowiecki, T. (2005) Temperature removal of templating agent from MCM-41 silica materials. Thermochim. Acta, 434 (1–2), 2–8.

Saab, A.P., Laub, M., Srdanov, V.I., and Stucky, G.D. (1998) Oxidized thin films of C_{60}: a new humidity-sensing material. Adv. Mater., 10 (6), 462–465.

Sadler, A.J., Horsch, J.G., Lawson, E.Q., Harmatz, D., Brandau, D.T., and Middaugh, C.R. (1984) Near-infrared photoacoustic spectroscopy of proteins. Anal. Biochem., 138, 44–51.

Santosa, E., te Lintel Hekkert, S., Harren, F.J.M., and Parker, D.H. (1999) The $\Delta v = 2$ CO laser: photoacoustic trace gas detection. AIP Conf. Proc., 463, 612–614.

Sarma, T.V.K., Sastry, C.V.R., and Santhamma, C. (1987) The photoacoustic spectra of substituted benzenes in the near infrared region – I. Spectrochim. Acta A, 43 (8), 1059–1065.

Saucy, D.A., Cabaniss, G.E., and Linton, R.W. (1985) Surface reactivities of polynuclear aromatic adsorbates on alumina and silica particles using infrared photoacoustic spectroscopy. Anal. Chem., 57 (4), 876–879.

Saucy, D.A., Simko, S.J., and Linton, R.W. (1985) Comparison of photoacoustic and attenuated total reflectance sampling depths in the infrared region. Anal. Chem., 57 (4), 871–875.

Schäfer, S., Mashni, M., Sneider, J., Miklós, A., Hess, P., Pitz, H., Pleban, K.-U., and Ebert, V. (1998a) Sensitive detection of methane with a 1.65 μm diode laser by photoacoustic and absorption spectroscopy. Appl. Phys. B, 66 (4), 511–516.

Schäfer, S., Miklós, A., Pusel, A., and Hess, P. (1998b) Absolute measurement of gas concentrations and saturation behavior in pulsed photoacoustics. Chem. Phys. Lett., 285 (3–4), 235–239.

Schendzielorz, A., Hanh, B.D., Neubert, R.H.H., and Wartewig, S. (1999) Penetration studies of Clotrimazole from semisolid formulation using step-scan FT-IR photoacoustic spectroscopy. *Pharm. Res.*, **16** (1), 42–45.

Schiewe-Langgartner, F., Wiesner, G., Gruber, M., and Hobbhahn, J. (2005) Expositions des Personals gegenüber Sevofluran. *Anaesthetist*, **54** (7), 667–672.

Schilt, S., Thévenaz, L., Courtois, E., and Robert, P.A. (2002) Ethylene spectroscopy using a quasi-room-temperature quantum cascade laser. *Spectrochim. Acta A*, **58** (11), 2533–2539.

Schilt, S., Thévenaz, L., Niklès, M., Emmenegger, L., and Hüglin, C. (2004a) Ammonia monitoring at trace level using photoacoustic spectroscopy in industrial and environmental applications. *Spectrochim. Acta A*, **60** (14), 3259–3268.

Schilt, S., Vicet, A., Werner, R., Mattiello, M., Thévenaz, L., Salhi, A., Rouillard, Y., and Koeth, J. (2004b) Application of antimonide diode lasers in photoacoustic spectroscopy. *Spectrochim. Acta A*, **60** (14), 3431–3436.

Schilt, S., Besson, J.-P., and Thévenaz, L. (2006) Near-infrared laser photoacoustic detection of methane: the impact of molecular relaxation. *Appl. Phys. B*, **82** (2), 319–329.

Schmidt, P., Roda, J., Nováková, V., and Pastor, J.M. (1997) Analysis of the chemical composition of poly(ε-caprolactam)-block-polybutadiene copolymers by photoacoustic FTIR spectroscopy and by FT Raman spectroscopy. *Angew. Makromol. Chem.*, **245** (4268), 113–123.

Schmidt, P., Kolařík, J., Lednický, F., Dybal, J., Lagarón, J.M., and Pastor, J.M. (2000a) Phase structure, composition and orientation of PC/PSAN blends studied by Raman spectroscopy, confocal Raman imaging spectroscopy and polarised PA-FTIR spectroscopy. *Polymer*, **41** (11), 4267–4279.

Schmidt, P., Raab, M., Kolařík, J., and Eichorn, K.J. (2000b) Comparison of two modern infrared spectroscopic methods for the determination of orientation in drawn bulk polymers. *Polym. Test.*, **19** (2), 205–212.

Schmidt, P., Baldrian, J., Ščudla, J., Dybal, J., Raab, M., and Eichhorn, K.-J. (2001) Structural transformation of polyethylene phase in oriented polyethylene/polypropylene blends: a hierarchical approach. *Polymer*, **42** (12), 5321–5326.

Schmidt, P., Dybal, J., Ščudla, J., Raab, M., Kratochvíl, J., Eichhorn, K.-J., Lopez Quintana, S., and Pastor, J.M. (2002) Structure of polypropylene/polyethylene blends assessed by polarised PA-FTIR spectroscopy, polarised FT Raman spectroscopy and confocal Raman microscopy. *Macromol. Symp.*, **184**, 107–122.

Schmohl, A., Miklós, A., and Hess, P. (2002) Detection of ammonia by photoacoustic spectroscopy with semiconductor lasers. *Appl. Opt.*, **41** (9), 1815–1823.

Schramm, D.U., Sthel, M.S., da Silva, M.G., Carneiro, L.O., Souza, A.P., and Vargas, H. (2003a) Diesel engines gas emissions monitored by photoacoustic spectroscopy. *Rev. Sci. Instrum.*, **74** (1), 513–515.

Schramm, D.U., Sthel, M.S., da Silva, M.G., Carneiro, L.O., Junior, A.J.S., Souza, A.P., and Vargas, H. (2003b) Application of laser photoacoustic spectroscopy for the analysis of gas samples emitted by diesel engines. *Infrared Phys. Technol.*, **44** (4), 263–269.

Schüle, G., Schmitz, B., and Steiner, R. (1999a) Spectral IR tissue diagnostics with photothermal detection. *Proc. SPIE Int. Soc. Opt. Eng.*, **3565**, 134–138.

Schüle, G., Schmitz, B., and Steiner, R. (1999b) Spectral photothermal tissue diagnostics with step-scan FT-IR technique. *AIP Conf. Proc.*, **463**, 615–617.

Scotoni, M., Rossi, A., Bassi, D., Buffa, R., Iannotta, S., and Boschetti, A. (2006) Simultaneous detection of ammonia, methane and ethylene at 1.63 µm with diode laser photoacoustic spectroscopy. *Appl. Phys. B*, **82** (3), 495–500.

Seaverson, L.M., McClelland, J.F., Burnet, G., Andregg, J.W., and Iles, M.K. (1985) Investigation of water and hydroxyl groups associated with coal fly ash by thermal desorption and Fourier transform infrared photoacoustic spectroscopies. *Appl. Spectrosc.*, **39** (1), 38–45.

Segura, Y., Chmielarz, L., Kustrowski, P., Cool, P., Dziembah, R., and Vansant, E.F. (2006) Preparation and characterization of vanadium oxide deposited on thermally stable mesoporous titania. *J. Phys. Chem. B*, **110** (2), 948–955.

Seiter, M. and Sigrist, M.W. (2000) Trace-gas sensor based on mid-IR difference-frequency generation in PPLN with saturated output power. *Infrared Phys. Technol.*, **41** (5), 259–269.

Shaw, M.D., Karunakaran, C., and Tabil, L.G. (2009) Physicochemical characteristics of densified untreated and steam exploded poplar wood and wheat straw grinds. *Biosyst. Eng.*, **103** (2), 198–207.

Shen, Y.C., MacKenzie, H.A., Lindberg, J., and Lu, Z.H. (1999) Time-resolved photoacoustics for glucose concentration measurement: theory and experiment. *Proc. SPIE Int. Soc. Opt. Eng.*, **3863**, 167–171.

Shoval, S., Yariv, S., and Michaelian, K.H. (1999) Hydroxyl stretching bands 'A' and 'Z' in Raman and IR spectra of kaolinites. *Clay Miner.*, **34** (4), 551–563.

Shoval, S., Yariv, S., Michaelian, K.H., Lapides, I., Boudeuille, M., and Panczer, G. (1999) A fifth OH-stretching band in IR spectra of kaolinites. *J. Colloid Interface Sci.*, **212** (2), 523–529.

Shoval, S., Yariv, S., Michaelian, K.H., Boudeulle, M., and Panczer, G. (2001) Hydroxyl-stretching bands in curve-fitted micro-Raman, photoacoustic and transmission infrared spectra of dickite from St. Claire, Pennsylvania. *Clays Clay Miner.*, **49** (4), 347–354.

Shoval, S., Yariv, S., Michaelian, K.H., Boudeulle, M., and Panczer, G. (2002a) Hydroxyl stretching bands in polarized micro-Raman spectra of oriented single crystal Keokuk kaolinite. *Clays Clay Miner.*, **50** (1), 56–62.

Shoval, S., Michaelian, K.H., Boudeulle, M., Panczer, G., Lapides, I., and Yariv, S. (2002b) Study of thermally treated dickite by curve-fitted transmission infrared, photoacoustic and micro-Raman spectra. *J. Therm. Anal. Calorim.*, **69** (1), 205–225.

Sieg, A., Guy, R.H., and Begoña Delgado-Charro, M. (2005) Noninvasive and minimally invasive methods for transdermal glucose monitoring. *Diabetes Technol. Ther.*, **7** (1), 174–197.

Siew, D.C.W., Heilmann, C., Easteal, A.J., and Cooney, R.P. (1999) Solution and film properties of sodium caseinate/glycerol and sodium caseinate/polyethylene glycol edible coating systems. *J. Agric. Food Chem.*, **47** (8), 3432–3440.

Sigrist, M.W., Bohren, A., Calasso, I.G., Naegele, M., and Romann, A. (2000a) Sensitive and selective monitoring of trace gases by laser photoacoustic spectroscopy. *Proc. SPIE Int. Soc. Opt. Eng.*, **3916**, 286–294.

Sigrist, M.W., Bohren, A., Calasso, I.G., Nägele, M., Romann, A., and Seiter, M. (2000b) Laser spectroscopic sensing of air pollutants. *Proc. SPIE Int. Soc. Opt. Eng.*, **4063**, 17–25.

Sikdar, D., Katti, D.R., and Katti, K.S. (2006) A molecular model for ε-caprolactam-based intercalated polymer clay nanocomposite: integrating modeling and experiments. *Langmuir*, **22** (18), 7738–7747.

Sikdar, D., Katti, K.S., and Katti, D.R. (2008) Molecular interactions alter clay and polymer structure in polymer clay nanocomposites. *J. Nanosci. Nanotechnol.*, **8** (4), 1638–1657.

Simms, J.R. and Yang, C.Q. (1994) Infrared spectroscopy studies of the petroleum pitch carbon fiber – II. The distribution of the oxidation products between the surface and the bulk. *Carbon*, **32** (4), 621–626.

Sivakesava, S. and Irudayaraj, J. (2000) Analysis of potato chips using FTIR photoacoustic spectroscopy. *J. Sci. Food Agric.*, **80** (12), 1805–1810.

Skřínský, J., Jacečková, R., Grigorová, E., Střižík, M., Kubát, P., Herecová, L., Nevrlý, V., Zelinger, Z., and Civiš, S. (2009) Allan variance for optimal signal averaging–monitoring by diode-laser and CO_2 laser photo-acoustic spectroscopy. *J. Mol. Spectrosc.*, **256** (1), 99–101.

Solomon, P.R. and Carangelo, R.M. (1982) FTIR analysis of coal. 1. Techniques and determination of hydroxyl concentrations. *Fuel*, **61** (7), 663–669.

Sowa, M.G. and Mantsch, H.H. (1993a) Phase modulated-phase resolved photoacoustic FT-IR study of calcified tissues. *Proc. SPIE Int. Soc. Opt. Eng.*, **2089**, 128–129.

Sowa, M.G. and Mantsch, H.H. (1993b) Photothermal infrared spectroscopy: applications to medicine. *J. Mol. Struct.*, **300** (C), 239–244.

Sowa, M.G. and Mantsch, H.H. (1994a) FT-IR step-scan photoacoustic phase analysis and depth profiling of calcified tissue. *Appl. Spectrosc.*, **48** (3), 316–319.

Sowa, M.G. and Mantsch, H.H. (1994b) FT-IR photoacoustic depth profiling spectroscopy of enamel. *Calcif. Tissue Int.*, **54** (6), 481–485.

Sowa, M.G., Wang, J., Schultz, C.P., Ahmed, M.K., and Mantsch, H.H. (1995) Infrared spectroscopic investigation of in vivo and ex vivo human nails. *Vib. Spectrosc.*, **10** (1), 49–56.

Spencer, N.D. (1986) Listening in on solid materials. *Chemtech*, **16** (6), 378–384.

Spencer, P., Byerley, T.J., Eick, J.D., and Witt, J.D. (1992) Chemical characterization of the dentin/adhesive interface by Fourier transform infrared photoacoustic spectroscopy. *Dent. Mater.*, **8** (1), 10–15.

Spencer, P., Trylovich, D.J., and Cobb, C.M. (1992) Chemical characterization of lased root surfaces using Fourier transform infrared photoacoustic spectroscopy. *J. Periodontol.*, **63** (7), 633–636.

Spencer, P., Cobb, C.M., McCollum, M.J., and Wieliczka, D.M. (1996) The effects of CO_2 laser and Nd:YAG with and without water/air surface cooling on tooth root structure: correlation between FTIR spectroscopy and histology. *J. Periodont. Res.*, **31** (7), 453–462.

Spencer, P., Payne, J.M., Cobb, C.M., Reinisch, L., Peavy, G.M., Drummer, D.D., Suchman, D.L., and Swafford, J.R. (1999) Effective laser ablation of bone based on the absorption characteristics of water and proteins. *J. Periodont.*, **70** (1), 68–74.

St.-Germain, F.G.T. and Gray, D.G. (1987) Photoacoustic Fourier transform infrared spectroscopy study of mechanical pulp brightening. *J. Wood Chem. Technol.*, **7** (1), 33–50.

Sudiyani, Y., Imamura, Y., Doi, S., and Yamauchi, S. (2003) Infrared spectroscopic investigations of weathering effects on the surface of tropical wood. *J. Wood Sci.*, **49** (1), 86–92.

Szakáll, M., Bozóki, Z., Mohácsi, A., Varga, A., and Szabo, G. (2004) Diode laser based photoacoustic water vapor detection system for atmospheric research. *Appl. Spectrosc.*, **58** (7), 792–798.

Teramae, N. and Tanaka, S. (1981) Structure elucidation of textile fabrics by Fourier transform infrared photoacoustic spectroscopy. *Spectrosc. Lett.*, **14** (10), 687–694.

Teramae, N. and Tanaka, S. (1984) Fourier transform infrared photoacoustic spectroscopy of film-like samples. *Bunseki Kagaku*, **33**, E397–E400.

Teramae, N. and Tanaka, S. (1985a) Subsurface layer detection by Fourier transform infrared photoacoustic spectroscopy. *Appl. Spectrosc.*, **39** (5), 797–799.

Teramae, N. and Tanaka, S. (1985b) Effect of heat from rear surface of a film sample on spectral features in Fourier transform infrared photoacoustic spectroscopy. *Anal. Chem.*, **57** (1), 95–99.

Teramae, N. and Tanaka, S. (1987) Fourier transform infrared photoacoustic spectroscopy of films, in *Fourier Transform Infrared Characterization of Polymers* (ed. H. Ishida), Plenum, New York, pp. 315–340.

Teramae, N., Hiroguchi, M., and Tanaka, S. (1982) Fourier transform infrared photoacoustic spectroscopy of polymers. *Bull. Chem. Soc. Jpn.*, **55** (7), 2097–2100.

Tranchart, S., Hadj Bachir, I., Huet, T.R., Olafsson, A., Destombes, J.-L., Naïm, S., and Fayt, A. (1999) High-resolution laser photoacoustic spectroscopy of OCS in the 12 000–13 000 cm^{-1} region. *J. Mol. Spectrosc.*, **196** (2), 265–273.

Urban, M.W. (1987) Photoacoustic Fourier transform infrared spectroscopy: a new method for characterization of coatings. *J. Coat. Technol.*, **59** (745), 29–34.

Urban, M.W. (1989a) Recent advances in coatings characterization by photoacoustic FT-IR spectroscopy. *Prog. Org. Coat.*, **16** (4), 321–353.

Urban, M.W. (1989b) Recent developments in the analysis of polymers and coatings by photoacoustic FT-IR spectroscopy. *Polym. Mater. Sci. Eng.*, **61**, 132–136.

Urban, M.W. (1999) Multi-dimensional spectroscopy of polymer films; surface and interfacial interactions. *Macromol. Symp.*, **141**, 15–31.

Urban, M.W. and Koenig, J.L. (1986) Depth-profiling studies of double-layer PVF$_2$-on-PET films by Fourier transform infrared photoacoustic spectroscopy. *Appl. Spectrosc.*, **40** (7), 994–998.

Utamapanya, S., Klabunde, K.J., and Schlup, J.R. (1991) Nanoscale metal oxide particles/clusters as chemical reagents. Synthesis and properties of ultrahigh surface area magnesium hydroxide and magnesium oxide. *Chem. Mater.*, **3** (1), 175–181.

van der Ven, L.G.J. and Leuverink, R. (2002) Durability prediction of p-urethane clearcoats using infrared P(hoto) A(coustic) S(pectroscopy). *Macromol. Symp.*, **187**, 845–852.

Van Der Voort, P., van Welzenis, R., de Ridder, M., Brongersma, H.H., Baltes, M., Mathieu, M., van de Ven, P.C., and Vansant, E.F. (2002) Controlled deposition of iron oxide on the surface of zirconia by the molecular designed dispersion of Fe(acac)$_3$: a spectroscopic study. *Langmuir*, **18** (11), 4420–4425.

van de Weert, M., van't Hof, R., van der Weerd, J., Heeren, R.M.A., Posthuma, G., Hennink, W.E., and Crommelin, D.J.A. (2000) Lysozome distribution and conformation in a biodegradable polymer matrix as determined by FTIR techniques. *J. Control. Release*, **68** (1), 31–40.

van de Weert, M., Haris, P.I., Hennink, W.E., and Crommelin, D.J.A. (2001) Fourier transform infrared spectrometric analysis of protein conformation: effect of sampling method and stress factors. *Anal. Biochem.*, **297**, 160–169.

van Herpen, M.M.J.W., Li, S., Bisson, S.E., te Lintel Hekkert, S., and Harren, F.J.M. (2002a) Tuning and stability of a continuous-wave mid-infrared high-power single resonant optical parametric oscillator. *Appl. Phys. B*, **75** (2–3), 329–333.

van Herpen, M.M.J.W., Li, S., Bisson, S.E., and Harren, F.J.M. (2002b) Photoacoustic trace gas detection of ethane using a continuously tunable, continuous-wave optical parametric oscillator based on periodically poled lithium niobate. *Appl. Phys. Lett.*, **81** (7), 1157–1159.

van Herpen, M., te Lintel Hekkert, S., Bisson, S.E., and Harren, F.J.M. (2002c) Wide single-mode tuning of a 3.0–3.8-μm, 700-mW, continuous-wave Nd:YAG-pumped optical parametric oscillator based on periodically poled lithium niobate. *Opt. Lett.*, **27** (8), 640–642.

van Herpen, M.M.J.W., Bisson, S.E., and Harren, F.J.M. (2003) Continuous-wave operation of a single-frequency optical parametric oscillator at 4–5 μm based on periodically poled LiNbO$_3$. *Opt. Lett.*, **28** (24), 2497–2499.

van Herpen, M.M.J.W., Ngai, A.K.Y., Bisson, S.E., Hackstein, J.H.P., Woltering, E.J., and Harren, F.J.M. (2006) Optical parametric oscillator-based photoacoustic detection of CO$_2$ at 4.23 μm allows real-time monitoring of the respiration of small insects. *Appl. Phys. B*, **82** (4), 665–669.

Varga, A., Bozóki, Z., Szakáll, M., and Szabó, G. (2006) Photoacoustic system for on-line process monitoring of hydrogen sulfide (H$_2$S) concentration in natural gas streams. *Appl. Phys. B*, **85** (2–3), 315–321.

Varlashkin, P.G. and Low, M.J.D. (1986) FT-IR photothermal beam deflection spectroscopy of solids submerged in liquids. *Appl. Spectrosc.*, **40** (8), 1170–1176.

Veres, A., Bozóki, Z., Mohácsi, A., Szakáll, M., and Szabo, G. (2003) External cavity diode laser based photoacoustic detection of CO$_2$ at 1.43 μm: the effect of molecular relaxation. *Appl. Spectrosc.*, **57** (8), 900–905.

Vidrine, D.W. (1980) Photoacoustic Fourier transform infrared spectroscopy of solid samples. *Appl. Spectrosc.*, **34** (3), 314–319.

Vidrine, W. (1981) Photoacoustic Fourier transform infrared spectroscopy of solids. *Proc. SPIE Int. Soc. Opt. Eng.*, **289**, 355–360.

Vidrine, D.W. (1984) Photoacoustic and reflection FT-IR spectrometry: photometric approximations for practical quantitative analysis. *Prepr. Pap. Am. Chem. Soc. Div. Fuel Chem.*, **25**, 147–148.

Vidrine, D.W. and Lowry, S.R. (1983) Photoacoustic Fourier transform IR spectroscopy and its application to polymer analysis. *Adv. Chem. Ser. (Polym. Charact.)*, **203**, 595–613.

von Lilienfeld-Toal, H., Weidenmüller, M., Xhelaj, A., and Mäntele, W. (2005) A novel approach to non-invasive glucose measurement by mid-infrared spectroscopy: the combination of quantum cascade lasers (QCL) and photoacoustic detection. *Vib. Spectrosc.*, **38** (1–2), 209–215.

Wahls, M.W.C., Kenttä, E., and Leyte, J.C. (2000) Depth profiles in coated paper: experimental and simulated FT-IR photoacoustic difference magnitude spectra. *Appl. Spectrosc.*, **54** (2), 214–220.

Wang, H.P., Eyring, E.M., and Huai, H. (1991) Photoacoustic enhancement of surface IR modes in zeolite channels. *Appl. Spectrosc.*, **45** (5), 883–885.

Wang, J., Sowa, M., Mantsch, H.H., Bittner, A., and Heise, H.M. (1996) Comparison of different infrared measurement techniques in the clinical analysis of biofluids. *Trends Analyt. Chem.*, **15** (7), 286–296.

Wang, N. and Low, M.J.D. (1989) Spectroscopic studies of carbons. XV. The pyrolysis of a lignin. DOE/PC/79920-10, *DE*90 006239, pp. 1–14.

Wang, N. and Low, M.J.D. (1990a) Spectroscopic studies of carbons. XVIII. The charring of rice hulls. *Mater. Chem. Phys.*, **26** (2), 117–130.

Wang, N. and Low, M.J.D. (1990b) Spectroscopic studies of carbons. XIX. The charring of sucrose. *Mater. Chem. Phys.*, **26** (5), 465–481.

Wang, N. and Low, M.J.D. (1991) Spectroscopic studies of carbon. XXI. An infrared study of the charring of coconut shell. *Mater. Chem. Phys.*, **27** (4), 359–374.

Wentrup-Byrne, E., Rintoul, L., Smith, J.L., and Fredericks, P.M. (1995) Comparison of vibrational spectroscopic techniques for the characterization of human gallstones. *Appl. Spectrosc.*, **49** (7), 1028–1036.

Wentrup-Byrne, E., Rintoul, L., Gentner, J.M., Smith, J.L., and Fredericks, P.M. (1997) Photoacoustic spectroscopy vs. chemical characterization of human gallstones. *Mikrochim. Acta*, **14** (Suppl.), 615–616.

Wetzel, D.L. and Carter, R.O. (1998) Synchrotron powered FT-IR microspectroscopic incremental probing of photochemically degraded polymer films. *AIP Conf. Proc.*, **430**, 567–570.

White, R.L. (1985) Analysis of thin-layer chromatographic adsorbates by Fourier transform infrared photoacoustic spectroscopy. *Anal. Chem.*, **57** (9), 1819–1822.

Wolff, M. and Harde, H. (2000) Photoacoustic spectrometer based on a DFB-diode laser. *Infrared Phys. Technol.*, **41** (5), 283–286.

Workman, J.J. (2001) Infrared and Raman spectroscopy in paper and pulp analysis. *Appl. Spectrosc. Rev.*, **36** (2–3), 139–168.

Xu, Z.H., Butler, I.S., and St.-Germain, F.G.T. (1986) FT-IR photoacoustic spectra of gaseous group VIB metal chalcocarbonyl complexes. *Appl. Spectrosc.*, **40** (7), 1004–1009.

Xu, Z., Wang, Z., Kung, J., Woods, J.R., Wu, X.A., Kotlyar, L.S., Sparks, B.D., and Chung, K.H. (2005) Separation and characterization of foulant material in coker gas oils from Athabasca bitumen. *Fuel*, **84** (6), 661–668.

Yamada, O., Yasuda, H., Soneda, Y., Kobayashi, M., Makino, M., and Kaiho, M. (1996) The use of step-scan FT-IR/PAS for the study of structural changes in coal and char particles during gasification. *Prepr. Am. Chem. Soc. Div. Fuel Chem.*, **41** (1), 93–97.

Yamauchi, S., Sudiyani, Y., Imamura, Y., and Doi, S. (2004) Depth profiling of weathered tropical wood using Fourier transform infrared photoacoustic spectroscopy. *J. Wood Sci.*, **50** (5), 433–438.

Yang, C.Q. (1990) Evaluation of photoacoustic and diffuse reflectance infrared spectroscopy for the near-surface analysis of polymeric fibers and films. *Polym. Mater. Sci. Eng.*, **62**, 903–910.

Yang, C.Q. (1991a) Characterizing ester crosslinkages in cotton cellulose with FT-IR photoacoustic spectroscopy. *Textile Res. J.*, **61** (5), 298–305.

Yang, C.Q. (1991b) Mechanism of esterification between polycarboxylic acid and cotton cellulose studied by FT-IR spectroscopy. *Polym. Mater. Sci. Eng.*, **64**, 372–374.

Yang, C.Q. (1991c) Comparison of photoacoustic and diffuse reflectance infrared spectroscopy as near-surface analysis techniques. *Appl. Spectrosc.*, **45** (1), 102–108.

Yang, C.Q. (1992) Infrared spectroscopic analysis of textile materials degradation using photoacoustic detection. *Ind. Eng. Chem. Res.*, **31** (2), 617–621.

Yang, C.Q. and Bresee, R.R. (1987) Studies of sized cotton yarns by FT-IR photoacoustic spectroscopy. *J. Coated Fabrics*, **17** (2), 110–128.

Yang, C.Q. and Fateley, W.G. (1987) Fourier-transform infrared photoacoustic spectroscopy evaluated for near-surface characterization of polymeric materials. *Anal. Chim. Acta*, **194** (C), 303–309.

Yang, C.Q. and Freeman, J.M. (1991a) Photo-oxidation of cotton cellulose studied by FT-IR photoacoustic spectroscopy. *Appl. Spectrosc.*, **45** (10), 1695–1698.

Yang, C.Q. and Freeman, J.M. (1991b) Studies of thermal oxidation and degradation of cotton cellulose using FT-IR photoacoustic spectroscopy. *Polym. Mater. Sci. Eng.*, **64**, 33–35.

Yang, C.Q. and Kottes Andrews, B.A. (1991) Infrared spectroscopic studies of the nonformaldehyde durable press finishing of cotton fabrics by use of polycarboxylic acids. *J. Appl. Polym. Sci.*, **43** (9), 1609–1616.

Yang, C.Q. and Martin, L.K. (1994) Photo- and thermal-oxidation of the nonwoven polypropylene fabric studied by FT-IR photoacoustic spectroscopy. *J. Appl. Polym. Sci.*, **51** (3), 389–397.

Yang, C.Q. and Simms, J.R. (1993) Infrared spectroscopy studies of the petroleum pitch carbon fiber – I. The raw materials, the stabilization, and carbonization processes. *Carbon*, **31** (3), 451–459.

Yang, C.Q. and Simms, J.R. (1995) Comparison of photoacoustic, diffuse reflectance and transmission infrared spectroscopy for the study of carbon fibres. *Fuel*, **74** (4), 543–548.

Yang, C.Q., Ellis, T.J., Bresee, R.R., and Fateley, W.G. (1985) Depth profiling of FT-IR photoacoustic spectroscopy and its application for polymeric material studies. *Polym. Mater. Sci. Eng.*, **53**, 169–175.

Yang, C.Q., Bresee, R.R., and Fateley, W.G. (1987) Near-surface analysis and depth profiling by FT-IR photoacoustic spectroscopy. *Appl. Spectrosc.*, **41** (5), 889–896.

Yang, C.Q., Perenich, T.A., and Fateley, W.G. (1989) Studies of foam finished cotton fabrics using FT-IR photoacoustic spectroscopy. *Textile Res. J.*, **59** (10), 562–568.

Yang, C.Q., Bresee, R.R., and Fateley, W.G. (1990) Studies of chemically modified poly(ethylene terephthalate) fibers by FT-IR photoacoustic spectroscopy and X-ray photoelectron spectroscopy. *Appl. Spectrosc.*, **44** (6), 1035–1039.

Yang, D.-Q. and Sacher, E. (2001) Argon ion treatment of the Dow Cyclotene 3022 surface and its effect on the adhesion of evaporated copper. *Appl. Surf. Sci.*, **173** (1–2), 30–39.

Yang, D.-Q., Martinu, L., Sacher, E., and Sadough-Vanini, A. (2001) Nitrogen plasma treatment of the dow Cyclotene 3022 surface and its reaction with evaporated copper. *Appl. Surf. Sci.*, **177** (1–2), 85–95.

Yang, H. and Irudayaraj, J. (1999) Estimation of particle size by DRIFTS and FTIR-PAS. *Part. Sci. Technol.*, **17** (4), 269–282.

Yang, H. and Irudayaraj, J. (2000a) Characterization of semisolid fats and edible oils by Fourier transform infrared photoacoustic spectroscopy. *J. Am. Oil Chem. Soc.*, **77** (3), 291–295.

Yang, H. and Irudayaraj, J. (2000b) Depth profiling Fourier transform analysis of cheese package using generalized two-dimensional photoacoustic correlation spectroscopy. *Trans. Am. Soc. Agric. Eng.*, **43** (4), 953–961.

Yang, H. and Irudayaraj, J. (2001a) Comparison of near infrared, Fourier transform infrared, and Fourier transform-Raman methods for determining olive pomace oil adulteration in extra virgin olive oil. *J. Am. Oil Chem. Soc.*, **78** (9), 889–895.

Yang, H. and Irudayaraj, J. (2001b) Characterization of beef and pork using Fourier transform infrared photoacoustic spectroscopy. *Lebensm. Wiss. Technol.*, **34** (6), 402–409.

Yang, H., Irudayaraj, J., and Sakhamuri, S. (2001) Characterization of edible coatings and microorganisms on food surfaces using Fourier transform infrared photoacoustic spectroscopy. *Appl. Spectrosc.*, **55** (5), 571–583.

Yang, X. and Noda, C. (1999) Overtone transitions of $^{12}CO_2$ beyond the Venus bands. *J. Mol. Spectrosc.*, **195** (2), 256–262.

Yaniger, S.I., Riseman, S.M., Frigo, T., and Eyring, E.M. (1982) Infrared photoacoustic spectroscopy of conducting polymers. II. P-doped polyacetylene. *J. Chem. Phys.*, **76** (8), 4298–4299.

Yaniger, S.I., Rose, D.J., McKenna, W.P., and Eyring, E.M. (1984a) Photoacoustic infrared spectroscopy of doped and undoped poly(p-phenylene). *Macromolecules*, **17** (12), 2579–2583.

Yaniger, S.I., Rose, D.J., McKenna, W.P., and Eyring, E.M. (1984b) Infrared photoacoustic spectroscopy of conducting polymers III. Polyparaphenylene and its derivatives. *Appl. Spectrosc.*, **38** (1), 7–11.

Yariv, S., Nasser, A., Michaelian, K.H., Lapides, I., Deutsch, Y., and Lahav, N. (1994) Thermal treatment of the kaolinite/$CsCl/H_2O$ intercalation complex. *Thermochim. Acta*, **234** (C), 275–285.

Yoshino, K., Fukuyama, A., Yokoyama, H., Meada, K., Fons, P.J., Yamada, A., Niki, S., and Ikari, T. (1999) Piezoelectric photoacoustic spectra of $CuInSe_2$ thin film grown by molecular beam epitaxy. *Thin Solid Films*, **343–344**, 591–593.

Zachmann, G. (1984) FT-IR spectroscopy of solid surfaces. *J. Mol. Struct.*, **115** (C), 465–468.

Zegadi, A., Slifkin, M.A., and Tomlinson, R.D. (1994) A photoacoustic spectrometer for measuring subgap absorption spectra of semiconductors. *Rev. Sci. Instrum.*, **65** (7), 2238–2243.

Zegadi, A., Al-Saffar, I.S., Yakushev, M.V., and Tomlinson, R.D. (1995) Photoacoustic spectroscopy use in the analysis of ion-implanted $CuInSe_2$ single crystals. *Rev. Sci. Instrum.*, **66** (8), 4095–4101.

Zegadi, A., Yakushev, M.V., Ahmed, E., Pilkington, R.D., Hill, A.E., and Tomlinson, R.D. (1996) A photoacoustic study on the effect of Se content on defect levels in $CuInSe_2$ single crystals. *Sol. Energy Mater. Sol. Cells*, **41–42**, 295–305.

Zelinger, Z., Střižik, M., Kubát, P., and Civiš, S. (2000) Quantitative analysis of trace mixtures of toluene and xylenes by CO_2 laser photoacoustic spectrometry. *Anal. Chim. Acta*, **422** (2), 179–185.

Zelinger, Z., Střižik, M., Kubát, P., Jaňour, Z., Berger, P., černý, A., and Engst, P. (2004) Laser remote sensing and photoacoustic spectrometry applied in air pollution investigation. *Opt. Lasers Eng.*, **42** (4), 403–412.

Zelinger, Z., Střižík, M., Kubát, P., Civiš, S., Grigorová, E., Janečková, R., Zavila, O., Nevrlý, V., Herecová, L., Bailleux, S., Horka, V., Ferus, M., Skřínský, J., Kozubková, M., Drábková, S., and Jaňour, Z. (2009) Dispersion of light and heavy pollutants in urban scale models: CO_2 laser photoacoustic studies. *Appl. Spectrosc.*, **63** (4), 430–436.

Zeninari, V., Tikhomirov, B.A., Ponomarev, Y.N., and Courtois, D. (1998) Preliminary results on photoacoustic study of the relaxation of vibrationally excited ozone (v_3). *J. Quant. Spectrosc. Radiat. Transf.*, **59** (3–5), 369–375.

Zeninari, V., Kapitanov, V.A., Coutois, D., and Ponomarev, Y.N. (1999) Design and characteristics of a differential Helmholtz resonant photoacoustic cell for infrared gas detection. *Infrared Phys. Technol.*, **40** (1), 1–23.

Zeninari, V., Tikhomirov, B.A., Ponomarev, Y.N., and Courtois, D. (2000) Photoacoustic measurements of the vibrational relaxation of the selectively excited ozone (v_3) molecule in pure ozone and its binary mixtures with O_2, N_2, and noble gases. *J. Chem. Phys.*, **112** (4), 1835–1843.

Zerlia, T. (1985) Fourier transform infrared photoacoustic spectroscopy of raw coal. *Fuel*, **64** (9), 1310–1312.

Zerlia, T. (1986) Depth profile study of large-sized coal samples by Fourier transform infrared photoacoustic spectroscopy. *Appl. Spectrosc.*, **40** (2), 214–217.

Zerlia, T., Marini, A., Berbenni, V., Massarotti, V., Giordano, F., La Manna, A., Bettinetti, G.P., and Margheritis, C. (1989) Solid state interaction study on the system polyvinylpyrrolidone-XL/trimethoprim. *Solid State Ionics*, **32–33** (1), 613–624.

Zhang, P. and Urban, M.W. (2004) Photoacoustic FT-IR depth imaging of polymeric surfaces: overcoming IR diffraction limits. *Langmuir*, **20** (24), 10691–10699.

Zhang, W., Wu, Z., and Yu, Q. (2007) Photoacoustic spectroscopy for fast and sensitive ammonia detection. *Chin. Opt. Lett.*, **5** (11), 677–679.

Zhao, Y. and Urban, M.W. (2000) Phase separation and surfactant stratification in styrene/n-butyl acrylate copolymer and latex blend films. 17. A spectroscopic study. *Macromolecules*, **33** (6), 2184–2191.

Appendix 1: Glossary

Amplitude modulation	Modulation of radiation from an interferometer, laser, etc. by a chopper, shutter, or other device capable of completely interrupting the beam
Depth profiling	Characterization of a (solid) sample at selected distances from its surface
Optical absorption length	Distance over which incident intensity is attenuated to $1/e$ of its original value; reciprocal of optical absorption coefficient β. Symbol: μ_β or l_β
Optically opaque	Most of the incident light is absorbed within a distance that is small compared with the physical length (l) of the sample ($\mu_\beta < l$)
Optically transparent	Light is absorbed throughout the length of the sample; some light is transmitted through the sample ($\mu_\beta > l$)
Optoacoustic	An alternative term for photoacoustic, implying the conversion of (absorbed) light into sound. Acronym: *OA*
Phase modulation	Modulation of radiation, typically by slightly varying beam position; in FT-IR spectroscopy, an interferometer mirror is oscillated about its equilibrium position in step-scan mode
Photoacoustic	Effect in which acoustic (pressure) waves are generated by the periodic absorption of radiation. Acronym: *PA*
Photopyroelectric	Production of a temperature-dependent voltage is caused by the absorption of modulated radiation. Acronym: *PPE*
Photothermal	Effect in which thermal waves are generated by the absorption of incident radiation. Acronym: *PT*
Photothermal radiometry	Infrared detection of the temperature perturbation in a sample caused by the absorption of light. Acronym: *PTR*
Piezoelectric	An effect in which a strain is converted to a voltage
Rapid-scan	An FT-IR scan mode in which the motion of the moving mirror is rapid and continuous; named to contrast with step-scan
Saturation	Observed spectrum is independent of the optical absorption coefficient β

Step-scan	An FT-IR scan mode in which the position of the moving mirror is varied in a stepwise fashion
Thermal conductivity	Proportionality constant between the rate of heat flow and a temperature gradient. Symbol: κ
Thermal diffusion length	Active thermal length; the distance over which the amplitude of a thermal wave decays to $1/e$ of its original magnitude. Symbol: μ
Thermal diffusivity	Ratio of thermal conductivity (κ) and thermal capacitance (ρC, where ρ is density and C is heat capacity). Symbol: α or D.
Thermal effusivity	Measure of thermal impedance or the ability of a sample to exchange heat with the environment; mathematically equal to $\sqrt{(\rho C \kappa)}$. Symbol: e
Thermally thick	Heat generated only in a region near the surface of the sample can escape; the physical length of the sample is greater than the thermal diffusion length [$\mu_s < l$, where $\mu_s = (\alpha/\pi f)^{1/2}$].
Thermally thin	Heat is conducted throughout the length of the sample; the thermal diffusion length is greater than the physical length of the sample ($\mu_s > l$).

Appendix 2: Literature Guide – Solids and Liquids

The following table lists 884 references on PA infrared spectra of solids and liquids that were published in the period beginning in 1978 and ending in 2009. Bibliographic data consists of authors' names and publication details (year of publication; name of journal, conference proceedings, or book; volume number; pagination). A majority of these publications are discussed within the main part of this book, where the titles of the articles are given in the reference lists.

Presentation or discussion of experimental results for a particular sample type is indicated by an 'x' in the appropriate column. The correspondence between most of the columns and the sections in Chapter 7 is straightforward. Articles discussing silica are cited under the catalysts and/or clays and minerals classifications according to their emphasis. References on carbonyl compounds are entered in the inorganics category, although some samples contained both organic and inorganic ligands.

Articles describing PA infrared spectra of gases are listed in Appendix 3. As mentioned in Chapter 7, a subset (mostly consisting of work published between 1994 and 2009) of the extensive literature on this subject is discussed in this book. The entire literature on PA spectra of gases is beyond the scope of the present publication; the Wiley-Interscience Chemical Analysis volume edited by M. W. Sigrist in 1994 or the proceedings of the fifteen International Conferences on Photoacoustic and Photothermal Phenomena should be consulted for more information on this subject. Similarly, publications discussing only theoretical aspects of PA infrared spectroscopy and those that describe instrumentation but do not present spectra of condensed-phase samples are not included. A few reviews and articles in less-accessible journals have also been omitted. Hence this listing is extensive but not comprehensive.

Photoacoustic IR Spectroscopy. 2nd Ed., Kirk H. Michaelian
Copyright © 2010 WILEY-VCH Verlag GmbH & Co. KGaA, Weinheim
ISBN: 978-3-527-40900-6

Citation	Carbons
Abu-Zeid, M. E., Nofal, E. E., Marafi, M. A., Tahseen, L. A., Abdul-Rasoul, F. A., Ledwith, A. (1984) *J. Appl. Polym. Sci.*, **29**, 2431–2442	
Abu-Zeid, M.E.; Tahseen, L. A.; Anani, A. A. (1985a) *Colloids Surf.* **16**, 301–307	
Abu-Zeid, M. E., Nofal, E. E., Tahseen, L. A., Abdul-Rasoul, F. A. (1985b) *J. Appl. Polym. Sci.*, **30**, 3791–3800	
Adams, M. J., Beadle, B. C., Kirkbright, G. F. (1978) *Anal. Chem.*, **50**, 1371–1374	
Adams, M. J. (1982) *Prog. Analyt. Atom. Spectrosc.*, **5**, 153–204	
Adamsons, K. (2000) *Prog. Polym. Sci.*, **25**, 1363–1409	
Ahmed, E., Zegadi, A., Hill, A. E., Pilkington, R. D., Tomlinson, R. D. (1995) *Thin Solid Films*, **268**, 144–151	
Ahmed, E., Zegadi, A., Hill, A. E., Pilkington, R. D., Tomlinson, R. D., Ahmed, W. (1998) *J. Mater. Process. Technol.*, **77**, 260–265	
Ahmed, E., Pilkington, R. D., Hill, A. E., Ali, N., Ahmed, W., Hassan, I. U. (2006) *Thin Solid Films*, **515**, 239–244	
Aho, M., Kortelainen, P., Rantanen, J., Linna, V. (1988) *J. Anal. Appl. Pyrolysis*, **15**, 297–306	
Airoldi, C., Alcântara, E. F. C., Nakamura, O., daPaixão, F. J., Vargas, H. (1993) *J. Mater. Chem.*, **3**, 479–482	
Allen, D. T., Palen, E. (1989) *J. Aerosol Sci.*, **20**, 441–455	x
Allen, G. M., Wu, G. L., Prentice, S. A. (1992) *J. Appl. Polym. Sci.*, **44**, 213–224	
Allen, R. W., Wheelock, T. D. (1996) *Fuel Sci. Technol. Int.*, **14**, 577–578	
d'Almeida, J. R. M., Cella, N., Monteiro, S. N., Miranda, L. C. M. (1998) *J. Appl. Polym. Sci.*, **69**, 1335–1341	
Almeida, E., Balmayore, M., Santos, T. (2002) *Prog. Org. Coat.*, **44**, 233–242	
Almeida, S. H., Kawano, Y. (1998) *Polym. Degrad. Stab.*, **62**, 291–297	
Amer, N. M. (1984) *U. S. Pat. Appl.*545338	
Anderson, M. S. (2000) *Appl. Spectrosc.*, **54**, 349–352	
Ando, T., Inoue, S., Ishii, M., Kamo, M., Sato, Y., Yamada, O., Nakano, T. (1993) *J. Chem. Soc. Faraday Trans.*, **89**, 749–751	x
Angle, C. W., Donini, J. C., Hamza, H. A. (1988) *Colloids Surfaces*, **30**, 373–385	
Annyas, J., Bićanić, D., Schouten, F. (1999) *Appl. Spectrosc.*, **53**, 339–343	
Antonialli Junior, W. F., Lima, S. M., Andrade, L. H. C., Súarez, Y. R. (2007) *Genet. Mol. Res.*, **6**, 492–499	
Antonialli, W. F., Súarez, Y. R., Isida, T., Andrade, L. H. C., Lima, S. M. (2008) *Genet. Mol. Res.*, **7**, 559–566	
Arakawa, E. T., Lavrik, N. V., Datskos, P. G. (2003) *Appl. Opt.*, **42**, 1757–1762	
Ardeleanu, M., Morisset, R., Bertrand, L. (1992) In: *Photoacoustic and Photothermal Phenomena III*, 81–84	
Arévalo, F., Saavedra, R., Paulraj, M. (2008) *J. Phys.: Conf. Ser.*, **134**, 012019	
Armenta, S., Moros, J., Garrigues, S., de laGuardia, M. (2006) *Anal. Chim. Acta*, **567**, 255–261	

Catalysts	Clays, minerals	Food products	Hydrocarbons, fuels	Inorganics	Medical, biological	Organics	Polymers	Textiles	Wood, paper
							x		
	x					x			
							x		
	x		x	x					
		x	x	x	x				
							x		
				x					
				x					
				x					
			x						
				x		x			
							x		
			x						
							x		
							x		
							x		
			x						
							x		
			x						
					x				
					x				
					x				
					x				
					x				
				x					
		x							

Citation	Carbons

Ashizawa, K. (1989) *J. Pharm. Sci.*, **78**, 256–260

Ashkenazy, R., Gottlieb, L., Yannai, S. (1997) *Biotechnol. Bioeng.*, **55**, 1–10

Baby, A. R., Lacerda, A. C. L., Kawano, Y., Velasco, M.V.R., Kaneko, T.M. (2007) *Pharmazie*, **62**, 727–731

Bain, C. D., Davies, P.B., Ong, T. H. (1992) In: *Photoacoustic and Photothermal Phenomena III*, 158–160

Bajic, S. J., Luo, S., Jones, R. W., McClelland, J. F. (1995) *Appl. Spectrosc.* **49**, 1000–1005

Bajic, S. J., McClelland, J. F., Jones, R. W. (1997) *Mikrochim. Acta [Suppl.]*, **14**, 611–612

Bajic, S. J., Jones, R. W., McClelland, J. F., Hames, B. R., Meglen, R. R. (1998) *AIP Conf. Proc.*, **430**, 466–469

Bajic, S. J., Jones, R. W., McClelland, J. F. (2005) *Appl. Spectrosc.*, **59**, 1420–1426

Bandyopadhyay, S., Massoth, F. E., Pons, S., Eyring, E. M. (1985) *J. Phys. Chem.*, **89**, 2560–2564

Barsan, M. M., Butler, I. S., Gilson, D. F. R., Moyer, R. O., Zhou, W., Wu, H., Udovic, T. J. (2008) *J. Phys. Chem. A*, **112**, 6936–6938

Barton, T. F., Price, T., Becker, K., Dillard, J. G. (1991) *Colloids Surf.*, **53**, 209–222

Bauer, D. R., Paputa Peck, M. C., Carter, R. O. (1987) *J. Coatings Technol.*, **59**, 103–109

Belton, P. S., Tanner, S. F. (1983) *Analyst*, **108**, 591–596

Belton, P. S. (1984) In: *Biophysical Methods in Food Research*, 123–135

Belton, P. S., Saffa, A. M., Wilson, R. H. (1987a) *Analyst*, **112**, 1117–1120

Belton, P. S., Wilson, R. H., Saffa, A. M. (1987b) *Anal. Chem.*, **59**, 2378–2382

Belton, P. S., Saffa, A. M., Wilson, R. H., Ince, A. D. (1988a) *Food Chem.*, **28**, 53–61

Belton, P. S., Saffa, A. M., Wilson, R. H. (1988b) *Proc. SPIE Int. Soc. Opt. Eng.*, **917**, 72–77   x

Belton, P. S., Saffa, A. M., Wilson, R. H. (1988c) In: *Analytical Applications of Spectroscopy*, 245–250

Bensaad, S., Jasse, B., Noel, C. (1999) *Polymer*, **40**, 7295–7301

Bensebaa, F., Kotlyar, L., Pleizier, G., Sparks, B., Deslandes, Y., Chung, K. (2000) *Surf. Interface Anal.*, **30**, 207–211

Bensebaa, F., Majid, A., Deslandes, Y. (2001) *Spectrochim. Acta A*, **57**, 2695–2702

Benziger, J. B., McGovern, S. J., Royce, B. S. H. (1985) In: *Catalyst Characterization Science*, 449–463

Berbenni, V., Marini, A., Bruni, G., Bini, M., Magnone-Grato, A., Villa, M. (1996) *Appl. Spectrosc.*, **50**, 871–879

Berghmans, P. A., Muir, I. J., Adams, F. C. (1990) *Surf. Interface Anal.*, **16**, 575–579

Bertrand, L., Monchalin, J.-P., Leputre, F. (1982) *Appl. Opt.*, **21**, 248–252

Bertrand, L. (1988) *Appl. Spectrosc.*, **42**, 134–138

Bhardwaj, N. K., Dang, V. Q., Nguyen, K. L. (2006) *Anal. Chem.*, **78**, 6818–6825

Bhardwaj, N. K., Nguyen, K. L. (2007) *Coll. Surf. A: Physicochem. Eng. Aspects*, **301**, 323–328

Bhattacharjee, D., Engineer, R. (1996) *J. Cell Plast.*, **32**, 260–273

Appendix 2: Literature Guide – Solids and Liquids | 299

Catalysts	Clays, minerals	Food products	Hydrocarbons, fuels	Inorganics	Medical, biological	Organics	Polymers	Textiles	Wood, paper
					x				
					x				
					x				
						x			
				x					
				x					
									x
						x			
x									
				x					
	x					x			
							x		
		x							
		x							
		x							
		x							
		x							
		x							
					x				
							x		
			x						
	x		x						
x	x								
					x				
	x								
	x								
	x						x		
									x
									x
							x		

Citation	Carbons

Bhowmik, R., Katti, K. S., Verma, D., Katti, D. R. (2007) *Mater. Sci. Eng. C*, **27**, 352–371

Bićanić, D., Krüger, S., Torfs, P., Bein, B., Harren, F. (1989) *Appl. Spectrosc.*, **43**, 148–153

Bićanić, D., Zuidberg, B., Jalink, H., Miklós, A., Hartmans, K., Van Es, A. (1990) *Appl. Spectrosc.*, **44**, 263–265

Bićanić, D., Jalink, H., Chirtoc, M., Sauren, H., Lubbers, M., Quist, J., Gerkema, E., vanAsselt, K., Miklós, A., Sólyom, A., Angeli, Gy.Z., Helander, P., Vargas, H. (1992) In: *Photoacoustic and Photothermal Phenomena III*, 20–27

Bićanić, D., Chirtoc, M., Chirtoc, I., Favier, J. P., Helander, P. (1995) *Appl. Spectrosc.*, **49**, 1485–1489

Bićanić, D., Fink, T., Franko, M., Močnik, G., van de Bovenkamp, P., van Veldhuizen, B., Gerkema, E. (1999) *AIP Conf. Proc.*, **463**, 637–639

Biresaw, G., Mohamed, A., Gordon, S. H., Harry-O'kuru, R. E., Carriere, C. C. (2008) *J. Appl. Polym. Sci.*, **110**, 2932–2941

Bjarnestad, S., Dahlman, O. (2002) *Anal. Chem.*, **74**, 5851–5858

Blank, R. E., Wakefield, T. (1979) *Anal. Chem.*, **51**, 50–54

Boccaccio, T., Bottino, A., Capannelli, G., Piaggio, P. (2002) *J. Membrane Sci.*, **210**, 315–329

Bordeleau, A., Rousset, G., Bertrand, L., Crine, J. P. (1986) *Can. J. Phys.*, **64**, 1093–1097

Bordeleau, A., Bertrand, L., Sacher, E. (1987) *Spectrochim. Acta, Part A*, **43**, 1189–1190

Bouzerar, R., Amory, C., Zeinert, A., Benlahsen, M., Racine, B., Durand-Drouhin, O., Clin, M. (2001) *J. Non-Cryst. Solids*, **281**, 171–180   x

Bowen, J. M., Compton, S. V., Blanche, M. S. (1989) *Anal. Chem.*, **61**, 2047–2050

Bozec, L., Hammiche, A., Pollock, H. M., Conroy, M., Chalmers, J. M., Everall, N. J.; Turin, L. (2001) *J. Appl. Phys.*, **90**, 5159–5165

Brienne, S. H. R., Bozkurt, M. M. V., Rao, S. R., Xu, Z., Butler, I. S., Finch, J. A. (1996) *Appl. Spectrosc.*, **50**, 521–527

Britcher, L., Barnes, T. J., Griesser, H. J., Prestidge, C. A. (2008) *Langmuir*, **24**, 7625–7627

Budevska, B. O., Manning, C. J. (1996) *Appl. Spectrosc.*, **50**, 939–947

Burggraf, L. W., Leyden, D. E. (1981) *Anal. Chem.*, **53**, 759–764

Burggraf, L. W., Leyden, D. E., Chin, R. L., Hercules, D. M. (1982) *J. Catal.*, **78**, 360–379

Burt, J. A., Michaelian, K. H., Zhang, S. L. (1997) *Mikrochim. Acta [Suppl.]*, **14**, 173–174

Butler, I. S., Xu, Z. H., Werbowyj, R. S., St.-Germain, F. (1987a) *Appl. Spectrosc.*, **41**, 149–153

Butler, I. S., Xu, Z. H., Darensbourg, D. J., Pala, M. (1987b) *J. Raman Spectrosc.*, **18**, 357–363

Butler, I. S., Li, H., Gao, J. P. (1991) *Appl. Spectrosc.*, **45**, 223–226

Butler, I. S., Gilson, D. F. R., Lafleur, D. (1992) *Appl. Spectrosc.*, **46**, 1605–1607

Caires, A. R. L., Teixeira, M. R. O., Súarez, Y. R., Andrade, L. H. C., Lima, S. M. (2008) *Appl. Spectrosc.*, **62**, 1044–1047

Calheiros, F. C., Braga, R. R., Kawano, Y., Ballester, R. Y. (2004) *Dent. Mater.*, **20**, 939–946

Cappitelli, F., Vicini, S., Piaggio, P., Abbruscato, P., Princi, E., Casadevall, A., Nosanchuk, J.D., Zanardini, E. (2005) *Macromol. Biosci.*, **5**, 49–57

Cardamone, J. M., Gould, J. M., Gordon, S. H. (1987) *Textile Res. J.*, **57**, 235–239

Appendix 2: Literature Guide – Solids and Liquids

Catalysts	Clays, minerals	Food products	Hydrocarbons, fuels	Inorganics	Medical, biological	Organics	Polymers	Textiles	Wood, paper
				x					
						x			
		x							
		x			x				
						x			
		x							
							x		
									x
				x					
							x		
							x		
				x					
	x					x			
						x	x		
	x								
	x								
							x		
x									
	x								
				x					
				x					
				x					
				x					
		x							
					x				
					x			x	
									x

Citation	Carbons
Carter, E. A., Fredericks, P. M., Church, J. S. (1996) *Textile Res. J.*, **66**, 787–794	
Carter, R. O., Bauer, D. R. (1987) *Polym. Mater. Sci. Eng.*, **57**, 875–879	
Carter, R. O., Paputa Peck, M. C. (1989a) *Appl. Spectrosc.*, **43**, 468–473	x
Carter, R. O., Paputa Peck, M. C., Bauer, D. R. (1989b) *Polym. Degrad. Stability*, **23**, 121–134	
Carter, R. O., Paputa Peck, M. C., Samus, M. A., Killgoar, P. C. (1989c) *Appl. Spectrosc.*, **43**, 1350–1354	x
Carter, R. O., Wright, S. L. (1991) *Appl. Spectrosc.*, **45**, 1101–1103	x
Carter, R. O. (1992) *Appl. Spectrosc.*, **46**, 219–224	
Carter, R. O., McCallum, J. B. (1994) *Polym. Degrad. Stab.*, **45**, 1–10	
Carvalho de Araújo, S., Kawano, Y. (2002) *J. Appl. Polym. Sci.*, **85**, 199–208	
Castleden, S. L., Kirkbright, G. F., Menon, K. R. (1980) *Analyst*, **105**, 1076–1081	
Castleden, S. L., Kirkbright, G. F., Long, S. E. (1982) *Can. J. Spectrosc.*, **27**, 244–248	
Cella, N., Vargas, H., Galembeck, E., Galembeck, R., Miranda, L. C. M. (1989) *J. Polym. Sci. Polym. Lett.*, **27**, 313–320	
Chalmers, J. M., Stay, B. J., Kirkbright, G. F., Spillane, D. E. M., Beadle, B. C. (1981) *Analyst*, **106**, 1179–1186	
Chalmers, J. M., Mackenzie, M. W., Willis, H. A. (1984) *Appl. Spectrosc.*, **38**, 763–773	
Chalmers, J. M., Wilson, J. (1988a) *Mikrochim. Acta*, **2**, 1099–1111	x
Chalmers, J. M., Mackenzie, M. W. (1988b) In: *Advances in Applied Fourier Transform Infrared Spectroscopy*, 105–188	x
Changwen, D., Linker, R., Shaviv, A. (2007) *Appl. Spectrosc.*, **61**, 1063–1067	
Chatzi, E. G., Urban, M. W., Ishida, H., Koenig, J. L. (1986) *Polymer*, **27**, 1850–1854	
Chatzi, E. G., Urban, M. W., Ishida, H., Koenig, J. L., Laschewski, A., Ringsdorf, H. (1988) *Langmuir*, **4**, 846–855	
Cherian, X. M., Satyamoorthy, P., Andrew, J. J., Bhattacharya, S. K. (1994) *Macromol. Rep.*, **A31**, 261–270	
Chia, C. H., Zakaria, S., Nguyen, K. L., Dang, V. Q., Duong, T. D. (2009) *Mater. Chem. Phys.*, **113**, 768–772	
Chien, P.-L., Markuszewski, R., McClelland, J. F. (1985a) *Preprints, ACS Div. Fuel Chem.*, **30**, 13–20	
Chien, P.-L., Markuszewski, R., Araghi, H. G., McClelland, J. F. (1985b) *Proc. 1985 Conf. Coal Sci.*, 818–821	
Choquet, M., Rousset, G., Bertrand, L. (1985) *Proc. SPIE Int. Soc. Opt. Eng.*, **553**, 224–225	
Choquet, M., Rousset, G., Bertrand, L. (1986) *Can. J. Phys.*, **64**, 1081–1085	
Chuang, T. J., Coufal, H., Träger, F. (1983) *J. Vac. Sci. Technol. A*, **1**, 1236–1239	
Church, J. S., Evans, D. J. (1995) *J. Appl. Polym. Sci.*, **57**, 1585–1594	
Cimadevilla, J. L. G., Álvarez, R., Pis, J. J. (2003) *Vib. Spectrosc.*, **31**, 133–141	
Cimadevilla, J. L. G., Álvarez, R., Pis, J. J. (2005) *Fuel Process. Technol.*, **87**, 1–10	

Appendix 2: Literature Guide – Solids and Liquids | 303

Catalysts	Clays, minerals	Food products	Hydrocarbons, fuels	Inorganics	Medical, biological	Organics	Polymers	Textiles	Wood, paper
								x	
							x		
							x		
							x		
							x		
							x		
							x		
							x		
							x		
		x							
					x				
							x		
							x		
							x		
							x		
				x			x		
	x			x					
								x	
							x		
							x			x
									x
			x						
			x						
	x								
	x								
				x					
								x	
			x						
			x						

Citation	Carbons
Ciurczak, E. (1999) *Pharm. Technol.*, **23**, 46–48	
Cody, G. D., Larsen, J. W., Siskin, M. (1989) *Energy Fuels*, **3**, 544–551	
Coelho, T. M., Nogueira, E. S., Steimacher, A., Medina, A. N., Weinand, W. R., Lima, W. M., Baesso, M. L., Bento, A. C. (2006) *J. Appl. Phys.*, **100**, 094312-1–6	
Cook, L. E., Luo, S. Q., McClelland, J. F. (1991) *Appl. Spectrosc.*, **45**, 124–126	
Cooper, A. I., Howdle, S. M., Hughes, C., Jobling, M., Kazarian, S. G., Poliakoff, M., Shepherd, L. A., Johnston, K. P. (1993) *Analyst*, **118**, 1111–1116	
Cooper, E. A., Urban, M. W., Provder, T. (1988) *Polym. Mater. Sci. Eng.*, **59**, 316–320	
Dang, V. Q., Bhardwaj, N. K., Hoang, V., Nguyen, K. L. (2007) *Carb. Polym.*, **68**, 489–494	
Davidson, R. S., King, D. (1983) *J. Textile Inst.*, **74**, 382–384	
Davidson, R. S., Fraser, G. V. (1984) *J. Soc. Dyers Colour.*, **100**, 167–170	
Davis, C. R., Snyder, R. W., Egitto, F. D., D'Couto, G. C., Babu, S. V. (1994) *J. Appl. Phys.*, **76**, 3049–3051	
Débarre, D., Boccara, A. C., Fournier, D. (1981) *Appl. Opt.*, **20**, 4281–4286	
Débarre, D., Boccara, A. C., Fournier, D. (1984) *Proc. Int. Conf. Photoacoustic Effect*, 147–153	
DeBellis, A. D., Low, M. J. D. (1987) *Infrared Phys.*, **27**, 181–191	
DeBellis, A. D., Low, M. J. D. (1988) *Infrared Phys.*, **28**, 225–237	x
de Jong, S. J., De Smedt, S. C., Wahls, M. W. C., Demeester, J., Kettenes-van den Bosch, J. J., Hennink, W. E. (2000) *Macromolecules*, **33**, 3680–3686	
Delgado, A. H., Paroli, R. M., Beaudoin, J. J. (1996) *Appl. Spectrosc.*, **50**, 970–976	
Delprat, P., Gardette, J.-L. (1993) *Polymer*, **34**, 933–937	
Deng, Z., Spear, J. D., Rudnicki, J. D.; McLarnon, F. R., Cairns, E. J. (1996) *J. Electrochem. Soc.*, **143**, 1514–1521	
De Oliveira, M. G., Pessoa, O., Vargas, H., Galembeck, F. (1988) *J. Appl. Polym. Sci.*, **35**, 1791–1802	
De Queiroz, A. A. A., Vitolo, M., De Oliveira, R. C., Higa, O. Z. (1996) *Radiat. Phys. Chem.*, **47**, 873–880	
Dias, D. T., Medina, A. N., Baesso, M. L., Bento, A. C., Porto, M. F., Rubira, A. F. (2002) *Brazilian J. Phys.*, **32**, 523–530	
Dias, D. T., Medina, A. N., Baesso, M. L., Bento, A. C., Porto, M. F., Muniz, E. C., Rubira, A. F. (2003) *Rev. Sci. Instrum.*, **74**, 325–327	
Dias, D. T., Medina, A. N., Baesso, M. L., Bento, A. C. (2005) *Appl. Spectrosc.*, **59**, 173–180	
Dick, R. J., Heater, K. J., McGinniss, V. D., McDonald, W. F., Russell, R. E. (1994) *J. Coat. Technol.*, **66**, 23–38	
Di Renzo, M., Ellis, T. H., Sacher, E., Stangel, I. (2001) *Biomaterials*, **22**, 787–792	
Di Renzo, M., Ellis, T. H., Sacher, E., Stangel, I. (2001) *Biomaterials*, **22**, 793–797	
Dittmar, R. A., Chao, J. L., Palmer, R. A. (1991) *Appl. Spectrosc.*, **45**, 1104–1110	
Dittmar, R. M., Palmer, R. A., Carter, R. O. (1994) *Appl. Spectrosc. Rev.*, **29**, 171–231	

Appendix 2: Literature Guide – Solids and Liquids | 305

Catalysts	Clays, minerals	Food products	Hydrocarbons, fuels	Inorganics	Medical, biological	Organics	Polymers	Textiles	Wood, paper
					x				
			x						
					x				
							x		
				x			x		
							x		
									x
								x	
								x	
							x		
				x					
				x					
				x	x	x			
				x		x			
					x				
	x			x					
							x		
				x					
							x		
							x		
							x		
							x		
							x		
							x		
					x				
					x				
							x		
							x		

Citation	Carbons
Dóka, O., Bićanić, D., Szücs, M., Lubbers, M. (1998) *Appl. Spectrosc.*, **52**, 1526–1529	
Dokolas, P., Qiao, G. G., Solomon, D. H. (2002) *J. Appl. Polym. Sci.*, **83**, 898–915	
Dolby, P. A., McIntyre, R. (1991) *Polymer*, **32**, 586–589	
Donini, J. C., Michaelian, K. H. (1984) *Infrared Phys.*, **24**, 157–163	
Donini, J. C., Michaelian, K. H. (1985) *Proc. SPIE Int. Soc. Opt. Eng.*, **553**, 344–345	
Donini, J. C., Michaelian, K. H. (1986) *Infrared Phys.*, **26**, 135–140	
Donini, J. C., Michaelian, K. H. (1988) *Appl. Spectrosc.*, **42**, 289–292	
Drapcho, D. L., Curbelo, R., Jiang, E. Y., Crocombe, R. A., McCarthy, W. J. (1997) *Appl. Spectrosc.*, **51**, 453–460	
Du, C., Linker, R., Shaviv, A. (2008) *Geoderma*, **143**, 85–90	
Du, C., Zhou, J., Wang, H., Chen, X., Zhu, A., Zhang, J. (2009) *Vib. Spectrosc.*, **49**, 32–37	
Dubois, M., Enguehard, F., Bertrand, L., Choquet, M., Monchalin, J.-P. (1994) *J. Phys. IV* **C7**, 377–380	x
Duerst, R. W., Mahmoodi, P. (1984) *Polym. Prepr.*, **25**, 194–195	
Duerst, R. W., Mahmoodi, P., Duerst, M. D. (1987) In: *Fourier Transform Infrared Characterization of Polymers*, 113–122	
Eckhardt, H., Chance, R. R. (1983) *J. Chem. Phys.*, **79**, 5698–5704	
Eichhorn, K.-J., Hopfe, I., Poetschke, P., Schmidt, P. (2000) *J. Appl. Polym. Sci.*, **75**, 1194–1204	
Einsiedel, H., Kreiter, M., Leclerc, M., Mittler-Neher, S. (1998) *Opt. Mater.*, **10**, 61–68	
El Shafei, G. M., Mokhtar, M. M. (1995) *Coll. Surf. A*, **94**, 267–277	
Esumi, K., Nichina, S., Sakurada, S., Meguro, K., Honda, H. (1987) *Carbon*, **25**, 821–825	x
Evanson, K. W., Urban, M. W. (1990) *Polym. Mater. Sci. Eng.*, **62**, 895–899	
Evanson, K. W., Urban, M. W. (1991) *J. Appl. Polym. Sci.*, **42**, 2287–2296	
Eyring, E. M., Riseman, S. M., Massoth, F. E. (1984) *ACS Symp. Ser.*, **248**, 399–410	
Factor, A., Tilley, M. G., Codella, P. J. (1991) *Appl. Spectrosc.*, **45**, 135–138	
Fainchtein, R., Stoyanov, B. J., Murphy, J. C., Wilson, D. A., Hanley, D. F. (2000) *Proc. SPIE Int. Soc. Opt. Eng.*, **3916**, 19–33	
Fairbrother, J. E. (1983) *Pharm. J.*, **230**, 326–329	
Fan, M., Brown, R. C. (2001) *Fuel*, **80**, 1545–1554	x
Fang, T.-T., Wu, M.-S., Tsai, J.-D. (2002) *J. Am. Ceram. Soc.*, **85**, 2984–2988	
Fangxin, L., Jinlong, Y., Tianpeng, Z. (1997) *Phys. Rev. B*, **55**, 8847–8851	
Fathallah, M., Rezig, B., Zouaghi, M., Amer, N. M., Roger, J. P., Boccara, A. C., Fournier, D. (1988) In: *Photoacoustic and Photothermal Phenomena*, 260–262	
Faubel, W., Heissler, S., Palmer, R. A. (2003) *Rev. Sci. Instrum.*, **74**, 331–333	
Favier, J. P., Bićanić, D., van de Bovenkamp, P., Chirtoc, M., Helander, P. (1996) *Anal. Chem.*, **68**, 729–733	
Favier, J. P., Bićanić, D., Cozijnsen, J., van Veldhuizen, B., Helander, P. (1998) *J. Amer. Oil Chem. Soc.*, **75**, 359–362	

Catalysts	Clays, minerals	Food products	Hydrocarbons, fuels	Inorganics	Medical, biological	Organics	Polymers	Textiles	Wood, paper
	x								
							x		
							x		
						x	x		
	x								
	x		x						
	x								
							x		
	x								
	x								
							x		
							x		
							x		
							x		
							x		
							x		
				x					
							x		
							x		
x									
							x		
					x				
			x			x		x	x
				x					
				x					
				x					
		x							
		x							

Citation	Carbons

Fernandes, E. G., Tramidi, C., Di Gregorio, G. M., Angeloni, G., Chiellini, E. (2008) *J. Appl. Polym. Sci.*, **110**, 1606–1612

Fischer, D., Lappan, U., Hopfe, I., Eichhorn, K.-J., Lunkwitz, K. (1998) *Polymer*, **39**, 573–582

Forsskåhl, I., Kenttä, E., Kyyrönen, P., Sundström, O. (1995) *Appl. Spectrosc.*, **49**, 163–170

Forsythe, J. S., Hill, D. J. T., Logothetis, A. L., Seguchi, T., Whittaker, A. K. (1998) *Radiat. Phys. Chem.*, **53**, 657–667

Foster, N. S., Amonette, J. E., Autrey, T., Ho, J. T. (2001) *Sens. Actuators, B*, **77**, 620–624

Foster, N. S., Valentine, N. B., Thompson, S. E., Johnson, T. J., Amonette, J. E. (2004) *Proc. SPIE Int. Soc. Opt. Eng.*, **5269**, 172–182

Fournier, D., Boccara, A. C., Badoz, J. (1982) *Appl. Opt.*, **21**, 74–76

Fowkes, F. M., Huang, Y. C., Shah, B. A., Kulp, M. J., Lloyd, T. B. (1988) *Colloids. Surf.*, **29**, 243–261

Franceschini, M. A., Gratton, E., Hueber, D., Fantini, S. (1999) *Proc. SPIE Int. Soc. Opt. Eng.*, **3597**, 526–531

Friesen, W. I., Michaelian, K. H. (1986) *Infrared Phys.*, **26**, 235–242

Friesen, W. I., Michaelian, K. H. (1991) *Appl. Spectrosc.*, **45**, 50–56

Friesen, W. I., Michaelian, K. H., Long, Y., Dabros, Y. (2005) *Energy Fuels*, **19**, 1109–1115

Friesen, W. I., Michaelian, K. H., Long, Y., Dabros, Y. (2005) *Prepr. Pap.-Am. Chem. Soc., Div. Fuel Chem.*, **50**, 228–229

Fukuyama, A., Akashi, Y., Suemitsu, M., Ikari, T. (2000) *J. Cryst. Growth*, **210**, 255–259

Gaboury, S. R., Urban, M. W. (1988) *Polym. Prepr.*, **29**, 356–357

Gaboury, S. R.; Urban, M. W. (1989) *Polym. Mater. Sci. Eng.*, **60**, 875–879

Gac, W. (2007) *Appl. Catal. B: Environ.*, **75**, 107–117

Gac, W., Derylo-Marczewska, A., Pasieczna-Patkowska, S., Popivnyak, N., Zukocinski, G. (2007) *J. Mol. Catal. A: Chem.*, **268**, 15–23

Gagarin, S. G., Gladun, T. G., Friesen, W. I., Michaelian, K. H. (1993) *Coke Chem.*, 9–15

Gagarin, S. G., Friesen, W. I., Michaelian, K. H., Gladun, T. G. (1994) *Solid Fuel Chem.*, **28**, 35–42

Gagarin, S. G., Friesen, W. I., Michaelian, K. H. (1995a) *Coke Chem.*, 6–16

Gagarin, S. G., Friesen, W. I., Michaelian, K. H. (1995b) *Coke Chem.*, 23–28

Gao, Y., Choudhury, N. R., Dutta, N., Matisons, J., Reading, M., Delmotte, L. (2001) *Chem. Mater.*, **13**, 3644–3652

Gao, Y., Choudhury, N. R., Matisons, J., Schubert, U., Moraru, B. (2002) *Chem. Mater.*, **14**, 4522–4529

Garbassi, F., Occhiello, E. (1987) *Anal. Chim. Acta*, **197**, 1–42

Garcia, J. A., Mandelis, A., Marinova, M., Michaelian, K. H., Afrashtehfar, S. (1998) *Appl. Spectrosc.*, **52**, 1222–1229

Garcia, J. A., Mandelis, A., Marinova, M., Michaelian, K. H., Afrashtehfar, S. (1999) *AIP Conf. Proc.*, **463**, 395–397

Gardella, J. A., Eyring, E. M., Klein, J. C., Carvalho, M. B. (1982) *Appl. Spectrosc.*, **36**, 570–573

Catalysts	Clays, minerals	Food products	Hydrocarbons, fuels	Inorganics	Medical, biological	Organics	Polymers	Textiles	Wood, paper
							x		
							x		
									x
							x		
			x						
					x				
				x					
				x					
					x				
	x		x						
			x						
			x						
			x						
				x					
							x		
							x		
x									
x									
			x						
			x						
			x						
			x						
				x			x	x	
				x			x	x	
							x		
									x
									x
	x								

Citation

Gardella, J. A., Jiang, D.-Z., McKenna, W. P., Eyring, E. M. (1983a) *Appl. Surf. Sci.*, **15**, 36–49

Gardella, J. A., Jiang, D.-Z., Eyring, E. M. (1983b) *Appl. Spectrosc.*, **37**, 131–133

Gardella, J. A., Grobe, G. L., Hopson, W. L., Eyring, E. M. (1984) *Anal. Chem.*, **56**, 1169–1177

Gentzis, T., Goodarzi, F., McFarlane, R. A. (1992) *Org. Geochem.*, **18**, 249–258

Gerson, D. J. (1984a) *Appl. Spectrosc.*, **38**, 436–437

Gerson, D. J., McClelland, J. F., Veysey, S., Markuszewski, R. (1984b) *Appl. Spectrosc.*, **38**, 902–904

Gillis-D'Hamers, I., Philippaerts, J., Van Der Voort, P., Vansant, E. (1990) *J. Chem. Soc. Faraday Trans.*, **86**, 3747–3750

Gillis-D'Hamers, I., Vrancken, K. C., Vansant, E. F., De Roy, G. (1992) *J. Chem. Soc. Faraday Trans.*, **88**, 2047–2050

Gladun, T.G.; Gagarin, S.G. (2002) *Russ. J. Appl. Chem.* **75**, 1345–1348

Gomes Lage, L., Gomes Delgado, P., Kawano, Y. (2004) *Eur. Polym. J.*, **40**, 1309–1316

Gómez Rivas, J., Sprik, R. (2004) *Rev. Sci. Instrum.*, **75**, 281–283

Gonon, L., Vasseur, O. J., Gardette, J.-L. (1999) *Appl. Spectrosc.*, **53**, 157–163

Gonon, L., Mallegol, J., Commereuc, S., Verney, V. (2001) *Vib. Spectrosc.*, **26**, 43–49

Goodall, R. A., David, B., Bartley, J. P. (1997) *Proc. Sixth Australas. Archaeom. Conf.*, 117–118

Goodarzi, F., McFarlane, R. A. (1991) *Int. J. Coal Geol.*, **19**, 283–301

Gordon, S. H. (1987) *Appl. Spectrosc.*, **41**, 195–199

Gordon, S. H., Greene, R. V., Freer, S. N.; James, C. (1990) *Biotech. Appl. Biochem.*, **12**, 1–10

Gordon, S. H., Schudy, R. B., Wheeler, B. C., Wicklow, D. T., Greene, R. V. (1997) *Int. J. Food Microbiol.*, **35**, 179–186

Gordon, S. H., Wheeler, B. C., Schudy, R. B., Wicklow, D. T., Greene, R. V. (1998) *J. Food Prot.*, **61**, 221–230

Gordon, S. H., Jones, R. W., McClelland, J. F., Wicklow, D. T., Greene, R. V. (1999) *J. Agric. Food Chem.*, **47**, 5267–5272

Gosselin, F., DiRenzo, M., Ellis, T. H., Lubell, W. D. (1996) *J. Org. Chem.*, **61**, 7980–7981

Gotter, B., Faubel, W., Neubert, R. H. H. (2008) *Skin. Pharmacol. Physiol.*, **21**, 156–165

Graf, R. T., Koenig, J. L., Ishida, H. (1987) *Polym. Sci. Technol.*, **36**, 1–32

Graham, J. A., Grim, W. M., Fateley, W. G. (1985) In: *Fourier Transform Infrared Spectroscopy*, **4**, 345–392

Graves, D. J., Luo, S. (1994) *Biochem. Biophys. Res. Comm.*, **205**, 618–624

Greene, R. V., Freer, S. N., Gordon, S. H. (1988) *FEMS Microbiol. Lett.*, **52**, 73–78

Greene, R. V., Gordon, S. H., Jackson, M. A., Bennett, G. A., McClelland, J. F., Jones, R. W. (1992) *J. Agric. Food Chem.*, **40**, 1144–1149

Gregoriou, V. G., Daun, M., Schauer, M. W., Chao, J. L., Palmer, R. A. (1993) *Appl. Spectrosc.*, **47**, 1311–1316

Gregoriou, V. G., Hapanowicz, R. (1996) *Prog. Nat. Sci.*, **6** [Suppl.], S10–S13

Gregoriou, V. G., Hapanowicz, R. (1997) *Macromol. Symp.*, **119**, 101–111

Catalysts	Clays, minerals	Food products	Hydrocarbons, fuels	Inorganics	Medical, biological	Organics	Polymers	Textiles	Wood, paper
x				x					
x									
							x		
			x						
							x		
			x						
	x								
	x								
			x						
							x		
				x					
							x		
							x		
	x								
			x						
									x
					x				
		x			x				
		x			x				
		x			x				
						x			
					x				
							x		
					x				
					x				
		x			x				
							x		
							x		
							x		

Citation	Carbons

Gregoriou, V. G., Rodman, S. E. (2002) *Anal. Chem.*, **74**, 2361–2369

Gremlich, H.-U. (1999) *Biotechnol. Bioeng.*, **61**, 179–187

Griffiths, P. R., de Haseth, J. A. (1986) Fourier Tranform Infrared Spectrometry, 312–337   x

Grobe, G. L., Gardella, J. A., Hopson, W. L., McKenna, W. P. (1987) *J. Biomed. Mater. Res.*, **21**, 211–229

Grosse, P. (1990) *Vib. Spectrosc.*, **1**, 187–198

Gundjian, M., Cole, K. C. (2000) *J. Appl. Polym. Sci.*, **75**, 1458–1473

Gurnagul, N., St-Germain, F. G. T., Gray, D. G. (1986) *J. Pulp Paper Sci.*, **12**, J156–J159

Gurton, K. P., Dahmani, R., Ligon, D., Bronk, B. V. (2005) *Appl. Opt.*, **44**, 4096–4101

Gurton, K. P., Felton, M., Dahmani, R., Ligon, D. (2007) *Appl. Opt.*, **46**, 6323–6329

Haas, U., Seiler, H. (1984) *Z. Naturforsch.*, **39a**, 1242–1249

Haas, U., Jäger, M. (1986) *J. Food Sci.*, **51**, 1087–1088

Halttunen, M., Tenhunen, J., Saarinen, T., Stenius, P. (1999) *Vib. Spectrosc.*, **19**, 261–269

Hammiche, A., Pollock, H. M., Reading, M., Claybourn, M., Turner, P. H., Jewkes, K. (1999) *Appl. Spectrosc.*, **53**, 810–815

Hammiche, A., Bozec, L., Conroy, M., Pollock, H. M., Mills, G., Weaver, J. M. R., Price, D. M., Reading, M., Hourston, D. J., Song, M. (2000) *J. Vac. Sci. Technol. B*, **18**, 1322–1332

Hamza, H. A., Michaelian, K. H., Andersen, N. E. (1983) *Proc. 1983 Conf. Coal Sci.*, 248–251

Han, H.-S., Chon, H. (1994) *Microporous Mater.*, **3**, 331–335

Hanh, B. D., Neubert, R. H. H., Wartewig, S., Christ, A., Hentzsch, C. (2000) *J. Pharm. Sci.*, **89**, 1106–1113

Hanh, B. D., Neubert, R. H. H., Wartewig, S., Lasch, J. (2001) *J. Controlled Release*, **70**, 393–398

Hannigan, J., Greig, F., Freeborn, S. S., MacKenzie, H. A. (1999) *Meas. Sci. Technol.*, **10**, 93–99

Harbour, J. R., Hopper, M. A., Marchessault, R. H., Dobbin, C. J., Anczurowski, E. (1985) *J. Pulp Paper Sci.*, **11**, J42–J47

Harpeness, R., Gedanken, A., Weiss, A. M., Slifkin, M. A. (2003) *J. Mater. Chem.*, **13**, 2603–2606

Harris, M., Pearson, G. N., Willetts, D. V., Ridley, K., Tapster, P. R., Perrett, B. (2000) *Appl. Opt.*, **39**, 1032–1041

Hartwig, A., Hunnekuhl, J., Vitr, G., Dieckhoff, S., Vohwinkel, F., Hennemann, O.-D. (1997) *J. Appl. Polym. Sci.*, **64**, 1091–1096

Harvey, T. J., Henderson, A., Gazi, E., Clarke, N. W., Brown, M., Correia Faria, E., Snook, R. D., Gardner, P. (2007) *Analyst*, **132**, 292–295

Hauser, M., Oelichmann, J. (1988) *Mikrochim. Acta*, **1**, 39–43

Heinrich, G., Güsten, H., Ache, H. J. (1986) *Appl. Spectrosc.*, **40**, 363–368

Helander, P. (1993) *Meas. Sci. Technol.*, **4**, 178–185

Helwig, E., Sandner, B., Gopp, U., Vogt, F., Wartewig, S., Henning, S. (2001) *Biomaterials*, **22**, 2695–2702

Henderson, G., Bryant, M. F. (1980) *Anal. Chem.*, **52**, 1787–1790

Catalysts	Clays, minerals	Food products	Hydrocarbons, fuels	Inorganics	Medical, biological	Organics	Polymers	Textiles	Wood, paper
							x		
							x		
x				x		x	x		
					x				
	x						x		
							x		
									x
	x		x		x				
						x			
						x			
									x
							x		
							x		
			x						
				x					
					x				
					x				
			x						
									x
x				x					
						x			
							x		
					x				
		x			x			x	
				x					
					x				
		x							

Citation	Carbons
Herres, W., Zachmann, G. (1984) *LaborPraxis*, 632–638	
Hess, A., Kemnitz, E. (1997) *J. Catal.*, **149**, 449–457	
Hewes, J. D., Curran, S., Leone, E. A. (1994) *J. Appl. Polym. Sci.*, **53**, 291–295	
Highfield, J. G., Moffat, J. B. (1984a) *Appl. Catal. A: General*, **149**, 373–389	
Highfield, J. G., Moffat, J. B. (1984b) *J. Catal.*, **89**, 185–195	
Highfield, J. G., Moffat, J. B. (1985a) *Appl. Spectrosc.*, **39**, 550–552	
Highfield, J. G., Moffat, J. B. (1985b) *J. Catal.*, **95**, 108–119	
Highfield, J. G., Moffat, J. B. (1986) *J. Catal.*, **98**, 245–258	
Hocking, M. B., Syme, D. T., Axelson, D. E., Michaelian, K. H. (1990a) *J. Polym. Sci.: Part A: Polym. Chem.*, **28**, 2949–2968	
Hocking, M. B., Syme, D. T., Axelson, D. E., Michaelian, K. H. (1990b) *J. Polym. Sci.: Part A: Polym. Chem.*, **28**, 2969–2982	
Hocking, M. B., Syme, D. T., Axelson, D. E., Michaelian, K. H. (1990c) *J. Polym. Sci.: Part A: Polym. Chem.*, **28**, 2983–2996	
Hocking, M. B., Klimchuk, K. A., Lowen, S. (2000) *J. Polym. Sci., Part A: Polym. Chem.*, **38**, 3128–3145	
Hocking, M. B., Klimchuk, K. A., Lowen, S. (2001) *J. Polym. Sci., Part A: Polym. Chem.*, **39**, 1960–1977	
Hofmann, G. R., Faubel, W., Ache, H. J. (1996) *Prog. Nat. Sci.*, **6** [Suppl.], S630–S633	
Honda, F., Imada, Y., Nakajima, K. (1988) *Hyomen Kagaku*, **9**, 356–361	
Hosomi, T., Suematsu, H., Fjellvåg, H., Karppinen, M., Yamauchi, H. (1999) *J. Mater. Chem.*, **9**, 1141–1148	
Hou, R., Wu, J., Soloway, R. D., Guo, H., Zhang, Y., Du, Y., Liu, F., Xu, G. (1988) *Mikrochim. Acta*, **2**, 133–136	
Hövel, H., Grosse, P., Theiss, W. (1992) *J. Non-Cryst. Solids*, **145**, 159–163	
Howdle, S. M., Ramsay, J. M., Cooper, A. I. (1994) *J. Polym. Sci.: Part B: Polym. Phys.*, **32**, 541–549	
Hu, Z., Vansant, E. F. (1995) *Carbon*, **33**, 1293–1300	x
Huang, C.-Y. F., Yuan, C.-J., Luo, S., Graves, D. J. (1994) *Biochem.*, **33**, 5877–5883	
Hunnicutt, M. L., Harris, J. M., Lochmüller, C. H. (1985) *J. Phys. Chem.*, **89**, 5246–5250	
Huvenne, J. P., Lacroix, B. (1988) *Spectrochim. Acta, Part A*, **44**, 109–113	
Imhof, R. E., McKendrick, A. D., Xiao, P. (1995) *Rev. Sci. Instrum.*, **66**, 5203–5213	
Irudayaraj, J., Yang, H. (2000) *Appl. Spectrosc.*, **54**, 595–600	
Irudayaraj, J., Sivakesava, S., Kamath, S., Yang, H. (2001a) *J. Food Sci.*, **66**, 1416–1421	
Irudayaraj, J., Yang, H. (2001b) *J. Food Eng.*, **55**, 25–33	
Irudayaraj, J., Yang, H., Sivakesava, S. (2002) *J. Mol. Struct.*, **606**, 181–188	
Jackson, R. S., Michaelian, K. H., Homes, C. C. (2001) *OSA Technical Digest*, 161–163	x
Janáky, C., Visy, C. (2008) *Synth. Met.*, **158**, 1009–1014	
Jasse, B. (1989) *J. Macromol. Sci.-Chem.*, **A26**, 43–67	

Appendix 2: Literature Guide – Solids and Liquids | 315

Catalysts	Clays, minerals	Food products	Hydrocarbons, fuels	Inorganics	Medical, biological	Organics	Polymers	Textiles	Wood, paper
			x				x		
x							x		
x									
x									
x									
x									
x									
							x		
							x		
							x		
							x		
							x		
	x					x			
				x					
				x					
					x				
				x					
				x			x		
			x						
					x				
	x					x			
					x				
							x		
		x							
		x							
		x							
		x							
	x								
				x			x		
	x						x		

Citation	Carbons

Jawhari, T., Quintanilla, L., Pastor, J. M. (1994) *J. Appl. Polym. Sci.*, **51**, 463–471

Jiang, E. Y., Palmer, R. A., Chao, J. L. (1995) *J. Appl. Phys.*, **78**, 460–469

Jiang, E. Y., Palmer, R. A. (1997a) *Anal. Chem.*, **69**, 1931–1935

Jiang, E. Y., Palmer, R. A., Barr, N. E., Morosoff, N. (1997b) *Appl. Spectrosc.*, **51**, 1238–1244

Jiang, E. Y., McCarthy, W. J., Drapcho, D. L., Crocombe, R. A. (1997c) *Appl. Spectrosc.*, **51**, 1736–1740

Jiang, E. Y., Drapcho, D. L., McCarthy, W. J., Crocombe, R. A. (1998) *AIP Conf. Proc.*, **430**, 381–384

Jiang, E. Y., McCarthy, W. J., Drapcho, D. L. (1998) *Spectroscopy*, **13**, 21–40

Jiang, E. Y. (1999) *Appl. Spectrosc.*, **53**, 583–587

Jianjun, Z., Xiangguong, Y., Yingli, B., Kaiji, Z. (1992) *Catal. Today*, **13**, 555–558

Jin, Q., Kirkbright, G. F., Spillane, D. E. M. (1982) *Appl. Spectrosc.*, **36**, 120–124

Johgo, A., Ozawa, E., Ishida, H., Shoda, K. (1987) *J. Mater. Sci. Lett.*, **6**, 429–430

Jones, R. W., McClelland, J. F. (1996) *Appl. Spectrosc.*, **50**, 1258–1263

Jones, R. W., McClelland, J. F. (2001) *Appl. Spectrosc.*, **55**, 1360–1367   x

Jones, R. W., McClelland, J. F. (2002) *Appl. Spectrosc.*, **56**, 409–418

Jurdana, L. E., Ghiggino, K. P., Leaver, I. H., Barraclough, C. G., Cole-Clarke, P. (1994) *Appl. Spectrosc.*, **48**, 44–49

Jurdana, L. E., Ghiggino, K. P., Leaver, I. H.; Cole-Clarke, P. (1995) *Appl. Spectrosc.*, **49**, 361–366

Kammer, S., Albinsky, K., Sandner, B., Wartewig, S. (1999) *Polymer*, **40**, 1131–1137

Kannan, A. G., Choudhury, N. R., Dutta, N. K. (2007) *Polymer*, **48**, 7078–7086

Kano, Y., Akiyama, S., Kobayashi, S. (1996) *J. Adhesion*, **55**, 261–272

Kanstad, S. O., Nordal, P.-E. (1977) *Int. J. Quantum Chem.*, **12** [Suppl. 2], 123–130

Kanstad, S. O.; Nordal, P.-E. (1978a) *Opt. Commun.*, **26**, 367–371

Kanstad, S. O.; Nordal, P.-E. (1978b) *Powder Technol.*, **22**, 133–137

Kanstad, S. O.; Nordal, P.-E. (1979) *Infrared Phys.*, **19**, 413–422

Kanstad, S. O., Nordal, P.-E. (1980a) *Phys. Technol.*, **11**, 142–147

Kanstad, S. O., Nordal, P.-E. (1980b) *Appl. Surf. Sci.*, **5**, 286–295

Kanstad, S. O., Nordal, P.-E., Hellgren, L., Vincent, L. (1981) *Naturwiss.*, **68**, 47–48

Kaplanová, M., Katuščáková, G. (1992) In: *Photoacoustic and Photothermal Phenomena III*, 180–182

Katti, K. S., Urban, M. W. (2003) *Polymer*, **44**, 3319–3325

Kawano, Y., Denofre, S., Gushikem, Y. (1994) *Vib. Spectrosc.*, **7**, 293–302

Kemsley, E. K., Tapp, H. S., Scarlett, A. J., Miles, S. J., Hammond, R., Wilson, R. H. (2001) *J. Agric. Food Chem.*, **49**, 603–609

Kendall, D. S., Leyden, D. E., Burggraf, L. W., Pern, F. J. (1982) *Appl. Spectrosc.*, **36**, 436–440

Khalil, O. S. (2004) *Diabetes Technol. Therapeutics*, **6**, 660–697

Kieffer, A., Hartwig, A., Schmidt-Naake, G., Hennemann, O.-D. (1998) *Acta Polym.*, **49**, 720–724

Appendix 2: Literature Guide – Solids and Liquids | 317

Catalysts	Clays, minerals	Food products	Hydrocarbons, fuels	Inorganics	Medical, biological	Organics	Polymers	Textiles	Wood, paper
							x		
							x		
							x		
							x		
							x		
							x		
							x		
					x				
x									
		x							x
				x					
							x		
							x		
							x		
					x			x	
					x			x	
							x		
							x		
							x		
				x					
				x		x			
				x		x			
				x					
				x					
				x					
					x				
									x
							x		
	x			x					
					x				
	x					x			
					x				
							x		

Citation	Carbons

Kiland, B. R., Urban, M. W., Ryntz, R. A. (2000a) *Polymer*, **41**, 1597–1606

Kiland, B. R., Urban, M. W., Ryntz, R. A. (2000b) *Polymer*, **42**, 337–344

Kim, H., Urban, M. W. (2000) *Langmuir*, **16**, 5382–5390

Kimura, T., Kato, Y., Yoshida, Z., Nitani, N. (1996) *J. Nucl. Sci. Technol.*, **33**, 519–521

Kinney, J. B., Staley, R. H., Reichel, C. L., Wrighton, M. S. (1981) *J. Amer. Chem. Soc.*, **103**, 4273–4275

Kinney, J. B., Staley, R. H. (1983a) *Anal. Chem.*, **55**, 343–348

Kinney, J. B., Staley, R. H. (1983b) *J. Phys. Chem.*, **87**, 3735–3740

Kirkbright, G. F. (1978) *Optica Pura y Aplicada*, **11**, 125–136

Klimchuk, K. A., Hocking, M. B., Lowen, S. (2000) *J. Polym. Sci., Part A: Polym. Chem.*, **38**, 3146–3160

Koenig, J. L. (1985) *Pure Appl. Chem.*, **57**, 971–976

Kowalska, S., Krupczyńska, K., Buszewski, B. (2006) *Biomed. Chromatogr.*, **20**, 4–22

Krishnan, K. (1981a) *Appl. Spectrosc.*, **35**, 549–557

Krishnan, K., Hill, S. L., Witek, H., Knecht, J. (1981b) *Proc. SPIE Int. Soc. Opt. Eng.*, **289**, 96–98

Krishnan, K., Sill, S., Hobbs, J. P., Sung, C. S. P. (1982) *Appl. Spectrosc.*, **36**, 257–259

Kuo, M.-L., McClelland, J. F., Luo, S., Chien, P.-L., Walker, R. D., Hse, C.-Y. (1988) *Wood Fiber Sci.*, **20**, 132–145

Kuśtrowski, P., Chmielarz, L., Dziembaj, R., Cool, P., Vansant, E. F. (2005) *J. Phys. Chem. A*, **109**, 330–336

Kuśtrowski, P., Chmielarz, L., Dziembaj, R., Cool, P., Vansant, E. F. (2005) *J. Phys. Chem. B*, **109**, 11552–11558

Kuwahata, H., Muto, N., Uehara, F. (2000) *Jpn. J. Appl. Phys.*, **39**, 3169–3171

Lai, E. P., Chan, B. L., Hadjmohammadi, M. (1985) *Appl. Spectrosc. Rev.*, **21**, 179–210

Larsen, J. W. (1988) *Preprints, ACS Div. Fuel Chem.*, **33**, 400–406

Laufer, G., Huneke, J. T., Royce, B. S. H., Teng, Y. C. (1980) *Appl. Phys. Lett.*, **37**, 517–519

Laufer, J., Elwell, C., Delpy, D., Beard, P. (2005) *Phys. Med. Biol.*, **50**, 4409–4428

Leprince, O., Harren, F. J. M., Buitnik, J., Alberda, M., Hoekstra, F. A. (2000) *Plant Physiol.*, **122**, 597–608

Lerner, B., Perkins, J. H., Pariente, G. L., Griffiths, P. R. (1989) *Proc. SPIE Int. Soc. Opt. Eng.*, **1145**, 476–477

Letzelter, N. S., Wilson, R. H., Jones, A. D., Sinnaeve, G. (1995) *J. Sci. Food Agric.*, **67**, 239–245

Lewis, L. N. (1982) *J. Organometall. Chem.*, **234**, 355–365

Li, D.-M., Zhao, Z.-X., Liu, S.-Q., Liu, G.-F., Shi, T.-S., Liu, X.-X. (2000) *Synth. Commun.*, **30**, 4017–4026

Li, G., Burggraf, L. W., Baker, W. P. (2000) *Appl. Phys. Lett.*, **76**, 1122–1124

Li, H., Butler, I. S. (1992) *Appl. Spectrosc.*, **46**, 1785–1789

Li, H., Butler, I. S. (1993) *Appl. Spectrosc.*, **47**, 218–221

Catalysts	Clays, minerals	Food products	Hydrocarbons, fuels	Inorganics	Medical, biological	Organics	Polymers	Textiles	Wood, paper
							x		
							x		
							x		
				x					
	x			x		x			
x	x					x			
x									
	x			x	x	x			x
							x		
	x						x		
					x				
	x		x		x		x		x
							x		
							x		
									x
	x								
	x								
				x					
	x						x		
			x						
				x					
					x				
					x				
							x		
		x							
				x					
						x			
						x			
				x					
				x					

Citation	Carbons
Li, Y.-X., Klabunde, K. J. (1991) *Langmuir*, **7**, 1388–1393	
Li, Y.-X., Schlup, J. R., Klabunde, K. J. (1991) *Langmuir*, **7**, 1394–1399	
Liang, J.-J., Hawthorne, F. C., Swainson, I. P. (1998) *Can. Mineral.*, **36**, 1017–1027	
Liebman, S. A., Pesce-Rodriguez, R. A., Matthews, C. N. (1995) *Adv. Space Res.*, **15**, 71–80	
Lima, J. A. P., Massunaga, M. S. O., Cardoso, S. L., Vargas, H., de Melo Monte, M. B., Duarte, A. C. P., do Amaral, M. R., de Souza-Barros, F. (2003) *Rev. Sci. Instrum.*, **74**, 773–775	
Lin, J. W. P., Dudek, L. P. (1985) *J. Polymer Sci., Polym. Chem. Ed.*, **23**, 1589–1597	
Linker, R. (2008) *Appl. Spectrosc.*, **62**, 248–250	
Linton, R. W., Miller, M. L., Maciel, G. E., Hawkins, B. L. (1985) *Surf. Interf. Anal.*, **7**, 196–203	
Liu, D.K., Wrighton, M. S., McKay, D. R., Maciel, G. E. (1984) *Inorg. Chem.*, **23**, 212–220	
Lloyd, L. B., Yeates, R. C., Eyring, E. M. (1982) *Anal. Chem.*, **54**, 549–552	
Lochmüller, C. H., Wilder, D. R. (1980a) *Anal. Chim Acta*, **116**, 19–24	
Lochmüller, C. H., Wilder, D. R. (1980b) *Anal. Chim Acta*, **118**, 101–108	
Lochmüller, C. H., Thompson, M. M., Kersey, M. T. (1987) *Anal. Chem.*, **59**, 2637–2638	
Lopes, M. H., Pascoal Neto, C., Barros, A. S., Rutledge, D., Delgadillo, I., Gil, A. M. (2000) *Biopolymers (Biospectrosc.)*, **57**, 344–351	
Lopes, M. H., Barros, A. S., Pascoal Neto, C., Rutledge, D., Delgadillo, I., Gil, A. M. (2001) *Biopolymers (Biospectrosc.)*, **62**, 268–277	
López Quintana, S., Schmidt, P., Dybal, J., Kratochvíl, J., Pastor, J. M., Merino, J. C. (2002) *Polymer*, **43**, 5187–5195	
Lotta, T. I., Tulkki, A. P., Virtanen, J. A., Kinnunen, P. K. J. (1990) *Chem. Phys. Lipids*, **52**, 11–27	
Low, M. J. D., Parodi, G. A. (1978) *Spectrosc. Lett.*, **11**, 581–588	x
Low, M. J. D., Parodi, G. A. (1980a) *Infrared Phys.*, **20**, 333–340	x
Low, M. J. D., Parodi, G. A. (1980b) *Spectrosc. Lett.*, **13**, 151–158	
Low, M. J. D., Parodi, G. A. (1980c) *Spectrosc. Lett.*, **13**, 663–669	x
Low, M. J. D., Parodi, G. A. (1980d) *Appl. Spectrosc.*, **34**, 76–80	
Low, M. J. D., Parodi, G. A. (1980e) *J. Mol. Struct.*, **61**, 119–124	
Low, M. J. D., Lacroix, M. (1982a) *Infrared Phys.*, **22**, 139–147	x
Low, M. J. D., Lacroix, M., Morterra, C. (1982b) *Appl. Spectrosc.*, **36**, 582–584	
Low, M. J. D., Lacroix, M., Morterra, C., Severdia, A.G. (1982c) *Amer. Lab.*, 16–27	
Low, M. J. D., Severdia, A. G. (1982d) *J. Mol. Struct.*, **80**, 209–212	
Low, M. J. D., Lacroix, M., Morterra, C. (1982e) *Spectrosc. Lett.*, **15**, 57–64	
Low, M. J. D., Morterra, C., Lacroix, M. (1982f) *Spectrosc. Lett.*, **15**, 159–164	
Low, M. J. D., Morterra, C., Severdia, A. G. (1982g) *Spectrosc. Lett.*, **15**, 415–421	x
Low, M. J. D., Parodi, G. A. (1982h) *J. Photoacoustics*, **1**, 131–144	x
Low, M. J. D. (1983a) *Spectrosc. Lett.*, **16**, 913–922	x
Low, M. J. D., Morterra, C. (1983b) *Carbon*, **21**, 275–281	x

Appendix 2: Literature Guide – Solids and Liquids | 321

Catalysts	Clays, minerals	Food products	Hydrocarbons, fuels	Inorganics	Medical, biological	Organics	Polymers	Textiles	Wood, paper
				x		x			
				x		x			
	x								
						x	x		
				x					
							x		
	x								
	x								
				x					
						x	x		
x	x								
x	x								
x	x								
					x				
					x		x		
							x		
					x				
x	x								
				x	x		x		
					x				
	x		x						
	x								
					x	x			
	x				x		x		
	x					x			
	x								
						x			
						x			
x									
				x	x			x	

Citation	Carbons
Low, M. J. D., Arnold, T. H., Severdia, A. G. (1983c) *Infrared Phys.*, **23**, 199–206	
Low, M. J. D. (1984a) *Spectrosc. Lett.*, **17**, 455–461	x
Low, M. J. D., Morterra, C., Severdia, A. G. (1984b) *Mater. Chem. Phys.*, **10**, 519–528	
Low, M. J. D. (1985a) *Spectrosc. Lett.*, **18**, 619–625	x
Low, M. J. D., Morterra, C. (1985b) *Carbon*, **23**, 311–316	x
Low, M. J. D., Tascon, J. M. D. (1985) *Phys. Chem. Minerals*, **12**, 19–22	
Low, M. J. D. (1986a) *Appl. Spectrosc.*, **40**, 1011–1019	x
Low, M. J. D., Morterra, C., Khosrofian, J. M. (1986b) *IEEE Trans. UFFC*, **33**, 573–584	x
Low, M. J. D., Morterra, C. (1986c) *IEEE Trans. UFFC*, **33**, 585–589	x
Low, M. J. D., Morterra, C. (1987a) *Appl. Spectrosc.*, **41**, 280–287	x
Low, M. J. D., DeBellis, A. D. (1987b) *Spectrosc. Lett.*, **20**, 213–219	
Low, M. J. D., Glass, A. S. (1989) *Spectrosc. Lett.*, **22**, 417–429	x
Low, M. J. D., Morterra, C. (1989) In: *Structure and Reactivity of Surfaces*, 601–609	x
Low, M. J. D., Politou, A. S., Varlashkin, P. G., Wang, N. (1990a) *Spectrosc. Lett.*, **23**, 527–531	x
Low, M. J. D., Wang, N. (1990b) *Spectrosc. Lett.*, **23**, 983–990	x
Low, M. J. D. (1993) *Spectrosc. Lett.*, **26**, 453–459	
Lowry, S. R., Mead, D. G., Vidrine, D. W. (1981) *Proc. SPIE Int. Soc. Opt. Eng.*, **289**, 102–104	
Lowry, S. R., Mead, D. G., Vidrine, D. W. (1982) *Anal. Chem.*, **54**, 546–548	
Ludwig, B. W., Urban, M. W. (1993) *Polymer*, **34**, 3376–3379	
Ludwig, B. W., Urban, M. W. (1994) *Polymer*, **35**, 5130–5137	
Ludwig, B. W., Urban, M. W. (1994) *J. Coat. Technol.*, **66**, 59–67	
Ludwig, B. W., Urban, M. W. (1998) *Polymer*, **39**, 5899–5912	
Luo, S., Liao, C. X., McClelland, J. F., Graves, D. J. (1987) *Int. J. Pept. Protein Res.*, **29**, 728–733	
Luo, S., Huang, C.-Y. F., McClelland, J. F., Graves, D. J. (1994) *Anal. Biochem.*, **216**, 67–76	
Luong, J. H. T., Hrapovic, S., Liu, Y., Yang, D.-Q., Sacher, E., Wang, D., Kingston, C. T., Enright, G. D. (2005) *J. Phys. Chem. B*, **109**, 1400–1407	x
Lynch, B. M., MacEachern, A. M., MacPhee, J. A., Nandi, B. N., Hamza, H., Michaelian, K. H. (1983) *Proc. 1983 Conf. Coal Sci.*, 653–654	
Lynch, B. M., Lancaster, L.-I., Fahey, J. T. (1986) *Preprints, ACS Div. Fuel Chem.*, **31**, 43–48	
Lynch, B. M., Lancaster, L.-I., MacPhee, J. A. (1987a) *Fuel*, **66**, 979–983	
Lynch, B. M., Lancaster, L.-I., MacPhee, J. A. (1987b) *Preprints, ACS Div. Fuel Chem.*, **32**, 138–145	
Lynch, B. M., MacPhee, J. A., Martin, R. R. (1987c) *Proc. Int. Conf. Coal. Sci.*, 19–22	
Lynch, B. M., Lancaster, L.-I., MacPhee, J. A. (1988) *Energy Fuels*, **2**, 13–17	
Lynch, B. M., MacPhee, J. A. (1989) In: *Chemistry of Coal Weathering*, 83–106	
Mackenzie, M. W., Sellors, J. (1988) *Polym. Degrad. Stab.*, **22**, 303–312	
Mahmoodi, P., Duerst, R. W., Meiklejohn, R. A. (1984) *Appl. Spectrosc.*, **38**, 437–438	

Appendix 2: Literature Guide – Solids and Liquids

Catalysts	Clays, minerals	Food products	Hydrocarbons, fuels	Inorganics	Medical, biological	Organics	Polymers	Textiles	Wood, paper
					x				
				x		x	x	x	x
	x								
			x						
	x								
				x		x			x
							x		
			x						
	x					x			
	x					x			
							x		
							x		
							x		
							x		
				x					
				x					
				x					
				x					
				x					
				x					
			x						
				x					
				x					
								x	
								x	

Citation	Carbons

Majid, A., Argue, S., Boyko, V., Pleizier, G., L'Ecuyer, P., Tunney, J., Lang, S. (2003) *Coll. Surf. A: Physicochem. Eng. Aspects*, **224**, 33–44

Mallégol, J., Lemaire, J., Gardette, J.-L. (2000) *Prog.Org. Coat.*, **39**, 107–113

Mallégol, J., Gonon, L., Lemaire, J., Gardette, J.-L. (2001) *Polym. Degrad. Stab.*, **72**, 191–197

Maniar, P. D., Navrotsky, A., Rabinovich, E. M., Ying, J. Y., Benziger, J. B. (1990) *J. Non-Cryst. Solids*, **124**, 101–111.

Manning, C. J., Palmer, R. A., Chao, J. L. (1991) *Rev. Sci. Instrum.*, **62**, 1219–1229

Manning, C. J., Dittmar, R. M., Palmer, R. A.; Chao, J. L. (1992a) *Infrared Phys.*, **33**, 53–62

Manning, C. J., Palmer, R. A., Chao, J. L., Charbonnier, F. (1992b) *J. Appl. Phys.*, **71**, 2433–2440

Manning, C. J., Charbonnier, F., Chao, J. L., Palmer, R. A. (1992c) In: *Photoacoustic and Photothermal Phenomena III*, 161–164

Manzanares, C., Blunt, V. M., Peng, J. (1993) *Spectrochim. Acta, Part A*, **49**, 1139–1152

Marchand, H., Cournoyer, A., Enguehard, F., Bertrand, L. (1997) *Opt. Eng.*, **36**, 312–320

Martin, B. L., Wu, D., Tabatabai, L., Graves, D. J. (1990) *Arch. Biochem. Biophys.*, **276**, 94–101

Martin, B. L., Li, B., Liao, C., Rhode, D. J. (2000) *Arch. Biochem. Biophys.*, **380**, 71–77

Martin, L. K., Yang, C. Q. (1994) *J. Environ. Polym. Degrad.*, **2**, 153–160

Martin, M. A., Childers, J. W., Palmer, R. A. (1987) *Appl. Spectrosc.*, **41**, 120–126

| Masujima, T., Yoshida, H., Kawata, H., Amemiya, Y., Katsura, T., Ando, M., Nanba, T., Fukui, K., Watanabe, M. (1989) *Rev. Sci. Instrum.*, **60**, 2318–2320 | x |

McAskill, N. A. (1987) *Appl. Spectrosc.*, **41**, 313–317

McClelland, J. F. (1983) *Anal. Chem.*, **55**, 89A–105A

McClelland, J. F., Luo, S., Jones, R. W., Seaverson, L. M. (1991) *Proc. SPIE Int. Soc. Opt. Eng.*, **1575**, 226–227

McClelland, J. F., Luo, S., Jones, R. W., Seaverson, L. M. (1992) In: *Photoacoustic and Photothermal Phenomena III*, 113–124

McClelland, J. F., Jones, R. W., Luo, S., Seaverson, L. M. (1993) In: *Practical Sampling Techniques for Infrared Analysis*, 107–144

McClelland, J. F., Jones, R. W., Ochiai, S. (1994) *Proc. SPIE Int. Soc. Opt. Eng.*, **2089**, 302–303

McClelland, J. F., Jones, R. W., Bajic, S. J., Power, J. F. (1997) *Mikrochim. Acta [Suppl.]*, **14**, 613–614

McClelland, J. F., Bajic, S. J., Jones, R. W., Seaverson, L. M. (1998) In: *Modern Techniques in Applied Molecular Spectroscopy*, 221–265

McClelland, J. F., Jones, R. W., Bajic, S. J. (2001a) In: *Handbook of Vibrational Spectroscopy*, vol. 2, 1231–1251

| McClelland, J. F., Jones, R. W. (2001b) *Lubr. Eng.*, **57**, 17–21 | x |

McClelland, J. F., Jones, R. W., Luo, S. (2003) *Rev. Sci. Instrum.*, **74**, 285–290

McClelland, J. F. (2007) *Rev. Sci. Instrum.*, **78**, 050901

McDonald, W. F., Goettler, H., Urban, M. W. (1989a) *Appl. Spectrosc.*, **43**, 1387–1393

Catalysts	Clays, minerals	Food products	Hydrocarbons, fuels	Inorganics	Medical, biological	Organics	Polymers	Textiles	Wood, paper
	x		x						
							x		
							x		
	x								
							x		
							x		
					x				
					x				
						x			
							x		
				x					
				x					
								x	
	x				x				
			x						
			x				x		
							x		
	x						x		
							x		
							x		
			x						
							x		
			x		x		x	x	x
							x		

Citation	Carbons
McDonald, W. F., Urban, M. W. (1989b) *Polym. Mater. Sci. Eng.*, **60**, 739–743	
McDonald, W. F., Urban, M. W. (1991) *Composites*, **22**, 307–318	
McFarlane, R. A., Gentzis, T., Goodarzi, F., Hanna, J. V., Vassallo, A. M. (1993) *Int. J. Coal Geol.*, **22**, 119–147	
McGovern, S. J., Royce, B. S. H., Benziger, J. B. (1984) *Appl. Surf. Sci.*, **18**, 401–413	
McGovern, S. J., Royce, B. S. H., Benziger, J. B. (1985a) *J. Appl. Phys.*, **57**, 1710–1718	
McGovern, S. J., Royce, B. S. H., Benziger, J. B. (1985b) *Appl. Opt.*, **24**, 1512–1514	x
McKenna, W. P., Bandyopadhyay, S., Eyring, E. M. (1984) *Appl. Spectrosc.*, **38**, 834–837	
McKenna, W. P., Gale, D. J., Rivett, D. E.; Eyring, E. M. (1985a) *Spectrosc. Lett.*, **18**, 115–122	
McKenna, W. P., Eyring, E. M. (1985b) *J. Mol. Catal.*, **29**, 363–369	
McKenna, W. P., Higgins, B. E., Eyring, E. M. (1985c) *J. Mol. Catal.*, **31**, 199–206	
McQueen, D. H., Wilson, R., Kinnunen, A. (1995a) *Trends Anal. Chem.*, **14**, 482–492	
McQueen, D. H., Wilson, R., Kinnunen, A., Jensen, E. P. (1995b) *Talanta*, **42**, 2007–2015	
Mead, D. G., Lowry, S. R., Vidrine, D. W., Mattson, D. R. (1979) Fourth Int. Conf. Infrared Millimeter Waves Appl., 231	
Mead, D. G., Lowry, S. R., Anderson, C. R. (1981) *Int. J. Infrared Millimeter Waves*, **2**, 23–34	
Mehicic, M., Kollar, R. G., Grasselli, J. G. (1981) *Proc. SPIE Int. Soc. Opt. Eng.*, **289**, 99–101	
Mendelovici, E., Frost, R. L., Kloprogge, J. T. (2001) *J. Coll. Interf. Sci.*, **238**, 273–278	
Michaelian, K. H., Friesen, W. I. (1985) *Proc. SPIE Int. Soc. Opt. Eng.*, **553**, 260–261	
Michaelian, K. H. (1987a) *Infrared Phys.* **27**, 287–296	
Michaelian, K. H., Bukka, K., Permann, D. N. S. (1987b) *Can. J. Chem.*, **65**, 1420–1423	
Michaelian, K. H. (1989a) *Appl. Spectrosc.*, **43**, 185–190	x
Michaelian, K. H. (1989b) *Infrared Phys.*, **29**, 87–100	x
Michaelian, K. H. (1990a) *Infrared Phys.*, **30**, 181–186	
Michaelian, K. H. (1990b) In: *Vibrational Spectra and Structure*, **18**, 81–126	
Michaelian, K. H., Friesen, W. I. (1990c) *Fuel*, **69**, 1271–1275	
Michaelian, K. H. (1991a) *Appl. Spectrosc.*, **45**, 302–304	
Michaelian, K. H., Yariv, S., Nasser, A. (1991b) *Can. J. Chem.*, **69**, 749–754	
Michaelian, K. H., Birch, J. R. (1991c) *Infrared Phys.*, **31**, 527–537	
Michaelian, K. H., Ogunsola, O. I., Bartholomew, R. J. (1995a) *Can. J. Appl. Spectrosc.*, **40**, 94–99	
Michaelian, K. H., Friesen, W. I., Zhang, S. L., Gentzis, T., Crelling, J. C., Gagarin, S. G. (1995b) In: *Coal Science*, 255–258	
Michaelian, K. H., Akers, K. L., Zhang, S. L., Yariv, S., Lapides, I. (1997) *Mikrochim. Acta [Suppl.]*, **14**, 211–212	
Michaelian, K. H., Lapides, I., Lahav, N., Yariv, S., Brodsky, I. (1998) *J. Coll. Interf. Sci.*, **204**, 389–393	

Catalysts	Clays, minerals	Food products	Hydrocarbons, fuels	Inorganics	Medical, biological	Organics	Polymers	Textiles	Wood, paper
							x		
			x						
x	x								
	x								
	x								
x									
								x	
x									
x									
		x							
		x							
			x						
				x					
x			x				x		
	x								
	x								
			x						
	x								
	x		x				x		
	x		x						
	x								
	x		x						
			x						
			x						
	x								
			x						
			x						
			x						
	x								
	x								

Citation	Carbons
Michaelian, K. H., Zhang, S. L., Hall, R. H., Bulmer, J. T. (2001a) *Can. J. Anal. Sci. Spectrosc.*, **46**, 10–22	
Michaelian, K. H. (2001b) In: *Frontiers in Science and Technology*	
Michaelian, K. H., Jackson, R. S., Homes, C. C. (2001c) *Rev. Sci. Instrum.*, **72**, 4331–4336	x
Michaelian, K. H., Hall, R. H., Bulmer, J. T. (2002a) *J. Therm. Anal. Calorim.*, **69**, 135–147	x
Michaelian, K. H., Hall, R. H., Bulmer, J. T. (2003b) *Spectrochim. Acta, Part A*, **59**, 895–903	
Michaelian, K. H. (2003) *Rev. Sci. Instrum.*, **74**, 659–662	
Michaelian, K. H., Hall, R. H., Bulmer, J. T. (2003a) *Spectrochim. Acta, Part A*, **59**, 811–824	
Michaelian, K. H., Hall, R. H., Bulmer, J. T. (2003c) *Spectrochim. Acta, Part A*, **59**, 2971–2984	
Michaelian, K. H. (2007) *Rev. Sci. Instrum.*, **78**, 051301-1-12	x
Michaelian, K. H., May, T. E., Hyett, C. (2008) *Rev. Sci. Instrum.*, **79**, 014903-1-5	x
Michaelian, K. H., Billinghurst, B. E., Shaw, J. M., Lastovka, V. (2009) *Vib. Spectrosc.*, **49**, 28–31	
Mikula, R. J., Axelson, D. E., Michaelian, K. H. (1985) In: *Proc. 1985 Int. Conf. Coal Sci.*, 495–498	
Millar, G. J., McCann, G. F., Hobbis, C. M., Bowmaker, G. A., Cooney, R. P. (1994) *J. Chem. Soc. Faraday Trans.*, **90**, 2579–2584	
Miller, J. E., Reagan, B. M. (1989) *J. Amer. Inst. Conservation*, **28**, 97–115	
Miller, R. M. (1988) *Proc. SPIE Int. Soc. Opt. Eng.*, **917**, 56–71	
Minato, H., Ishido, Y. (2001) *Rev. Sci. Instrum.*, **72**, 2889–2892	
Mistry, M. K., Choudhury, N. R., Dutta, N. K., Knott, R., Shi, Z., Holdcroft, S. (2008) *Chem. Mater.*, **20**, 6857–6870	
Moffat, J. B., Highfield, J. G. (1984a) *Stud. Surf. Sci. Catal.*, **19**, 77–84	
Moffat, J. B., Highfield, J. G. (1984b) *Preprints, ACS Div. Fuel Chem.*, **29**, 254–260	
Moffat, J. B. (1989) *J. Mol. Catal.*, **52**, 169–191	
Mohamed, A., Rayas-Duarte, P., Gordon, S. H., Xu, J. (2004) *Food Chem.*, **87**, 195–203	
Mohamed, A., Gordon, S. H., Biresaw, G. (2007) *Polym. Degrad. Stabil.*, **92**, 1177–1185	
Mohamed, M. M., Vansant, E. F. (1995) *J. Mater. Sci.*, **30**, 4834–4838	
Mohamed, M. M. (1995) *Spectrochim. Acta, Part A*, **51**, 1–9	
Mohamed, M. M., Vansant, E. F. (1995) *Coll. Surf. A*, **96**, 253–260	
Mohamed, M. M. (2003) *J. Colloid Interface Sci.*, **265**, 106–114	
Mojumdar, S. C., Raki, L. (2005) *J. Thermal Anal. Calorim.*, **82**, 89–95	
Monchalin, J.-P., Gagné, J.-M., Parpal, J.-L., Bertrand, L. (1979) *Appl. Phys. Lett.*, **35**, 360–363	
Monchalin, J.-P., Bertrand, L., Rousset, G., Lepoutre, F. (1984) *J. Appl. Phys.*, **56**, 190–210	
Mongeau, B., Rousset, G., Bertrand, L. (1986) *Can. J. Phys.*, **64**, 1056–1058	
Moore, P., Poslusny, M., Daugherty, K. E., Venables, B. J., Okuda, T. (1990) *Appl. Spectrosc.*, **44**, 326–328	
Morey, M., Davidson, A., Stucky, G. (1996) *Microporous Mater.*, **6**, 99–104	
Morterra, C., Low, M. J. D. (1982a) *Spectrosc. Lett.*, **15**, 689–697	x

Catalysts	Clays, minerals	Food products	Hydrocarbons, fuels	Inorganics	Medical, biological	Organics	Polymers	Textiles	Wood, paper
			x						
	x		x						
	x						x		
			x						
			x						
			x						
			x						
	x								
	x								
			x						
			x						
x							x		
								x	
								x	
				x					
				x		x			
x									
x									
x									
		x							
							x		
x									
x									
x									
x									
							x		
	x								
	x								
				x					
									x
x									

Citation	Carbons
Morterra, C., Low, M. J. D., Severdia, A. G. (1982b) *Infrared Phys.*, **22**, 221–227	x
Morterra, C., Low, M. J. D. (1983) *Carbon*, **21**, 283–288	x
Morterra, C., Low, M. J. D., Severdia, A. G. (1984) *Carbon*, **22**, 5–12	x
Morterra, C., Low, M. J. D. (1985a) *Mater. Chem. Phys.*, **12**, 207–233	x
Morterra, C., Low, M. J. D. (1985b) *Carbon*, **23**, 301–310	x
Morterra, C., Low, M. J. D. (1985c) *Carbon*, **23**, 335–341	x
Morterra, C., Low, M. J. D. (1985d) *Carbon*, **23**, 525–530	x
Morterra, C., Low, M. J. D. (1985e) *Langmuir*, **1**, 320–326	x
Morterra, C., Low, M. J. D. (1985f) *Mater. Chem. Phys.*, **12**, 207–233	
Morterra, C., O'Shea, M. L., Low, M. J. D. (1988) *Mater. Chem. Phys.*, **20**, 123–144	x
Moyer, D. J. D., Wightman, J. P. (1989) *Surf. Interf. Anal.*, **14**, 496–504	x
Muraishi, S. (1984) *Bunko Kenkyu*, **33**, 269–270	
Mutuku, J. N., Karanja, P. K. (2003) *J. Polym. Mater.*, **20**, 445–452	
Nadler, M. P., Nissan, R. A., Hollins, R. A. (1988) *Appl. Spectrosc.*, **42**, 634–642	
Nakamura, O., Lowe, R. D., Mitchem, L., Snook, R. D. (2001) *Anal. Chim. Acta*, **427**, 63–73	
Nakanaga, T., Matsumoto, M., Kawabata, Y., Takeo, H., Matsumura, C. (1989) *Chem. Phys. Lett.*, **160**, 129–133	
Nasreddine, V., Halla, J., Reven, L. (2001) *Macromolecules*, **34**, 7403–7410	
Natale, M., Lewis, L. N. (1982) *Appl. Spectrosc.*, **36**, 410–413	
Nelson, J. H., MacDougall, J. J., Baglin, F. G., Freeman, D. W., Nadler, M., Hendrix, J. L. (1982) *Appl. Spectrosc.*, **36**, 574–576	x
Neubert, R., Collin, B., Wartewig, S. (1997) *Vib. Spectrosc.*, **13**, 241–244	
Neubert, R., Collin, B., Wartewig, S. (1997) *Pharm. Res.*, **14**, 946–948	
Neubert, R., Raith, K., Raudenkolb, S., Wartewig, S. (1998) *Anal. Commun.*, **35**, 161–164	
Nguyen, T. T. (1989) *J. Appl. Polym. Sci.*, **38**, 765–768	
Nishikawa, Y., Kimura, K., Matsuda, A., Kenpo, T. (1992) *Appl. Spectrosc.*, **46**, 1695–1698	
Nishio, E., Abe, I., Ikuta, N., Koga, J., Okabayashi, H., Nishikida, K. (1991) *Appl. Spectrosc.*, **45**, 496–497	
Niu, B.-J., Martin, L. R., Tebelius, L. K., Urban, M. W. (1996) *ACS Symp. Ser.*, **648**, 301–331	
Niu, B.-J., Urban, M. W. (1998) *J. Appl. Polym. Sci.*, **70**, 1321–1348	
Noda, I., Story, G. M., Dowrey, A. E., Reeder, R. C., Marcott, C. (1997) *Macromol. Symp.*, **119**, 1–13	
Noda, K., Tsuji, M., Takahara, A., Kajiyama, T. (2002) *Polymer*, **43**, 4055–4062	
Nordal, P.-E., Kanstad, S. O. (1977a) *Opt. Commun.*, **22**, 185–189	
Nordal, P.-E., Kanstad, S. O. (1977b) *Int. J. Quantum Chem.*, **12** [Suppl. 2], 115–121	
Nordal, P.-E., Kanstad, S. O. (1978) *Opt. Commun.*, **24**, 95–99	
Norton, G. A., McClelland, J. F. (1997) *Miner. Eng.*, **10**, 237–240	
Notingher, I., Imhof, R. E., Xiao, P., Pascut, F. C. (2003) *Rev. Sci. Instrum.*, **74**, 346–348	

Appendix 2: Literature Guide – Solids and Liquids | 331

Catalysts	Clays, minerals	Food products	Hydrocarbons, fuels	Inorganics	Medical, biological	Organics	Polymers	Textiles	Wood, paper
	x			x					
			x						
							x		
							x		
							x		
	x					x			
					x				
						x			
							x		
				x					
		x							
					x				
					x				
					x				
							x		
				x		x	x		
	x			x					
							x		
							x		
							x		
							x		
		x		x					
				x		x			
				x					
	x								
							x		

Citation	Carbons
Notingher, I., Imhof, R. E., Xiao, P., Pascut, F. C. (2003) *Appl. Spectrosc.*, **57**, 1494–1501	
Nowicki, W., Sikorska, A., Zachara, S. (1995) *Spectrosc. Lett.*, **28**, 81–88	
Ochiai, S. (1985) *Toso Kogaku*, **20**, 192–195	
Oelichmann, J. (1989) *Fresenius Z. Anal. Chem.*, **333**, 353–359	x
Oh, W., Nair, S. (2004) *J. Phys. Chem. B*, **108**, 8766–8769	
Oh, W., Nair, S. (2005) *Appl. Phys. Lett.*, **87**, 151912-1-3	
Ohkoshi, M. (2002) *J. Wood Sci.*, **48**, 394–401	
Olson, E. S., Diehl, J. W., Froehlich, M. L. (1988) *Fuel*, **67**, 1053–1061	
O'Shea, M. L., Low, M. J. D., Morterra, C. (1989) *Mater. Chem. Phys.*, **23**, 499–516	x
O'Shea, M. L., Morterra, C., Low, M. J. D. (1990a) *Mater. Chem. Phys.*, **25**, 501–521	x
O'Shea, M. L., Morterra, C., Low, M. J. D. (1990b) *Mater. Chem. Phys.*, **26**, 193–209	x
O'Shea, M. L., Morterra, C., Low, M. J. D. (1991a) *Mater. Chem. Phys.*, **27**, 155–179	x
O'Shea, M. L., Morterra, C., Low, M. J. D. (1991b) *Mater. Chem. Phys.*, **28**, 9–31	x
Paez, C. A., Lozada, O., Castellanos, N. J., Martinez, F. O., Ziarelli, F., Agrifoglio, G., Paez-Moro, E. A., Arzoumanian, J. (2009) *J. Mol. Catal. A: Chem.*, **299**, 53–59	
Palmer, R. A., Smith, M. J. (1986) *Can. J. Phys.*, **64**, 1086–1092	
Palmer, R. A., Smith, M. J., Manning, C. J., Chao, J. L., Boccara, A. C., Fournier, D. (1988) In: *Photoacoustic and Photothermal Phenomena*, 50–52	x
Palmer, R. A., Dittmar, R. M. (1993a) *Thin Solid Films*, **223**, 31–38	
Palmer, R. A., Chao, J. L., Dittmar, R. M., Gregoriou, V. G., Plunkett, S. E. (1993) *Appl. Spectrosc.*, **47**, 1297–1310	
Palmer, R. A. (1993) *Spectroscopy*, **8**, 26–36	
Palmer, R. A., Jiang, E. Y., Chao, J. L. (1994) *Proc. SPIE Int. Soc. Opt. Eng.*, **2089**, 250–251	
Palmer, R. A., Boccara, A. C., Fournier, D. (1996) *Prog. Nat. Sci.*, **6** [Suppl.], S3–S9	
Palmer, R. A., Jiang, E. Y., Chao, J. L. (1997) *Mikrochim. Acta [Suppl.]*, **14**, 591–594	
Pan, B., Ren, J., Yue, Q., Liu, B., Zhang, J., Yang, S. (2009) *Polym. Composites*, **30**, 147–153	
Pandey, G. C. (1989) *Indian J. Text. Res.*, **14**, 160–163	
Pandey, K. K., Theagarajan, K. S. (1997) *Holz Roh- Werkstoff*, **55**, 383–390	
Pandey, K. K., Vuorinen, T. (2008) *Polym. Degrad. Stab.*, **93**, 2138–2146	
Pandurangi, R. S., Seehra, M. S. (1990) *Anal. Chem.*, **62**, 1943–1947	
Pandurangi, R. S., Seehra, M. S. (1991) *Appl. Spectrosc.*, **45**, 673–676	
Pandurangi, R. S., Seehra, M. S. (1992) *Appl. Spectrosc.*, **46**, 1719–1723	
Papendorf, U., Riepe, W. (1989) In: *Proc. 1989 Int. Conf. Coal Science*, **2**, 1111–1113	x
Paputa Peck, M. C., Samus, M. A., Killgoar, P. C., Carter, R. O. (1991) *Rubber Chem. Technol.*, **64**, 610–621	
Parker, J. R., Waddell, W. H. (1996) *J. Elastomers Plast.*, **28**, 140–160	
Pasieczna, S., Ryczkowski, J. (2003) *J. Phys. IV*, **109**, 65–71	

Appendix 2: Literature Guide – Solids and Liquids | 333

Catalysts	Clays, minerals	Food products	Hydrocarbons, fuels	Inorganics	Medical, biological	Organics	Polymers	Textiles	Wood, paper
							x		
				x					
							x		
							x		
x									
x									
									x
			x						
x									
				x		x			
				x					
							x		
							x		
							x		
							x		
				x					
							x		
	x						x		
								x	
									x
									x
	x								
	x								
	x								
							x		
							x		
x									

Citation	Carbons
Pasieczna, S., Ryczkowski, J. (2007) *Appl. Surf. Sci.*, **253**, 5910–5913	
Pawsey, S., Yach, K., Reven, L. (2002) *Langmuir*, **18**, 5205–5212	
Pekel, N., Guven, O. (1998) *J. Appl. Polym. Sci.*, **69**, 1669–1674	
Pekel, N., Güven, O. (2002) *J. Appl. Polym. Sci.*, **85**, 2750–2756	
Pennington, B. D., Ryntz, R. A., Urban, M. W. (1999) *Polymer* **40**, 4795–4803	
Pennington, B. D., Urban, M. W. (1999) *J. Adhes. Sci. Technol.*, **13**, 19–34	
Peoples, M. E., Smith, M. J., Palmer, R. A. (1987) *Appl. Spectrosc.*, **41**, 1257–1259	x
Perkins, J. A., Griffiths, P. R. (1989) *Proc. SPIE Int. Soc. Opt. Eng.*, **1145**, 360–361	
Pesce-Rodriguez, R. A.; Fifer, R. A. (1991) *Appl. Spectrosc.*, **45**, 417–419	
Philippaerts, J., Vansant, E. F., Peeters, G., Vanderheyden, E. (1987) *Anal. Chim. Acta*, **195**, 237–246	
Philippaerts, J., Vanderheyden, E., Vansant, E. F. (1988a) *Mikrochim. Acta*, **2**, 145–148	
Philippaerts, J., Vanderheyden, E., Vansant, E. F. (1988b) In: *Photoacoustic and Photothermal Phenomena*, 33–34	
Phillips, M. J., Duncanson, P., Wilson, K., Darr, J. A., Griffiths, D. V., Rehman, I. (2005) *Adv. Appl. Ceram.*, **104**, 261–267	
Pichler, A., Sowa, M. G. (2004) *Appl. Spectrosc.*, **58**, 1228–1235	
Pichler, A., Sowa, M. G. (2005) *Appl. Spectrosc.*, **59**, 164–172	
Pichler, A., Sowa, M. G. (2005) *Vib. Spectrosc.*, **39**, 163–168	
Pichler, A., Sowa, M. G. (2005) *J. Mol. Spectrosc.*, **229**, 231–237	
Piyakis, K., Sacher, E., Domingue, A., Pireaux, J.-J., Leclerc, G., Bertrand, P., Lhoest, J. B. (1995) *Appl. Surf. Sci.*, **84**, 227–235	
Politou, A. S., Morterra, C., Low, M. J. D. (1990a) *Carbon*, **28**, 529–538	x
Politou, A. S., Morterra, C., Low, M. J. D. (1990b) *Carbon*, **28**, 855–865	x
Politou, A. S., Morterra, C., Low, M. J. D. (1991) *Polym. Degrad. Stab.*, **32**, 331–356	x
Pollock, H. M., Hammiche, A. (2001) *J. Phys. D: Appl. Phys.*, **34**, R23–R53	
Porter, M. D., Karweik, D. H., Kuwana, T., Theis, W. B., Norris, G. B., Tiernan, T. D. (1984) *Appl. Spectrosc.*, **38**, 11–16	
Poslusny, M., Daugherty, K. E. (1988) *Appl. Spectrosc.*, **42**, 1466–1469	
Poulet, P., Chambron, J., Unterreiner, R. (1980) *J. Appl. Phys.*, **51**, 1738–1742	
Poulin, S., Yang, D. Q., Sacher, E., Hyett, C., Ellis, T. H. (2000) *Appl. Surf. Sci.*, **165**, 15–22	
Power, M. D., Chappelow, C. C., Pinzino, C. S., Eick, J. D. (1999) *J. Appl. Polym. Sci.*, **74**, 1577–1583	
Prasad, R. L., Prasad, R., Bhar, G. C., Thakur, S. N. (2002) *Spectrochim. Acta, Part A*, **58**, 3093–3102	
Prasad, R. L., Thakur, S. N., Bhar, G. C. (2002) *Pramana*, **59**, 487–496	
Praschak, D., Bahners, T., Schollmeyer, E. (1998) *Appl. Phys. A*, **66**, 69–75	
Quaschning, V., Deutsch, J., Druska, P., Niclas, H.-J., Kemnitz, E. (1998) *J. Catal.*, **177**, 164–174	

Catalysts	Clays, minerals	Food products	Hydrocarbons, fuels	Inorganics	Medical, biological	Organics	Polymers	Textiles	Wood, paper
x									
				x		x			
							x		
							x		
							x		
							x		
							x		
							x		
					x				
	x								
							x		
							x		
				x					
							x		
							x		
							x		
							x		
							x		
							x		
				x					
									x
							x		
							x		
					x				
			x						
							x		
x									

Citation	Carbons
Quintanilla, L., Pastor, J. M. (1994) *Polymer*, **35**, 5241–5246	
Raveh, A., Martinu, L., Domingue, A., Wertheimer, M. R., Bertrand, L. (1992) In: *Photoacoustic and Photothermal Phenomena III*, 151–154	x
Reading, M., Grandy, D., Hammiche, A., Bozec, L., Pollock, H. M. (2002) *Vib. Spectrosc.*, **29**, 257–260	
Reddy, K. T. R., Chalapathy, R. B. V., Slifkin, M. A., Weiss, A. W., Miles, R. W. (2001) *Thin Solid Films*, **387**, 205–207	
Reddy, K. T. R., Slifkin, M. A., Weiss, A. M. (2001) *Opt. Mater.*, **16**, 87–91	
Renugopalakrishnan, V., Bhatnagar, R. S. (1984) *J. Am. Chem. Soc.*, **106**, 2217–2219	
Renugopalakrishnan, V., Horowitz, P. M., Glimcher, M. J. (1985) *J. Biol. Chem.*, **260**, 11406–11413	
Renugopalakrishnan, V., Chandrakasan, G., Moore, S., Hutson, T. B., Berney, C. V., Bhatnager, R. S. (1989) *Macromolecules*, **22**, 4121–4124	
Rintoul, L., Panayiotou, H., Kokot, S., George, G., Cash, G., Frost, R., Bui, T., Fredericks, P. (1998) *Analyst*, **123**, 571–577	
Riseman, S. M., Eyring, E. M. (1981a) *Spectrosc. Lett.*, **14**, 163–185	x
Riseman, S. M., Yaniger, S. I., Eyring, E. M., MacInnes, D., MacDiarmid, G., Heeger, A. J. (1981b) *Appl. Spectrosc.*, **35**, 557–559	
Riseman, S. M., Massoth, F. E., Dhar, G. M., Eyring, E. M. (1982) *J. Phys. Chem.*, **86**, 1760–1763	
Riseman, S. M., Bandyopadhyay, S., Massoth, F. E., Eyring, E. M. (1985) *Appl. Catal.*, **16**, 29–37	
Rocha, S. M., Goodfellow, B. J., Delgadillo, I., Neto, C. P., Gil, A. M. (2001) *Int. J. Biol. Macromol.*, **28**, 107–119	
Rockley, M. G. (1979) *Chem. Phys. Lett.*, **68**, 455–456	
Rockley, M. G. (1980a) *Appl. Spectrosc.*, **34**, 405–406	x
Rockley, M. G., Devlin, J. P. (1980b) *Appl. Spectrosc.*, **34**, 407–408	
Rockley, M. G., Richardson, H. H., Davis, D. M. (1980c) *Ultrasonics Symp. Proc.*, **2**, 649–651	
Rockley, M. G. (1980d) *Chem. Phys. Lett.*, **75**, 370–372	
Rockley, M. G., Davis, D. M., Richardson, H. H. (1980e) *Science*, **210**, 918–920	
Rockley, M. G., Davis, D. M., Richardson, H. H. (1981) *Appl. Spectrosc.*, **35**, 185–186	
Rockley, M. G., Richardson, H. H., Davis, D. M. (1982) *J. Photoacoustics*, **1**, 145–149	
Rockley, M. G., Woodard, M., Richardson, H. H., Davis, D. M., Purdie, N., Bowen, J. M. (1983) *Anal. Chem.*, **55**, 32–34	
Rockley, M. G., Ratcliffe, A. E., Davis, D. M., Woodard, M. K. (1984b) *Appl. Spectrosc.*, **38**, 553–556	x
Rockley, N. L., Woodard, M. K., Rockley, M. G. (1984) *Appl. Spectrosc.*, **38**, 329–334	
Rockley, N. L., Rockley, M. G. (1987) *Appl. Spectrosc.*, **41**, 471–475	
Rodriguez-Cabello, J. C., Santos, J., Merino, J. C., Pastor, J. M. (1996) *J. Polym. Sci., Part B: Polym. Phys.*, **34**, 1243–1255	
Röhl, R., Childers, J. W., Palmer, R. A. (1982) *Anal. Chem.*, **54**, 1234–1236	
Rosenthal, R. J., Lowry, S. R. (1987b) *Mikrochim. Acta*, **2**, 291–302	

Catalysts	Clays, minerals	Food products	Hydrocarbons, fuels	Inorganics	Medical, biological	Organics	Polymers	Textiles	Wood, paper
							x		
							x		
				x					
				x					
					x				
					x				
					x				
							x		
							x		
x									
x									
					x		x		
							x		
					x		x		
			x						
	x								
				x					
					x				
				x					
	x								
		x				x			
						x	x		
	x								
				x			x		
							x		

Citation	Carbons
Rosenthal, R. J., Carl, R. T., Beauchaine, J. P., Fuller, M. P. (1988) *Mikrochim. Acta*, **2**, 149–153	
Roush, P. B., Oelichmann, J. (1988) *Mikrochim. Acta*, **1**, 49–52	
Royce, B. S. H., Teng, Y. C., Enns, J. (1980) *Ultrason. Symp. Proc.*, 652–657	
Royce, B. S. H., Teng, Y. C., Ors, J. A. (1981) *Ultrason. Symp. Proc.*, 784–787	
Royce, B. S. H., Benziger, J. B. (1986) *IEEE Trans. UFFC*, **33**, 561–572	x
Royce, B. S. H., Alexander, J. (1988) In: *Photoacoustic and Photothermal Phenomena*, 9–18	
Ryczkowski, J. (1994) *Proc. SPIE Int. Soc. Opt. Eng.*, **2089**, 182–183	
Ryczkowski, J. (2004) *Vib. Spectrosc.*, **34**, 247–252	
Ryczkowski, J., Goworek, J., Gac, W., Pasieczna, S., Borowiecki, T. (2005) *Thermochim. Acta*, **434**, 2–8	
Ryczkowski, J. (2007) *Vib. Spectrosc.*, **43**, 203–209	
Ryczkowski, J. (2007) *Catal. Today*, **124**, 11–20	
Saab, A. P., Laub, M., Srdanov, I., Stucky, G. D. (1998) *Adv. Mater.*, **10**, 462–465	x
Saadallah, F., Yacoubi, N., Genty, F., Alibert, C. (2002) *Appl. Opt.*, **41**, 7561–7568	
Sadler, A. J., Horsch, J. G., Lawson, E. Q., Harmatz, D., Brandau, D. T., Middaugh, C. R. (1984) *Anal. Biochem.*, **138**, 44–51	
Saffa, A. M., Michaelian, K. H. (1994a) *Proc. SPIE Int. Soc. Opt. Eng.*, **2089**, 566–567	
Saffa, A. M., Michaelian, K. H. (1994b) *Appl. Spectrosc.*, **48**, 871–874	
Sakthivel, R., Prescott, H., Kemnitz, E. (2004) *J. Mol. Catal. A: Chem.*, **223**, 137–142	
Salazar-Rojas, E. M., Urban, M. W. (1987) *Preprints, ACS Div. Polym. Chem.*, **28**, 1–2	
Salazar-Rojas, E. M., Urban, M. W. (1989) *Prog. Org. Coat.*, **16**, 371–386	
Salnick, A. O., Faubel, W. (1995) *Appl. Spectrosc.*, **49**, 1516–1524	
Salnick, A., Faubel, W., Klewe-Nebenius, H., Vendl, A., Ache, H.-J. (1995) *Corros. Sci.*, **37**, 741–767	
Salnick, A., Faubel, W. (1996) *Prog. Nat. Sci.*, **6** [Suppl.], S14–S17	
Sarma, T. V. K., Sastry, C. V. R., Santhamma, C. (1987) *Spectrochim. Acta, Part A*, **43**, 1059–1065	
Saucy, D. A., Simko, S. J., Linton, R. W. (1985a) *Anal. Chem.*, **57**, 871–875	
Saucy, D. A., Cabaniss, G. E., Linton, R. W. (1985b) *Anal. Chem.*, **57**, 876–879	
Sawatari, N., Fukuda, M., Taguchi, Y., Tanaka, M. (2005) *J. Appl. Polym. Sci.*, **97**, 682–690	
Schendzielorz, A., Hanh, B. D., Neubert, R. H. H., Wartewig, S. (1999) *Pharm. Res.*, **16**, 42–45	
Schiewe-Langgartner, Wiesner, G., Gruber, M., Hobbhahn, J. (2005) *Anaesthetist*, **54**, 667–672	
Schmidt, P., Roda, J., Nováková, V., Pastor, J. M. (1997) *Angew. Makromol. Chem.*, **245**, 113–123	
Schmidt, P., Raab, M., Kolařík, J., Eichorn, K. J. (2000a) *Polym. Test.*, **19** (2), 205–212.	
Schmidt, P., Kolařík, J., Lednický, F., Dybal, J., Lagarón, J. M., Pastor, J. M. (2000b) *Polymer*, **41**, 4267–4279.	
Schmidt, P., Dybal, J., Ščudla, J., Raab, M., Kratochvíl, J., Eichhorn, K.-J., Lopez Quintana, S., Pastor, J. M. (2002a) *Macromol. Symp.*, **184**, 107–122	
Schmidt, P., Baldrian, J., Ščudla, J., Dybal, J., Raab, M., Eichhorn, K.-J. (2002b) *Polymer*, **42**, 5321–5326	

Appendix 2: Literature Guide – Solids and Liquids | **339**

Catalysts	Clays, minerals	Food products	Hydrocarbons, fuels	Inorganics	Medical, biological	Organics	Polymers	Textiles	Wood, paper
					x				
					x		x		
x			x	x			x		
							x		
x									
	x								
x						x			
x									
x									
x									
x									
				x					
					x				
	x								
	x								
x									
							x		
	x								
	x								
	x								
						x			
							x		
						x			
							x		
				x					
				x					
							x		
							x		
							x		
							x		
							x		

Citation	Carbons

Schrijnemakers, K., Van DerVoort, P., Vansant, E. F. (1999) *Phys. Chem. Chem. Phys.*, **1**, 2569–2572

Schroeder, M. A., Fifer, R. A., Miller, M. S., Pesce-Rodriguez, R. A., McNesby, C. J. S., Singh, G., Widder, J. M. (2001) *Combust. Flame*, **126**, 1577–1598

Schüle, G., Schmitz, B., Steiner, R. (1998) *Proc. SPIE Int. Soc. Opt. Eng.*, **3565**, 134–138

Schüle, G., Schmitz, B., Steiner, R. (1999) *AIP Conf. Proc.*, **463**, 615–617

Schwerha, D. J., Orr, C.-S., Chen, B. T., Soderholm, S. C. (2002) *Anal. Chim. Acta*, **457**, 257–264

Seaverson, L. M., McClelland, J. F., Burnet, G., Anderegg, J. W., Iles, M. K. (1985) *Appl. Spectrosc.*, **39**, 38–45

Seehra, M. S., Pandurangi, R. (1989) *J. Phys.: Condens. Matter*, **1**, 5301–5304

Seehra, M. S., Roy, P., Raman, A., Manivannan, A. (2004) *Solid State Comm.*, **130**, 597–601

Segura, Y., Chmielarz, L., Kustrowski, P., Cool, P., Dziembaj, R., Vansant, E. F. (2006) *J. Phys. Chem. B*, **110**, 948–955

Shaw, M. D., Karunakaran, C., Tabil, L. G. (2009) *Biosystems Eng.*, **103**, 198–207

Shelat, K. J., Dutta, N. K., Choudhury, N. R. (2008) *Langmuir*, **24**, 5464–5473

Shen, Y. C., MacKenzie, H. A., Lindberg, J., Lu, Z. H. (1999) *Proc. SPIE Int. Soc. Opt. Eng.*, **3863**, 167–171

Shoval, S., Yariv, S., Michaelian, K. H., Boudeuille, M., Panczer, G. (1999a) *Clay Miner.*, **34**, 551–563

Shoval, S., Yariv, S., Michaelian, K. H., Lapides, I., Boudeuille, M., Panczer, G. (1999b) *J. Coll. Interf. Sci.*, **212**, 523–529

Shoval, S., Yariv, S., Michaelian, K. H., Boudeuille, M., Panczer, G. (2001) *Clays Clay Miner.*, **49**, 347–354

Shoval, S., Yariv, S., Michaelian, K. H., Boudeuille, M., Panczer, G. (2002a) *Clays Clay Miner.*, **50**, 56–62

Shoval, S., Michaelian, K. H., Boudeuille, M., Panczer, G., Lapides, I., Yariv, S. (2002b) *J. Thermal Anal. Calorim.*, **69**, 205–225

Sieg, A., Guy, R. H., Delgado-Charro, M. M. (2005) *Diabetes Technol. Therapeutics*, **7**, 174–197

Siesler, H. W. (1991) *Polym. Mater. Sci. Eng.*, **64**, 30

Siew, D. C. W., Heilmann, C., Easteal, A. J., Cooney, R. P. (1999) *J. Agric. Food Chem.*, **47**, 3432–3440

Sikdar, D., Katti, D. R., Katti, K. S. (2006) *Langmuir*, **22**, 7738–7747

Sikdar, D., Katti, K. S., Katti, D. R. (2008) *J. Nanosci. Nanotchnol.*, **8**, 1638–1657

Simms, J. R., Yang, C. Q. (1994) *Carbon*, **32**, 621–626   x

Sivakesava, S., Irudayaraj, J. (2000) *J. Sci. Food Agric.*, **80**, 1805–1810

Small, R. D., Ors, J. A. (1984) *Org. Coat. Appl. Polym. Sci. Proc.*, **48**, 678–686

Smith, M. J., Palmer, R. A. (1987) *Appl. Spectrosc.*, **41**, 1106–1113

Smith, M. J., Manning, C. J., Palmer, R. A., Chao, J. L. (1988a) *Appl. Spectrosc.*, **42**, 546–555

Smith, M. J., Palmer, R. A. (1988b) In: *Photoacoustic and Photothermal Phenomena*, 211–213

Catalysts	Clays, minerals	Food products	Hydrocarbons, fuels	Inorganics	Medical, biological	Organics	Polymers	Textiles	Wood, paper
	x								
						x			
				x					
				x					
	x		x						
	x								
				x					
x									
									x
							x		
					x				
	x								
	x								
	x								
	x								
	x								
					x				
							x		
		x							
	x						x		
	x						x		
		x							
							x		
						x			
				x					

Citation	Carbons

Snook, R. D. (2000) *Analyst*, **125**, 45–50

Solomon, P. R., Carangelo, R. M. (1982) *Fuel*, **61**, 663–669

Sowa, M. G., Mantsch, H. H. (1993) *Proc. SPIE Int. Soc. Opt. Eng.*, **2089**, 128–129

Sowa, M. G., Mantsch, H. H. (1993) *J. Mol. Struct.*, **300**, 239–244

Sowa, M. G., Mantsch, H. H. (1994) *Appl. Spectrosc.*, **48**, 316–319

Sowa, M. G., Mantsch, H. H. (1994) *Calcif. Tissue Int.*, **54**, 481–485

Sowa, M. G., Wang, J., Schultz, C. P., Ahmed, M. K., Mantsch, H. H. (1995) *Vib. Spectrosc.*, **10**, 49–56

Sowa, M. G., Fischer, D., Eysel, H. H., Mantsch, H. H. (1996) *J. Mol. Struct.*, **379**, 77–85

Sparks, B. D., Kotlyar, L. S., O'Carroll, J. B., Chung, K. H. (2003) *J. Pet. Sci. Eng.*, **39**, 417–430

Spencer, N. D. (1986) *CHEMTECH*, **16**, 378–384

Spencer, P., Trylovich, D. J., Cobb, C. M. (1992) *J. Periodontol.*, **63**, 633–636

Spencer, P., Byerley, T. J., Eick, J. D., Witt, J. D. (1992) *Dent. Mater.*, **8**, 10–15

Spencer, P., Cobb, C. M., McCollum, M. H., Wieliczka, D. M. (1996) *J. Periodont. Res.*, **31**, 453–462

Spencer, P., Payne, J. M., Cobb, C. M., Reinisch, L., Peavy, G. M., Drummer, D. D., Sichman, D. L., Swafford, J. R. (1999) *J. Periodont.*, **70**, 68–74

Stegge, J. M., Urban, M. W. (2001) *Polymer*, **42**, 5479–5484

Stevenson, C. M., Abdelrahim, I. M., Novak, S. W. (2001) *J. Archaeol. Sci.*, **28**, 109–115

St-Germain, F. G. T., Gray, D. G. (1987) *J. Wood Chem. Technol.*, **7**, 33–50

Story, G. M., Marcott, C. (1998) *AIP Conf. Proc.*, **430**, 513–515

Story, G. M., Marcott, C., Noda, I. (1994) *Proc. SPIE Int. Soc. Opt. Eng.*, **2089**, 242–243

Stout, P. J., Crocombe, R. A. (1994) *Proc. SPIE Int. Soc. Opt. Eng.*, **2089**, 300–301

Su, C., Puls, R. W. (2004) *Environ. Sci. Technol.*, **38**, 5224–5231

Sudiyani, Y., Imamura, Y., Doi, S., Yamauchi, S. (2003) *J. Wood Sci.*, **49**, 86–92

Sweterlitsch, J. J., Jones, R. W., Hsu, D. K., McClelland, J. F. (2004) *Appl. Spectrosc.*, **58**, 1420–1423   x

Szurkowski, J., Wartewig, S. (1999a) *Instrument. Sci. Technol.*, **27**, 311–317

Szurkowski, J., Wartewig, S. (1999b) *AIP Conf. Proc.*, **463**, 618–620

Szurkowski, J., Pawelska, I., Wartewig, S., Pogorzelski, S. (2000) *Acta Phys. Polon. A*, **97**, 1073–1082

Szurkowski, J. (2001) *Bull. Env. Contamination Toxicology*, **66**, 683–690

Tabor, R., Lepovitz, J., Potts, W., Latham, D., Latham, L. (1997) *J. Cell. Plast.*, **33**, 372–399

Tanaka, K., Gotoh, T., Yoshida, N., Nonomura, S. (2002) *J. Appl. Phys.*, **91**, 125–128

Tanaka, S., Teramae, N. (1984) *Preprints, ACS Div. Polym. Chem.*, **25**, 190–191

Tandon, P., Raudenkolb, S., Neubert, R. H. H., Rettig, W., Wartewig, S. (2001) *Chem. Phys. Lipids*, **109**, 37–45

Teng, Y. C., Royce, B. S. H. (1982) *Appl. Opt.*, **21**, 77–80

Catalysts	Clays, minerals	Food products	Hydrocarbons, fuels	Inorganics	Medical, biological	Organics	Polymers	Textiles	Wood, paper
					x		x		
			x						
					x				
					x				
					x				
					x				
					x				
							x		
	x		x						
x									
					x				
					x				
					x				
					x				
							x		
	x								
									x
					x		x		
							x		
							x		
	x			x					
									x
		x							
		x							
		x							
									x
							x		
				x					
							x		
					x				
							x		

Citation	Carbons
Teramae, N., Tanaka, S. (1981) *Spectrosc. Lett.*, **14**, 687–694	
Teramae, N., Hiroguchi, M., Tanaka, S. (1982a) *Bull. Chem. Soc. Jpn.*, **55**, 2097–2100	
Teramae, N., Yamamoto, T., Hiroguchi, M., Matsui, T., Tanaka, S. (1982) *Chem. Lett.*, 37–40	
Teramae, N., Tanaka, S. (1984) *Bunseki Kagaku*, **33**, E397–E400	
Teramae, N., Tanaka, S. (1985a) *Appl. Spectrosc.*, **39**, 797–799	
Teramae, N., Tanaka, S. (1985b) *Anal. Chem.*, **57**, 95–99	
Teramae, N., Tanaka, S. (1987) In: *Fourier Transform Infrared Characterization of Polymers*, 315–340	
Teramae, N., Tanaka, S. (1988) *Mikrochim. Acta*, **2**, 159–162	
Thompson, M. M., Palmer, R. A. (1988) *Appl. Spectrosc.*, **42**, 945–951	
Thompson, M. M., Palmer, R. A. (1988) *Anal. Chem.*, **60**, 1027–1032	x
Thompson, S. E., Foster, N. S., Johnson, T. J., Valentine, N. B., Amonette, J. E. (2003) *Appl. Spectrosc.*, **57**, 893–899	
Tiefenthaler, A. M., Urban, M. W. (1988) *Polym. Mater. Sci. Eng.*, **59**, 311–315	
Tiefenthaler, A. M., Urban, M. W. (1989) *Composites*, **20**, 145–150	
Tiefenthaler, A. M., Urban, M. W. (1989) *Composites*, **20**, 585–588	
Träger, F., Coufal, H., Chuang, T. J. (1982) *Phys. Rev. Lett.*, **49**, 1720–1723	
Träger, F., Chuang, T. J., Coufal, H. (1983) *J. Electron Spectrosc. Related Phenom.*, **30**, 19–24	
Tran, N. D., Choudhury, N. R., Dutta, N. K. (2005) *Polym. Int.*, **54**, 513–525	
Tsuge, A., Uwamino, Y., Ishizuka, T. (1988) *Appl. Spectrosc.*, **42**, 168–169	
Tu, L., Kruger, D., Wagener, J. B., Carstens, P. A. B. (1997) *J. Adhes.*, **62**, 187–211	
Uotila, J., Kauppinen, J. (2008) *Appl. Spectrosc.*, **62**, 655–660	
Urban, M. W., Koenig, J. L. (1985) *Appl. Spectrosc.*, **39**, 1051–1056	
Urban, M. W., Koenig, J. L. (1986a) *Appl. Spectrosc.*, **40**, 513–519	
Urban, M. W., Koenig, J. L. (1986b) *Appl. Spectrosc.*, **40**, 851–856	
Urban, M. W., Koenig, J. L. (1986c) *Appl. Spectrosc.*, **40**, 994–998	
Urban, M. W., Chatzi, E. G., Perry, B. C., Koenig, J. L. (1986d) *Appl. Spectrosc.*, **40**, 1103–1107	
Urban, M. W. (1987) *J. Coatings Technol.*, **59**, 29–34	
Urban, M. W., Koenig, J. L. (1988) *Anal. Chem.*, **60**, 2408–2412	
Urban, M. W. (1989a) *Prog. Org. Coat.*, **16**, 321–353	
Urban, M. W., Gaboury, S. R. (1989) *Macromolecules*, **22**, 1486–1487	
Urban, M. W., Evanson, K. W. (1990) *Polym. Commun.*, **31**, 279–282	
Urban, M. W., Salazar-Rojas, E. M. (1990) *J. Polym. Sci., Part A: Polym. Chem.*, **28**, 1593–1613	
Urban, M. W. (1991) *Polym. Mater. Sci. Eng.*, **64**, 31–32	
Urban, M. W. (1997) *Prog. Org. Coat.*, **32**, 215–229	
Urban, M. W. (1999) *Macromol. Symp.*, **141**, 15–31	
Urban, M. W. (2000a) *Prog. Org. Coat.*, **40**, 195–202	

Appendix 2: Literature Guide — Solids and Liquids | 345

Catalysts	Clays, minerals	Food products	Hydrocarbons, fuels	Inorganics	Medical, biological	Organics	Polymers	Textiles	Wood, paper
								x	
							x		
							x		
							x		
							x		
							x		
							x		
						x			
				x					
					x				
								x	
							x		
							x		
				x					
				x					
							x		
	x			x					
							x		
		x					x		
	x								
	x					x			
	x					x			
							x		
							x		
							x		
	x								
							x		
							x		
							x		
							x		
							x		
							x		
							x		
							x		

Citation	Carbons
Urban, M. W. (2000b) *Eur. Coat. J.*, 58–59	
Urban, M. W., Allison, C. L., Johnson, G. L., DiStefano, F. (1999) *Appl. Spectrosc.*, **53**, 1520–1527	
Utamapanya, S., Klabunde, K. J., Schlup, J. R. (1991) *Chem. Mater.*, **3**, 175–181	
van Dalen, G. (2000) *Appl. Spectrosc.*, **54**, 1350–1356	
van der Ven, L. G. J., Leuverink, R. (2002) *Macromol. Symp.*, **187**, 845–852	
Van Der Voort, P., Swerts, J., Vrancken, K. C., Vansant, E. F., Geladi, P., Grobet, P. (1993) *J. Chem. Soc. Faraday Trans.*, **89**, 63–68	
Van Der Voort, P., van Welzenis, R., de Ridder, M., Brongersma, H. H., Baltes, M., Mathieu, M., van de Ven, P. C., Vansant, E. F. (2002) *Langmuir*, **18**, 4420–4425	
van de Weert, M., van't Hof, R., van der Weerd, J., Heeren, R. M. A., Posthuma, G., Hennink, W. E., Crommelin, D. J. A. (2000) *J. Controlled Release*, **68**, 31–40	
van de Weert, M., Haris, P. I., Hennink, W. E., Crommelin, D. J. A. (2001) *Anal. Biochem.*, **297**, 160–169	
Van Neste, C. W., Senesac, L. R., Yi, D., Thundat, T. (2008) *Appl. Phys. Lett.*, **92**, 134102-1-3	
Van Neste, C. W., Senesac, L. R., Thundat, T. (2008) *Appl. Phys. Lett.*, **92**, 234102-1-3	
von Lilienfeld-Toal, H., Weidenmüller, M., Xhelaj, A., Mäntele, W. (2005) *Vib. Spectrosc.*, **38**, 209–215	
Vargas, H., Miranda, L. C. M. (1988) *Phys. Rep.*, **161**, 43–101	
Varlashkin, P. G., Low, M. J. D., Parodi, G. A., Morterra, C. (1986a) *Appl. Spectrosc.*, **40**, 636–641	x
Varlashkin, P. G., Low, M. J. D. (1986b) *Appl. Spectrosc.*, **40**, 1170–1176	x
Verma, D., Katti, K., Katti, D. (2006) *Spectrochim. Acta, Part A*, **64**, 1051–1057	
Verma, D., Katti, K., Katti, D. (2006) *J. Biomed. Mater. Res. A*, **77**, 59–66	
Vidrine, D. W. (1980) *Appl. Spectrosc.*, **34**, 314–319	
Vidrine, D. W. (1981) *Proc. SPIE Int. Soc. Opt. Eng.*, **289**, 355–360	
Vidrine, D.W. (1982) In: *Fourier Transform Infrared Spectroscopy* **3**, 125–148	
Vidrine, D. W., Lowry, S. R. (1983) *Adv. Chem. Ser. (Polym. Charact.)*, **203**, 595–613	
Vidrine, D. W. (1984) *Polym. Prepr.*, **25**, 147–148	
Vreugdenhil, A. J., Donley, M. S., Grebasch, N. T., Passinault, R. J. (2001) *Prog. Org. Coat.*, **41**, 254–260	
Waddell, W. H., Parker, J. R. (1992) *Rubber World*, **207**, 29–35	
Wahls, M. W. C., Toutenhoofd, J. P., Leyte-Zuiderweg, L. H., de Bleijser, J., Leyte, J. C. (1997) *Appl. Spectrosc.*, **51**, 552–557	x
Wahls, M. W. C., Leyte, J. C. (1998) *J. Appl. Phys.*, **83**, 504–509	
Wahls, M. W. C., Weisman, J. L., Jesse, W. J., Leyte, J. C. (1998b) *AIP Conf. Proc.*, **430**, 392–394	x
Wahls, M. W. C., Leyte, J. C. (1998c) *Appl. Spectrosc.*, **52**, 123–127	
Wahls, M. W. C., Kenttä, E., Leyte, J. C. (2000) *Appl. Spectrosc.*, **54**, 214–220	
Wahls, M. W. C., Leyte, J. C. (2001) In: *Frontiers in Science and Technology*	
Wang, H., Phifer, E. B., Palmer, R. A. (1998) *Fresenius J. Anal. Chem.*, **362**, 34–40	

Catalysts	Clays, minerals	Food products	Hydrocarbons, fuels	Inorganics	Medical, biological	Organics	Polymers	Textiles	Wood, paper
							x		
							x		
				x					
					x			x	
							x		
	x								
x				x					
					x				
					x				
						x			
						x			
					x				
			x			x			
	x				x				
				x	x				
					x				
					x		x		
			x		x		x		
x							x		
							x		
							x		
							x		
							x		
							x		
							x		
							x		
							x		
									x
							x		x
							x		

Citation	Carbons
Wang, H. P., Eyring, E. M., Huai, H. (1991) *Appl. Spectrosc.*, **45**, 883–885	
Wang, J., Ahmed, M. K., Sowa, M. G., Mantsch, H. H. (1994) *Proc. SPIE Int. Soc. Opt. Eng.*, **2089**, 492–493	
Wang, J., Sowa, M., Mantsch, H. H., Bittner, A., Heise, H. M. (1996) *Trends Anal. Chem.*, **15**, 286–296	
Wang, N., Low, M. J. D. (1989) DOE/PC/79920–10, DE90 006239, 1–14	x
Wang, N., Low, M.J.D. (1990a) *Mater. Chem. Phys.* **26**, 117–130	x
Wang, N.; Low, M. J. D. (1990b) *Mater. Chem. Phys.*, **26**, 465–481	x
Wang, N., Low, M. J. D. (1991) *Mater. Chem. Phys.*, **27**, 359–374	x
Wang, W., Qu, B. (2003) *Polym. Degrad. Stabil.*, **81**, 531–537	
Wang, Z. Y., Ng, C. F., Lim, P. K., Leung, H. W., Wu, J. G., Chen, H. F. (1994) *J. Phys.: Condens. Matter*, **6**, 7207–7215	
Wen, Q., Michaelian K. H. (2008) *Opt. Lett.*, **33**, 1875–1877	x
Wentrup-Byrne, E., Rintoul, L., Smith, J. L., Fredericks, P. M. (1995) *Appl. Spectrosc.*, **49**, 1028–1036	
Wentrup-Byrne, E., Rintoul, L., Gentner, J. M., Smith, J. L., Fredericks, P. M. (1997) *Mikrochim. Acta [Suppl.]*, **14**, 615–616	
Wetzel, D. L., Carter, R. O. (1998) *AIP Conf. Proc.*, **430**, 567–570	
White, R. L. (1985) *Anal. Chem.*, **57**, 1819–1822	
Will, F. G., McDonald, R. S., Gleim, R. D., Winkle, M. R. (1983) *J. Chem. Phys.*, **78**, 5847–5852	
Will, F. G. (1985) *J. Electrochem. Soc.*, **132**, 518–519	
Willett, J. L. (1998) *Polym. Preprints*, **39**, 112–113	
Wilson, R. H. (1990) *TrAC, Trends Anal. Chem.*, **9**, 127–131	
Woo, S. I., Hill, C. G. (1985) *J. Mol. Catal.*, **29**, 209–229	
Workman, J. J. (2001) *Appl. Spectrosc. Rev.*, **36**, 139–168	
Xie, R., Qu, B. (2001) *Polym. Degrad. Stab.*, **71**, 395–402	
Xu, G., Gong, M., Shi, W. (2005) *Polym. Adv. Technol.*, **16**, 473–479	
Xu, Z. H., Butler, I. S., St.-Germain, F. G. T. (1986) *Appl. Spectrosc.*, **40**, 1004–1009	
Xu, Z., Butler, I. S., Wu, J., Xu, G. (1988) *Mikrochim. Acta*, **2**, 171–174	
Xu, Z., Wang, J., Kung, J., Woods, J. R., Wu, X. A., Kotlyar, L. S., Sparks, B. D., Chung, K. H. (2005) *Fuel*, **84**, 661–668	
Yamada, O., Yasuda, H., Soneda, Y., Kobayashi, M., Makino, M., Kaiho, M. (1996) *Preprints, ACS Div. Fuel Chem.*, **41**, 93–97	
Yamauchi, S., Sudiyani, Y., Imamura, Y., Doi, S. (2004) *J. Wood Sci.*, **50**, 433–438	
Yan, B., Gremlich, H.-U., Moss, S., Coppola, G. M., Sun, Q., Liu, L. (1999) *J. Comb. Chem.*, **1**, 46–54	
Yang, C. Q., Ellis, T. J., Bresee, R. R., Fateley, W. G. (1985) *Polymer. Mater. Sci. Eng.*, **53**, 169–175	
Yang, C. Q., Fateley, W. G. (1986a) *J. Mol. Struct.*, **141**, 279–281	

Appendix 2: Literature Guide – Solids and Liquids | 349

Catalysts	Clays, minerals	Food products	Hydrocarbons, fuels	Inorganics	Medical, biological	Organics	Polymers	Textiles	Wood, paper
x									
					x				
					x				
							x		
						x			
							x		
					x				
					x				
							x		
	x								
							x		
							x		
							x		
		x							
x									
									x
							x		
							x		
				x					
				x					
	x		x						
			x						
									x
						x	x		
							x	x	
	x								

Citation	Carbons
Yang, C. Q., Fateley, W. G. (1986b) *J. Mol. Struct.*, **146**, 25–39	
Yang, C. Q., Fateley, W. G. (1986c) *Polym. Mater. Sci. Eng.*, **54**, 404–410	
Yang, C. Q., Bresee, R. R. (1987a) *J. Coated Fabrics*, **17**, 110–128	
Yang, C. Q., Bresee, R. R., Fateley, W. G. (1987b) *Appl. Spectrosc.*, **41**, 889–896	
Yang, C. Q., Fateley, W. G. (1987c) *Anal. Chim. Acta*, **194**, 303–309	
Yang, C. Q., Bresee, R. R., Fateley, W. G. (1987d) *ACS Symp. Ser.*, **340**, 214–232	
Yang, C. Q., Perenich, T. A., Fateley, W. G. (1989) *Textile Res. J.*, **59**, 562–568	
Yang, C. Q. (1990) *Polym. Mater. Sci. Eng.*, **62**, 903–910	
Yang, C. Q., Bresee, R. R., Fateley, W. G. (1990) *Appl. Spectrosc.*, **44**, 1035–1039	
Yang, C. Q. (1991a) *Appl. Spectrosc.*, **45**, 102–108	
Yang, C. Q. (1991b) *Textile Res. J.*, **61**, 298–305	
Yang, C. Q. (1991c) *Polym. Mater. Sci. Eng.*, **64**, 372–374	
Yang, C. Q. (1992) *Ind. Eng. Chem. Res.*, **31**, 617–621	
Yang, C. Q., Freeman, J. M. (1991) *Polym. Mater. Sci. Eng.*, **64**, 33–35	
Yang, C. Q., Freeman, J. M. (1991b) *Appl. Spectrosc.*, **45**, 1695–1698	
Yang, C. Q., Kottes Andrews, B. A. (1991) *J. Appl. Polym. Sci.*, **43**, 1609–1616	
Yang, C. Q., Simms, J. R. (1993) *Carbon*, **31**, 451–459	x
Yang, C. Q., Martin, L. K. (1994) *J. Appl. Polym. Sci.*, **51**, 389–397	
Yang, C. Q., Simms, J. R. (1995) *Fuel*, **74**, 543–548	x
Yang, D.-Q., Sacher, E. (2001a) *Appl. Surf. Sci.*, **173**, 30–39	
Yang, D.-Q., Martinu, L., Sacher, E., Sadough-Vanini, A. (2001b) *Appl. Surf. Sci.*, **177**, 85–95	
Yang, D.-Q., Rochette, J.-F., Sacher, E. (2005a) *J. Phys. Chem. B*, **109**, 4481–4484	x
Yang, D.-Q., Meunier, M., Sacher, E. (2005b) *J. Appl. Phys.*, **98**, 114310-1-6	
Yang, H., Irudayaraj, J. (1999) *Part. Sci. Technol.*, **17**, 269–282	
Yang, H., Irudayaraj, J. (2000a) *J. Am. Oil Chem. Soc.*, **77**, 291–295	
Yang, H., Irudayaraj, J. (2000b) *Trans. Am. Soc. Agric. Eng.*, **43**, 953–961	
Yang, H., Irudayaraj, J., Sakhamuri, S. (2001a) *Appl. Spectrosc.*, **55**, 571–583	
Yang, H., Irudayaraj, J. (2001b) *J. Am. Oil Chem. Soc.*, **78**, 889–895	
Yang, H., Irudayaraj, J. (2001c) *Lebensm.-Wiss. u.-Technol.*, **34**, 402–409	
Yang, H., Irudayaraj, J. (2002) *J. Pharm. Pharmacol.*, **54**, 1247–1255	
Yang, X., Hamza, H., Czarnecki, J. (2004) *Energy Fuels*, **18**, 770–777	
Yaniger, S. I., Riseman, S. M., Frigo, T., Eyring, E. M. (1982) *J. Chem. Phys.*, **76**, 4298–4299	
Yaniger, S. I., Rose, D. J., McKenna, W. P., Eyring, E. M. (1984a) *Macromolecules*, **17**, 2579–2583	
Yaniger, S. I., Rose, D. J., McKenna, W. P., Eyring, E. M. (1984b) *Appl. Spectrosc.*, **38**, 7–11	
Yariv, S., Nasser, A., Michaelian, K. H., Lapides, I., Deutsch, Y., Lahav, N. (1994) *Thermochim. Acta*, **234**, 275–285	

Catalysts	Clays, minerals	Food products	Hydrocarbons, fuels	Inorganics	Medical, biological	Organics	Polymers	Textiles	Wood, paper
	x								
							x		
								x	
							x	x	
							x	x	
								x	
								x	
							x	x	
							x		
							x	x	
								x	
								x	
								x	
								x	
								x	
								x	
								x	
						x			
						x			
	x								
		x							
		x							
		x							
		x							
		x							
		x							
		x							
			x						
							x		
							x		
							x		
	x								

Citation	Carbons

Yasumoto, M., Ogawa, H., Sei, N., Yamada, K. (2004) *Proc. FEL Conf.*, 703–705

Yeboah, S. A., Griffiths, P. R., Krishnan, K., Kuehl, D. (1981) *SPIE*, **289**, 105–107

Ying, J. Y., Benziger, J. B. (1992) *J. Non-Cryst. Solids*, **147–148**, 222–231

Ying, J. Y., Benziger, J. B., Gleiter, H. (1993) *Phys. Rev. B*, **48**, 1830–1836

Yokoyama, Y., Kosugi, M., Kanda, H., Ozasa, M., Hodouchi, K. (1984) *Bunseki Kagaku*, **33**, E1–E7

Yoshino, K., Fukuyama, A., Yokoyama, H., Meada, K., Fons, P. J., Yamada, A., Niki, S., Ikari, T. (1999) *Thin Solid Films*, **343–344**, 591–593

Yu, X., Song, H., Su, Q. (2003) *J. Mol. Struct.*, **644**, 119–123

Yue-e, F., Xia, Z., Wu, G. X. (1997) *Radiat. Phys. Chem.*, **50**, 487–491

Yue-e, F., Jun, J., Chaoxiong, M. (1999) *Radiat. Phys. Chem.*, **54**, 159–163

Zachmann, G. (1984) *J. Mol. Struct.*, **115**, 465–468

Zegadi, A., Slifkin, M. A., Tomlinson, R. D. (1994) *Rev. Sci. Instrum.*, **65**, 2238–2243

Zegadi, A., Al-Saffar, I. S., Yakushev, M. V., Tomlinson, R. D. (1995) *Rev. Sci. Instrum.*, **66**, 4095–4101

Zegadi, A., Yakushev, M. V., Ahmed, E., Pilkington, R. D., Hill, A. E., Tomlinson, R. D. (1996) *Sol. Energy Mater. Sol. Cells*, **41–42**, 295–305

Zerlia, T. (1985) *Fuel*, **64**, 1310–1312

Zerlia, T. (1986) *Appl. Spectrosc.*, **40**, 214–217

Zerlia, T., Girelli, A. (1988a) *Mikrochim. Acta*, **2**, 175–178

Zerlia, T., Carimati, A., Marengo, S., Martinengo, S., Zanderighi, L. (1988b) *Struct. React. Surf.*, **48**, 943–953

Zerlia, T., Marini, A., Berbenni, V., Massarotti, V., Giordano, F., La Manna, A., Bettinetti, G. P., Margheritis, C. (1989) *Solid State Ionics*, **32–33**, 613–624

Zhang, G., Wang, Q., Yu, X., Su, D., Li, Z., Zhang, G. (1991) *Spectrochim. Acta, Part A*, **47**, 737–741

Zhang, J., Zhong, Q., Wu, N., Wang, J., Zhao, J., Xu, Y., Guo, S., Ying, C. (1996) *Phys. Lett. A*, **215**, 291–295

Zhang, P., Urban, M. W. (2004) *Langmuir*, **20**, 10691–10699

Zhang, S. L., Michaelian, K. H., Burt, J. A. (1997) *Opt. Eng.*, **36**, 321–325

Zhang, W. R., Lowe, C., Smith, R. (2009) *Prog. Org. Coat.*, **65**, 469–476

Zhao, Y., Urban, M. W. (2000) *Macromolecules*, **33**, 2184–2191

Zhou, W., Xie, S., Qian, S., Wang, G., Qian, L., Sun, L., Tang, D., Liu, Z. (2000) *J. Phys. Chem. Solids*, **61**, 1165–1169   x

Catalysts	Clays, minerals	Food products	Hydrocarbons, fuels	Inorganics	Medical, biological	Organics	Polymers	Textiles	Wood, paper
							x		
						x	x		
				x					
				x					
				x					
				x					
				x					
							x		
							x		
							x		
				x					
				x					
				x					
			x						
			x						
x			x						
x									
					x		x		
						x			
				x					
							x		
	x								
							x		
							x		

Appendix 3: Literature Guide – Gases

The following 13 tables list references on PA infrared spectra of gases. These tables are organized according to chemical class or, in some cases, the number of atoms per molecule. As in Appendix 2, bibliographic data consists of authors' names and publication details. Many of these publications are discussed within the main part of this book, where titles of the articles are also given in the reference lists.

Presentation or discussion of experimental results for a particular compound is indicated by an 'x' in the appropriate column. Formulas of isotopomers (fully or partially deuterated compounds) are given explicitly in place of this designation.

Articles that describe spectra of more than one compound appear in several tables as appropriate. This organization differs from that in Appendix 2 (where each reference appears once) and is necessitated by the large number of gases included in Appendix 3.

The very substantial body of literature on PA infrared spectra of gases precludes preparation of comprehensive reference lists. Instead, this guide provides interested readers with a means for approaching this literature. Additional references can be found by searching various databases, particularly for articles by the more prominent researchers in this field.

Table A3.1 Acids.

Citation	CH_2O_2	$C_2H_4O_2$	$C_3H_6O_2$
	Formic acid	Acetic acid	Propanoic acid
Kaestle, R., Sigrist, M. W. (1996) *Appl. Phys. B*, **63**, 389–397		CD_3COOD	
Kaestle, R., Sigrist, M. W. (1996) *Spectrochim. Acta, Part A*, **52**, 1221–1228		x, CD_3COOD	x
Merker, U., Engels, P., Madeja, F., Havenity, M., Urban, W. (1999) *Rev. Sci. Instrum.*, **70**, 1933–1938	x		

Photoacoustic IR Spectroscopy. 2nd Ed., Kirk H. Michaelian
Copyright © 2010 WILEY-VCH Verlag GmbH & Co. KGaA, Weinheim
ISBN: 978-3-527-40900-6

Table A3.2 Alcohols.

Citation	CH$_4$O Methanol	C$_2$H$_6$O Ethanol	C$_2$H$_6$O$_2$ Ethylene glycol	C$_3$H$_8$O$_2$ 1,3-Propanediol	C$_4$H$_{10}$O$_2$ 1,4-Butanediol	C$_6$H$_6$O Phenol
Barbieri, S., Pellaux, J.-P., Studemann, E., Rosset, D. (2002) *Rev. Sci. Instrum.*, **73**, 2458–2461		x				
Bijnen, F.G.C.; Zuckermann, H.; Harren, F.J.M.; Reuss, J. (1998) *Appl. Opt.*, **37**, 3345–3353		x				
Bohren, A., Sigrist, M.W. (1997) *Infrared Phys. Technol.*, **38**, 423–435	x	x				
Brienne, S. H. R., Bozkurt, M. M. V., Rao, S. R., Xu, Z., Butler, I. S., Finch, J. A. (1996) *Appl. Spectrosc.*, **50**, 521–527		x				
Busse, G., Bullemer, B. (1978) *Infrared Phys*, **18**, 631–634	x					
Elia, A., DiFranco, C., Lugara, P. M., Scamarcio, G. (2006) *Sensors*, **6**, 1411–1419	x					
Hofstetter, D., Beck, M., Faist, J., Nagele, M., Sigrist, M. (2001) *Opt. Lett.*, **26**, 887–889	x					
Howard, D.L., Kjaergaard, H.G. (2006) *J. Phys. Chem. A*, **110**, 10245–10250			x	x	x	
Huang, J., Urban, M. W. (1993) *J. Chem. Phys*, **98**, 5259–5268		x				
Ishiuchi, S-i., Fujii, M., Robinson, T. W., Miller, B. J., Kjaergaard, H. G. (2006) *J. Phys. Chem. A*, **110**, 7345–7354						x
Kreuzer, L.B. (1971) *J. Appl. Phys*, **42**, 2934–2943	x					
Li, C.-P.; Davis, J. (1979) *Appl. Opt.*, **18**, 3541–3543	x					
Moeckli, M.A.; Hilbes, C.; Sigrist, M.W. (1998) *Appl. Phys. B*, **67**, 449–458	x					

Reference	Marker
Moraes, J.C.S.; Telles, E.M.; Cruz, F.C.; Scalabrin, A.; Pereira, D.; Carelli, G.; Ioli, N.; Moretti, A.; Strumia, F. (1992) *Int. J. Infrared Millimeter Waves*, **13**, 1801–1823	$^{13}CD_3OH$
Naegele, M., Sigrist, M. W. (2000) *Appl. Phys. B*, **70**, 895–901	x
Oomens, J., Zuckermann, H., Persijn, S., Parker, D. H., Harren, F. J. M. (1998) *Appl. Phys. B*, **67**, 459–466	x
Pereira, D., Telles, E. M., Moraes, J. C. S., Scalabrin, A., Carelli, G., Ioli, N., Moretti, A., Strumia, F. (1992) *Proc. SPIE Int. Soc. Opt. Eng*, **1929**, 250–251	$^{13}CD_3OH$
Persijn, S. T., Veltman, R. H., Oomens, J., Harren, F. J. M., Parker, D. H. (1999) *AIP Conf. Proc.*, **463**, 609–611	x
Persijn, S. T., Veltman, R. H., Oomens, J., Harren, F. J. M., Parker, D. H. (2000) *Appl. Spectrosc.*, **54**, 62–71	x
Pushkarsky, M., Webber, M., Patel, C. K. N. (2005) *Proc. SPIE Int. Soc. Opt. Eng.*, **5732**, 93–107	x
Repond, P., Sigrist, M. W. (1994) *J. Phys. IV*, **4**, C7-523-526	x
Repond, P., Sigrist, M. W. (1996) *Appl. Opt.*, **35**, 4065–4085	x
Rey, J. M., Schramm, D., Hahnloser, D., Marinov, D., Sigrist, M. W. (2008) *Meas. Sci. Technol*, **19**, 075602	x
Sigrist, M. W. (1999) *AIP Conf. Proc.*, **463**, 203–207	x
Sigrist, M. W., Bohren, A., Calasso, I. G., Nägele, M., Romann, A., Seiter, M. (2000) *Proc. SPIE Int. Soc. Opt. Eng.*, **3916**, 286–294	x
Sigrist, M. W., Bohren, A., Calasso, I. G., Nägele, M., Romann, A., Seiter, M. (2000) *Proc. SPIE Int. Soc. Opt. Eng*, **4063**, 17–25	x
Skřínský, J., Janečková, R., Grigorová, E., Stčížik, M., Kubát, P., Herecová, L., Nevrlý, V., Zelinger, Z., Civiš, S. (2009) *J. Mol. Spectrosc.*, **256**, 99–101	x

Table A3.2 Continued

Citation	CH$_4$O	C$_2$H$_6$O	C$_2$H$_6$O$_2$	C$_3$H$_8$O$_2$	C$_4$H$_{10}$O$_2$	C$_6$H$_6$O
	Methanol	Ethanol	Ethylene glycol	1,3-Propanediol	1,4-Butanediol	Phenol
Telles, E. M., Moraes, J. C. S., Scalabrin, A., Pereira, D., Carelli, G., Ioli, N., Moretti, A., Strumia, F. (1994) *IEEE J. Quantum Elect.*, **30**, 2946–2949	^{13}CD$_3$OH					
Telles, E. M., Cruz, F. C., Scalabrin, A., Pereira, D. (2001) *Int. J. Infrared Millimeter Waves*, **22**, 521–530		x				
Tittel, F. K., Richter, D., Fried, A. (2003) *Topics Appl. Phys.*, **89**, 445–516	x					
Viscovini, R. C., Scalabrin, A., Pereira, D. (1997) *Int. J. Infrared Millimeter Waves*, **17**, 1821–1838	x					
Viscovini, R. C., Scalabrin, A., Pereira, D. (1997) *IEEE J. Quantum Elect.*, **33**, 916–918	x					
Viscovini, R. C., Scalabrin, A., Pereira, D. (2000) *Int. J. Infrared Millimeter Waves*, **21**, 621–632	^{13}CD$_3$OD					
Wysocki, G., Lewicki, R., Curl, R. F., Tittel, F. K., Diehl, L., Capasso, F., Troccoli, M., Hofler, G., Bour, D., Corzine, S., Maulini, R., Giovannini, M., Faist, J. (2008) *Appl. Phys. B*, **92**, 305–311		x				
Zelinger, Z., Strizik, M., Kubat, P., Janour, Z., Berger, P., Cerny, A., Engst, P. (2004) *Opt. Lasers Eng.*, **42**, 403–412		x				
Zelinger, Z., Střížík, M., Kubát, P., Civiš, S., Grigorová, E., Janečková, R., Zavila, O., Nevrlý, V., Herecova, L., Bailleux, S., Horka, V., Ferus, M., Skřínský, J., Kozubkova, M., Drabková, S., Janour, Z. (2009) *Appl. Spectrosc.*, **63**, 430–436	x	x				

Table A3.3 Aldehydes.

Citation	CH$_2$O Formaldehyde	CH$_2$S Thio-formaldehyde	C$_2$H$_4$O Acetaldehyde	C$_3$H$_4$O Acrolein
Angelmahr, M., Miklos, A., Hess, P. (2006) Appl. Phys. B, **85**, 285–288	x			
Bijnen, F. G. C., Zuckermann, H., Harren, F. J. M., Reuss, J. (1998) Appl. Opt., **37**, 3345–3353			x	
Cihelka, J., Matulková, I., Civiš, S. (2009) J. Mol. Spectrosc., **256**, 68–74	x			
Darwish, A. M., Petersen, J. C., Izatt, J. R. (1992) Proc. SPIE Int. Soc. Opt. Eng., **1929**, 168–169		x		
Fischer, C., Sigrist, M. W. (2002) Appl. Phys. B, **75**, 305–310	x			
Horstjann, M., Bakhirkin, Y. A., Kosterev, A. A., Curl, R. F., Tittel, F. K., Wong, C. M., Hill, C. J., Yang, R. Q. (2004) Appl. Phys. B, **79**, 799–803	x			
Oomens, J., Zuckermann, H., Persijn, S., Parker, D. H., Harren, F. J. M. (1998) Appl. Phys. B, **67**, 459–466			x	
Persijn, S. T., Veltman, R. H., Oomens, J., Harren, F. J. M., Parker, D. H. (1999) AIP Conf. Proc., **463**, 609–611			x	
Persijn, S. T., Veltman, R. H., Oomens, J., Harren, F. J. M., Parker, D. H. (2000) Appl. Spectrosc., **54**, 62–71			x	
Seiter, M., Sigrist, M. W. (2000) Infrared Phys. Technol., **41**, 259–269	x			
Sigrist, M. W., Bernegger, S., Meyer, P. L. (1989) Infrared Phys., **29**, 805–814	x		x	x
Sigrist, M. W., Bohren, A., Calasso, I. G., Nägele, M., Romann, A., Seiter, M. (2000) Proc. SPIE Int. Soc. Opt. Eng., **4063**, 17–25				x
So, S., Koushanfar, F., Kosterev, A., Tittel, F. K. (2007) Proc. Int. Conf. Inform. Proc. Sensor Networks, 226–237	x			
Spagnolo, V., Elia, A., Franco, C. D., Lugarà, P. M., Vitiello, M. S., Scamarcio, G. (2009) Proc. SPIE Int. Soc. Opt. Eng., **7222**, 72220K	x			

Table A3.4 Alkanes.

Citation	CH$_4$ Methane	C$_2$H$_6$ Ethane	C$_3$H$_8$ Propane	C$_4$H$_{10}$ Butane	C$_5$H$_{12}$ Pentane	C$_5$H$_{12}$ Isopentane
Ageev, B. G., Ponomarev, Yu. N., Sapozhnikova, V. A. (1998) *Appl. Phys. B*, **67**, 467–473	x					
Baxter, G. W., Barth, H.-D., Orr, B.J. (1998) *Appl. Phys. B*, **66**, 653–657	x					
Berrou, A., Raybaut, M., Godard, A., Lefebvre, M. (2010) *Appl. Phys. B*, **98**, 217–230	x					
Besson, J.-P., Schilt, S., Thevenaz, L. (2004) *Spectrochim. Acta, Part A*, **60**, 3449–3456	x					
Besson, J.-P., Schilt, S., Thevenaz, L. (2006) *Spectrochim. Acta, Part A* **63**, 899–904	x					
Bijnen, F. G. C., Brugman, T., Harren, F. J. M., Reuss, J. (1992) *Photoacoustic and Photothermal Phenomena III*, 34–37		x				
Bijnen, F. G. C., Harren, F. J. M., Hackstein, J. H. P., Reuss, J. (1996) *Appl. Opt.*, **35**, 5357–5368	x					
Bohren, A., Sigrist, M. W. (1997) *Infrared Phys. Technol.*, **38**, 423–435						x
Boschetti, A., Bassi, D., Iacob, E., Iannota, S., Ricci, L., Scotoni, M. (2002) *Appl. Phys. B*, **74**, 273–278	x					
Civis, S., Horka, V., Cihelka, J., Simecek, T., Hulicius, E., Oswald, J., Pangrac, J., Vicet, A., Rouillard, Y., Salhi, A., Alibert, C., Werner, R., Koeth, J. (2005) *Appl. Phys. B*, **81**, 857–861	x					
Claspy, P. C., Pao, Y-H., Kwong, S., Nodov, E. (1976) *Appl. Opt*, **15**, 1506–1509	x					
Firebaugh, S. L. (2001) *J. Microelectromech. Syst.*, **10**, 232–237			x			
Fischer, C., Sigrist, M. W., Yu, Q., Seiter, M. (2001) *Opt. Lett.*, **26**, 1609–1611	x					
Fischer, C., Sigrist, M. W. (2002) *Appl. Phys. B*, **75**, 305–310	x	x				
Fischer, C., Sorokin, E., Sorokina, I. T., Sigrist, M. W. (2005) *Opt. Lasers Eng.*, **43**, 573–582	x					

Fischer, C., Bartlome, R., Sigrist, M. W. (2006) *Appl. Phys. B*, **85**, 289–294

Grossel, A., Zeninari, V., Joly, L., Parvitte, B., Courtois, D., Durry, G. (2006) *Spectrochim. Acta, Part A*, **63**, 1021–1028

Grossel, A., Zeninari, V., Parvitte, B., Joly, L., Courtois, D. (2007) *Appl. Phys. B*, **88**, 483–492

Horka, V., Civis, S., Xu, L.-H., Lees, R. M. (2005) *Analyst*, **130**, 1148–1154

Kapitanov, V. A., Zeninari, V., Parvitte, B., Courtois, D., Ponomarev, Yu. N. (2002) *Spectrochim. Acta, Part A*, **58**, 2397–2404

Kapitanov, V. A., Ponomarev, Yu. N., Tyryshkin, I. S., Rostov, A. P. (2007) *Spectrochim. Acta, Part A*, **66**, 811–818

Karbach, A., Roeper, J., Hess, P. (1984) *Rev. Sci. Instrum.*, **55**, 892–895

Karbach, A., Hess, P. (1985) *J. Chem. Phys.*, **83**, 1075–1084

Karbach, A., Hess, P. (1985) *J. Appl. Phys.*, **58**, 3851–3855

Karbach, A., Hess, P. (1986) *J. Chem. Phys.*, **84**, 2945–2952

Kauppinen, J., Wilcken, K., Kauppinen, I., Koskinen, V. (2004) *Microchem. J.*, **76**, 151–159

Koskinen, V., Fonsen, J., Kauppinen, J., Kauppinen, I. (2006) *Vib. Spectrosc.*, **42**, 239–242

Koskinen, V., Fonsen, J., Roth, K., Kauppinen, J. (2008) *Vib. Spectrosc.*, **48**, 16–21

Kosterev, A. A., Bakhirkin, Yu. A., Curl, R. F., Tittel, F. K. (2002) *Opt. Lett.*, **27**, 1902–1904

Kosterev, A. A., Bakhirkin, Yu. A., Tittel, F. K., McWhorter, S., Ashcraft, B. (2008) *Appl. Phys. B*, **92**, 103–109

Kreuzer, L. B. (1971) *J. Appl. Phys.*, **42**, 2934–2943

Kuhnemann, F., Schneider, K., Hecker, A., Martis, A. A. E., Urban, W., Schiller, S., Mlynek, J. (1998) *Appl. Phys. B*, **66**, 741–745

Kuusela, T., Kauppinen, J. (2007) *Appl. Spectrosc. Rev.*, **42**, 443–474

Table A3.4 Continued

Citation	CH$_4$ Methane	C$_2$H$_6$ Ethane	C$_3$H$_8$ Propane	C$_4$H$_{10}$ Butane	C$_5$H$_{12}$ Pentane	C$_5$H$_{12}$ Isopentane
Kuusela, T., Peura, J., Matveev, B. A., Remennyy, M. A., Stus', N. M. (2009) *Vib. Spectrosc.*, **51**, 289–293	x		x			
Lawton, S. A., Bragg, S. L. (1984) *Appl. Opt.*, **23**, 3042–3044	x					
Liang, G.-C., Liu, H.-H., Kung, A. H., Mohacsi, A., Miklos, A., Hess, P. (2000) *J. Phys. Chem. A*, **104**, 10179–10183	x					
Miklos, A., Hess, P., Romolini, A., Spada, C., Lancia, A., Kamm, S., Schaefer, S. (1999) *AIP Conf. Proc.*, **463**, 217–219	x					
Miklos, A., Lim, C.-H., Hsiang, W.-W., Liang, G.-C., Kung, A. H., Schmohl, A., Hess, P. (2002) *Appl. Opt.*, **41**, 2985–2993	x					
Mueller, F., Popp, A., Kuehnemann, F., Schiller, S. (2003) OSA TOPS, (no page numbers)		x				
Müller, F., Popp, A., Kühnemann, F., Schiller, S. (2003) *Opt. Express*, **11**, 2820–2825		x				
Ng, J., Kung, A. H., Miklos, A., Hess, P. (2004) *Opt. Lett.*, **29**, 1206–1208	x					
Ngai, A. K. Y., Persijn, S. T., VonBasum, G., Harren, F. J. M. (2006) *Appl. Phys. B*, **85**, 173–180	x					
Ngai, A. K. Y., Persijn, S. T., Lindsay, I. D., Kosterev, A. A., Gross, P., Lee, C. J., Cristescu, S. M., Tittel, F. K., Boller, K.-J., Harren, F. J. M. (2007) *Appl. Phys. B*, **89**, 123–128		x				
Oomens, J., Bisson, S., Harting, M., Kulp, T., Harren, F. J. M. (2000) *Proc. SPIE Int. Soc. Opt. Eng.*, **3916**, 295–300	x	x		x	x	
Osada, T., Rom, H. B., Dahl, P. (1998) *Trans. Am. Soc. Agric. Eng.*, **41**, 1109–1114	x					
Osada, T., Fukumoto, Y. (2001) *Water Sci. Technol.*, **44**, 79–86	x					

Persijn, S. T., Santosa, E., Harren, F. J. M. (2002) *Appl. Phys. B*, **75**, 335–342

Rosengren, L.-G. (1973) *Infrared Phys.*, **13**, 109–121

Santosa, E., te Lintel Hekkert, S., Harren, F. J. M., Parker, D. H. (1999) *AIP Conf. Proc.*, **463**, 612–614

Schaefer, S., Mashni, M. Sneider, J., Miklos, A., Hess, P., Pitz, H., Pleban, K.-U., Ebert, V. (1998) *Appl. Phys. B*, **66**, 511–516

Schilt, S., Besson, J.-Ph., Thevenaz, L. (2005) *J. Phys. IV*, **125**, 7–10

Schilt, S., Besson, J.-Ph., Thevenaz, L. (2006) *Appl. Phys. B*, **82**, 319–329

Scotoni, M., Roosi, A., Bassi, D., Buffa, R., Iannotta, S., Boschetti, A. (2006) *Appl. Phys. B*, **82**, 495–500

Sell, J. A. (1985) *Appl. Opt.*, **24**, 152–153

Sigrist, M. W. (2003) *Rev. Sci. Instrum.*, **74**, 486–490

So, S., Koushanfar, F., Kosterev, A., Tittel, F. K. (2007) *Proc. Int. Conf. Inform. Proc. Sensor Networks*, 226–237

Song, K., Cha, H. K., Kapitanov, V. A., Ponomarev, Yu. N., Rostov, A. P., Courtois, D., Parvitte, B., Zeninari, V. (2002) *Appl. Phys. B*, **75**, 215–227

Uotila, J., Koskinen, V., Kauppinen, J. (2005) *Vib. Spectrosc.*, **38**, 3–9

Uotila, J., Kauppinen, J. (2008) *Appl. Spectrosc.*, **62**, 655–660

vanHerpen, M., te Lintel Hekkert, S., Bisson, S. E., Harren, F. J. M. (2002) *Opt. Lett.*, **27**, 640–642

Wilcken, K., Kauppinen, J. (2003) *Appl. Spectrosc.*, **57**, 1087–1092

Zeninari, V., Parvitte, B., Courtois, D., Kapitanov, V. A., Ponomarev, Yu. N. (2003) *Infrared Phys. Technol.*, **44**, 253–261

Table A3.5 Alkenes and alkynes.

Citation	C_2H_4 Ethylene	C_3H_6 Propylene	C_5H_8 Isoprene	C_2H_2 Acetylene
Ageev, B. G., Ponomarev, Yu. N., Sapozhnikova, V. A. (1998) *Appl. Phys. B*, **67**, 467–473	x			
Bijnen, F. G. C., Zuckermann, H., Harren, F. J. M., Reuss, J. (1998) *Appl. Opt.*, **37**, 3345–3353	x			
Boschetti, A., Bassi, D., Iacob, E., Iannotta, S., Ricci, L., Scotoni, M. (2002) *Appl. Phys. B*, **74**, 273–278	x			
Besson, J.-P., Schilt, S., Thevenaz, L. (2006) *Spectrochim. Acta, Part A* **63**, 899–904	x			
Calasso, I. G., Sigrist, M. W. (1999) *Rev. Sci. Instrum*, **70**, 4569–4578	x			
Cristescu, S., Dumitras, D. C., Dutu, D. C. A. (2000) *Proc. SPIE Int. Soc. Opt. Eng.*, **4070**, 457–464	x			
Dahnke, H., Kahl, J., Schüler, G., Boland, W., Urban, W., Kühnemann, F. (2000) *Appl. Phys. B*, **70**, 275–280.			x	
Fournier, D., Boccara, A. C., Amer, N. M., Gerlach, R. (1980) *Appl. Phys. Lett.*, **37**, 519–521	x			
Gondal, M.A. (1997) *Appl. Opt.*, **36**, 3195–3201	x			
Haub, J. G., Johnson, M. J., Orr, B. J. (1993) *J. Opt. Soc. Am. B: Opt. Phys.*, **10**, 1765–1777				x
Hornberger, Ch., Konig, M., Rai, S. B., Demtroder, W. (1995) *Chem. Phys.*, **190**, 171–177				x
Kania, P., Civis, S. (2003) *Spectrochim. Acta, Part A*, **59**, 3063–3074				x
Kuhnemann, F., Wolfertz, M., Arnold, S., Lagemann, M., Popp, A., Schuler, G., Jux, A., Boland, W. (2002) *Appl. Phys. B*, **75**, 397–403			x	
Li, J. S., Gao, X., Li, W., Cao, Z., Deng, L., Zhao, W., Huang, M., Zhang, W. (2006) *Spectrochim. Acta, Part A*, **64**, 338–342				x
Li, J. S., Gao, X., Fang, L., Zhang, W., Cha, H. (2007) *Opt. Laser Technol.*, **39**, 1144–1149				x

Marinov, D., Sigrist, M. W. (2003) *Photochem. Photobiol. Sci.*, **2**, 774–778

Moeckli, M. A., Hilbes, C., Sigrist, M. W. (1998) *Appl. Phys. B*, **67**, 449–458

Mueller, F., Popp, A., Kuehnemann, F., Schiller, S. (2003) *OSA TOPS*, (no page numbers)

Müller, F., Popp, A., Kühnemann, F., Schiller, S. (2003) *Opt. Express*, **11**, 2820–2825

Olafsson, A., Hammerich, M., Henningsen, J. (1992) *Appl. Opt.*, **31**, 2657–2668

Oomens, J., Zuckermann, H., Persijn, S., Parker, D. H., Harren, F. J. M. (1998) *Appl. Phys. B*, **67**, 459–466

Oomens, J., Bisson, S., Harting, M., Kulp, T., Harren, F. J. M. (2000) *Proc. SPIE Int. Soc. Opt. Eng.*, **3916**, 295–300

Perlmutter, P., Shtrikman, S., Slatkine, M. (1979) *Appl. Opt.*, **18**, 2267–2274

Persijn, S. T., Veltman, R. H., Oomens, J., Harren, F. J. M., Parker, D. H. (1999) *AIP Conf. Proc.*, **463**, 609–611

Persijn, S. T., Veltman, R. H., Oomens, J., Harren, F. J. M., Parker, D. H. (2000) *Appl. Spectrosc.*, **54**, 62–71

Petra, N., Zweck, J., Kosterev, A. A., Minkoff, S. E., Thomazy, D. (2009) *Appl. Phys. B*, **94**, 673–680

Portnov, A., Ganot, Y., Rosenwaks, S., Bar, I. (2005) *J. Mol. Struct.*, **744–747**, 107–115

Puskarsky, M. B., Dunayevskiy, I. G., Prasanna, M., Tsekoun, A. G., Go, R., Patel, C. K. N. (2006) *Proc. Natl. Acad. Sci. U.S.A.*, **103**, 19630–19634

Radak, B. B., Pastirk, I., Ristic, G. S., Petkovska, L. T. (1998) *Infrared Phys. Technol.*, **39**, 7–13

Repond, P., Sigrist, M. W. (1994) *J. Phys. IV*, **4**, C7-523–526

Repond, P., Sigrist, M. W. (1996) *Appl. Opt.*, **35**, 4065–4085

Santosa, E., te Lintel Hekkert, S., Harren, F. J. M., Parker, D. H. (1999) *AIP Conf. Proc.*, **463**, 612–614

Schaefer, S., Miklos, A., Hess, P. (1997) *Appl. Opt.*, **36**, 3202–3211

Schaefer, S., Miklos, A., Pusel, A., Hess, P. (1998) *Chem. Phys. Lett.*, **285**, 235–239

Table A3.5 Continued

Citation	C$_2$H$_4$ Ethylene	C$_3$H$_6$ Propylene	C$_5$H$_8$ Isoprene	C$_2$H$_2$ Acetylene
Schilt, S., Thevenaz, L., Courtois, E., Robert, P. A. (2002) *Spectrochim. Acta, Part A,* **58**, 2533–2539	x			
Schilt, S., Kosterev, A. A., Tittel, F. K. (2009) *Appl. Phys. B,* **95**, 813–824	x			
Schramm, D. U., Sthel, M. S., daSilva, M. G., Carneiro, L. O., Junior, A. J. S., Souza, A. P., Vargas, H. (2003) *Infrared Phys. Technol.,* **44**, 263–269	x			
Scotoni, M., Roosi, A., Bassi, D., Buffa, R., Iannotta, S., Boschetti, A. (2006) *Appl. Phys. B,* **82**, 495–500	x			
Sell, J. A. (1984) *Appl. Opt.,* **23**, 1586–1597	x			
Sigrist, M. W., Bernegger, S., Meyer, P. L. (1989) *Infrared Phys.,* **29**, 805–814	x	x		
Sigrist, M. W. (1999) *AIP Conf. Proc.,* **463**, 203–207	x			
Sigrist, M. W. (2003) *Rev. Sci. Instrum.,* **74**, 486–490	x			x
Sigrist, M. W., Bohren, A., Calasso, I. G., Nägele, M., Romann, A., Seiter, M. (2000) *Proc. SPIE Int. Soc. Opt. Eng.,* **3916**, 286–294	x			
Sigrist, M. W., Bohren, A., Calasso, I. G., Nägele, M., Romann, A., Seiter, M. (2000) *Proc. SPIE Int. Soc. Opt. Eng.,* **4063**, 17–25	x	x	x	
So, S., Koushanfar, F., Kosterev, A., Tittel, F. K. (2007) *Proc. Int. Conf. Inform. Proc. Sensor Networks,* 226–237				x
Zelinger, Z., Strizik, M., Kubat, P., Janour, Z., Berger, P., Cerny, A., Engst, P. (2004) *Opt. Lasers Eng.,* **42**, 403–412	x	x		
Zeninari, V., Kapitanov, V. A., Courtois, D., Ponomarev, Yu. N. (1999) *Infrared Phys. Technol.,* **40**, 1–23	x			

Table A3.6 Aromatics.

Citation	C_6H_6 Benzene	C_7H_8 Toluene	C_8H_8 Styrene	C_8H_{10} Xylene
Beenen, A., Niessner, R. (1998) *Analyst*, **123**, 543–545	x	x		
Beenen, A., Niessner, R. (1999) *Appl. Spectrosc.*, **53**, 1040–1044	x	x		x
Bohren, A., Sigrist, M. W. (1997) *Infrared Phys. Technol.*, **38**, 423–435	x			
Maravić, D. S., Trtica, M. S., Milhanić, Š. S., Radak, B. B. (2006) *Anal. Chim. Acta*, **555**, 259–262		x		x
Mohácsi, A., Bozóki, Z., Niessner, R. (2001) *Sens. Actuators, B*, **79**, 127–131	x			
Naegele, M., Sigrist, M. W. (2000) *Appl. Phys. B*, **70**, 895–901	x			
Radak, B. B., Petkovska, M. T., Trtica, M. S., Miljanic, S. S., Petkovska, L. T. (2004) *Anal. Chim. Acta*, **505**, 67–71	x			
Repond, P., Sigrist, M. W. (1994) *J. Phys. IV*, **4**, C7-523–526		x		
Repond, P., Sigrist, M. W. (1996) *Appl. Opt.*, **35**, 4065–4085		x		
Sigrist, M. W., Bernegger, S., Meyer, P. L. (1989) *Infrared Phys.*, **29**, 805–814	x	x		x
Sigrist, M. W., Bohren, A., Calasso, I. G., Nägele, M., Romann, A., Seiter, M. (2000) *Proc. SPIE Int. Soc. Opt. Eng.*, **3916**, 286–294	x	x		x
Sigrist, M. W., Bohren, A., Calasso, I. G., Nägele, M., Romann, A., Seiter, M. (2000) *Proc. SPIE Int. Soc. Opt. Eng.*, **4063**, 17–25	x	x		x
Wright, R. S., Howe, G. B., Jayanty, R. K. M. (1998) *J. Air Waste Manage. Assoc.*, **48**, 1077–1084			x	
Zelinger, Z., Střížik, M., Kubát, P., Civiš, S. (2000) *Anal. Chim. Acta*, **422**, 179–185		x		x

Table A3.7 Chlorides.

Citation	CH$_3$Cl	C$_2$HCl	C$_2$HCl$_3$	C$_2$H$_3$Cl	C$_2$H$_5$Cl
	Methyl chloride	Monochloro-acetylene	Trichloro-ethylene	Vinyl chloride	Ethyl chloride
Fischer, C., Sigrist, M. W. (2002) *Appl. Phys. B*, **75**, 305–310	x				
Marchetti, S., Simili, R., Carelli, G., Moretti, A., Strumia, F. (1998) *Infrared Phys. Technol.*, **39**, 89–92					x
Nela, M., Niskanen, K., Vaittinen, O., Halonen, L., Buerger, H., Polanz, O. (2002) *Mol. Phys.*, **100**, 655–665		x			
Radak, B. B., Petkovska, M. T., Trtica, M. S., Miljanic, S. S., Petkovska, L.T. (2004) *Anal. Chim. Acta*, **505**, 67–71			x		
Sigrist, M. W., Bohren, A., Calasso, I. G., Nägele, M., Romann, A., Seiter, M. (2000) *Proc. SPIE Int. Soc. Opt. Eng.*, **4063**, 17–25	x			x	

Table A3.8 Diatomics.

Citation	H$_2$	N$_2$	O$_2$	CO	HF	HCl
	Hydrogen	Nitrogen	Oxygen	Carbon monoxide	Hydrogen fluoride	Hydrogen chloride
Besson, J.-P., Schilt, S., Thevenaz, L. (2004) *Spectrochim. Acta, Part A*, **60**, 3449–3456						x
Besson, J.-P., Schilt, S., Thevenaz, L. (2006) *Spectrochim. Acta, Part A*, **63**, 899–904						x
Besson, J.-P., Schilt, S., Sauser, F., Rochat, E., Hamel, P., Sandoz, F., Nikles, M., Thevenaz, L. (2006) *Appl. Phys. B*, **85**, 343–348						x
Cattaneo, H., Laurila, T., Hernberg, R. (2006) *Appl. Phys. B*, **85**, 337–341			x			
Fischer, C., Sigrist, M. W. (2002) *Appl. Phys. B*, **75**, 305–310						x
Fischer, C., Sorokin, E., Sorokina, I. T., Sigrist, M. W. (2005) *Opt. Lasers Eng.*, **43**, 573–582				x		

Table A3.8 *Continued*

Citation	H$_2$ Hydrogen	N$_2$ Nitrogen	O$_2$ Oxygen	CO Carbon monoxide	HF Hydrogen fluoride	HCl Hydrogen chloride
Horka, V., Civis, S., Xu, L.-H., Lees, R. M. (2005) *Analyst*, **130**, 1148–1154				x		
Horstjann, M., Bakhirkin, Y. A., Kosterev, A. A., Curl, R. F., Tittel, F. K., Wong, C. M., Hill, C. J., Yang, R. Q. (2004) *Appl. Phys. B*, **79**, 799–803						x
Kania, P., Civis, S. (2003) *Spectrochim. Acta, Part A*, **59**, 3063–3074				x		
Kosterev, A. A., Bakhirkin, Y. A., Tittel, F. K., Blaser, S., Bonetti, Y., Hvozdara, L. (2004) *Appl. Phys. B*, **78**, 673–676				x		
Kosterev, A. A., Bakhirkin, Y. A., Tittel, F. K. (2005) *Appl. Phys. B*, **80**, 133–138				x		
Lawton, S. A., Bragg, S. L. (1984) *Appl. Opt.*, **23**, 3042–3044				x		
Miklos, A., Feher, M. (1996) *Infrared Phys. Technol.*, **37**, 21–27	x		x			
Schley, R. S., Telschow, K. L. (2001) *Proc. SPIE Int. Soc. Opt. Eng.*, **4205**, 18–24					x	
Sigrist, M. W. (2003) *Rev. Sci. Instrum.*, **74**, 486–490				x		
So, S., Koushanfar, F., Kosterev, A., Tittel, F. K. (2007) *Proc. Int. Conf. Inform. Proc. Sensor Networks*, 226–237				x		
Willing, B., Muralt, P., Oehler, O. (1999) *AIP Conf. Proc.*, **463**, 277–79				x		
Wolff, M., Harde, H. (2000) *Infrared Phys. Technol.*, **41**, 283–286					x	
Wolff, M., Groninga, H., Harde, H. (2004) *Appl. Spectrosc.*, **58**, 552–554					x	
Zeninari, V., Kapitanov, V. A., Courtois, D., Ponomarev, Yu. N. (1999) *Infrared Phys. Technol.*, **40**, 1–23		x				

Table A3.9 Haloalkanes.

Citation	CHClF$_2$ Freon 22	CHCl$_2$F Dichlorofluoromethane	CH$_3$Br Bromomethane	CClF$_3$ Freon 13	CCl$_2$F$_2$ Freon 12	CCl$_3$F Freon 11	C$_2$HF$_5$ Freon 125
Fischer, C., Sigrist, M. W. (2002) *Appl. Phys. B*, **75**, 305–310			x				
Kosterev, A., Wysocki, G., Bakhirkin, Y., So, S., Lewicki, R., Fraser, M., Tittel, F., Curl, R. F. (2008) *Appl. Phys. B*, **90**, 165–176							x
Lewicki, R., Wysocki, G., Kosterev, A. A., Tittel, F. K. (2007) *Opt. Express*, **15**, 7357–7366							x
Li, C.-P., Davis, J. (1979) *Appl. Opt.*, **18**, 3541–3543		x					
McCurdy, M. R., Bakhirkin, Y., Wysocki, G., Lewicki, R., Tittel, F. K. (2007) *J. Breath Res.*, **1**, 014001 (12 pp)							x
Phillips, M. C., Myers, T. L., Wojcik, M. D., Cannon, B. D. (2007) *Opt. Lett.*, **32**, 1177–1179							x
So, S., Koushanfar, F., Kosterev, A., Tittel, F. K. (2007) *Proc. Int. Conf. Inform. Proc. Sensor Networks*, 226–237							x
Troccoli, M., Diehl, L., Bour, D. P., Corzine, S. W., Yu, N., Wang, C. Y., Belkin, M. A., Höfler, G., Lewicki, R., Wysocki, G., Tittel, F. K., Capasso, F. (2008) *J. Lightwave Technol.*, **26**, 3534–3555.							x
Wojcik, M. D., Phillips, M. C., Cannon, B. D., Taubman, M. S. (2006) *Appl. Phys. B*, **85**, 307–313				x	x	x	
Zelinger, Z., Jancik, I., Engst, P. (1992) *Appl. Opt.*, **31**, 6974–6975	x						x
Zelinger, Z., Strizik, M., Kubat, P., Janour, Z., Berger, P., Cerny, A., Engst, P. (2004) *Opt. Lasers Eng.*, **42**, 403–412							x

Table A3.10 Hydrogen sulfide and selenide.

Citation	H$_2$S	H$_2$Se
	Hydrogen sulfide	Hydrogen selenide
Hao, L-y., Han, J-x., Shi, Q., Zhang, J-h., Zheng, J-j., Zhu, Q-s. (2000) *Rev. Sci. Instrum.*, **71**, 1975–1980		x
Hao, L-y., Qiang, S., Wu, G-r., Qi, L., Feng, D., Zhu, Q-s. (2002) *Rev. Sci. Instrum.*, **73**, 2079–2085		HDSe
Kadibelban, R., Hess, P. (1982) *Appl. Opt.*, **21**, 61–63	x	

Table A3.11 Nitrogen compounds.

Citation	NH$_3$	N$_2$O	NO	NO$_2$
	Ammonia	Nitrous oxide	Nitric oxide	Nitrogen dioxide
Ageev, B. G., Ponomarev, Yu. N., Sapozhnikova, V. A. (1998) *Appl. Phys. B*, **67**, 467–473	x			
Berrou, A., Raybaut, M., Godard, A., Lefebvre, M. (2010) *Appl. Phys. B*, **98**, 217–230		x		
Besson, J.-P., Schilt, S., Rochat, E., Thevenaz, L. (2006) *Appl. Phys. B*, **85**, 323–328	x			
Bozoki, Z., Mohacsi, A., Szabo, G., Bor, Z., Erdelyi, M., Chen, W., Tittel, F. K. (2002) *Appl. Spectrosc.*, **56**, 715–719	x			
Brewer, R. J., Bruce, C. W. (1978) *Appl. Opt.*, **17**, 3746–3749	x			
Civis, S., Horka, V., Cihelka, J., Simecek, T., Hulicius, E., Oswald, J., Pangrac, J., Vicet, A., Rouillard, Y., Salhi, A., Alibert, C., Werner, R., Koeth, J. (2005) *Appl. Phys. B*, **81**, 857–861	x			
Claspy, P. C., Pao, Y-H., Kwong, S., Nodov, E. (1976) *Appl. Opt.*, **15**, 1506–1509			x	x
Corriveau, R., Monchalin, J.-P., Bertrand, L. (1984) *Appl. Opt.*, **23**, 372–373			x	
Costopoulos, D., Miklos, A., Hess, P. (2002) *Appl. Phys. B*, **75**, 385–389		x		
Cristescu, S., Dumitras, D. C., Dutu, D. C. A. (2000) *Proc. SPIE Int. Soc. Opt. Eng.*, **4070**, 457–464	x			

Table A3.11 Continued

Citation	NH$_3$ Ammonia	N$_2$O Nitrous oxide	NO Nitric oxide	NO$_2$ Nitrogen dioxide
Danworaphong, S., Calasso, I. G., Beveridge, A., Diebold, G. J., Gmachl, C., Capasso, F., Sivco, D. L., Cho, A.Y. (2003) Appl. Opt., 27, 5561–5565		x		
Elia, A., DiFranco, C., Lugara, P. M., Scamarcio, G. (2006) Sensors, 6, 1411–1419	x		x	
Feher, M., Jiang, Y., Maier, J. P., Miklos, A. (1994) Appl. Opt., 33, 1655–1658	x			
Filho, M. B., da Silva, M. G., Sthel, M. S., Schramm, D. U., Vargas, H., Miklos, A., Hess, P. (2006) Appl. Opt., 45, 4966–4971	x			
Fischer, C., Sigrist, M. W. (2002) Appl. Phys. B, 75, 305–310				x
Fischer, C., Sorokin, E., Sorokina, I. T., Sigrist, M. W. (2005) Opt. Lasers Eng., 43, 573–582		x		
Grossel, A., Zeninari, V., Joly, L., Parvitte, B., Durry, G., Courtois, G. (2007) Infrared Phys. Technol., 51, 95–101			x	
Grossel, A., Zeninari, V., Parvitte, B., Joly, L., Courtois, D. (2007) Appl. Phys. B, 88, 483–492		x		
Hofstetter, D., Beck, M., Faist, J., Nagele, M., Sigrist, M. (2001) Opt. Lett., 26, 887–889	x			
Horka, V., Civis, S., Xu, L.-H., Lees, R. M. (2005) Analyst, 130, 1148–1154	x	x		
Huszar, H., Pogany, A., Bozoki, Z., Mohacsi, A., Horvath, L., Szabo, G. (2008) Sens. Actuators, B, 134, 1027–1033	x			
Kania, P., Civis, S. (2003) Spectrochim. Acta, Part A, 59, 3063–3074	x			
Kondo, T., Mitsui, T., Kitagawa, M., Nakae, Y. (2000) Digest. Diseases Sci., 45, 2054–2057		x		
Kosterev, A. A., Tittel, F. K. (2004) Appl. Opt., 43, 6213–6217	x			
Kosterev, A. A., Bakhirkin, Y. A., Tittel, F. K. (2005) Appl. Phys. B, 80, 133–138		x		
Lawton, S. A., Bragg, S. L. (1984) Appl. Opt., 23, 3042–3044	x			

Table A3.11 *Continued*

Citation	NH$_3$ Ammonia	N$_2$O Nitrous oxide	NO Nitric oxide	NO$_2$ Nitrogen dioxide
Lewicki, R., Wysocki, G., Kosterev, A. A., Tittel, F. K. (2007) *Appl. Phys. B*, **87**, 157–162	x			
Lima, J. P., Vargas, H., Miklos, A., Angelmahr, M., Hess, P. (2006) *Appl. Phys. B*, **85**, 279–284		x		x
Marinov, D., Sigrist, M. W. (2003) *Photochem. Photobiol. Sci.*, **2**, 774–778	x			
Max, E., Rosengren, L.-G. (1974) *Opt. Comm.*, **11**, 422–426	x			
McCurdy, M. R., Bakhirkin, Y., Wysocki, G., Lewicki, R., Tittel, F. K. (2007) *J. Breath Res.*, **1**, 014001 (12 pp)	x			
Miklos, A., Bozoki, Z., Jiang, Y., Feher, M. (1994) *Appl. Phys. B*, **58**, 483–492	x			
Miklos, A., Feher, M. (1996) *Infrared Phys. Technol.*, **37**, 21–27	x		x	
Miklos, A., Hess, P., Mohacsi, A., Sneider, J., Kamm, S., Schaefer, S. (1999) *AIP Conf. Proc.*, **463**, 126–128	x			
Mitsui, T., Miyamura, M., Matsunami, A., Kitagawa, K., Arai, N. (1997) *Clin. Chem.*, **43**, 1993–1995		x		
Mitsui, T., Kondo, T. (1999) *Digest. Diseases Sci.*, **44**, 1216–1219		x		
Mitsui, T., Kato, N., Kondo, T. (2000) *Digest. Diseases Sci.*, **45**, 1002–1005		x		
Mitsui, T., Kondo, T. (2004) *Clin. Chim. Acta*, **345**, 129–133		x		
Mukherjee, A., Dunayevskiy, I., Prasanna, M., Go, R., Tsekoun, A., Wang, X., Fan, J., Patel, C. K. N. (2008) *Appl. Opt.*, **47**, 1543–1548	x			
Mukherjee, A., Prasanna, M., Lane, M., Go, R., Dunayevskiy, I., Tsekoun, A., Patel, C. K. N. (2008) *Appl. Opt.*, **47**, 4884–4887	x			x
Osada, T., Rom, H. B., Dahl, P. (1998) *Trans. Am. Soc. Agric. Eng.*, **41**, 1109–1114		x		
Osada, T., Fukumoto, Y. (2001) *Water Sci. Technol.*, **44**, 79–86	x	x		

Table A3.11 *Continued*

Citation	NH$_3$ Ammonia	N$_2$O Nitrous oxide	NO Nitric oxide	NO$_2$ Nitrogen dioxide
Paldus, B. A., Spence, T. G., Zare, R. N., Oomens, J., Harren, F. J. M., Parker, D. H., Gmachl, C., Cappasso, F., Sivco, D. L., Baillargeon, J. N., Hutchinson, A. L., Cho, A. Y. (1999) *Opt. Lett.*, **24**, 178–180	x			
Patel, C. K. N., Kerl, R. J., Burkhardt, E. G. (1977) *Phys. Rev. Lett.*, **38**, 1204–1207			x	
Patel, C. K. N. (1978) *Phys. Rev. Lett.*, **40**, 535–538			x	
Peng, Y., Zhang, W., Li, L., Yu, Q. (2009) *Spectrochim. Acta, Part A*, **74**, 924–927	x			
Petkovska, L. T., Miljanic, S. S. (1997) *Infrared Phys. Technol.*, **38**, 331–336	x			
Petra, N., Zweck, J., Kosterev, A. A., Minkoff, S. E., Thomazy, D. (2009) *Appl. Phys. B*, **94**, 673–680	x			
Pogány, A., Mohácsi, A., Varga, A., Bozóki, Z., Galbács, Z., Horváth, L., Szabó, G. (2009) *Environ. Sci. Technol.*, **43**, 826–830	x			
Pushkarsky, M. B., Webber, M. E., Baghdassarian, O., Narasimhan, L. R., Patel, C. K. N. (2002) *Appl. Phys. B*, **75**, 391–396	x			
Pushkarsky, M. B., Webber, M. E., Patel, C. K. N. (2003) *Appl. Phys. B*, **77**, 381–385	x			
Pushkarsky, M., Webber, M., Patel, C. K. N. (2005) *Proc. SPIE Int. Soc. Opt. Eng.*, **5732**, 93–107	x			
Pushkarsky, M., Tsekoun, A., Dunayevskiy, I. G., Go, R., Patel, C. K. N. (2006) *Proc. Nat. Acad. Sci. U.S.A.*, **103**, 10846–10849				x
Radak, B. B., Pastirk, I., Ristic, G. S., Petkovska, L. T. (1998) *Infrared Phys. Technol.*, **39**, 7–13	x			
Repond, P., Sigrist, M. W. (1996) *Appl. Opt.*, **35**, 4065–4085	x			
Rey, J. M., Schramm, D., Hahnloser, D., Marinov, D., Sigrist, M. W. (2008) *Meas. Sci. Technol.*, **19**, 075602	x			
Romann, A., Sigrist, M. W. (2002) *Appl. Phys. B*, **75**, 377–383			x	

Table A3.11 *Continued*

Citation	NH$_3$	N$_2$O	NO	NO$_2$
	Ammonia	Nitrous oxide	Nitric oxide	Nitrogen dioxide
Schiewe-Langgartner, F., Wiesner, G., Gruber, M., Hobbhahn, J. (2005) *Anesthetist*, **54**, 667–672		x		
Schilt, S., Thevenaz, L., Nikles, M., Emmenegger, L., Huglin, C. (2004) *Spectrochim. Acta, Part A*, **60**, 3259–3268	x			
Schmohl, A., Miklos, A., Hess, P. (2002) *Appl. Opt.*, **41**, 1815–1823	x			
Schramm, D. U., Sthel, M. S., daSilva, M. G., Cameiro, L. O., Souza, A. P., Vargas, H. (2003) *Rev. Sci. Instrum.*, **74**, 513–515		x		x
Schramm, D. U., Sthel, M. S., daSilva, M. G., Cameiro, L. O., Junior, A. J. S., Souza, A. P., Vargas, H. (2003) *Infrared Phys. Technol.*, **44**, 263–269		x		x
Scotoni, M., Roosi, A., Bassi, D., Buffa, R., Iannotta, S., Boschetti, A. (2006) *Appl. Phys. B*, **82**, 495–500	x			
Sigrist, M. W. (2003) *Rev. Sci. Instrum.*, **74**, 486–490	x	x	x	
Sigrist, M. W., Bohren, A., Calasso, I. G., Nägele, M., Romann, A., Seiter, M. (2000) *Proc. SPIE Int. Soc. Opt. Eng.*, **4063**, 17–25	x			
Sigrist, M. W., Naegele, M., Romann, A. (2001) *Proc. SPIE Int. Soc. Opt. Eng.*, **4419**, 14–17	x			
So, S., Koushanfar, F., Kosterev, A., Tittel, F. K. (2007) *Proc. Int. Conf. Inform. Proc. Sensor Networks*, 226–237	x	x		
Solyom, A. M., Angeli, G. Z., Bicanic, D. D., Lubbers, M. (1992) *Analyst*, **117**, 379–382	x			
Spagnolo, V., Elia, A., Franco, C. D., Lugarà, P. M., Vitiello, M. S., Scamarcio, G. (2009) *Proc. SPIE Int. Soc. Opt. Eng.*, **7222**, 72220K			x	
Tittel, F. K., Richter, D., Fried, A. (2003) *Top. Appl. Phys.*, **89**, 445–516	x			
Tittel, F. K., Bakhirkin, Yu., Kosterev, A. A., Wysocki, G. (2006) *Rev. Laser Eng.*, **34**, 275–284	x			

Table A3.11 Continued

Citation	NH$_3$ Ammonia	N$_2$O Nitrous oxide	NO Nitric oxide	NO$_2$ Nitrogen dioxide
Tittel, F. K., Bakhirkin, Y., Kosterev, A., Lewicki, R., So, S., Wysocki, G., Curl, R. F. (2008) *Proc. SPIE Int. Soc. Opt. Eng.*, **6900**, 69000Z-1–7			x	
Webber, M. E., Pushkarsky, M., Patel, C. K. N. (2003) *Appl. Opt.*, **42**, 2119–2126	x			
Webber, M. E., MacDonald, T., Pushkarsky, M. B., Patel, C. K. N., Zhao, Y., Marcillac, N., Mitloehner, F. M. (2005) *Meas. Sci. Technol.*, **16**, 1547–1553	x			
Weidmann, D., Kosterev, A. A., Tittel, F. K., Ryan, N., McDonald, D. (2004) *Opt. Lett.*, **29**, 1837–1839	x			
Wysocki, G., Lewicki, R., Curl, R. F., Tittel, F. K., Diehl, L., Capasso, F., Troccoli, M., Hofler, G., Bour, D., Corzine, S., Maulini, R., Giovannini, M., Faist, J. (2008) *Appl. Phys. B*, **92**, 305–311			x	
Zelinger, Z., Jancik, I., Engst, P. (1992) *Appl. Opt.*, **31**, 6974–6975	x			
Zhang, W., Wu, Z., Yu, Q. (2007) *Chin. Opt. Lett.*, **5**, 677–679	x			

Table A3.12 Triatomics (first group).

Citation	CO$_2$ Carbon dioxide	H$_2$O Water	O$_3$ Ozone	HCN Hydrogen cyanide
Ageev, B. G., Ponomarev, Yu. N., Sapozhnikova, V. A. (1998) *Appl. Phys. B*, **67**, 467–473	x			
Beenen, A., Niessner, R. (1998) *Analyst*, **123**, 543–545		x		
Besson, J.-P., Schilt, S., Thevenaz, L. (2004) *Spectrochim. Acta, Part A*, **60**, 3449–3456		x		
Besson, J.-P., Schilt, S., Thevenaz, L. (2006) *Spectrochim. Acta, Part A*, **63**, 899–904		x		

Table A3.12 Continued

Citation	CO$_2$ Carbon dioxide	H$_2$O Water	O$_3$ Ozone	HCN Hydrogen cyanide
Besson, J.-P., Schilt, S., Sauser, F., Rochat, E., Hamel, P., Sandoz, F., Nikles, M., Thevenaz, L. (2006) *Appl. Phys. B*, **85**, 343–348		x		
Bijnen, F. G. C., Harren, F. J. M., Hackstein, J. H. P., Reuss, J. (1996) *Appl. Opt.*, **35**, 5357–5368	x	x		
Bijnen, F. G. C., Zuckermann, H., Harren, F. J. M., Reuss, J. (1998) *Appl. Opt.*, **37**, 3345–3353	x	x		
Bohren, A., Sigrist, M. W. (1997) *Infrared Phys. Technol.*, **38**, 423–435	x			
Booth, D. T., Sowa, S. (2001) *J. Arid Environ.*, **48**, 35–39	x			
Bozóki, Z., Sneider, J., Gingl, Z., Mohácsi, Á., Szakáll, M., Bor, Z., Szabó, G. (1999) *Meas. Sci. Technol.*, **10**, 999–1003		x		
Brienne, S. H. R., Bozkurt, M. M. V., Rao, S. R., Xu, Z., Butler, I. S., Finch, J. A. (1996) *Appl. Spectrosc.*, **50**, 521–527	x	x		
Calasso, I. G., Sigrist, M. W. (1997) *Appl. Opt.*, **36**, 3212–3216	x			
Claspy, P. C., Pao, Y-H., Kwong, S., Nodov, E. (1976) *Appl. Opt.*, **15**, 1506–1509		x		
Cohen, Y., Bar, I., Rosenwaks, S. (1996) *J. Mol. Spectrosc.*, **180**, 298–304		D$_2$O		
Cristescu, S., Dumitras, D. C., Dutu, D. C. A. (2000) *Proc. SPIE Int. Soc. Opt. Eng.*, **4070**, 457–464			x	
Elia, A., DiFranco, C., Lugara, P. M., Scamarcio, G. (2006) *Sensors*, **6**, 1411–1419	x	x	x	
Firebaugh, S. L. (2001) *J. Microelectromech. Syst.*, **10**, 232–237	x			
Fischer, C., Sigrist, M. W. (2002) *Appl. Phys. B*, **75**, 305–310			x	
Fischer, C., Sorokin, E., Sorokina, I. T., Sigrist, M. W. (2005) *Opt. Lasers Eng.*, **43**, 573–582	x	x		
Fournier, D., Boccara, A. C., Amer, N. M., Gerlach, R. (1980) *Appl. Phys. Lett.*, **37**, 519–521		x		

Table A3.12 *Continued*

Citation	CO₂ Carbon dioxide	H₂O Water	O₃ Ozone	HCN Hydrogen cyanide
German, K. R., Gornall, W. S. (1981) *J. Opt. Soc. Am.*, **71**, 1452–1457				x
Grossel, A., Zeninari, V., Parvitte, B., Joly, L., Courtois, D. (2007) *Appl. Phys. B*, **88**, 483–492		x		
Hofstetter, D., Beck, M., Faist, J., Nagele, M., Sigrist, M. (2001) *Opt. Lett.*, **26**, 887–889	x			
Horka, V., Civis, S., Xu, L.-H., Lees, R. M. (2005) *Analyst*, **130**, 1148–1154	x			
Hubner, M., Sigrist, M. W., Demartines, N., Gianella, M., Clavien, P. A., Hahnloser, D. (2008) *Patient Safety Surgery*, **2** (5 pages)		x		
Kapitanov, V. A., Zeninari, V., Parvitte, B., Courtois, D., Ponomarev, Yu. N. (2002) *Spectrochim. Acta, Part A*, **58**, 2397–2404		x		
Kerr, E. L., Atwood, J. G. (1968) *Appl. Opt.*, **7**, 915–921	x	x		
Koskinen, V., Fonsen, J., Roth, K., Kauppinen, J. (2008) *Vib. Spectrosc.*, **48**, 16–21	x			
Kosterev, A. A., Mosely, T. S., Tittel, F. K. (2006) *Appl. Phys. B*, **85**, 295–300				x
Kosterev, A., Wysocki, G., Bakhirkin, Y., So, S., Lewicki, R., Fraser, M., Tittel, F., Curl, R. F. (2008) *Appl. Phys. B*, **90**, 165–176	x			
Kuusela, T., Peura, J., Matveev, B. A., Remennyy, M. A., Stus', N. M. (2009) *Vib. Spectrosc.*, **51**, 289–293	x			
Laurila, T., Cattaneo, H., Koskinen, V., Kauppinen, J., Hernberg, R. (2005) *Opt. Express*, **13**, 2453–2458	x			
Lawton, S. A., Bragg, S. L. (1984) *Appl. Opt.*, **23**, 3042–3044	x			
Lecoutre, M., Rohart, F., Huet, T. R., Maki, A. G. (2000) *J. Mol. Spectrosc.*, **203**, 158–164				x
Lewicki, R., Wysocki, G., Kosterev, A. A., Tittel, F. K. (2007) *Appl. Phys. B*, **87**, 157–162	x			
Li, J. S., Liu, K., Zhang, W. J., Chen, W. D., Gao, X. M. (2008) *J. Quant. Spectrosc. Radiat. Transfer*, **109**, 1575–1585	x			

Table A3.12 *Continued*

Citation	CO$_2$ Carbon dioxide	H$_2$O Water	O$_3$ Ozone	HCN Hydrogen cyanide
Li, J., Liu, K., Zhang, W., Chen, W., Gao, X. (2008) *Opt. Appl.*, **38**, 341–352	x			
Liu, W., Sun, Z., Ranheimer, M., Forsling, W., Tang, H. (1999) *Analyst*, **124**, 361–365	x			
Liu, K., Li, J., Wang, L., Tan, T., Zhang, W., Gao, X., Chen, W., Tittel, F. K. (2009) *Appl. Phys. B*, **94**, 527–533		x		
Ludwig, B. W., Urban, M. W. (1994) *J. Coat. Technol.*, **66**, 59–67	x			
Marinov, D., Sigrist, M. W. (2003) *Photochem. Photobiol. Sci.*, **2**, 774–778	x			
Moeckli, M. A., Hilbes, C., Sigrist, M. W. (1998) *Appl. Phys. B*, **67**, 449–458	x	x	x	
Naegele, M., Sigrist, M. W. (2000) *Appl. Phys. B*, **70**, 895–901	x	x		
Ngai, A. K. Y., Persijn, S. T., VonBasum, G., Harren, F. J. M. (2006) *Appl. Phys. B*, **85**, 173–180	x			
Ngai, A. K. Y., Persijn, S. T., Lindsay, I. D., Kosterev, A. A., Gross, P., Lee, C. J., Cristescu, S. M., Tittel, F. K., Boller, K.-J., Harren, F. J. M.*Appl. Phys. B*, **89**, 123–128		x		
Oehler, O. (1995) *Sens. Rev.*, **15**, 14–16	x			
Oomens, J., Zuckermann, H., Persijn, S., Parker, D. H., Harren, F. J. M. (1998) *Appl. Phys. B*, **67**, 459–466		x		
Paldus, B. A., Spence, T. G., Zare, R. N., Oomens, J., Harren, F. J. M., Parker, D. H., Gmachl, C., Cappasso, F., Sivco, D. L., Baillargeon, J. N., Hutchinson, A. L., Cho, A. Y. (1999) *Opt. Lett.*, **24**, 178–180		x		
Perlmutter, P., Shtrikman, S., Slatkine, M. (1979) *Appl. Opt.*, **18**, 2267–2274	x			
Persijn, S. T., Veltman, R. H., Oomens, J., Harren, F. J. M., Parker, D. H. (1999) *AIP Conf. Proc.*, **463**, 609–611	x			
Persijn, S. T., Veltman, R. H., Oomens, J., Harren, F. J. M., Parker, D. H. (2000) *Appl. Spectrosc.*, **54**, 62–71	x			

Table A3.12 Continued

Citation	CO$_2$ Carbon dioxide	H$_2$O Water	O$_3$ Ozone	HCN Hydrogen cyanide
Persijn, S. T., Santosa, E., Harren, F. J. M. (2002) *Appl. Phys. B*, **75**, 335–342	x			
Ponomarev, Yu. N. (2004) *Spectrochim. Acta, Part A*, **60**, 3469–3476		x		
Repond, P., Sigrist, M. W. (1996) *Appl. Opt.*, **35**, 4065–4085	x		x	
Rey, J. M., Schramm, D., Hahnloser, D., Marinov, D., Sigrist, M. W. (2008) *Meas. Sci. Technol.*, **19**, 075602	x	x		
Rey, J. M., Sigrist, M. W. (2008) *Sens. Actuators, B*, **135**, 161–165		x		
Riddle, A., Selker, M. (2006) *Appl. Phys. B*, **85**, 329–336		x		
Romann, A., Sigrist, M. W. (2002) *Appl. Phys. B*, **75**, 377–383	x			
Samura, K., Hashimoto, S., Kawasaki, M., Hayashida, A., Kagi, E., Ishiwata, T., Matsumi, Y. (2002) *Appl. Opt.*, **41**, 2349–2354		x		
Schley, R. S., Telschow, K. L. (2001) *Proc. SPIE Int. Soc. Opt. Eng.*, **4205**, 18–24		x		
Sigrist, M. W., Bernegger, S., Meyer, P. L. (1989) *Infrared Phys.*, **29**, 805–814	x	x	x	
Sigrist, M. W. (1999) *AIP Conf. Proc.*, **463**, 203–207	x	x		
Sigrist, M. W., Bohren, A., Calasso, I. G., Nägele, M., Romann, A., Seiter, M. (2000) *Proc. SPIE Int. Soc. Opt. Eng.*, **4063**, 17–25	x	x	x	
Sigrist, M. W. (2003) *Rev. Sci. Instrum.*, **74**, 486–490	x	x		
Skřínský, J., Janečková, R., Grigorová, E., Střižík, M., Kubát, P., Herecová, L., Nevrlý, V., Zelinger, Z., Civiš, S. (2009) *J. Mol. Spectrosc.*, **256**, 99–101			x	
So, S., Koushanfar, F., Kosterev, A., Tittel, F. K. (2007) *Proc. Int. Conf. Inform. Proc. Sensor Networks*, 226–237	x	x		x
Sokabe, N., Hammerich, M., Pedersen, T., Olafsson, A., Henningsen, J. (1992) *J. Mol. Spectrosc.*, **152**, 420–433			x	

Table A3.12 *Continued*

Citation	CO$_2$ Carbon dioxide	H$_2$O Water	O$_3$ Ozone	HCN Hydrogen cyanide
Song, K., Cha, H. K., Kapitanov, V. A., Ponomarev, Yu. N., Rostov, A. P., Courtois, D., Parvitte, B., Zeninari, V. (2002) *Appl. Phys. B*, **75**, 215–227		x		
Szakall, M., Bozoki, Z., Mohacsi, A., Varga, A., Szabo, G. (2004) *Appl. Spectrosc.*, **58**, 792–798		x		
Tittel, F. K., Richter, D., Fried, A. (2003) *Top. Appl. Phys.*, **89**, 445–516	x	x		
Tittel, F. K., Bakhirkin, Yu., Kosterev, A. A., Wysocki, G. (2006) *Rev. Laser Eng.*, **34**, 275–284		x		
Tittel, F. K., Bakhirkin, Y., Kosterev, A., Lewicki, R., So, S., Wysocki, G., Curl, R. F. (2008) *Proc. SPIE Int. Soc. Opt. Eng.*, **6900**, 69000Z-1–7		x		
Uotila, J., Koskinen, V., Kauppinen, J. (2005) *Vib. Spectrosc.*, **38**, 3–9	x			
Weidmann, D., Kosterev, A. A., Tittel, F. K., Ryan, N., McDonald, D. (2004) *Opt. Lett.*, **29**, 1837–1839	x	x		
Willing, B., Muralt, P., Oehler, O. (1999) *AIP Conf. Proc.*, **463**, 277–279	x			
Wolff, M., Harde, H. (2003) *Infrared Phys. Technol.*, **44**, 51–55	x			
Wysocki, G., Kosterev, A. A., Tittel, F. K. (2006) *Appl. Phys. B*, **85**, 301–306	x			
Yang, X., Noda, C. (1999) *J. Mol. Spectrosc.*, **195**, 256–262	x			
Zelinger, Z., Strizik, M., Kubat, P., Janour, Z., Berger, P., Cerny, A., Engst, P. (2004) *Opt. Lasers Eng.*, **42**, 403–412			x	
Zeninari, V., Tikhomirov, B. A., Ponomarev, Yu. N., Courtois, D. (1998) *J. Quant. Spectrosc. Radiat. Transfer*, **59**, 369–375			x	
Zeninari, V., Tikhomirov, B. A., Ponomarev, Yu. N., Courtois, D. (2000) *J. Chem. Phys.*, **112**, 1835–1843			x	
Zeninari, V., Vicet, A., Parvitte, B., Joly, L., Durry, G. (2004) *Infrared Phys. Technol.*, **45**, 229–237	x			

Table A3.13 Triatomics (second group).

Citation	SO$_2$ Sulfur dioxide	OCS Carbonyl sulfide	CS$_2$ Carbon disulfhide
Brienne, S. H. R., Bozkurt, M. M. V., Rao, S. R., Xu, Z., Butler, I. S., Finch, J. A. (1996) *Appl. Spectrosc.*, **50**, 521–527		x	x
Fischer, C., Sigrist, M. W. (2002) *Appl. Phys. B*, **75**, 305–310		x	
Gondal, M.A. (1997) *Appl. Opt.*, **36**, 3195–3201	x		
Horka, V., Civis, S., Xu, L.-H., Lees, R. M. (2005) *Analyst*, **130**, 1148–1154		x	
Kuusela, T., Peura, J., Matveev, B. A., Remennyy, M. A., Stus', N. M. (2009) *Vib. Spectrosc.*, **51**, 289–293	x		
Lecoutre, M., Hadj Bachir, I., Huet, T. R., Fayt, A. (2002) *J. Mol. Spectrosc.*, **216**, 472–480		x	
Mukherjee, A., Prasanna, M., Lane, M., Go, R., Dunayevskiy, I., Tsekoun, A., Patel, C. K. N. (2008) *Appl. Opt.*, **47**, 4884–4887	x		
Olafsson, A., Hansen, G. I., Loftsdottir, A. S., Jakobsson, S. (1999) *AIP Conf. Proc.*, **463**, 208–210		x	
Schramm, D. U., Sthel, M. S., daSilva, M. G., Cameiro, L. O., Souza, A. P., Vargas, H. (2003) *Rev. Sci. Instrum.*, **74**, 513–515	x		
Schramm, D. U., Sthel, M. S., daSilva, M. G., Cameiro, L. O., Junior, A. J. S., Souza, A. P., Vargas, H. (2003) *Infrared Phys. Technol.*, **44**, 263–269	x		
Tranchart, S., Hadj Bachir, I., Huet, T. R., Olafsson, A., Destombes, J.-L., Naim, S., Fayt, A. (1999) *J. Mol. Spectrosc.*, **196**, 265–273		x	
Zúñiga, J., Bastida, A., Alacid, M., Requena, A. (2000) *J. Chem. Phys.*, **113**, 5695–5704		x	

Index

a

Alumina 15, 98, 172, 202–206, 208, 209
Amide I and II bands 100, 103, 119, 183, 184, 186, 191, 194, 240, 241, 245, 258–260
Ammonium sulfate 9, 10, 18
Amplitude modulation 12, 34–36, 79–82, 107–111, 117, 163, 251
– definition 293
Amplitude spectrum (see magnitude spectrum)
Asbestos 15, 247, 248, 251, 255
Attenuated total reflectance (ATR) 21, 29, 31, 60, 61, 64, 94, 95, 103, 151, 166, 178, 186, 191, 195–198, 201, 227, 231, 237, 238, 243, 245, 256, 259, 260

b

Biology and biochemistry 100, 101, 182–187, 193, 194, 239, 240

c

Cantilever-enhanced detection 47–50
Carbon black 17, 21, 28–30, 32–34, 50, 80, 83, 85, 88, 90–92, 128, 129, 134–136, 151, 152, 179, 238, 257
Carbon-filled rubber 30, 32, 37, 87, 152
Carbons 2, 12, 16, 17, 27, 58, 141, 143–156
– 1600-cm^{-1} band 148–150, 155, 168
Carbonyls 176–179, 225, 226
Catalysts 11, 12, 16, 17, 21, 27, 56, 98, 99, 106, 148, 202–209
Cellulose 100, 145–147, 149, 150, 161, 183, 230–232, 257, 262, 263
Clays and minerals 12, 15, 32, 33, 35, 40, 53, 54, 72, 73, 75–77, 81, 82, 87, 100, 101, 103, 134–136, 138, 157, 158, 247–256
CO laser excitation 3, 11, 40, 63, 64, 210–213, 256

CO_2 laser excitation 3, 8–11, 13–15, 17, 38–40, 51, 57, 58, 61, 63, 64, 99, 139, 189, 191, 210, 213–218, 242, 243, 245, 248
Coals 16, 17, 19, 20, 22, 27, 101, 105, 106, 108–110, 131, 150, 153, 157–165
Cokes 87, 143, 150, 154–156
Corrosion 27
Curve fitting 164–167, 185, 252, 253, 255

d

Depth profiling 1, 2, 12, 16, 21, 35, 38, 83, 93, 94, 103–120, 162, 163, 179, 190, 191, 195, 197–202, 230, 234–236, 243–245, 256, 257, 260, 261, 263
– definition 293
Diffuse reflectance 29, 31, 95, 100, 103, 151, 152, 157, 158, 160, 162, 188, 191, 197, 204, 227, 228, 231, 238–240, 243, 253, 259, 263
Diode laser excitation 43–45, 49, 50, 58, 59, 210, 219–221, 223, 224
Dispersive PA spectroscopy 11–13, 25–29, 161, 171, 176, 177, 182, 187, 202, 237, 238, 247, 255
Drugs (see pharmaceuticals)

f

Far-infrared PA spectroscopy 4, 16, 20, 174–176, 249–251, 254, 255
Films 53, 58, 105, 106, 115, 116, 133, 141, 152, 153, 197, 198, 260
Food products 26, 63–65, 97, 101–103, 105, 106, 112, 113, 118–120, 145, 149, 150, 211–213, 237–246

g

Gases 8, 9, 11, 15, 42–50, 57, 177, 187, 194, 209–227

Gas-microphone cell
– microsampling 92–96
– noise reduction 31
– resonance 9, 21, 197, 212, 215–217, 257
– signal generation 2, 14, 29–31
– signal-to-noise ratios 93
Generalized two-dimensional correlation 107, 117–120, 244
Glassy carbon 30, 32, 34, 87

h

Hair 92, 94, 105, 191, 192, 259
HF laser excitation 15, 255
Hydrocarbon fuels 2, 97, 156–170, 173, 256
Hydrocarbons 9, 12–16, 19, 26, 53, 57, 58, 61, 63, 64, 85, 86, 97, 98, 144–148, 151–153, 171–176, 203, 204, 210–216, 218–225

i

Inorganics 9, 10, 16–22, 54, 55, 98, 99, 114, 115, 139–141, 176–182, 204
In-phase (real) spectrum 60, 111, 116, 137–139, 152, 179, 190, 235, 244, 251

l

Lignin 145, 149, 150, 228, 229
Linearization of PA spectra 132–136, 190
Lipids 51, 101, 183, 185, 189, 190, 213, 242, 244

m

Magnitude (amplitude) spectrum 60, 82, 111, 127, 132–136, 191, 193, 234–236, 244, 245, 255
Medical and dental applications 12, 64, 105, 116, 133, 187–194
Metals 2, 10, 14, 17, 59, 151, 202, 203
Microspectroscopy 92–96, 177, 198, 254
Mirage effect 56

n

Near-infrared PA spectroscopy 3, 12, 13, 20, 25, 26, 28, 54, 55, 58, 97, 159–161, 171, 172, 176–182, 187, 191, 197, 202, 218–221, 231, 232, 237, 238, 248–251
Normalization of PA spectra 32–35, 143, 144
Nuclear magnetic resonance spectroscopy 13, 186

o

Open-membrane cell 14
Optothermal window spectroscopy 62–65, 241–243

p

Paper 227, 230–236
Particle size effects 12, 13, 19, 20, 97, 99, 158, 232
Pharmaceuticals 17, 20, 26, 39, 97, 99, 101, 187–189
Phase analysis 137–141
Phase modulation 35–37, 72–77, 82–86, 94, 106, 107, 111, 112, 163, 190
– definition 293
Phase rotation 37, 94, 107, 113–116, 190, 260
Phase spectrum 37, 38, 60, 81, 82, 94, 107, 111–113, 132, 190, 193, 235, 244, 255
Photoacoustic (PA) infrared spectroscopy
– advantages 1, 12, 20, 157, 172, 182, 246, 247
– general description 1–3
– history 7–22
– review articles 4, 5, 37, 38, 171, 195, 210
– surface sensitivity 2, 21, 261, 262
Photopyroelectric detection 243
Photothermal beam deflection spectroscopy 56–59, 139, 143, 146, 150
Photothermal radiometry 233, 234
– definition 293
Piezoelectric detection 15, 36, 38, 51–56, 62, 63, 172, 181, 189
– definition 293
Polymers 2, 16, 20, 21, 37, 39–42, 50, 94–96, 101, 105, 106, 112–120, 130, 133, 134, 138–140, 145, 148, 149, 183, 185, 194–201, 230, 244, 257, 260–263
Power (modulus) spectrum 133
Proteins 12, 26, 64, 97, 101, 103, 105, 106, 119, 120, 182, 183, 185, 186, 190, 191, 194, 237, 238, 241–245, 259

q

Quadrature (imaginary) spectrum 60, 111, 112, 116, 137–139, 141, 152, 179, 190, 235, 244, 251, 252
Quantitative analysis 12, 13, 19, 96–103, 171, 183, 184, 187, 188, 203, 206, 215, 237, 238, 242
Quantum cascade laser excitation 41–47, 220–223
Quartz-enhanced detection 42–47

r

Raman spectroscopy 97, 149, 150, 167, 177, 187, 201, 207, 221, 229, 245, 251, 254

Rapid-scan PA spectroscopy
- as experimental method   29–31
- definition   293
- depth profiling in   93, 94, 104–107, 163, 197, 198, 259, 260
Rare earths   12
Reflection-absorption spectroscopy   10, 14, 15
Reverse mirage PA spectroscopy   59–61
Rosencwaig-Gersho theory   33, 179

s
Sample heating   169, 177, 243, 248
Saturation   98–100, 102, 111, 137, 151, 178, 183, 191, 204, 242, 248, 251, 252, 260, 263
- definition   293
Signal averaging   127–132, 164, 250
Signal demodulation by
- digital signal processor (DSP)   36, 71–74, 94, 111, 112
- lock-in amplifier   12, 33, 34, 36, 52, 53, 56, 74–77, 80–83, 85, 86, 108, 112, 113, 137, 216, 235, 238, 260
Signal generation   2, 31
Silica   11, 13, 16, 19, 26, 98, 100, 103, 172, 202–208
Step-scan PA spectroscopy   34–38, 60, 82, 93, 94, 106, 137, 188, 190, 193, 235, 244, 251, 252
- definition   294
Synchrotron radiation   87–92, 128–130

t
Textiles   197, 257–263
Thermal conductivity   12, 155, 204, 233
- definition   294

- of paper   233
- of wool and hair   259
Thermal diffusion length   31, 62, 95, 99, 102, 104, 107, 108, 115, 195, 235, 259, 261
- definition   294
Thermal diffusivity   31, 104, 107, 155
- definition   294
- of coal   109
- of foods   243
- of grease   94
- of human skin   189
- of paper   233
- of polyethylene terephthalate   116
- of polymers   199
- of wood products   227, 229, 230
Thermal effusivity   62, 155
- definition   294
Thermally thick   33, 62, 102, 188
- definition   294
Thermally thin   33, 95
- definition   294
Transmission spectroscopy   20, 21, 29, 31, 33, 53, 95, 100, 141, 151–153, 157, 158, 160, 166, 171, 172, 176, 178, 180, 190, 191, 196, 197, 204, 227, 238, 242, 248, 251, 252, 254–256, 260, 263

w
Wood   161, 227–230

x
X-ray photoelectron spectroscopy   262

z
Zeolites   16, 22, 204–206